Current Developments in Biotechnology and Bioengineering

Current Developments in Biotechnology and Bioengineering

Technologies for Production of Nutraceuticals and Functional Food Products

Edited by

Amit Kumar Rai
Institute of Bioresources and Sustainable Development, Sikkim Centre, Tadong, India;
Institute of Bioresources and Sustainable Development, Regional Centre, Gangtok, India

Sudhir P. Singh
Center of Innovative and Applied Bioprocessing, Mohali, India

Ashok Pandey
Centre for Innovation and Translational Research, CSIR-Indian Institute of Toxicology Research, Lucknow, India

Christian Larroche
Chemical and Biochemical Engineering Laboratory, Institute Pascal, University Clermont Auvergne, Clermont Ferrand, France

Carlos Ricardo Soccol
Bioprocess Engineering and Biotechnology Department, Federal University of Paraná (UFPR), Curitiba, Brazil

Series Editor

Ashok Pandey

Elsevier
Radarweg 29, PO Box 211, 1000 AE Amsterdam, Netherlands
The Boulevard, Langford Lane, Kidlington, Oxford OX5 1GB, United Kingdom
50 Hampshire Street, 5th Floor, Cambridge, MA 02139, United States

Copyright © 2022 Elsevier B.V. All rights reserved.

No part of this publication may be reproduced or transmitted in any form or by any means, electronic or mechanical, including photocopying, recording, or any information storage and retrieval system, without permission in writing from the publisher. Details on how to seek permission, further information about the Publisher's permissions policies and our arrangements with organizations such as the Copyright Clearance Center and the Copyright Licensing Agency, can be found at our website: www.elsevier.com/permissions.

This book and the individual contributions contained in it are protected under copyright by the Publisher (other than as may be noted herein).

Notices

Knowledge and best practice in this field are constantly changing. As new research and experience broaden our understanding, changes in research methods, professional practices, or medical treatment may become necessary.

Practitioners and researchers must always rely on their own experience and knowledge in evaluating and using any information, methods, compounds, or experiments described herein. In using such information or methods they should be mindful of their own safety and the safety of others, including parties for whom they have a professional responsibility.

To the fullest extent of the law, neither the Publisher nor the authors, contributors, or editors, assume any liability for any injury and/or damage to persons or property as a matter of products liability, negligence or otherwise, or from any use or operation of any methods, products, instructions, or ideas contained in the material herein.

British Library Cataloguing-in-Publication Data
A catalogue record for this book is available from the British Library

Library of Congress Cataloging-in-Publication Data
A catalog record for this book is available from the Library of Congress

ISBN: 978-0-12-823506-5

For Information on all Elsevier publications
visit our website at https://www.elsevier.com/books-and-journals

Publisher: Susan Dennis
Acquisitions Editor: Katie Hammon
Editorial Project Manager: Maria Elaine D Desamero
Production Project Manager: R. Vijay Bharath
Cover Designer: Greg Harris

Typeset by MPS Limited, Chennai, India

Contents

List of contributors   xiii
About the editors   xix
Preface   xxiii

1. **Microbial bioprocesses for production of nutraceuticals and functional foods**   1
   LORENI CHIRING PHUKON, SUDHIR P. SINGH, ASHOK PANDEY AND AMIT KUMAR RAI

   1.1 Introduction   1
   1.2 Microbial production of nutraceuticals and functional foods   2
   1.3 Conclusions and perspectives   16
   References   16

2. **Microbial transformation for improving food functionality**   31
   ANTONIA TERPOU AND AMIT KUMAR RAI

   2.1 Microbes and food production over the centuries   31
   2.2 Bioactive compounds as functional food ingredients   33
   2.3 Health-promoting bioactive compounds synthesized *via* microbial transformation   34
   2.4 Trends in the delivery of bioactive compounds targeting enhanced food functionality   38
   2.5 Fermented food as potential for antiviral therapy   40
   2.6 Conclusions and perspectives   41
   References   41

3. Bioactive peptide production in fermented foods — 47

ANJA DULLIUS, GABRIELA RABAIOLI RAMA, MAIARA GIROLDI, MÁRCIA INÊS GOETTERT, DANIEL NEUTZLING LEHN AND CLAUCIA FERNANDA VOLKEN DE SOUZA

- 3.1 Introduction — 47
- 3.2 Bioactive peptides in fermented dairy products — 48
- 3.3 Bioactive peptides in fermented meat products — 52
- 3.4 Bioactive peptides in fermented vegetables — 55
- 3.5 Biotechnological progress for industrial production of bioactive peptides — 57
- 3.6 Conclusions and perspectives — 62
- References — 63

4. Probiotics in fermented products and supplements — 73

NEHA BALIYAN, MADHU KUMARI, POONAM KUMARI, KIRAN DINDHORIA, SRIJANA MUKHIA, SANJEET KUMAR, MAHESH GUPTA AND RAKSHAK KUMAR

- 4.1 Introduction — 73
- 4.2 Probiotic bacteria — 74
- 4.3 Probiotics in fermented foods — 80
- 4.4 Interactions between probiotics and components of fermented foods — 86
- 4.5 Impact of probiotics on the human gut microbiome — 87
- 4.6 Ethnic fermented probiotics products translated to industry and their advantages over nonfermented products — 90
- 4.7 Challenges in the probiotics industry and future prospects — 92
- 4.8 Conclusions and perspectives — 95
- Acknowledgments — 95
- References — 96

5. Fructooligosaccharides production and the health benefits of prebiotics — 109

CLARISSE NOBRE, LÍVIA S. SIMÕES, DANIELA A. GONÇALVES, PAULO BERNI AND JOSÉ A. TEIXEIRA

- 5.1 Introduction — 109
- 5.2 Fructans occurrence, chemical structure, technological properties, and market — 110
- 5.3 Fructooligosaccharides production — 112
- 5.4 Role of prebiotics in control of human diseases — 120
- 5.5 Conclusions and perspectives — 127
- Acknowledgments — 128
- References — 128

6. Production of food enzymes — 139

QINGHUA LI, GUOQIANG ZHANG AND GUOCHENG DU

- 6.1 Introduction — 139
- 6.2 Applications of food enzymes in food industry — 141
- 6.3 Current progress and challenges in food enzymes production — 144
- 6.4 Advanced technologies for food enzymes production — 147
- 6.5 Conclusions and perspectives — 151
- References — 151

7. Production of fibrinolytic enzymes during food production — 157

ALI MUHAMMED MOULA ALI, SRI CHARAN BINDU BAVISETTY, MARIA GULLO, SITTIWAT LERTSIRI, JOHN MORRIS AND SALVATORE MASSA

- 7.1 Introduction — 157
- 7.2 Mechanism of blood clotting and fibrinogenesis — 158
- 7.3 Mechanism of action of FEs: fibrinolysis and thrombolysis — 159
- 7.4 Classification of FEs — 160

7.5	Purification of FEs	160
7.6	Biochemical characterization of FEs	161
7.7	Sources of FEs	163
7.8	Production of FEs	170
7.9	Applications	174
7.10	Conclusions and perspectives	177
	References	177

8. Microbial production and transformation of polyphenols — 189
PUJA SARKAR, MD MINHAJUL ABEDIN, SUDHIR P. SINGH, ASHOK PANDEY AND AMIT KUMAR RAI

8.1	Introduction	189
8.2	Types of polyphenols and their health benefits	190
8.3	Microbial production and enhancement of polyphenols during fermentation	192
8.4	Effect of gut microbes on polyphenols	200
8.5	Conclusions and perspectives	200
	References	201

9. Bioprocess technologies for production of structured lipids as nutraceuticals — 209
SUZANA FERREIRA-DIAS, NATÁLIA OSÓRIO AND CARLA TECELÃO

9.1	Introduction	209
9.2	General aspects of lipids: definition, structure, and properties	210
9.3	The role of lipids in human nutrition	213
9.4	Production of structured lipids	215
9.5	Conclusions and perspectives	230
	List of symbols	230
	References	231

10. **Microbial fermentation for reduction of antinutritional factors**   239

EBENEZER JEYAKUMAR AND RUBINA LAWRENCE

    10.1 Introduction   239
    10.2 Nutritive and antinutritive properties in food   240
    10.3 Neutralizing antinutritional factors in fermented foods   248
    10.4 Conclusions and perspectives   254
    References   255

11. **Mycotoxins in foods: impact on health**   261

SAMUEL AYOFEMI OLALEKAN ADEYEYE

    11.1 Introduction   261
    11.2 Mycotoxins in foods   262
    11.3 Mycotoxins in foods: impact on health   264
    11.4 Economic implications of mycotoxins in foods   266
    11.5 Mitigation and control of mycotoxins in foods   266
    11.6 Conclusions and perspectives   268
    References   269

12. **Gut microbes: Role in production of nutraceuticals**   273

PALANISAMY ATHIYAMAN BALAKUMARAN, K. DIVAKAR, RAVEENDRAN SINDHU, ASHOK PANDEY AND PARAMESWARAN BINOD

    12.1 Introduction   273
    12.2 Role of the gut microbiome in human health   274
    12.3 Nutraceuticals and gut microbes   276
    12.4 Nutrigenomics: Gut microbes and health benefits   276
    12.5 Developments in nutraceuticals production using gut microbes   278
    12.6 Microbial strategies for the production of nutraceuticals   279
    12.7 Production of gamma-aminobutyric acid and hyaluronic acid by lactic acid bacteria   279

12.8	Vitamin B12 and folate production by *Propionibacteria*	280
12.9	Nutraceuticals production by *Bacillus subtilis*	281
12.10	Glutathione, carotenoids, and other nutraceuticals production by yeasts	281
12.11	Metabolic engineering of microbes for the production of nutraceuticals	282
12.12	Market trends in the production of nutraceuticals	287
12.13	Conclusions and perspectives	288
	Acknowledgment	289
	References	289

13. Bioprocessing of agri-food processing residues into nutraceuticals and bioproducts — 301

VINOD KUMAR, SUDESH K. YADAV, ANIL K. PATEL, BHUWAN B. MISHRA, VIVEK AHLUWALIA, LALITESH K. THAKUR AND JITENDRA KUMAR

13.1	Introduction	301
13.2	Agri-food processing residues and valorization	303
13.3	Conclusions and perspectives	314
	Acknowledgments	314
	References	314

14. Genetically modified microorganisms for enhancing nutritional properties of food — 323

PARDEEP KUMAR BHARDWAJ AND KASHMIR SINGH

14.1	Introduction	323
14.2	Techniques used for genetic modification	324
14.3	Advantages of genetic modification	324
14.4	Need for genetically modified food products	324
14.5	Microorganisms as source of food enzymes	325
14.6	Use of genetically modified microbes in food industry	325
14.7	Characteristics of genetically modified microorganisms	327
14.8	Products development using genetically modified microorganisms in food industry	330

14.9	Fermented food products and their health benefits	332
14.10	Conclusions and perspectives	333
	References	334

15. Exopolysaccharide producing microorganisms for functional food industry — 337

RWIVOO BARUAH, KUMARI RAJSHEE AND PRAKASH M. HALAMI

15.1	Introduction	337
15.2	Exopolysaccharides	338
15.3	Important exopolysaccharides producing bacterial genera	339
15.4	Fermented food containing exopolysaccharides	345
15.5	Technological advantages of exopolysaccharides	347
15.6	Health benefits of exopolysaccharides	347
15.7	Conclusions and perspectives	349
	Acknowledgments	349
	References	349
	Further reading	354

16. Microbial metabolites beneficial in regulation of obesity — 355

KHUSHBOO AND KASHYAP KUMAR DUBEY

16.1	Introduction	355
16.2	Composition of gut microbiota	356
16.3	Obesity and lipid metabolism	357
16.4	Pancreatic lipase	358
16.5	Antilipid effect of Cineromycin B	359
16.6	Gut–brain axis	359
16.7	Short-chain fatty acids	360
16.8	New players for obesity treatment	362
16.9	Orlistat: Food and Drug Administration–approved antiobesity drug	364
16.10	*Bacillus natto* probiotics	365

	16.11 Inhibitory effect of fermented skim milk	366
	16.12 Conclusions and perspectives	367
	References	367

17. Potential bovine colostrum for human and animal therapy — 377
MARIA GIOVANA BINDER PAGNONCELLI, FERNANDA GUILHERME DO PRADO, JULIANE MAYARA CASARIM MACHADO, ANDREIA ANSCHAU AND CARLOS RICARDO SOCCOL

	17.1 Introduction	377
	17.2 Bovine colostrum composition	378
	17.3 Passive immunity mechanism in human and different animal species	382
	17.4 Bovine colostrum application for human therapy	383
	17.5 Bovine colostrum application for animal therapy	386
	17.6 Conclusions and perspectives	389
	References	389

18. Colostrum new insights: products and processes — 397
ALESSANDRA CRISTINE NOVAK SYDNEY, ISADORA KANIAK IKEDA, MARIA CAROLINA DE OLIVEIRA RIBEIRO, EDUARDO BITTENCOURT SYDNEY, DÃO PEDRO DE CARVALHO NETO, SUSAN GRACE KARP, CRISTINE RODRIGUES AND CARLOS RICARDO SOCCOL

	18.1 Introduction	397
	18.2 New insights in colostrum production process and quality control	398
	18.3 Colostrum patents	402
	18.4 Commercial bovine colostrum products	414
	18.5 Conclusion and perspectives	417
	References	418

Index 423

List of contributors

Md Minhajul Abedin Institute of Bioresources and Sustainable Development, Regional Centre, Tadong, India

Samuel Ayofemi Olalekan Adeyeye Department of Food Technology, Hindustan Institute of Technology and Science, Hindustan University, Chennai, India

Vivek Ahluwalia Institute of Pesticide Formulation Technology, India

Ali Muhammed Moula Ali Department of Food Science and Technology, Faculty of Food-Industry, King Mongkut's Institute of Technology Ladkrabang, Bangkok, Thailand

Andreia Anschau Bioprocess Engineering and Biotechnology Department, Federal University of Technology—Paraná (UTFPR), Curitiba, Brazil

Palanisamy Athiyaman Balakumaran Microbial Processes and Technology Division, CSIR-National Institute for Interdisciplinary Science and Technology (CSIR-NIIST), Thiruvananthapuram, India

Neha Baliyan Biotechnology Division, CSIR-Institute of Himalayan Bioresource Technology, Palampur, India; Academy of Scientific and Innovative Research (AcSIR), CSIR- Human Resource Development Centre, Ghaziabad, India

Rwivoo Baruah Microbiology and Fermentation Technology Department, CSIR – Central Food Technological Research Institute, Mysuru, 570020, India

Sri Charan Bindu Bavisetty Department of Fermentation Technology, Faculty of Food-Industry, King Mongkut's Institute of Technology Ladkrabang, Bangkok, Thailand

Paulo Berni Centre of Biological Engineering (CEB), University of Minho, Braga, Portugal

Pardeep Kumar Bhardwaj Institute of Bioresources & Sustainable Development, Imphal, India

Parameswaran Binod Microbial Processes and Technology Division, CSIR-National Institute for Interdisciplinary Science and Technology (CSIR-NIIST), Thiruvananthapuram, India

Dão Pedro de Carvalho Neto Federal University of Paraná—(UFPR), Curitiba, Paraná, Brazil

Maria Carolina de Oliveira Ribeiro Federal University of Technology - Parana (UTFPR) - Campus Ponta Grossa, Ponta Grossa, Paraná, Brazil

Kiran Dindhoria Biotechnology Division, CSIR-Institute of Himalayan Bioresource Technology, Palampur, India; Academy of Scientific and Innovative Research (AcSIR), CSIR- Human Resource Development Centre, Ghaziabad, India

K. Divakar Department of Biotechnology, Sri Venkateswara College of Engineering, Chennai, India

Fernanda Guilherme do Prado Bioprocess Engineering and Biotechnology Department, Federal University of Paraná (UFPR), Curitiba, Brazil

Guocheng Du School of Biotechnology and Key Laboratory of Industrial Biotechnology, Ministry of Education, Jiangnan University, Wuxi, P.R. China; The Key Laboratory of Carbohydrate Chemistry and Biotechnology, Ministry of Education, Jiangnan University, Wuxi, P.R. China

Kashyap Kumar Dubey Bioprocess Engineering Laboratory, Department of Biotechnology, Central University of Haryana, Mahendergarh, India; Bioprocess Engineering Laboratory, School of Biotechnology, Jawaharlal Nehru University, New Delhi, India

Anja Dullius Food Biotechnology Laboratory, University of Vale do Taquari - Univates, Lajeado, RS, Brazil; Biotechnology Graduate Program, University of Vale do Taquari - Univates, Lajeado, RS, Brazil

Suzana Ferreira-Dias Instituto Superior de Agronomia, LEAF, Linking Landscape, Environment, Agriculture and Food, Universidade de Lisboa, Lisbon, Portugal

Maiara Giroldi Food Biotechnology Laboratory, University of Vale do Taquari - Univates, Lajeado, RS, Brazil; Biotechnology Graduate Program, University of Vale do Taquari - Univates, Lajeado, RS, Brazil

Márcia Inês Goettert Biotechnology Graduate Program, University of Vale do Taquari - Univates, Lajeado, RS, Brazil; Cell Culture Laboratory, University of Vale do Taquari - Univates, Lajeado, RS, Brazil

Daniela A. Gonçalves Centre of Biological Engineering (CEB), University of Minho, Braga, Portugal

Maria Gullo Department of Life Sciences, University of Modena and Reggio Emilia, Reggio Emilia, Italy

Mahesh Gupta Academy of Scientific and Innovative Research (AcSIR), CSIR- Human Resource Development Centre, Ghaziabad, India; Food and Nutraceutical Division, CSIR-Institute of Himalayan Bioresource Technology, Palampur, India

Prakash M. Halami Microbiology and Fermentation Technology Department, CSIR – Central Food Technological Research Institute, Mysuru, 570020, India

Isadora Kaniak Ikeda Federal University of Technology – Parana (UTFPR) - Campus Ponta Grossa, Ponta Grossa, Paraná, Brazil

Ebenezer Jeyakumar Department of Industrial Microbiology, Jacob Institute of Biotechnology and Bioengineering, Sam Higginbottom University of Agriculture, Technology and Sciences, Prayagraj, India

Susan Grace Karp Federal University of Paraná—(UFPR), Curitiba, Paraná, Brazil

Khushboo Bioprocess Engineering Laboratory, Department of Biotechnology, Central University of Haryana, Mahendergarh, India

Jitendra Kumar Institute of Pesticide Formulation Technology, India

Rakshak Kumar Biotechnology Division, CSIR-Institute of Himalayan Bioresource Technology, Palampur, India; Academy of Scientific and Innovative Research (AcSIR), CSIR- Human Resource Development Centre, Ghaziabad, India

Sanjeet Kumar Biotechnology Division, CSIR-Institute of Himalayan Bioresource Technology, Palampur, India

Vinod Kumar Center of Innovative and Applied Bioprocessing, India; Division of Fermentation & Microbial Biotechnology, CSIR-Indian Institute of Integrative Medicine (IIIM), Jammu, India

Madhu Kumari Academy of Scientific and Innovative Research (AcSIR), CSIR- Human Resource Development Centre, Ghaziabad, India; Food and Nutraceutical Division, CSIR-Institute of Himalayan Bioresource Technology, Palampur, India

Poonam Kumari Biotechnology Division, CSIR-Institute of Himalayan Bioresource Technology, Palampur, India

Rubina Lawrence Department of Industrial Microbiology, Jacob Institute of Biotechnology and Bioengineering, Sam Higginbottom University of Agriculture, Technology and Sciences, Prayagraj, India

Daniel Neutzling Lehn Food Biotechnology Laboratory, University of Vale do Taquari – Univates, Lajeado, RS, Brazil

Sittiwat Lertsiri Department of Biotechnology, Faculty of Science, Mahidol University, Bangkok, Thailand

Qinghua Li National Engineering Laboratory for Cereal Fermentation Technology, Jiangnan University, Wuxi, P.R. China; School of Biotechnology and Key Laboratory of Industrial Biotechnology, Ministry of Education, Jiangnan University, Wuxi, P.R. China

Juliane Mayara Casarim Machado Bioprocess Engineering and Biotechnology Department, Federal University of Technology—Paraná (UTFPR), Curitiba, Brazil

Salvatore Massa Department of Agricultural, Food and Environmental Sciences, University of Foggia, Foggia, Italy

Bhuwan B. Mishra Center of Innovative and Applied Bioprocessing, India

John Morris KMITL Research and Innovation Services, King Mongkut's Institute of Technology Ladkrabang, Bangkok, Thailand

Srijana Mukhia Biotechnology Division, CSIR-Institute of Himalayan Bioresource Technology, Palampur, India; Department of Microbiology, Guru Nanak Dev University, Amritsar, India

Clarisse Nobre Centre of Biological Engineering (CEB), University of Minho, Braga, Portugal

Natália Osório Instituto Politécnico de Setúbal, Escola Superior de Tecnologia do Barreiro, Lavradio, Portugal; Instituto Superior de Agronomia, Centro de Estudos Florestais, Universidade de Lisboa, Lisbon, Portugal

Maria Giovana Binder Pagnoncelli Department of Chemistry and Biology, Federal University of Technology—Paraná (UTFPR), Curitiba, Brazil

Ashok Pandey Centre for Innovation and Translational Research, CSIR-Indian Institute of Toxicology Research (CSIR-IITR), Lucknow, India; CSIR-Indian Institute of Toxicology Research, Lucknow, India

Anil K. Patel Department of Chemical and Biological Engineering, Korea University, Seoul, Republic of Korea

Loreni Chiring Phukon Institute of Bioresources and Sustainable Development, Regional Centre, Tadong, India

Amit Kumar Rai Institute of Bioresources and Sustainable Development, Regional Centre, Tadong, India; Institute of Bioresources and Sustainable Development, Regional Centre, Gangtok, India

Kumari Rajshee Microbiology and Fermentation Technology Department, CSIR – Central Food Technological Research Institute, Mysuru, 570020, India

Gabriela Rabaioli Rama Food Biotechnology Laboratory, University of Vale do Taquari – Univates, Lajeado, RS, Brazil; Biotechnology Graduate Program, University of Vale do Taquari – Univates, Lajeado, RS, Brazil

Cristine Rodrigues Federal University of Paraná—(UFPR), Curitiba, Paraná, Brazil

Puja Sarkar Institute of Bioresources and Sustainable Development, Regional Centre, Tadong, India

Lívia S. Simões Centre of Biological Engineering (CEB), University of Minho, Braga, Portugal

Raveendran Sindhu Microbial Processes and Technology Division, CSIR-National Institute for Interdisciplinary Science and Technology (CSIR-NIIST), Thiruvananthapuram, India

Kashmir Singh Department of Biotechnology, Panjab University, Chandigarh, India

Sudhir P. Singh Center of Innovative and Applied Bioprocessing, Mohali, India

Carlos Ricardo Soccol Federal University of Paraná—(UFPR), Curitiba, Paraná, Brazil; Bioprocess Engineering and Biotechnology Department, Federal University of Paraná (UFPR), Curitiba, Brazil

Alessandra Cristine Novak Sydney Federal University of Technology – Parana (UTFPR) - Campus Ponta Grossa, Ponta Grossa, Paraná, Brazil

Eduardo Bittencourt Sydney Federal University of Technology – Parana (UTFPR) - Campus Ponta Grossa, Ponta Grossa, Paraná, Brazil

Carla Tecelão Politécnico de Leiria, Escola Superior de Turismo e Tecnologia do Mar, MARE-Marine and Environmental Sciences Centre, Peniche, Portugal

José A. Teixeira Centre of Biological Engineering (CEB), University of Minho, Braga, Portugal

Antonia Terpou Department of Agricultural Development, Agri-food, and Natural Resources Management, School of Agricultural Development, Nutrition & Sustainability, National & Kapodistrian University of Athens, Psachna, Greece

Lalitesh K. Thakur Institute of Pesticide Formulation Technology, India

Claucia Fernanda Volken de Souza Food Biotechnology Laboratory, University of Vale do Taquari – Univates, Lajeado, RS, Brazil; Biotechnology Graduate Program, University of Vale do Taquari – Univates, Lajeado, RS, Brazil

Sudesh K. Yadav Center of Innovative and Applied Bioprocessing, India

Guoqiang Zhang National Engineering Laboratory for Cereal Fermentation Technology, Jiangnan University, Wuxi, P.R. China; School of Biotechnology and Key Laboratory of Industrial Biotechnology, Ministry of Education, Jiangnan University, Wuxi, P.R. China; Jiangsu Provisional Research Center for Bioactive Product Processing Technology, Jiangnan University, Wuxi, P.R. China

About the editors

Amit Kumar Rai

Dr. Amit Kumar Rai is presently working as a Scientist-C at the Institute of Bioresources and Sustainable Development, Tadong, Sikkim. He completed his doctorate from CSIR-Central Food Technological Research Institute, Mysore, His research interests are in the area of food biotechnology for the production of nutraceuticals and functional foods rich in bioactive peptides and isoflavones using microorganisms associated with traditional fermented foods of North East India. To his credit, Dr. Rai has three patents and 73 publications, including 48 papers, 2 books, and 23 book chapter. He has won several awards, including the Scientist of the year 2020 from National Environment Science Academy, New Delhi; Young Scientist Award 2016 from Association of Food Scientist and Technologists (India); Young Scientist Award 2015 by Association of Microbiologist of India; AU-CBT Research Excellence Award 2009 by the Biotech Research Society of India, Award. Dr. Amit is serving as an Associate Editor of *Frontiers in Food Science and Technology* (Section — Food Biotechnology) and an Editorial Board Member of *BMC Microbiology* (Section — Microbial Biotechnology).

Sudhir P. Singh

Dr. Sudhir P. Singh obtained PhD from the University of Lucknow, India, in 2011. Subsequently, he worked as a research associate and then as a project scientist at National Agri-Food Biotechnology Institute, Mohali, India. He joined the Center of Innovative and Applied Bioprocessing, Mohali, India, in 2015 as a Scientist-C. He has been working in the area of molecular biology and synthetic biology. His current research focus is toward the development of biocatalysts for biotransformation of agro-industrial by-products and residues into value-added biomolecules. He has published 60 research papers, 5 review articles, edited 6 books, and has 7 patents to his credit.

Ashok Pandey

Prof. Ashok Pandey is currently a distinguished scientist at the Centre for Innovation and Translational Research, CSIR-Indian Institute of Toxicology Research, Lucknow, India, and Executive Director (Honorary) at the Centre for Energy and Environmental Sustainability—India. Formerly, he was an eminent scientist at the Center of Innovative and Applied Bioprocessing, Mohali, and Chief Scientist & Head of Biotechnology Division and Centre for Biofuels at CSIR's National Institute for Interdisciplinary Science and Technology, Trivandrum. His major research and technological development interests are industrial and environmental biotechnology and energy biosciences, focusing on biomass to biofuels and chemicals, waste to wealth and energy, industrial enzymes, and so on. Prof. Pandey is adjunct/visiting professor/scientist in universities in France, Brazil, Canada, China, Korea, South Africa, and Switzerland and also in several universities several in India. He has around 1550 publications/communications, which include 16 patents, 90 books, and more than 800 papers and book chapters with an h-index of 112 and approximately 55,000 citations (Google Scholar). Prof. Pandey is the recipient of many national and international awards and honors, which include Distinguished Fellow, the Biotech Research Society, India; Highest Cited Researcher (Top 1% in the world), Clarivate Analytics, Web of Science (2020, 2019, and 2018); Top Scientist in Biotechnology (#1 in India and #8 in the world), Stanford University world ranking (2020); Fellow, World Society of Sustainable Energy Technologies (2020); Fellow, Indian Chemical Society (2020); Distinguished Scientist, VDGOOD Professional Association, India (2020); Distinguished Professor of Eminence with global impact in the area of Biotechnology, Precious Cornerstone University, Nigeria (2020); IconSWM Life-time Achievement Award 2019, International Society for Solid Waste Management, KIIT, Bhubaneshwar, India (2019); Yonsei Outstanding Scholar, Yonsei University, Seoul, Korea (2019); Life-Time Achievement Award from the Biotech Research Society, India (2018); Life-Time Achievement Award from Venus International Research Awards (2018), Most Outstanding Researcher Award from Career360 (2018), Life-Time Achievement Award from the International Society for Energy, Environment and Sustainability (2017); Fellow, Royal Society of Biology, UK (2016); Fellow, International Society for Energy, Environment and Sustainability (2016); Academician of European Academy of Sciences and Arts, Austria (2015); Fellow, National Academy of Sciences, India (2012); Fellow, Association of Microbiologists of India (2008); Honorary Doctorate degree from Université Blaise Pascal, France (2007); Fellow, International Organization of Biotechnology and Bioengineering (2007); Thomson Scientific India Citation Laureate Award, USA (2006); Fellow, the Biotech Research Society, India (2005); UNESCO Professor (2000); Raman Research Fellowship Award, CSIR (1995); GBF, Germany and CNRS, France Fellowships (1992), and Young Scientist Award (1989). Prof. Pandey is Founder President of the Biotech Research Society, India (www.brsi.in); Founder and International Coordinator of International Forum on Industrial Bioprocesses, France (www.ifibiop.org), Chairman of the International Society for Energy, Environment & Sustainability (www.isees.in), Editor-in-Chief of *Bioresource Technology* (http://ees.elsevier.com/bite/), Honorary Executive Advisor of *Journal of Energy and Environmental Sustainability* (www.jees.in), *Journal of Systems Microbiology and Biomanufacturing* (https://www.springer.com/journal/43393), *Journal of Environmental Sciences and Engineering* (http://neerijese.org/editorial-board/), Subject Editor, *Proceedings of National Academy of Sciences*, India (https://www.springer.com/life+sciences/journal/40011), and Associate Editor, *Biologia*—Section Cellular and Molecular Biology (https://www.springer.com/journal/11756/editors) and editorial board member of several international and Indian journals.

Christian Larroche

Prof. Christian Larroche is Director of Polytech Clermont-Ferrand, a graduate school of engineering of University Clermont-Auvergne, France. He is also member of the research laboratory Institut Pascal and of the laboratory of excellence ImobS3 at the same university. Prof. Larroche has strong research skills and expertise in the area of applied microbiology and biochemical engineering. He is an author of 220 documents, including 114 articles, 3 patents, 15 book chapters, and 24 books or journal special issues. He is a member of French Society for Process Engineering (SFGP), French Society of Biotechnology, and European Federation of Chemical Engineering. He is an executive of IBA-IFIBiop and an editor of *Bioresource Technology*.

Carlos Ricardo Soccol

Prof. Carlos Ricardo Soccol is the research group leader of Department of Bioprocess Engineering and Biotechnology at the Federal University of Paraná, Brazil. He graduated in Chemical Engineering (UFPR, 1979), Master in Food Technology (UFPR, 1986), and PhD in Genie Enzymatique, Microbiologie et Bioconversion (Université de Technologie de Compiègne, France, 1992) and carried out postdoctoral studies at Institut ORSTOM/IRD (Montpellier, 1994 and 1997) and at the Université de Provence et de la Méditerranée (Marseille, 2000). Prof. Soccol is HDR professor at Ecole d'Ingénieurs Supériure of Luminy, Marseille-France. He has experience in the areas of science and food technology, with emphasis on agro-industrial and agroalimentary biotechnology, acting in the following areas: bioprocess engineering and solid state fermentation, submerged fermentation, bioseparations, industrial bioprocesses, enzyme technology, tissue culture, bio-industrial projects, and bioproduction. Prof. Soccol received several national and international awards, which include Science & Technology award of the Govt. of Paraná (1996), Scopus/Elsevier award (2009), Dr. Honoris Causa, University Blaise Pascal-France (2010),

Outstanding Scientist—5th International Conference on Industrial Bioprocesses, Taipei, Taiwan (2012), Elected Titular Member of the Brazilian Academy of Sciences (2014). He has more than 1000 publications/communications, which include 17 books, 107 book chapters, 270 original research papers, 557 research communications in international and national conferences, and has registered 44 patents. His h-index is 83 with approximately 32,000 citations (Google Scholar).

Preface

The book entitled *"Technologies for Production of Nutraceuticals and Functional Food Products"* is a part of the comprehensive series on *Current Developments in Biotechnology and Bioengineering* (Editor-in-Chief: Ashok Pandey). Functional foods and nutraceuticals are gaining popularity as they have specific health benefits beyond their nutritional role. Globally growing demands for functional foods and value-added nutraceuticals for prevention and treatment of diseases have rendered them a multibillion dollar market. Microorganisms from different sources with specific enzyme-producing ability have been used for the production of nutraceuticals and fermented functional foods. The nutraceuticals are produced either by biotransformation reactions of food components or by microorganisms. Native and recombinant microorganisms can be used in developing processing technologies in the food, feed, and nutraceutical industries. Due to increasing awareness and the demand of specific nutraceuticals, microbial and enzyme technologies are gaining popularity globally. This book will provide comprehensive information on recent technologies related to high-value products for the functional food industry.

This book intends to cover a wide range of topic related to biotechnological approaches for the production of high-value nutraceuticals and fermented functional foods. The scientific information is contributed by the global experts in their specific areas and compiled in the book. The bioactive compounds derived from foods substrate covered in this book mainly include bioactive peptides, polyphenols, oligosaccharides, and prebiotics. This book includes chapters on food enzymes responsible for producing nutraceutical and fibrinolytic enzymes, which acts as therapeutic agents. Furthermore, there are chapters on important technologies related to bovine colostrums and their effect on human and animal therapy. Apart from the improvement of nutraceutical potential, there are chapters on the effect of microbial fermentation on reducing antinutritional factors and approaches for reducing mycotoxin in food. The strategic approach of microbial fermentation for the recovery of nutraceuticals from agri-food processing waste is also described. The scientific information related to recombinant microorganisms and their role in the production of nutraceutical and functional foods is an essential component in the book. The translational aspects of microbial bioprocess technologies are illustrated, by emphasizing the present status, current requirements, and future perspectives for the production of nutraceuticals and functional foods.

This book would be a unique source of knowledge with state-of-the-art technologies ideal for the researchers, graduate students, educators, and policy makers; each chapter presents detailed discussions and comprehensive knowledge with state-of-the-art information.

We highly appreciate authors' contribution from different countries presenting comprehensive scientific information on recent developments on the technologies for the development of functional foods and nutraceuticals. We strongly believe that the enriched scientific information offered in this book will be beneficial for different sections of scientific and academic communities. We gratefully acknowledge the contribution of the reviewers' efforts in the critical review of the chapters, which led to the scientific enrichment of this volume. We acknowledge Dr. Kostas Marinakis, Former Senior Book Acquisition Editor; Ms. Katie Hammon, Senior Acquisitions Editor; Desamero, Maria Elaine D. Desamero, Editorial Project Manager; Vijay Bharath Rajan, Project Manager, and the entire production team of Elsevier for their support and cooperation in publishing this book.

Amit Kumar Rai
Sudhir P. Singh
Ashok Pandey
Christian Larroche
Carlos Ricardo Soccol

Microbial bioprocesses for production of nutraceuticals and functional foods

Loreni Chiring Phukon[1], Sudhir P. Singh[2], Ashok Pandey[3], Amit Kumar Rai[1]

[1]INSTITUTE OF BIORESOURCES AND SUSTAINABLE DEVELOPMENT, REGIONAL CENTRE, TADONG, INDIA [2]CENTER OF INNOVATIVE AND APPLIED BIOPROCESSING, MOHALI, INDIA [3]CSIR-INDIAN INSTITUTE OF TOXICOLOGY RESEARCH, LUCKNOW, INDIA

1.1 Introduction

Unhealthy diets and lifestyles in this modern era have become one of the major reasons contributing to the emergence of life-threatening diseases and malnutrition worldwide. Awareness among the common people to combat the risk of diseases and proper intake of nutrients has focused on the consumption of functional foods and nutraceuticals with health beneficial properties [1,2]. Nutraceuticals are a structurally and functionally diverse group of bioactive compounds imparting nutrition as well as therapeutic activity [3]. In simple words, nutraceuticals can be defined as a hybrid of nutrition and pharmaceuticals, which helps in maintaining sound health [4,5]. Consumption of nutraceuticals helps in boosting up the immune and digestive system, prevents or controls several prevailing health concerns of the present generation such as cardiovascular diseases, obesity, cancer, inflammation, gastrointestinal diseases, hypertension, and aging-associated diseases, and it also imparts antimicrobial activities [3,6,7]. The global market of nutraceuticals reached a value of US$382.51 billion in 2019 from 171.8 billion in 2014, and it is estimated to reach a value of approximately US$722.49 billion by 2027 [3,8].

The increasing demand for nutraceuticals in the global market has led researchers to look for different sources, apart from the conventional production through the nutraceutical industries. Plant- and animal-based diets are rich sources of bioactive compounds with nutraceutical properties; however, the low abundance of raw material, limited capability to chemical modification, and sustainability have shifted the focus toward microbial fermentation, microbial-based genetic manipulation, and enzyme-catalyzed production of nutraceuticals [3,9]. Nutraceuticals, including micronutrients, vitamins, gamma-aminobutyric acid (GABA), pigments, oligosaccharides, lipids, polyphenols, and peptides, are some of the major health-promoting bioactive compounds trending in the global market.

Functional foods are similar to conventional foods in appearance and consumed as part of a normal diet but with added benefits of health-promoting and disease-preventing properties along with basic nutritional properties [10,11]. The functionality of conventional foods can be enhanced through microbial fermentation [12,13]. Increasing malnutrition and chronic diseases worldwide is the reason for the increase in the global market for functional food [14]. Bioactive compound-enriched foods provide health beneficial properties like antihypertensive, antiinflammatory, immunomodulatory, anticancer, antithrombotic, antioxidative, and antimicrobial properties [1,10,15]. Fermented foods consumed in different parts of the world include various fermented dairy and nondairy products such as *Kinema, Natto, Cheonggukjang, Tungrymbai, Tempeh, Thua-nao, Hawaijar, soy milk, soy sauce, Tofu, Kimchi, cheese, curd, sour-milk, Kefir, PaoCai,* and *Kumis* [2,11,12,15]. Various researches throughout the world are studying the underlying pathways involved in the addition of functionality of common food products upon traditional fermentation as well as to develop new fermentation techniques with enhanced functionality, hygiene process, increased yield, use of defined starter strains, and standardized fermentation processes. This chapter focuses on different bioprocess technologies involved in the development of various functional foods and nutraceuticals along with the underlying health beneficial properties of some of the functional foods and nutraceuticals.

1.2 Microbial production of nutraceuticals and functional foods

There are different types of nutraceuticals and functional foods produced using microbial processes (Fig. 1–1). The microorganism produces these bioactive compounds directly or via the transformation of food substrate into value-added compounds. The bioactive compounds exhibit different types of health benefits (Fig. 1–2). The different types of functional foods and nutraceuticals produced using microbial processes are discussed below.

1.2.1 Polyphenols

Polyphenols are phenolic secondary metabolites produced by plants and are the richest source of antioxidants in our day-to-day diet. Polyphenols are among the most widely available bioactive compounds in nature and have potential nutraceutical activities including antioxidative, antidiabetic, antiinflammatory, anticancer, antimicrobial, and neuroprotective properties [16,17]. Polyphenols range from simple molecules to highly complex molecules conjugated with hydroxyl-linked-sugar residues [18]. There are several thousands of phenolic compounds available in nature but based on the chemical structure and complexity, polyphenols are categorized as flavonoids and nonflavonoids [19,20]. Flavonoids are a group of polyphenols with a 15-Carbon backbone with two phenyl rings and an oxygenated heterocyclic ring and are further classified according to their structural diversity into flavanones, flavan-3-ols, flavanonols, isoflavones, flavonols, flavones, dihydroxyflavonols, and anthocyanidins, while nonflavonoids includes phenolic acid, lignans, and stilbenoids [16,21]. Although polyphenols are widely distributed in nature and regularly consumed with normal

FIGURE 1–1 A dendrogram illustrating a schematic pathway involved in the microbial-based production of bioactive compounds using different food source/substrate and technology.

diets, their bioavailability and poor absorption become a major constraint in availing the beneficial effects of this highly functional compound. Microorganisms play a vital role in transforming unavailable conjugated polyphenols to easily absorbable simple phenolic compounds [22]. Fermented food products with bioavailable polyphenol are among the food products currently in demand for nutraceutical properties.

Legumes are consumed worldwide daily and are a rich source of protein and bioactive compounds including polyphenols, vitamins, and lipids [13]. Fermented soybean products are a rich source of polyphenols that are consumed in most Asian countries. *Bacillus amyloliquefaciens* and *Bacillus licheniformis* isolated from traditional *Kinema* (fermented soybean

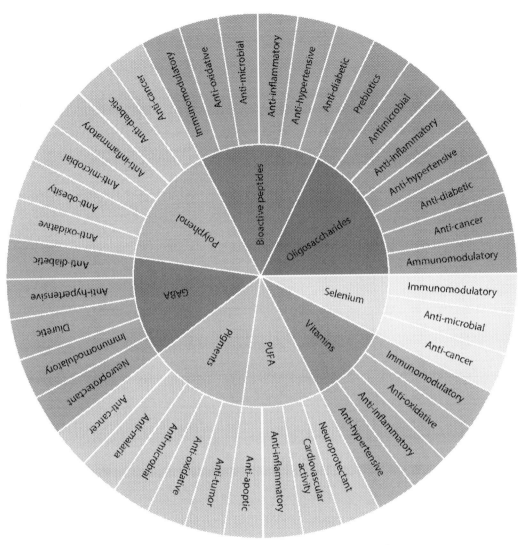

FIGURE 1-2 Sunburst chart illustrating the biological activities of bioactive compounds.

product from India) were used for fermentation of soybean resulting in the production of free polyphenol enriched bioactive soybean hydrolysate [13,23]. Fermentation of soybean by *Saccharomyces cerevisiae* ATCC 32052 resulted in the production of a high polyphenolic extract with potent antioxidative activity [24]. A steep rise in polyphenol content with high antioxidative activity was observed in Korean fermented soybean *Cheonggukjang* prepared through fermentation of soybean and hazelnut using *Bacillus subtilis* KCCM 11316 [25]. *Rhizopus oligosporus* strain NRRL 5905 was used for fermentation of soybean for production of an Indonesian soybean product *Tempeh* with high isoflavone aglycones [26]. Apart from fermented legumes, fermented cereals, and leaves are also consumed for their high phenolic contents. Fermentation of oats by *Monascus anka* and *B. subtilis* resulted in the generation of free polyphenols, enhancing the functionality of oats [27]. Metabolomics studies of sourdough fermented by *Lactobacillus plantarum* showed the importance of these microbes for improving the functionality of sourdough by releasing antioxidative compounds such as phenolic compounds and flavonoids [28]. A significant increase in nutritional contents and total phenolic content of Bamboo Tree (*Sasacoreana* Nakai) leaves was observed after fermentation by *B. subtilis* KCCM 11315 [29]. The study suggests the potential use of fermented leaves as a seasoning in various foods for enhancing the nutritional quality.

Along with solid diets, beverages are consumed all over the world from the ancient ages. Gaining knowledge and enhancing the functionality of traditional beverages along with the development of new functional beverages are the focus of researchers working in the field of functional foods. A potentially high antioxidative activity exhibiting functional soy beverage with high total phenolic contents and isoflavone aglycone was produced from fermented soy whey using *Lb. plantarum* B1−6 [30]. An antiobesity fermented beverage with increased polyphenolic content was prepared through fermentation of *Houttuynia cordata* leaves and green tea using *Lactobacillus paracasei*, oral administration of this product to obese mice showed increased lipolytic activity resulting in inhibition of lipogenesis of adipocytes and reduced body weight [31]. The agricultural waste product, mango peel, was used to improve the functionality of probiotic fermented *Kefir* milk using *Kefir* microorganisms resulting in the development of novel fermented milk enriched with free phenolic compounds from mango peel [32]. While in a similar study, high antioxidative activity exhibiting fermented milk with hydrolyzable polyphenols was prepared using pomegranate peel by fermenting with *Lactobacillus acidophilus* [33]. Two-step fermentation process using *Lactobacillus delbrueckii* ssp. *Bulgaricus*, *Streptococcus thermophilus*, and *Bifidobacterium longum* were employed for the production of a functional fermented soy milk-tea enriched with bioactive compounds from tea [34]. The resulting product showed high total phenolic compounds and isoflavones content, resulting in high antioxidant activity.

Microorganisms not only play an important role in the bioavailability of polyphenols through fermentation but it has also emerged as a cost-effective alternative source to plant-based production of polyphenols through the introduction of crucial plant genes for enhanced production. Genetic engineering of microbes serves as an eco-friendly approach for the production of plant-based polyphenols. The first microbial-based polyphenols produced from engineered microorganisms are naringenin and pinocembrin. *Escherichia coli*, *Streptomyces venezuelae*, *S. cerevisiae*, *Lactococcus lactis*, and *Corynebacterium glutamicum* are some of the most successful microbial hosts used for the production of microbial-based polyphenols [35,36]. *E. coli* and *S. cerevisiae* have been genetically engineered for the production of a numbers of polyphenols such as flavonols, flavanones, naringenin, and genistein [37]. Several studies on the genetic remodeling of *C. glutamicum* reported the production of polyphenols like hydroxybenzoic acid, naringenin, and noreugenin, and the reports suggest *C. glutamicum* as a potential host for microbial production of plant polyphenols [38−40]. Microbial-based production of stilbenes through genetic manipulation of *L. lactis* was also reported [41]. This finding shows the use of such microbial hosts for fast and enhanced production of microbial-based polyphenols having therapeutic potentials.

1.2.2 Selenium

Selenium (Se), an essential micronutrient for both humans and animals, got its name after Selene, the Greek goddess of the moon due to its dual nature of essentiality just like the bright and dark side of the moon [42]. Selenium has antioxidative properties and its deficiency causes oxidative stress resulting in various health issues such as infertility, cardiovascular diseases, Keshan disease, cancer, and neurological disorders [11,43,44]. Intake of a particular amount of selenium in the diet protects against free radicals, carcinogenic components, and pathogenic bacteria; however, it also has a dark side when the dosage exceeds the recommended amount resulting in promoting of disease such as type 2 diabetes and liver tumors [42,45,46]. Naturally, selenium occurs in two forms: organic and inorganic selenium, inorganic forms are present as mineral supplements such as selenite and selenite. The organic form of selenium includes Selenomethionine (SeMet) and Selenocysteine (SeCys), two seleno amino acids found in nuts, meat, fish, marine foods, eggs, and cereals [43,45,47,48].

Although selenium-rich foods are available in nature, less or excess dietary intake of selenium could harm the wellbeing rather than providing beneficial effects. To avoid the deficiency or toxicity of selenium intake, microbial-based selenium-rich supplements such as Se-yeast and Se-lactic acid bacteria (LAB)/*Bifidobacterium* with improved nutraceutical activity are recommended [46,48,49]. The cell wall of yeast cells consists of numerous groups of cell wall proteins and polysaccharides that are biological binders of selenium [50]. Yeast cells can be used for the production of selenium-enriched functional food as well as Se-yeast for nutrition supplement. Se-yeast is the product of aerobic fermentation of inorganic Se-enriched medium. Inorganic selenium is transformed into bioactive SeMet, the major seleno amino acid accumulated in the yeast cell wall [51]. Selenium metabolomics studies of Se-yeast identified about 60 different selenium metabolites [52]. Two yeast species commonly used for the production of Se-yeast are *Saccharomyces* and *Candida*, and *S. cerevisiae* is among the potential microorganisms to transform inorganic selenium to active form SeMet [53]. Various statistical approaches are applied to increase the productivity of Se-yeast, optimization of fermentation condition such as the concentration of selenium, glucose along with optimized temperature, and aeration for *S. cerevisiae* increased the biomass production of Se-yeast from 3.73 to 5.89 g/L [54]. Se-yeast is used for the direct dietary supplement as well as for the production of functional food such as functional ice cream, Se-enriched yeast dietary bakery, Se-enriched beer, and other cereal-based foods [51,55,56]. Se-enriched *Candida utilis* was grown using inorganic selenium, and their therapeutic properties were studied. Experimental supplementation of Se-enriched *C. utilis* to 40 rats significantly improved growth rate, antioxidative activity, and immune system compared to control [57]. In another study, rats supplemented with Se-enriched *C. utilis* showed enhanced antioxidant capacity [58]. Metastatic brain tumor of rats was subsequently decreased upon feeding of Se-yeast, suggesting its potentiality in the prevention of various metastatic diseases [49].

Probiotic Bifidobacteria and LAB have gained limelight in the food and nutraceutical field due to their involvement in the wellbeing of humans as probiotics, fermentative bacteria, and for the production of valuable biomolecules with health beneficial properties. The capability of some LAB and Bifidobacteria to accumulate and transform inorganic Se into a more active form of Se has interested researchers in Se-enriched bacterial fermentation for the production of selenium-enriched food and supplements [59]. In separate studies, different LAB strains were studied for biotransformation of inorganic selenium to organic selenium. Se-enriched *Lb. bulgaricus* generated by supplementation of inorganic selenium in growth medium showed high intracellular accumulation of selenium with an increase in selenium concentration [60]. Another study on two potentially probiotic strains, *Lb. acidophilus* CRL 636 and *Lactobacillus reuteri* CRL 1101, also showed the transformation of inorganic selenium to Se-nanoparticles (SeNPs) and intracellular accumulation as SeMet and SeCys [61]. Three LAB strains: *Lactobacillus brevis* CRL 2051, *Lb. plantarum* CRL 2030, and *Fructobacillustropaeoli* CRL 2034, was able to transform inorganic selenium to organic form from a study conducted with 96 LAB strains, the study reported the formation of SeNPs and intracellular accumulation of only SeCys [62]. The above studies suggest the potentiality of these LAB strains for application in the production of Se-enriched fermented foods and SeNPs. Absorption and distribution of selenium upon consumption of Se-enriched probiotic *B. longum* DD98 (Se-DD98), Se-yeast, and inorganic selenium was studied in rat models, and it was observed that organic selenium (Se-DD98 and Se-yeast) was easily absorbed and distributed in various organs than inorganic selenium [63]. Comparative analysis of the antibacterial activity of two LAB, *Lb. delbrueckii* ssp. *Bulgaricus* and *S. thermophilus* with and without selenium enrichment showed high antibacterial activity of Se-enriched LAB than LAB without enrichment against four pathogenic bacteria, *Salmonella typhimurium*, *E. coli*, *Staphylococcus aureus*, and *Listeria monocytogenes* [46]. In a similar study on the antimicrobial activity of two SeNPs-enriched probiotic strains *Lb. plantarum* and *Lactobacillus johnsonii* showed inhibition of *Candida albicans* [64]. Immunomodulatory effect of SeNPs-enriched probiotics, *Lb. brevis* and *Lb. plantarum* in two studies on tumor-induced mice by feeding SeNPs-enriched probiotics before as well as after tumor induction showed a significant rise in NK cytotoxicity, IFN-γ, and IL-17 levels along with a decrease in tumor volume and longer life span [65,66].

The development of selenium-enriched functional foods along with direct supplements of Se-yeast and Se-enriched *Bifidobacteria* and LAB is among the major approaches under consideration to meet the worldwide demand of this essential micronutrient. Bread produced through fermentation of dough using Se-yeast was found to contain bioavailable SeMet in the range of 50–200 μg per 100 g of dough approximately meeting up to 25%–100% of the daily recommended concentration of selenium for human consumption [56,67]. SeNPs-enriched yogurt with an enhanced concentration of selenium was produced by coculturing yogurt starter culture and selenium-enriched *Lb. brevis* [68]. In another study, a potential selenium-enriched beverage was developed through fermentation of blended juices by Se-enriched probiotics [69]. The development of Se-enriched functional foods is among the most promising bioprocess technologies to meet the global demand for selenium for human consumption. Studies on the generation of Se-enriched yeast, probiotic, and Se-enriched functional foods showed the potential application of these cultures and food products as nutraceuticals for combating various ailments related to the deficiency of selenium.

1.2.3 Vitamins

Vitamins are micronutrients, essential for the normal development, maintenance, and functioning of living beings. Vitamins act as cofactors/coenzymes in metabolic reactions and are categorized into fat-soluble (A, D, E, and K) and water-soluble (B and C) vitamins [70]. Vitamins are essential micronutrients for most animals while most plants and microbes can synthesize naturally [71]. Vitamins are consumed worldwide for their nutritive properties as well as for their anticancer, immunomodulatory, antiinflammatory, antioxidative, and antihypertensive properties [72–74]. Vitamins are commonly produced through chemical processes but the production of hazardous wastes and the use of nonrenewable chemicals during the chemical process has been one of the major driving forces for the need for biotechnological process-based production of vitamins [70]. Currently, microbial production of vitamins has emerged as a better alternative over chemical synthesis due to advantages such as flexibility on choice of microorganisms, renewable source of biomass, and eco-friendly process [75].

1.2.3.1 Riboflavin (B_2)

Riboflavin plays a key role in the cellular metabolism of an organism as it the precursor of two important coenzymes flavin mononucleotide and flavin adenine dinucleotide. The deficiency of riboflavin can cause common serious health issues like cancer, cardiovascular diseases, eye disorders, and neurological disorders [76]. Flavogenic fungi and bacteria are commonly used in the production of riboflavin on an industrial scale, *Ashbyagossypii* and *B. subtilis* are two main strains used industrially [70]. Various carbon sources such as whey, molasses, palm oil, corn-steep liquor, and other agricultural wastes have been used for the production of riboflavin using *A. gossypii* [75]. Genetic manipulation is a crucial step toward enhanced production and study of the pathways involved in the production of microbial ingredients. Metabolic engineering of *A. gossypii* resulted in overproduction of riboflavin up to 5.4-fold higher than the wild strains [77]. *Candida famata*, a riboflavin producing yeast under iron scarce, was engineered, and production of riboflavin was carried out under optimized conditions resulting in overproduction of riboflavin up to 16.4 g/L [78]. LAB has also been used for the production of riboflavin due to its applicability in the production of food and mainly due to the presence of complete riboflavin operon required for the biosynthesis of riboflavin. Two probiotic LAB strains, *Lb. plantarum* RYG-YYG-9049 and *Lb. plantarum* RYG-YYG-9049-M10 isolated from pickle showing resistance to roseoflavin (a toxic analog of riboflavin that causes mutations in LAB) showed high production of riboflavin during soymilk fermentation [76]. In similar studies, riboflavin content on soymilk was significantly increased upon fermentation using *Lb. plantarum* CRL 725 and *Lb. plantarum* BBC32B [79,80]. Riboflavin-enriched bread was produced using roseoflavin resistant *Lb. plantarum* strains and *Lactobacillus fermentum* PBCC11.5 in two different studies, there was a significant increase in final riboflavin content of bread with up to threefold and twofold increase, respectively [81,82].

The above studies suggest the possible use of the biotechnologically processes for overproduction of riboflavin on an industrial scale and shows the potentiality of genetic manipulation for enhanced production. The capability of LAB strains to produce high riboflavin during soymilk and bread fermentation shows the applicability of these strains for the production of riboflavin-enriched fermented foods.

1.2.3.2 Vitamin K

Vitamin K occurs in two forms, phylloquinone (PK) and menaquinone (MK), and is fat-soluble vitamins. PK occurs abundantly in plant-based foods but has less bioactivity, while MK is predominantly synthesized by microbes and has higher bioactivity than PK [83].

Vitamin K plays a vital role in several physiological activities such as cholesterol metabolism, growth regulation, homeostasis, calcium, and bone metabolism [84]. Most MKs has unsaturated 5-carbon prenyl side chains with varying length ranging from 4 to 13prenyl units, and MKs are classified based on the number of prenyl units MK4−13. MK provides numerous health benefits including prevention of cardiovascular diseases, anticancer activity, anticoagulation activity, and osteoporosis [85−87]. As MKs have multiple health benefitting properties and are synthesized mostly by fermenting microbes has led to the production of MK-rich fermented foods on an industrial scale [88]. Fermented dairy and soybean products are the richest sources of MK, and *Natto*, a Japanese fermented soybean product fermented by *B. subtilis* natto, has been reported to have the highest amount of MK7 [86].

B. subtilis natto is one of the main strains used for the production of MK industrially, and several studies have been done for enhanced production of MKs, such as optimization of fermentation conditions, genetic manipulation, and use of different substrates. Mutation of *B. subtilis* natto followed by statistical optimization of fermenting conditions resulted in an about 1.66-fold increase in MK production as compared to controlled [89]. Economically production of MK was successfully studied through the fermentation of wheat starch wastewater by *B. subtilis* W-17 [90]. Long chained MKs were produced using six *L. lactis* strains using different carbons source and fermentation conditions, and it was observed that aerobic fermentation of fructose and trehalose resulted in 3.7- and 5.2-fold increased production of MKs as compared to static fermentation of glucose [84]. Studies on the use of different strains symbolize the utility of these strains for enhanced production of safe and health beneficial MKs and MK-enriched food products.

1.2.4 Gamma-aminobutyric acid

Gamma-aminobutyric acid (GABA), a naturally occurring nonprotein amino acid widely produced by various plants, animals, and microorganisms, acts as a potent inhibitory neurotransmitter in the mammalian central nervous system [91−93]. Apart from being an inhibitory neurotransmitter, it has antidiabetic, antihypertensive, diuretic, immunomodulatory, and tranquilizing properties [91,94−96]. The health beneficial properties of GABA have promoted the consumption and demand of GABA-rich food supplements and fermented foods. Fermented food products are the richest source of GABA, and many food-grade microbes involved in the fermentation of traditional fermented foods have been reported to produce GABA [96,97]. LABs are the most dominating microbes found in fermented foods imparting enhanced functionality. LABs isolated from fermented food products have been reported for the production of GABA. A high GABA-producing probiotic strain *Lb. brevis* DSM 32386 isolated from alpine cheese showed 200% GABA production using synthetic medium [98], *Lb. brevis* VTCC-B-454 isolated from pickle showed high GABA production on fermentation of rice bran defatted extract [99], while in a similar study, an effective fed-batch fermentation method was optimized for large-scale production of GABA using *Lb. brevis* NCL912 isolated from a Chinese fermented vegetable *Paocai* [95]. Enhanced production of GABA up to 35.47 g/L was observed during batch fermentation of empty fruit bunch solution using engineered *C. glutamicum* strains [100]. GABA-enriched yogurt was developed by fermentation of milk and monosodium glutamate with optimized conditions using *S. thermophilus* fmb5 [101]. Low cost, improved functionality, high yield, and eco-friendly methods of production are the major objectives in the commercialization and production of any product on an industrial scale. Various studies are going on the development of microbial-based production techniques for cost-effective production of GABA-rich foods and supplements to meet the demands in the global market. Microbes isolated from traditional Chinese fermented food products were used for the production of GABA-enriched novel mulberry beverage, mulberry leaf powder, and fermented milk were in separate studies by coculturing *S. cerevisiae* SC12 and *Lb. plantarum* BC114, *Lactobacillus pentosus* SS6, and *S. thermophilus* GABA, respectively [96,97,102]. Eco-friendly and edible GABA-rich fermented soy protein (GABA-RFSP) biofilms suitable for food packaging were prepared through fermentation of soy protein isolate, chitosan, and glutamate by *Lb. brevis* [103]. Response surface methodology following composite central design was applied for the production of novel *Kung-Som* (a traditional Thai fermented shrimp) using autochthonous isolate *Lactobacillus futsaii* CS3 as a starter culture. A fourfold increase in GABA content was observed in novel *Kung-Som* as compared to controlled (without starter culture) and commercialized *Kung-Som* products [104].

Bioactivity of GABA- and GABA-enriched fermented foods has been demonstrated through in vitro and in vivo studies revealing the potentiality of the compound [94,105,106]. Oral administration of two GABA-producing *Lb. brevis* strains in diet-induced obese mice showed a reduction in symptoms of metabolic syndromes [105]. Antihypertensive and antioxidative activity of GABA-enriched *idli* (fermented food of India) fermented by *Aspergillus oryzae* NSK was observed in hypertensive rats, lower systolic blood pressure, and significantly higher glutathione reductase and superoxide dismutase activities were observed as compared to plain *idli*-fed hypertensive rats [106]. Suppressed adhesion of *E. coli* ATCC25922 and *Salmonella enterica* C29039 with increased expression of tight junction proteins and TGF-β cytokine involved in maintaining the integrity of epithelial barrier was observed in an in vitro study on Caco-2 cells with supernatant of GABA-producing *Lb. brevis* BGZLS10-17 isolated from artisanal *Zlatar* cheese suggesting its role as an antimicrobial and antiinflammatory agent [94].

The above studies demonstrated the possibility of the production of GABA-enriched fermented foods and supplements on an industrial scale through the use of food-grade GABA-producing isolates, optimization of fermentation conditions, genetic manipulation, and substrate selection. The health beneficial effects of GABA-enriched foods on experimental animal models project the applicability of these foods not just as a source of nutrition but also as a therapeutic agent for the wellbeing of humankind.

1.2.5 Pigments

Pigments are chemical compounds that absorb specific wavelengths of light and reflect certain wavelengths of light [107]. Varieties of pigments having different chemical composition, color, and functional properties are produced by plants, animals, and microorganisms. Microorganisms possesses an advantage over other sources of pigment owing to their high yield, ease of cultivation, cost-effectiveness, and multidisciplinary applicability of such pigments as a colorant in foods, cosmetics, textiles, and nutraceuticals exhibiting various antimicrobial, antitumor, antioxidant, anticancer, and antimalarial activity [107,108]. Carotenoids, monascins, flavins, phenazines, indigo, melanin, quinones, and prodigiosin are some of the major pigments produced by microorganisms that have wide application in the industrial world [109].

Monascus species producing red, orange, and yellow pigments are used for the production of fermented rice in Eastern Asia. Red mold rice, also known V as *Koji* in Japan, *Angkhak*, and *Honchi* in China is a famous *Monascus* fermented rice [110]. *Monascus* fermented red rice possesses bioactive metabolites with numerous health beneficial properties including antioxidative, antihypertensive, antihyperglycemic, antidiabetic, and antiobesity effects [111,112]. *Monascus* pigments (MPs) have wide application as a coloring agent in the food industry and nutraceutical industry for their therapeutic properties [113]. Low yield and high production cost are the major constraints in the large-scale production of MPs. Strain improvement, optimization of fermentation conditions, and use of cheaper raw materials are some of the strategies understudy for the cost-effective production of MPs. High yield and low-cost process for the production of MPs was investigated using a mutant strain *Monascus purpureus* M630 for fermentation of rice straw hydrolysate, where the mutant strain was able to produce up to 34.12 U/mL MPs under submerged fermentation (SF) condition [113]. High MP production was observed during SF of rice powder using two engineered strains *Monascus ruber* ACL501969-11 and *M. ruber* ACL438469-4 with overexpressed ATP-citrate lyase genes, production of MPs by engineered strains was much higher than the parent strain [114]. A novel red mold cheese with high soluble nitrogen content was developed using *M. ruber* NBRC 32318 during the ripening stage [112]. The antiobesity effect of red mold rice was demonstrated using 3T3-L1 cell line, the extract of red mold rice suppressed the growth and differentiation of preadipocytes resulting in inhibition of new adipocytes [111]. Effect of red mold rice extract on hamsters with induced tumor showed a significant decrease in tumor formation with a decreased level of oxidative stress and inflammation [115]. MPs not only have attractive colors but also have therapeutic values against chronic diseases.

Carotenoids are isopentenyl pyrophosphate (IPP) derived fat-soluble pigments, synthesized by plants, algae, fungi, and microorganisms as yellow, orange, and red pigments [116]. There are about 600 carotenoids pigments divided into two groups: Carotenes and Xanthophylls [117,118]. Carotenoids are grouped among the most important naturally produced pigment exhibiting health beneficial activities such as anticancer, antioxidant, immunomodulatory, antimicrobial, anticataract, and antiinflammatory activities, while some carotenoids act as a precursor molecule for the biosynthesis of vitamin A [117,119–121]. Carotenoids producing industries currently rely on chemical, plants, and microbial synthesis. Chemical- and plant-based syntheses of carotenoids have numerous pitfalls such as health deteriorating effects of synthetic carotenoids, production of hazardous wastes during chemical synthesis, the dependence of seasonal and geographical factors for extraction of plant-based carotenoid, low yield, and high production costs [120]. However, microbial production of carotenoids has the advantage over chemical and plant synthesis including the use of low-cost renewable substrates, less cultivation time, eco-friendly, strain improvement possibility, and process optimization resulting in high yield with low production cost [122]. A marine bacterium, *Gordonia terrae* TWRH01 isolated from Taiwan was investigated for the production of carotenoids in an optimized fermentation medium resulting in the production of a commercially important carotenoid pigment, echinenone [119]. Optimization of fermentation conditions for carotenoid production by coculturing *Lactobacillus casei* subsp. *casei* and *Rhodotorula rubra* GED2 using cheese whey-filtrate showed high carotenoid production with up to 60% β-carotene content [122]. *Rhodotorula glutinis* CBS 20 and *Rhodotorula gracilis* ATCC 10788 were used in separate studies for cost-effective production of high-value carotenoids using low-cost industrial wastewaters [123,124]. Optimized fermentation of agricultural waste by *Rhodotorula mucilaginosa* MTCC-1403 showed production of torularhodin, β-carotene, and torulene [125]. Microbial carotenoids have been demonstrated to high antimicrobial and antioxidative properties, which makes them a suitable candidate for the replacement of synthetic drugs. Carotenoids extract from a halophilic archaea, *Haloterrigena turkmenica* showed higher antioxidative activity than ascorbic acid, butylhydroxytoluene, and α-tocopherol [126]. Separate studies on the bioactivity of red yeast species, *R. glutinis* strains and *R. mucilaginosa* strains, showed high antioxidant activity as well as antimicrobial activity against pathogenic bacteria and fungi [127–129]. Biotechnological advantages of microbial production of pigment and high biological activity of microbial pigments as demonstrated by various studies threw light on various biotechnological aspects useful in enhanced production of pigments.

1.2.6 Bioactive peptides

Bioactive peptides are unique amino acid fragments of protein sequence having profound biological activity upon release from the parent protein through proteolysis [130–132]. Fermented food products such as fermented milk, soybean, cereals, fish, meat, and vegetables are rich sources of bioactive peptides [133] (Table 1–1). Microbial fermentation and enzymatic hydrolysis or a combination of both methods is generally employed for the production of bioactive peptides [157]. They have numerous biological properties such as antihypertensive, antimicrobial, antioxidative, antidiabetic, and immunomodulatory activities [158,159]. The therapeutic potential of peptides produced during the fermentation of varieties of protein-rich dairy and nondairy food products has been demonstrated through in vitro and in vivo studies [130,131,160].

Proteolysis of protein is the major phenomenon responsible for the formation of bioactive peptides. Lactic acid-producing microorganisms particularly LAB that have a well-characterized proteolytic system are the most dominating microorganisms in fermented dairy foods [157,161,162]. Starter culture, fermentation condition, and ripening duration play important roles in the production of milk-derived bioactive peptides. Peptidomics profiling of commercial Parmigiano-Reggiano (PR) cheese of three ripening durations, 12, 18, and 24 months showed the positive impact of ripening duration for release of bioactive peptides in PR cheese. The highest release of bioactive peptides (257) with previously assigned activity such as antihypertensive, antidiabetic, antioxidative, antimicrobial, anxiolytic, and immunomodulatory were found in 24-month ripened PR cheese [163]. Dutch-type cheese produced

Table 1-1 Bioactive peptides–rich functional foods and microbes involved in the production.

Fermented product	Fermenting microorganism	Bioactive peptides	Bioactivity	References
Prato cheese	*Lactococcus lactis* strains	β-CN (f193–209)	Antihypertensive	[134,135]
Yoghurt	*Lactobacillus helveticus* LH-B02	β-CN (f94–123)	Maintenance of intestinal homeostasis and antihypertensive	[136,137]
Yoghurt	*Lactobacillus delbrueckii* ssp. *bulgaricus*, *Streptococcus salivarius* ssp. *thermophilus*	LYQEPVLGPVRGPFPIIV, VLPVPQK, LQDKIHP, YPFPGPIPK,	Antihypertensive, antioxidative, antimicrobial	[138]
Yoghurt	*S. salivarius* subsp. *thermophilus* CH1, *Lb. delbrueckii* subsp. *bulgaricus* LB-12-DRI-VAC, *Lb. helveticus* CH 5, and *Lactobacillus acidophilus* 20552 ATCC	αS1-CN f(24–32), β-CN f(193–209)	Antihypertensive	[139]
Yoghurt	*Streptococcus thermophiles*, *Lb. delbrueckii* subsp. *bulgaricus* YF-L812, *Lb. helveticus* LH-B02	MQTDIMFTIGPA	Antihypertensive	[140]
Fermented milk	*Lactobacillus rhamnosus* MTCC 5945 (NS4)	VPP, IPP	Antihypertensive	[141]
Fermented milk	*Lactobacillus casei* PRA205	FSDI-PNPIGSENSEKTTMPLW	Antioxidative	[142]
Fermented milk	*Lc. lactis* SL6	LVYPFP, LPLP, VLPVPQK	Antihypertensive, antioxidative	[143]
Fermented milk	*Bifidobacterium bifidum* MF 20/5	VPP, IPP	Antihypertensive	[144]
Fermented honey milk	*Lb. helveticus* MTCC5463	AVPYPQR, DFGHIQYVAAYR, IHPFAQTQSLVYPPGP, RFFVAPFPEVFGK, PPPEVFGK, EMPFPKYPVEPF	Antihypertensive	[145]
Kefir	Kefir microorganisms	AASDISLLDAQSAPLR, LKGYGGVSLPEW, LKPTPEGDLE, LIVTQTMK	Antimicrobial, antihypertensive, antidiabetic, immunomodulatory	[146]
Whey	*Enterococcus faecalis* 2/28	E. SYFVDAQPKKKEEGN.K, N. ALPEEVIQHTFNLK.S, L. VGYGSADGVDYWIAKN.S	Antihypertensive, antioxidative	[147,148]
Fermented soymilk	*Lactobacillus plantarum* strain C2	AF, FI, IF	Antihypertensive	[149]
Soy sauce	*Aspergillus oryzae*	KFNKYGR, FPFPRPPHQK, GQSSRPQDRHQK, QRFDQRSPQ, EROFPFRPPHQK, EQPRPIPFFRPQPR	Angiogenic	[150]
Natto (fermented soybean)	*Bacillus subtilis*	LE, EW, SP, VE, VL, VT, EF	Antidiabetic	[151]
Cheonggukjang (fermented soybean)	*Bacillus licheniformis* B1	EAKPSFYLK, AIGIFVKPDTAV, PVNNNAWAYATNPVGK	Antibacterial, antioxidative	[152]
Fermented bitter bean	*Lactobacillus fermentum* ATCC9338	AKVGLKPGGFFVLK, GSTIK, HGDRPR, TAHDDYK, LLLSK	Antimicrobial, ant cancer	[153]
Fermented kenaf seed	*Lb. casei* ATCC334	VPP, IPP, LQP, LLP	Antihypertensive	[154]
Rye malt sourdough	*Lactobacillus reuteri* TMW 1-106, *Lb. reuteri* LTH 5448, *Lactobacillus rossiae* 34I, *Lactobacillus hammesii* DSM 16381, and *Lb. plantarum* FUA 3002	DPVAPLQRSGPEJ, PVAPQLSRGLL, VAPSRPTPR, DIIIPD, PRSGNVGESGLID, DPVAPLQRSGPE	Antihypertensive	[155]
Sourdough	*Lactobacillus alimentarius* 15M, *Lactobacillus brevis* 14G, *Lactobacillus hilgardii* 51B, and *Lactobacillus sanfranciscensis* strains	VSWYDNEYGYSTR, ISWYDNEYGYSAR	Antimicrobial	[156]
Wine	*Saccharomyces cerevisiae* strains			

through fermentation of milk by *L. lactis* strains showed high antihypertensive properties through the release of angiotensin-converting enzyme (ACE) inhibitory peptides during the ripening stage, a steep increase in ACE inhibition activity was observed with an increase in ripening time [164]. Similar studies, on peptide profile of Prato cheese, prepared using *Lactobacillus helveticus* LH-B02 as adjunct culture resulted in the release of antihypertensive peptides during the ripening stage [134,135]. Probiotic strains are consumed as a normal diet for maintaining a healthy gut micro-flora along with other health beneficial physiological properties. Cottage cheese produced using probiotic strains, *Lb. casei* and *Lactobacillus rhamnosus* GG showed antimicrobial activity against foodborne pathogen *L. monocytogenes* through the release of antioxidative peptides [165]. Coculturing of probiotic strain *Lb. bulgaricus* with a mixed starter culture of *L. lactis* strains for production of goat-milk cheese showed high antioxidative and antihypertensive activity [166]. The use of probiotic strains and longer ripening duration has a positive effect on the release of biologically active peptides, and consumption of bioactive peptide-rich fermented milk products has added benefits apart from being a nutritional source.

Soybean is a nutritious pulse rich in proteins, lipids, isoflavones, free sugars, vitamins, and minerals [131,167]. Soybean products are consumed worldwide while fermented soybean products are most commonly consumed in Asian countries owing to enhanced functionality after fermentation [15]. Fermentation of soybean with proteolytic microorganisms releases biologically active peptides with profound therapeutic activity [2,131]. *Natto*, a traditional fermented soybean product, has been reported to produce bioactive peptides with immunomodulatory properties [15]. Peptides released from the fermentation of soybean meal by *B. subtilis* SHZ, *B. subtilis* BS12, and *Lb. plantarum* strain RM10 on separate studies showed antioxidative, antifatigue activity, and antiinflammatory activity on animal models [168–170]. Peptide enriched fermented soybean seasoning (FSS) with high antihypertensive activity was developed through fermentation of steamed soybean and wheat by *Aspergillus sojae*. The antihypertensive property of FSS was investigated through oral administration of FSS to hypertensive mice [171,172]. A comparative study on the production of peptides by *B. amyloliquefaciens* XZ-173 under SF and solid-state fermentation (SSF) of soybean flour showed a higher generation of peptides under SSF as compared to SF [173]. Fermented soymilk produced using *Enterococcus faecium* strains has been reported to have antioxidative and antihypertensive properties [131]. Soymilk fermented with *Lb. plantarum* strain C2 (LP C2) resulted in the release of 20 bioactive peptides having antioxidative and antihypertensive activity [147]. Efficient production of bioactive peptides during soybean fermentation helps in improving the functionality of soy-based fermented food with immense functional property.

1.2.7 Oligosaccharides

Oligosaccharides are low molecular weight nondigestible carbohydrates consisting of 2–10 monosaccharide units exhibiting health beneficial properties such as prebiotics, antimicrobial, anticancer, antidiabetic, antihypertensive, antiinflammatory, and immunomodulatory activities [174–177]. Oligosaccharides consisting of monosaccharide units of glucose, fructose, xylose, and galactose are considered as the major classes of oligosaccharides with functional properties [178]. Although oligosaccharides are naturally found in milk, cereals, vegetables, and honey, due to its functional attributes and low availability from the natural sources, it is produced industrially through chemical and microbial processes [179]. Microbial process includes the use of microbial enzymes and whole-cell and has an advantage over chemical processes due to its cost-effective and eco-friendly processes.

Currently, there are more than 13 classes of functional oligosaccharides produced on a commercial scale, while the most common oligosaccharides include fructooligosaccharides (FOS), galactooligosaccharides (GOS), and xylooligosaccharides (XOS) [1,178,180]. Enhanced production of FOS was observed in separate studies through optimization of fermentation conditions using *Aspergillus ibericus* and either coculturing *Aureobasidium pullulans* or *S. cerevisiae* or in a two-step process using *A. pullulans* for FOS production followed by the metabolism of small saccharides by *S. cerevisiae* [181,182]. Statistical optimization of SSF using sugarcane bagasse by *A. oryzae* DIA-MF resulted in enhanced production of FOS [183]. A higher level of FOS was obtained with an increase in the concentration of sucrose, fermented by *Bacillus aryabhattai* GYC2-3 [184]. Enzymes from microbial sources are extensively used for the production of functional oligosaccharides. In separate studies, invertase from *S. cerevisiae* SAA-612, endo-β-1,4-D-xylanase from *Bacillus* sp., and β-galactosidase from probiotic *Pediococcus acidilactici* was used for production FOS, XOS, and GOS, respectively [185–187]. In a similar study, high purity XOS was produced from sugarcane bagasse using xylanase from *B. subtilis* KCX006 [188]. The diverse enzymatic system of microorganisms holds a promising position for the industrial use of whole microbial cells and their enzymes toward the production of high-quality oligosaccharides of functional interest.

1.2.8 Polyunsaturated fatty acids

Polyunsaturated fatty acids (PUFAs) are an important class of fatty acids categorized into two subgroups: omega-3 and omega-6 fatty acids, essential for growth and development [189]. Omega-3 fatty acids have been among the most demanding fatty acid in the global market due to their role in maintaining a healthy life such as the development of the brain, protection against cardiovascular diseases and neurological diseases, antiapoptotic and antiinflammatory properties [1,190]. Linoleic acid, eicosapentaenoic acid (EPA), docosahexaenoic acid (DHA), and α-linolenic acid (ALA) are the major omega-3 fatty acids exhibiting health beneficial properties [191,192]. The major sources of omega-3 fatty acids are fish, fish oils, and plant seeds but low productivity and unsuitable extraction processes have resulted in the favored use of microbes for the production of therapeutic PUFAs [189,193–195]. About 90% pure DHA was produced through hydrolysis of algal oils using the combined hydrolytic activity of lipases from *Candida Antarctica* and *Thermomyces lanuginosus* [196]. Genetic manipulation of yeast cells has become an important approach in the production of PUFAs in recent years [11]. Engineered oleaginous yeast *Yarrowia lipolytica* was used for the production of EPA, resulting in the production of 56.6% EPA [197]. The highest production of ALA up to 1.4 g/L was achieved under low-temperature fermentation in a fed-batch bioreactor by coculturing an engineered lipid-producing strain and *Y. lipolytica* [198]. Forest biomass, birch, and spruce were utilized for the production of high-value PUFA using an oleaginous marine microalgae *Phaeodactylum tricornutum*, resulting in the high production of EPA and DHA as compared to production without glucose [199].

1.3 Conclusions and perspectives

Biological activity of functional foods and nutraceutical compounds against prevailing chronic diseases of the generation such as cancer, cardiovascular, neurodegenerative, and immunological disease has resulted in the ever-increasing demand for these products in the global market. Many industries have set up for production of functional food products to meet the demand of consumers. Currently, industrial production of nutraceuticals and food supplements is based on chemical synthesis; however, chemical synthesis has numerous drawbacks such as the generation of hazardous wastes, health deteriorating compounds, and low yield with high production costs. Bioprocess technologies involving microorganisms and the enzymatic system of microbe has become a new trend that offers a platform for the cost-effective and eco-friendly production of nutraceuticals ingredients and the development of functional food with biological activities. Selection of potential microorganisms and cheap substrate, metabolic engineering of potent microorganisms, and optimization of fermentation conditions are among the major strategies involved toward enhanced production of microbial-based products. Development of effective and standardized microbial bioprocess will be helpful in meeting the global demand of eco-friendly production of various nutraceuticals and functional foods in industrial scale.

References

[1] R. Chourasia, L.C. Phukon, S.P. Singh, A.K. Rai, D. Sahoo, Role of enzymatic bioprocesses for the production of functional food and nutraceuticals, Biomass, Biofuels, Biochemicals, Elsevier Ltd., 2020, pp. 309–334. Available from: https://doi.org/10.1016/B978-0-12-819820-9.00015-6.

[2] A.K. Rai, J. Kumaraswamy, Health benefits of functional proteins in fermented foods, Health Benefits of Fermented Foods and Beverages, CRC Press, 2015, pp. 455–474. Available from: https://doi.org/10.13140/RG.2.1.2335.6963.

[3] J. Wang, S. Guleria, M.A.G. Koffas, Y. Yan, Microbial production of value-added nutraceuticals, Current Opinion in Biotechnology 37 (2016) 97–104. Available from: https://doi.org/10.1016/j.copbio.2015.11.003.

[4] L. Das, E. Bhaumik, U. Raychaudhuri, R. Chakraborty, Role of nutraceuticals in human health, Journal of Food Science and Technology 49 (2012) 173–183. Available from: https://doi.org/10.1007/s13197-011-0269-4.

[5] S. Ghaffari, N. Roshanravan, The role of nutraceuticals in prevention and treatment of hypertension: an updated review of the literature, Food Research International 128 (2020). Available from: https://doi.org/10.1016/j.foodres.2019.108749.

[6] M.A. Ballou, E.M. Davis, B.A. Kasl, Nutraceuticals: an alternative strategy for the use of antimicrobials, Vet Clin North Am Food Anim Pract 35 (2019) 507–534. Available from: https://doi.org/10.1016/j.cvfa.2019.08.004.

[7] P. Santhakumaran, S.M. Ayyappan, J.G. Ray, Nutraceutical applications of twenty-five species of rapid-growing green-microalgae as indicated by their antibacterial, antioxidant and mineral content, Algal Research 47 (2020). Available from: https://doi.org/10.1016/j.algal.2020.101878.

[8] Global nutraceutical market growth analysis report, 2020–2027. <https://www.grandviewresearch.com/industry-analysis/nutraceuticals-market>, 2020.

[9] Z.M. Zhang, X. Li Wu, G. Yuan Zhang, X. Ma, D.X. He, Functional food development: Insights from TRP channels, Journal of Functional Foods 56 (2019) 384–394. Available from: https://doi.org/10.1016/j.jff.2019.03.023.

[10] M. Gobbetti, R. Di Cagno, M. De Angelis, et al., Functional microorganisms for functional food quality, Critical Reviews in Food Science and Nutrition 50 (2010) 716–727. Available from: https://doi.org/10.1080/10408398.2010.499770.

[11] A.K. Rai, A. Pandey, D. Sahoo, Biotechnological potential of yeasts in functional food industry, Trends in Food Science and Technology 83 (2019) 129–137. Available from: https://doi.org/10.1016/j.tifs.2018.11.016.

[12] J. Kumar, N. Sharma, G. Kaushal, S. Samurailatpam, D. Sahoo, A.K. Rai, et al., Metagenomic insights into the taxonomic and functional features of kinema, a traditional fermented soybean product of Sikkim Himalaya, Frontiers in Microbiology 10 (2019). Available from: https://doi.org/10.3389/fmicb.2019.01744.

[13] A.K. Rai, S. Sanjukta, R. Chourasia, I. Bhat, P.K. Bhardwaj, D. Sahoo, Production of bioactive hydrolysate using protease, B-glucosidase and A-amylase of Bacillus spp. isolated from kinema, Bioresource Technology 235 (2017) 358–365. Available from: https://doi.org/10.1016/j.biortech.2017.03.139.

[14] J. Bogue, O. Collins, A.J. Troy, Market analysis and concept development of functional foods, Developing New Functional Food and Nutraceutical Products, Elsevier Ltd., 2017, pp. 29–45. Available from: https://doi.org/10.1016/B978-0-12-802780-6.00002-X.

[15] Z.H. Cao, J.M. Green-Johnson, N.D. Buckley, Q.Y. Lin, Bioactivity of soy-based fermented foods: a review, Biotechnology Advances 37 (2019) 223–238. Available from: https://doi.org/10.1016/j.biotechadv.2018.12.001.

[16] F. Cardona, C. Andrés-Lacueva, S. Tulipani, F.J. Tinahones, M.I. Queipo-Ortuño, Benefits of polyphenols on gut microbiota and implications in human health, The Journal of Nutritional Biochemistry 24 (2013) 1415–1422. Available from: https://doi.org/10.1016/j.jnutbio.2013.05.001.

[17] L. Marín, E.M. Miguélez, C.J. Villar, F. Lombó, Bioavailability of dietary polyphenols and gut microbiota metabolism: antimicrobial properties, BioMed Research International 2015 (2015) 1–18. Available from: https://doi.org/10.1155/2015/905215.

[18] L. Bravo, Polyphenols: chemistry, dietary sources, metabolism, and nutritional significance, Nutrition Reviews 56 (1998) 317–333. Available from: https://doi.org/10.1111/j.1753-4887.1998.tb01670.x.

[19] T.A.F. Corrêa, M.M. Rogero, N.M.A. Hassimotto, F.M. Lajolo, The two-way polyphenols-microbiota interactions and their effects on obesity and related metabolic diseases, Frontiers in Nutrition 6 (2019) 1–15. Available from: https://doi.org/10.3389/fnut.2019.00188.

[20] C. Manach, A. Scalbert, C. Morand, C. Rémésy, L. Jiménez, Polyphenols: food sources and bioavailability, American Journal of Clinical Nutrition 79 (2004) 727–747. Available from: https://doi.org/10.1093/ajcn/79.5.727.

[21] C. Papuc, G.V. Goran, C.N. Predescu, V. Nicorescu, G. Stefan, Plant polyphenols as antioxidant and antibacterial agents for shelf-life extension of meat and meat products: classification, structures, sources, and action mechanisms, Comprehensive Reviews in Food Science and Food Safety 16 (2017) 1243–1268. Available from: https://doi.org/10.1111/1541-4337.12298.

[22] K.S. Shivashankara, S.N. Acharya, Bioavailability of dietary polyphenols and the cardiovascular diseases, The Open Nutraceuticals Journal 3 (2014) 227–241. Available from: https://doi.org/10.2174/1876396001003010227.

[23] S. Sanjukta, S. Padhi, P. Sarkar, S.P. Singh, D. Sahoo, A.K. Rai, Production, characterization and molecular docking of antioxidant peptides from peptidome of kinema fermented with proteolytic *Bacillus* spp, Food Research International 141 (2021). Available from: https://doi.org/10.1016/j.foodres.2021.110161.

[24] A.M. Romero, M.M. Doval, M.A. Sturla, M.A. Judis, Antioxidant properties of polyphenol-containing extract from soybean fermented with *Saccharomyces cerevisiae*, European Journal of Lipid Science and Technology 106 (2004) 424–431. Available from: https://doi.org/10.1002/ejlt.200400953.

[25] W. Choi, N. Lee, U. Choi, Changes in the quality characteristics and antioxidative activities of Cheonggukjang prepared using hazelnut, The Korean Journal of Food and Nutrition 30 (2017) 1229–1234.

[26] M. Kuligowski, K. Pawłowska, I. Jasińska-Kuligowska, J. Nowak, Isoflavone composition, polyphenols content and antioxidative activity of soybean seeds during tempeh fermentation, CyTA − Journal of Food 15 (2016) 1–7. Available from: https://doi.org/10.1080/19476337.2016.1197316.

[27] G. Chen, Y. Liu, J. Zeng, X. Tian, Q. Bei, Z. Wu, Enhancing three phenolic fractions of oats (*Avena sativa* L.) and their antioxidant activities by solid-state fermentation with *Monascus anka* and *Bacillus subtilis*, Journal of Cereal Science 93 (2020). Available from: https://doi.org/10.1016/j.jcs.2020.102940.

[28] M. Ferri, D.I. Serrazanetti, A. Tassoni, M. Baldissarri, A. Gianotti, Improving the functional and sensorial profile of cereal-based fermented foods by selecting *Lactobacillus plantarum* strains via a metabolomics approach, Food Research International 89 (2016) 1095–1105. Available from: https://doi.org/10.1016/j.foodres.2016.08.044.

[29] H.-G. Jo, D.-S. Kim, H.-J. Shin, Changes of nutritional components, polyphenols, and antioxidant activities of domestic bamboo tree (*Sasa coreana* Nakai) leaves fermented with *Bacillus subtilis*, Korean Society for Biotechnology and Bioengineering Journal 32 (2017) 63–70. Available from: https://doi.org/10.7841/ksbbj.2017.32.1.63.

[30] Y. Xiao, L. Wang, X. Rui, W. Li, X. Chen, M. Jiang, et al., Enhancement of the antioxidant capacity of soy whey by fermentation with *Lactobacillus plantarum* B1–6, Journal of Functional Foods 12 (2015) 33–44. Available from: https://doi.org/10.1016/j.jff.2014.10.033.

[31] L.C. Wang, T.M. Pan, T.Y. Tsai, Lactic acid bacteria-fermented product of green tea and *Houttuynia cordata* leaves exerts anti-adipogenic and anti-obesity effects, Journal of Food and Drug Analysis 26 (2018) 973–984. Available from: https://doi.org/10.1016/j.jfda.2017.11.009.

[32] G.M. Vicenssuto, R.J.S. de Castro, Development of a novel probiotic milk product with enhanced antioxidant properties using mango peel as a fermentation substrate, Biocatalysis and Agricultural Biotechnology 24 (2020). Available from: https://doi.org/10.1016/j.bcab.2020.101564.

[33] C.L. Chan, R. You Gan, N.P. Shah, H. Corke, Enhancing antioxidant capacity of *Lactobacillus acidophilus*-fermented milk fortified with pomegranate peel extracts, Food Bioscience 26 (2018) 185–192. Available from: https://doi.org/10.1016/j.fbio.2018.10.016.

[34] D. Zhao, N.P. Shah, Antiradical and tea polyphenol-stabilizing ability of functional fermented soymilk-tea beverage, Food Chemistry 158 (2014) 262–269. Available from: https://doi.org/10.1016/j.foodchem.2014.02.119.

[35] A. Dudnik, P. Gaspar, A.R. Neves, J. Forster, Engineering of microbial cell factories for the production of plant polyphenols with health-beneficial properties, Current Pharmaceutical Design 24 (2018) 2208–2225. Available from: https://doi.org/10.2174/1381612824666180515152049.

[36] L. Milke, J. Aschenbrenner, J. Marienhagen, N. Kallscheuer, Production of plant-derived polyphenols in microorganisms: current state and perspectives, Applied Microbiology and Biotechnology 102 (2018) 1575–1585. Available from: https://doi.org/10.1007/s00253-018-8747-5.

[37] S. Chouhan, K. Sharma, J. Zha, S. Guleria, M.A.G. Koffas, Recent advances in the recombinant biosynthesis of polyphenols, Frontiers in Microbiology 8 (2017) 1–16. Available from: https://doi.org/10.3389/fmicb.2017.02259.

[38] N. Kallscheuer, J. Marienhagen, *Corynebacterium glutamicum* as platform for the production of hydroxybenzoic acids, Microbial Cell Factories 17 (2018) 1–13. Available from: https://doi.org/10.1186/s12934-018-0923-x.

[39] L. Milke, P. Ferreira, N. Kallscheuer, A. Braga, M. Vogt, J. Kappelmann, et al., Modulation of the central carbon metabolism of *Corynebacterium glutamicum* improves malonyl-CoA availability and increases plant polyphenol synthesis, Biotechnology and Bioengineering 116 (2019) 1380–1391. Available from: https://doi.org/10.1002/bit.26939.

[40] L. Milke, N. Kallscheuer, J. Kappelmann, J. Marienhagen, Tailoring *Corynebacterium glutamicum* towards increased malonyl-CoA availability for efficient synthesis of the plant pentaketide noreugenin, Microbial Cell Factories 18 (2019) 1–12. Available from: https://doi.org/10.1186/s12934-019-1117-x.

[41] P. Gaspar, A. Dudnik, A.R. Neves, J. Forster, Engineering *Lactococcus lactis* for stilbene production, in: 28th International Conference on Polyphenols, 2016.

[42] D. Constantinescu-Aruxandei, R.M. Frîncu, L. Capră, F. Oancea, Selenium analysis and speciation in dietary supplements based on next-generation selenium ingredients, Nutrients 10 (2018) 1–34. Available from: https://doi.org/10.3390/nu10101466.

[43] P. Adadi, N.V. Barakova, K.Y. Muravyov, E.F. Krivoshapkina, Designing selenium functional foods and beverages: a review, Food Research International 120 (2019) 708–725. Available from: https://doi.org/10.1016/j.foodres.2018.11.029.

[44] V. Gómez-Jacinto, F. Navarro-Roldán, I. Garbayo-Nores, C. Vílchez-Lobato, A.A. Borrego, T. García-Barrera, In vitro selenium bioaccessibility combined with in vivo bioavailability and bioactivity in Se-enriched microalga (*Chlorella sorokiniana*) to be used as functional food, Journal of Functional Foods 66 (2020). Available from: https://doi.org/10.1016/j.jff.2020.103817.

[45] C.R.B. Rocourt, W.H. Cheng, Selenium supranutrition: are the potential benefits of chemoprevention outweighed by the promotion of diabetes and insulin resistance? Nutrients 5 (2013) 1349–1365. Available from: https://doi.org/10.3390/nu5041349.

[46] J. Yang, J. Wang, K. Yang, M. Liu, Y. Qi, T. Zhang, et al., Antibacterial activity of selenium-enriched lactic acid bacteria against common food-borne pathogens in vitro, Journal of Dairy Science 101 (2018) 1930–1942. Available from: https://doi.org/10.3168/jds.2017-13430.

[47] M. Kieliszek, S. Błażejak, E. Kurek, Binding and conversion of selenium in *Candida utilis* ATCC 9950 yeasts in bioreactor culture, Molecules 22 (2017). Available from: https://doi.org/10.3390/molecules22030352.

[48] K.M. Kubachka, T. Hanley, M. Mantha, R.A. Wilson, T.M. Falconer, Z. Kassa, et al., Evaluation of selenium in dietary supplements using elemental speciation, Food Chemistry 218 (2017) 313–320. Available from: https://doi.org/10.1016/j.foodchem.2016.08.086.

[49] J.K. Wrobel, M.J. Seelbach, L. Chen, R.F. Power, M. Toborek, Supplementation with selenium-enriched yeast attenuates brain metastatic growth, Nutrition and Cancer 65 (2013) 563–570. Available from: https://doi.org/10.1080/01635581.2013.775315.

[50] M. Kieliszek, I. Gientka, Accumulation and metabolism of selenium by yeast cells, Applied Microbiology and Biotechnology 99 pp (2015) 5373–5382. Available from: https://doi.org/10.1007/s00253-015-6650-x.

[51] M.P. Rayman, The use of high-selenium yeast to raise selenium status: how does it measure up? British Journal of Nutrition 92 (2004) 557–573. Available from: https://doi.org/10.1079/bjn20041251.

[52] C. Arnaudguilhem, K. Bierla, L. Ouerdane, H. Preud'homme, A. Yiannikouris, R. Lobinski, Selenium metabolomics in yeast using complementary reversed-phase/hydrophilic ion interaction (HILIC) liquid chromatography-electrospray hybrid quadrupole trap/Orbitrap mass spectrometry, Analytica Chimica Acta 757 (2012) 26–38. Available from: https://doi.org/10.1016/j.aca.2012.10.029.

[53] H. Zare, H. Vahidi, P. Owlia, M.H. Khujin, A. Khamisabadi, Yeast enriched with selenium: a promising source of selenomethionine and seleno-proteins, Trends in Peptides and Protein Sciences 1 (2017) 130–134. Available from: https://doi.org/10.22037/tpps.v1i3.16778.

[54] H. Zare, P. Owlia, H. Vahidi, M. Hosseindokht, Khujin, simultaneous optimization of the production of organic selenium and cell biomass in *Saccharomyces cerevisiae* by Plackett-Burman and Box-Behnken design, Iranian Journal of Pharmaceutical Research 17 (2018) 1081–1092. Available from: http://www.ncbi.nlm.nih.gov/pmc/articles/pmc6094416/.

[55] M. Sánchez-Martínez, E.G.P. Da Silva, T. Pérez-Corona, C. Cámara, S.L.C. Ferreira, Y. Madrid, Selenite biotransformation during brewing. Evaluation by HPLC-ICP-MS, Talanta 88 (2012) 272–276. Available from: https://doi.org/10.1016/j.talanta.2011.10.041.

[56] O. Stabnikova, V. Ivanov, I. Larionova, V. Stabnikov, M.A. Bryszewska, J. Lewis, Ukrainian dietary bakery product with selenium-enriched yeast, LWT—Food Science and Technology 41 (2008) 890–895. Available from: https://doi.org/10.1016/j.lwt.2007.05.021.

[57] D. Wang, B. Yang, G. Wei, Z. Liu, C. Wang, Efficient preparation of selenium/glutathione-enriched *Candida utilis* and its biological effects on rats, Biological Trace Element Research 150 (2012) 249–257. Available from: https://doi.org/10.1007/s12011-012-9459-9.

[58] B. Yang, D. Wang, G. Wei, Z. Liu, X. Ge, Selenium-enriched *Candida utilis*: efficient preparation with l-methionine and antioxidant capacity in rats, Journal of Trace Elements in Medicine and Biology 27 (2013) 7–11. Available from: https://doi.org/10.1016/j.jtemb.2012.06.001.

[59] S.D. Pophaly, P. Poonam, H. Singh, et al., Selenium enrichment of lactic acid bacteria and bifidobacteria: a functional food perspective, Trends in Food Science and Technology 39 (2014) 135–145. Available from: https://doi.org/10.1016/j.tifs.2014.07.006.

[60] S.K. Xia, L. Chen, J.Q. Liang, Enriched selenium and its effects on growth and biochemical composition in *Lactobacillus bulgaricus*, Journal of Agricultural and Food Chemistry 55 (2007) 2413–2417. Available from: https://doi.org/10.1021/jf062946j.

[61] M. Pescuma, B. Gomez-Gomez, T. Perez-Corona, G. Font, Y. Madrid, F. Mozzi, Food prospects of selenium enriched-*Lactobacillus acidophilus* CRL 636 and *Lactobacillus reuteri* CRL 1101, Journal of Functional Foods 35 (2017) 466–473. Available from: https://doi.org/10.1016/j.jff.2017.06.009.

[62] F.G. Martínez, G. Moreno-Martin, M. Pescuma, Y. Madrid-Albarrán, F. Mozzi, Biotransformation of selenium by lactic acid bacteria: formation of seleno-nanoparticles and seleno-amino acids, Frontiers in Bioengineering and Biotechnology 8 (2020) 1–17. Available from: https://doi.org/10.3389/fbioe.2020.00506.

[63] Y. Zhou, H. Zhu, Y. Qi, C. Wu, J. Zhang, L. Shao, et al., Absorption and distribution of selenium following oral administration of selenium-enriched *Bifidobacterium longum* DD98, selenized yeast, or sodium selenite in rats, Biological Trace Element Research 197 (2019) 599–605. Available from: https://doi.org/10.1007/s12011-019-02011-y.

[64] E. Kheradmand, F. Rafii, M.H. Yazdi, A.A. Sepahi, A.R. Shahverdi, M.R. Oveisi, The antimicrobial effects of selenium nanoparticle-enriched probiotics and their fermented broth against *Candida albicans*, DARU, Journal of Pharmaceutical Sciences 22 (2014) 1–6. Available from: https://doi.org/10.1186/2008-2231-22-48.

[65] M.H. Yazdi, M. Mahdavi, E. Kheradmand, A.R. Shahverdi, The preventive oral supplementation of a selenium nanoparticle-enriched probiotic increases the immune response and lifespan of 4T1 breast cancer bearing mice, Arzneimittel-Forschung/Drug Research 62 (2012) 525–531. Available from: https://doi.org/10.1055/s-0032-1323700.

[66] M.H. Yazdi, M. Mahdavi, N. Setayesh, M. Esfandyar, A.R. Shahverdi, Selenium nanoparticle-enriched *Lactobacillus brevis* causes more efficient immune responses in vivo and reduces the liver metastasis in metastatic form of mouse breast cancer, DARU, Journal of Pharmaceutical Sciences 21 (2013) 1–9. Available from: https://doi.org/10.1186/2008-2231-21-33.

[67] D. Guardado-Félix, M.A. Lazo-Vélez, E. Pérez-Carrillo, D.E. Panata-Saquicili, S.O. Serna-Saldívar, Effect of partial replacement of wheat flour with sprouted chickpea flours with or without selenium on physicochemical, sensory, antioxidant and protein quality of yeast-leavened breads, LWT 129 (2020). Available from: https://doi.org/10.1016/j.lwt.2020.109517.

[68] Y. Deng, C. Man, Y. Fan, Z. Wang, L. Li, H. Ren, et al., Preparation of elemental selenium-enriched fermented milk by newly isolated *Lactobacillus brevis* from kefir grains, International Dairy Journal 44 (2015) 31–36. Available from: https://doi.org/10.1016/j.idairyj.2014.12.008.

[69] X. Xu, Y. Bao, B. Wu, F. Lao, X. Hu, J. Wu, Chemical analysis and flavor properties of blended orange, carrot, apple and Chinese jujube juice fermented by selenium-enriched probiotics, Food Chemistry 289 (2019) 250–258. Available from: https://doi.org/10.1016/j.foodchem.2019.03.068.

[70] C.G. Acevedo-Rocha, L.S. Gronenberg, M. Mack, F.M. Commichau, H.J. Genee, Microbial cell factories for the sustainable manufacturing of B vitamins, Current Opinion in Biotechnology 56 (2019) 18–29. Available from: https://doi.org/10.1016/j.copbio.2018.07.006.

[71] K. Yoshii, K. Hosomi, K. Sawane, J. Kunisawa, Metabolism of dietary and microbial vitamin b family in the regulation of host immunity, Frontiers in Nutrition 6 (2019) 1–12. Available from: https://doi.org/10.3389/fnut.2019.00048.

[72] J. Han, C. Zhao, J. Cai, Y. Liang, Comparative efficacy of vitamin supplements on prevention of major cardiovascular disease: systematic review with network *meta*-analysis, Complementary Therapies in Clinical Practice 39 (2020). Available from: https://doi.org/10.1016/j.ctcp.2020.101142.

[73] Q. Jiang, Natural forms of vitamin E: metabolism, antioxidant, and anti-inflammatory activities and their role in disease prevention and therapy, Free Radical Biology and Medicine 72 (2014) 76–90. Available from: https://doi.org/10.1016/j.freeradbiomed.2014.03.035.

[74] E.J. Vandamme, Production of vitamins, coenzymes and related biochemicals by biotechnological processes, Journal of Chemical Technology & Biotechnology 53 (1992) 313–327. Available from: https://doi.org/10.1002/jctb.280530402.

[75] S.A. Survase, I.B. Bajaj, R.S. Singhal, Biotechnological production of vitamins, Food Technology and Biotechnology 44 (2006) 381–396.

[76] Y.-Y. Ge, J.-R. Zhang, H. Corke, R.-Y. Gan, Screening and spontaneous mutation of pickle-derived *Lactobacillus plantarum* with overproduction of Riboflavin, related mechanism, and food application, Foods 9 (2020) 1–12. Available from: https://doi.org/10.3390/foods9010088.

[77] R. Ledesma-Amaro, C. Serrano-Amatriain, A. Jiménez, J.L. Revuelta, Metabolic engineering of riboflavin production in *Ashbya gossypii* through pathway optimization, Microbial Cell Factories 14 (2015) 1–8. Available from: https://doi.org/10.1186/s12934-015-0354-x.

[78] K. Dmytruk, O. Lyzak, V. Yatsyshyn, M. Kluz, V. Sibirny, C. Puchalski, et al., Construction and fed-batch cultivation of *Candida famata* with enhanced riboflavin production, Journal of Biotechnology 172 (2014) 11–17. Available from: https://doi.org/10.1016/j.jbiotec.2013.12.005.

[79] B. Bhushan, C.R. Kumkum, M. Kumari, J.J. Ahire, L.M.T. Dicks, V. Mishra, Soymilk bio-enrichment by indigenously isolated riboflavin-producing strains of *Lactobacillus plantarum*, LWT 119 (2020). Available from: https://doi.org/10.1016/j.lwt.2019.108871.

[80] M. Juarez del Valle, J.E. Laiño, G. Savoy, et al., Riboflavin producing lactic acid bacteria as a biotechnological strategy to obtain bio-enriched soymilk, Food Research International 62 (2014) 1015–1019. Available from: https://doi.org/10.1016/j.foodres.2014.05.029.

[81] V. Capozzi, V. Menga, A.M. Digesù, P. De Vita, D. Van Sinderen, L. Cattivelli, et al., Biotechnological production of vitamin B2-enriched bread and pasta, Journal of Agricultural and Food Chemistry 59 (2011) 8013–8020. Available from: https://doi.org/10.1021/jf201519h.

[82] P. Russo, V. Capozzi, M.P. Arena, G. Spadaccino, M.T. Dueñas, P. López, et al., Riboflavin-overproducing strains of *Lactobacillus fermentum* for riboflavin-enriched bread, Applied Microbiology and Biotechnology 98 (2014) 3691–3700. Available from: https://doi.org/10.1007/s00253-013-5484-7.

[83] U. Gröber, J. Reichrath, M.F. Holick, K. Kisters, Vitamin K: an old vitamin in a new perspective, Dermato-Endocrinology 6 (2014) 1–6. Available from: https://doi.org/10.4161/19381972.2014.968490.

[84] Y. Liu, E.O. Van Bennekom, Y. Zhang, T. Abee, E.J. Smid, Long-chain vitamin K2 production in *Lactococcus lactis* is influenced by temperature, carbon source, aeration and mode of energy metabolism, Microbial Cell Factories 18 (2019) 1–14. Available from: https://doi.org/10.1186/s12934-019-1179-9.

[85] E. Mahdinia, S.J. Mamouri, V.M. Puri, A. Demirci, A. Berenjian, Modeling of vitamin K (Menaquinoe-7) fermentation by *Bacillus subtilis* natto in biofilm reactors, Biocatalysis and Agricultural Biotechnology 17 (2019) 196–202. Available from: https://doi.org/10.1016/j.bcab.2018.11.022.

[86] B. Walther, M. Chollet, Menaquinones, bacteria, and foods: vitamin K2 in the diet, Vitamin K2—Vital for Health and Wellbeing, Intechopen, 2017. Available from: http://doi.org/10.5772/63712.

[87] L. Ren, C. Peng, X. Hu, Y. Han, H. Huang, Microbial production of vitamin K2: current status and future prospects, Biotechnology Advances 39 (2020). Available from: https://doi.org/10.1016/j.biotechadv.2019.107453.

[88] B. Walther, J.P. Karl, S.L. Booth, P. Boyaval, Menaquinones, bacteria, and the food supply: the relevance of dairy and fermented food products to vitamin K requirements, Advances in Nutrition 4 (2013) 463–473. Available from: https://doi.org/10.3945/an.113.003855.

[89] J. Song, H. Liu, L. Wang, J. Dai, Y. Liu, H. Liu, et al., Enhanced production of vitamin K2 from *Bacillus subtilis* (natto) by mutation and optimization of the fermentation medium, Brazilian Archives of Biology and Technology 57 (2014) 606–612. Available from: https://doi.org/10.1590/S1516-8913201402126.

[90] C. Zhang, H. Ren, C. Zhong, Economical production of vitamin K2 using wheat starch wastewater, Journal of Cleaner Production 270 (2020). Available from: https://doi.org/10.1016/j.jclepro.2020.122486.

[91] Y. Cui, K. Miao, S. Niyaphorn, X. Qu, Production of gamma-aminobutyric acid from lactic acid bacteria: a systematic review, International Journal of Molecular Sciences 21 (2020) 1–21. Available from: https://doi.org/10.3390/ijms21030995.

[92] L. Diez-Gutiérrez, L. San Vicente, L.J. Luis, Md.C. Villarán, M. Chávarri, Gamma-aminobutyric acid and probiotics: multiple health benefits and their future in the global functional food and nutraceuticals market, Journal of Functional Foods 64 (2020). Available from: https://doi.org/10.1016/j.jff.2019.103669.

[93] S. Siragusa, M. De Angelis, R. Di Cagno, C.G. Rizzello, R. Coda, M. Gobbetti, Synthesis of γ-aminobutyric acid by lactic acid bacteria isolated from a variety of Italian cheeses, Applied and Environmental Microbiology 73 (2007) 7283–7290. Available from: https://doi.org/10.1128/aem.01064-07.

[94] S.S. Bajic, J. Djokic, M. Dinic, K. Veljovic, N. Golic, S. Mihajlovic, et al., GABA-producing natural dairy isolate from artisanal zlatar cheese attenuates gut inflammation and strengthens gut epithelial barrier in vitro, Frontiers in Microbiology 10 (2019) 1–13. Available from: https://doi.org/10.3389/fmicb.2019.00527.

[95] H. Li, T. Qiu, G. Huang, Y. Cao, Production of gamma-aminobutyric acid by *Lactobacillus brevis* NCL912 using fed-batch fermentation, Microbial Cell Factories 9 (2010) 1–7. Available from: https://doi.org/10.1186/1475-2859-9-85.

[96] Q. Zhang, Q. Sun, X. Tan, S. Zhang, L. Zeng, J. Tang, et al., Characterization of γ-aminobutyric acid (GABA)-producing *Saccharomyces cerevisiae* and coculture with *Lactobacillus plantarum* for mulberry beverage brewing, Journal of Bioscience and Bioengineering 129 (2020) 447–453. Available from: https://doi.org/10.1016/j.jbiosc.2019.10.001.

[97] M. Han, W. Yan Liao, S. Mao Wu, X. Gong, C. Bai, Use of *Streptococcus thermophilus* for the in situ production of γ-aminobutyric acid-enriched fermented milk, Journal of Dairy Science 103 (2020) 98–105. Available from: https://doi.org/10.3168/jds.2019-16756.

[98] A. Mancini, I. Carafa, E. Franciosi, T. Nardin, B. Bottari, R. Larcher, et al., In vitro probiotic characterization of high GABA producing strain *Lactobacilluas brevis* DSM 32386 isolated from traditional 'wild' Alpine cheese, Annals of Microbiology 69 (2019) 1435–1443. Available from: https://doi.org/10.1007/s13213-019-01527-x.

[99] L.Q. Dat, T.T.K. Ngan, N.T.X. Nu, Gamma-amino butyric acid (GABA) synthesis of Lactobacillus in fermentation of defatted rice bran extract, AIP Conference Proceedings 1878 (2017) 1–7. Available from: https://doi.org/10.1063/1.5000213.

[100] K.A. Baritugo, H.T. Kim, Y. David, T.U. Khang, S.M. Hyun, K.H. Kang, et al., Enhanced production of gamma-aminobutyrate (GABA) in recombinant *Corynebacterium glutamicum* strains from empty fruit

bunch biosugar solution, Microbial Cell Factories 17 (2018) 1–12. Available from: https://doi.org/10.1186/s12934-018-0977-9.

[101] L. Chen, J. Alcazar, T. Yang, Z. Lu, Y. Lu, Optimized cultural conditions of functional yogurt for γ-aminobutyric acid augmentation using response surface methodology, Journal of Dairy Science 101 (2018) 10685–10693. Available from: https://doi.org/10.3168/jds.2018-15391.

[102] Y. Zhong, S. Wu, F. Chen, M. He, J. Lin, Isolation of high γ-aminobutyric acid-producing lactic acid bacteria and fermentation in mulberry leaf powders, Experimental and Therapeutic Medicine 18 (2019) 147–153. Available from: https://doi.org/10.3892/etm.2019.7557.

[103] Z. Zareie, F. Tabatabaei Yazdi, S.A. Mortazavi, et al., Development and characterization of antioxidant and antimicrobial edible films based on chitosan and gamma-aminobutyric acid-rich fermented soy protein, Carbohydrate Polymers 244 (2020). Available from: https://doi.org/10.1016/j.carbpol.2020.116491.

[104] C. Sanchart, O. Rattanaporn, D. Haltrich, P. Phukpattaranont, S. Maneerat, Enhancement of gamma-aminobutyric acid (GABA) levels using an autochthonous *Lactobacillus futsaii* CS3 as starter culture in Thai fermented shrimp (Kung-Som), World Journal of Microbiology and Biotechnology 33 (2017). Available from: https://doi.org/10.1007/s11274-017-2317-3.

[105] E. Patterson, P.M. Ryan, N. Wiley, I. Carafa, E. Sherwin, G. Moloney, et al., Gamma-aminobutyric acid-producing lactobacilli positively affect metabolism and depressive-like behaviour in a mouse model of metabolic syndrome, Scientific Reports 9 (2019) 1–15. Available from: https://doi.org/10.1038/s41598-019-51781-x.

[106] M. Zareian, E. Oskoueian, M. Majdinasab, B. Forghani, Production of GABA-enriched: Idli with ACE inhibitory and antioxidant properties using *Aspergillus oryzae*: the antihypertensive effects in spontaneously hypertensive rats, Food and Function 11 (2020) 4304–4313. Available from: https://doi.org/10.1039/c9fo02854d.

[107] C. Ramesh, N.V. Vinithkumar, R. Kirubagaran, C.K. Venil, L. Dufossé, Multifaceted applications of microbial pigments: current knowledge, challenges and future directions for public health implications, Microorganisms 7 (2019). Available from: https://doi.org/10.3390/microorganisms7070186.

[108] P.S. Nigam, J.S. Luke, Food additives: production of microbial pigments and their antioxidant properties, Current Opinion in Food Science 7 (2016) 93–100. Available from: https://doi.org/10.1016/j.cofs.2016.02.004.

[109] C.K. Venil, Z.A. Zakaria, W.A. Ahmad, Bacterial pigments and their applications, Process Biochemistry 48 (2013) 1065–1079. Available from: https://doi.org/10.1016/j.procbio.2013.06.006.

[110] D. Kim, S. Ku, Pigments and derivatives: a mini review, Molecules 23 (2018) 1–15. Available from: https://doi.org/10.3390/molecules23010098.

[111] W. Chen, B. Ho, C. Lee, C. Lee, T. Pan, Red mold rice prevents the development of obesity, dyslipidemia and hyperinsulinemia induced by high-fat diet, International Journal of Obesity 32 (2008) 1694–1704. Available from: https://doi.org/10.1038/ijo.2008.156.

[112] H. Kumura, T. Ohtsuyama, Y. Matsusaki, M. Taitoh, H. Koyanagi, K. Kobayashi, et al., Application of red pigment producing edible fungi for development of a novel type of functional cheese, Journal of Food Processing and Preservation 42 (2018). Available from: http://hdl.handle.net/2115/75658.

[113] J. Liu, Y. Luo, T. Guo, C. Tang, X. Chai, W. Zhao, et al., Cost-effective pigment production by *Monascus purpureus* using rice straw hydrolysate as substrate in submerged fermentation, Journal of Bioscience and Bioengineering 129 (2020) 229–236. Available from: https://doi.org/10.1016/j.jbiosc.2019.08.007.

[114] C. Long, X. Zeng, J. Xie, Y. Liang, J. Tao, Q. Tao, High-level production of *Monascus* pigments in *Monascus ruber* CICC41233 through ATP-citrate lyase overexpression, Biochemical Engineering Journal 146 (2019) 160–169. Available from: https://doi.org/10.1016/j.bej.2019.03.007.

[115] R. Tsai, B. Ho, T. Pan, Red mold rice mitigates oral carcinogenesis in 7, 12-dimethyl-1, 2-benz [a] anthracene-induced oral carcinogenesis in Hamster, Evidence-Based Complementary and Alternative Medicine: eCAM 2011 (2011). Available from: https://doi.org/10.1093/ecam/nep215.

[116] S.A.E. Heider, P. Peters-Wendisch, V.F. Wendisch, J. Beekwilder, T. Brautaset, Metabolic engineering for the microbial production of carotenoids and related products with a focus on the rare C50 carotenoids, Applied Microbiology and Biotechnology 98 (2014) 4355–4368. Available from: https://doi.org/10.1007/s00253-014-5693-8.

[117] K. Kirti, S. Amita, S. Priti, A.M. Kumar, S. Jyoti, Colorful world of microbes: carotenoids and their applications, Advances in Biology 2014 (2014) 1–13. Available from: https://doi.org/10.1155/2014/837891.

[118] M. Liu, X. Jiang, A. Chen, T. Chen, Y. Cheng, X. Wu, Transcriptome analysis reveals the potential mechanism of dietary carotenoids improving antioxidative capability and immunity of juvenile Chinese mitten crabs *Eriocheir sinensis*, Fish and Shellfish Immunology 104 (2020) 359–373. Available from: https://doi.org/10.1016/j.fsi.2020.06.033.

[119] W.L.C. Loh, K.C. Huang, H.S. Ng, J.C.W. Lan, Exploring the fermentation characteristics of a newly isolated marine bacteria strain, *Gordonia terrae* TWRH01 for carotenoids production, Journal of Bioscience and Bioengineering 130 (2020) 187–194. Available from: https://doi.org/10.1016/j.jbiosc.2020.03.007.

[120] L.C. Mata-Gomez, J.C. Montanez, A. Mendez-Zavala, C.N. Aguilar, Biotechnological production of carotenoids by yeasts: an overview, Microbial Cell Factories 13 (2014) 1–12. Available from: https://doi.org/10.1186/1475-2859-13-12.

[121] S. Ravikumar, G. Uma, R. Gokulakrishnan, Antibacterial property of halobacterial carotenoids against human bacterial pathogens, Journal of Scientific & Industrial Research 75 (2016) 253–257.

[122] E.D. Simova, G.I. Frengova, D.M. Beshkova, Effect of aeration on the production of carotenoid pigments by *Rhodotorula rubra-Lactobacillus casei* subsp. *casei* co-cultures in whey ultrafiltrate, Zeitschrift für Naturforschung C 58 (2003) 225–229. Available from: https://doi.org/10.1515/znc-2003-3-415.

[123] A.M. Kot, S. Błażejak, M. Kieliszek, I. Gientka, K. Piwowarek, Biocatalysis and agricultural biotechnology production of lipids and carotenoids by *Rhodotorula gracilis* ATCC 10788 yeast in a bioreactor using low-cost wastes, Biocatalysis and Agricultural Biotechnology 26 (2020). Available from: https://doi.org/10.1016/j.bcab.2020.101634.

[124] T. Schneider, S. Graeff-Hönninger, W.T. French, R. Hernandez, W. Claupein, W.E. Holmes, et al., Screening of industrial wastewaters as feedstock for the microbial production of oils for biodiesel production and high-quality pigments, Journal of Combustion 2012 (2012). Available from: https://doi.org/10.1155/2012/153410.

[125] R. Sharma, G. Ghoshal, Optimization of carotenoids production by *Rhodotorula mucilaginosa* (MTCC-1403) using agro-industrial waste in bioreactor: a statistical approach, Biotechnology Reports 25 (2020) 1–11. Available from: https://doi.org/10.1016/j.btre.2019.e00407.

[126] G. Squillaci, R. Parrella, V. Carbone, P. Minasi, F.L. Cara, et al., Carotenoids from the extreme halophilic Archaeon *Haloterrigena turkmenica*: identification and antioxidant activity, Extremophiles 21 (2017) 933–945. Available from: https://doi.org/10.1007/s00792-017-0954-y.

[127] T.M. Keceli, Z. Erginkaya, E. Turkkan, U. Kaya, Antioxidant and antibacterial effects of carotenoids extracted from *Rhodotorula glutinis* strains, Asian Journal of Chemistry 25 (2013) 42–46. Available from: https://doi.org/10.14233/ajchem.2013.12377.

[128] M.D. Moreira, M.M. Melo, J.M. Coimbra, K.C. dos Reis, R.F. Schwan, C.F. Silva, Solid coffee waste as alternative to produce carotenoids with antioxidant and antimicrobial activities, Waste Management 82 (2018) 93–99. Available from: https://doi.org/10.1016/j.wasman.2018.10.017.

[129] A.Y. Yoo, M. Alnaeeli, J.K. Park, Production control and characterization of antibacterial carotenoids from the yeast *Rhodotorula mucilaginosa* AY-01, Process Biochemistry 51 (2016) 463–473. Available from: https://doi.org/10.1016/j.procbio.2016.01.008.

[130] A.K. Rai, S. Sanjukta, R. Chourasia, I. Bhat, P.K. Bhardwaj, D. Sahoo, Production of bioactive hydrolysate using protease, B-glucosidase and A-amylase of *Bacillus* spp. isolated from kinema, Bioresource Technology 235 (2017) 358–365. Available from: https://doi.org/10.1016/j.biortech.2017.03.139.

[131] S. Sanjukta, A.K. Rai, Production of bioactive peptides during soybean fermentation and their potential health benefits, Trends in Food Science and Technology 50 (2016) 1–10. Available from: https://doi.org/10.1016/j.tifs.2016.01.010.

[132] D. Tagliazucchi, S. Martini, L. Solieri, Bioprospecting for bioactive peptide production by lactic acid bacteria isolated from fermented dairy food, Fermentation 5 (2019). Available from: https://doi.org/10.3390/fermentation5040096.

[133] C. Martinez-Villaluenga, E. Peñas, J. Frias, Bioactive peptides in fermented foods: production and evidence for health effects, Fermented Foods in Health and Disease Prevention, Academic Press, 2017, pp. 23–47. Available from: https://doi.org/10.1016/B978-0-12-802309-9.00002-9.

[134] D.P. Baptista, B.D. Galli, F.G. Cavalheiro, F. Negrão, M.N. Eberlin, M.L. Gigante, *Lactobacillus helveticus* LH-B02 favours the release of bioactive peptide during Prato cheese ripening, International Dairy Journal 87 (2018) 75–83. Available from: https://doi.org/10.1016/j.idairyj.2018.08.001.

[135] D.P. Baptista, F. Negrão, M.N. Eberlin, M.L. Gigante, Peptide profile and angiotensin-converting enzyme inhibitory activity of Prato cheese with salt reduction and *Lactobacillus helveticus* as an adjunct culture, Food Research International 133 (2020). Available from: https://doi.org/10.1016/j.foodres.2020.109190.

[136] P. Plaisancié, J. Claustre, M. Estienne, G. Henry, R. Boutrou, A. Paquet, et al., A novel bioactive peptide from yoghurts modulates expression of the gel-forming MUC2 mucin as well as population of goblet cells and Paneth cells along the small intestine, Journal of Nutritional Biochemistry 24 (2013) 213–221. Available from: https://doi.org/10.1016/j.jnutbio.2012.05.004.

[137] P. Plaisancié, R. Boutrou, M. Estienne, G. Henry, J. Jardin, A. Paquet, et al., β-Casein(94–123)-derived peptides differently modulate production of mucins in intestinal goblet cells, Journal of Dairy Research 82 (2015) 36–46. Available from: https://doi.org/10.1017/s0022029914000533.

[138] S. Taha, M. El Abd, C. De Gobba, M. Abdel-Hamid, E. Khalil, D. Hassan, Antioxidant and antibacterial activities of bioactive peptides in buffalo's yoghurt fermented with different starter cultures, Food Science and Biotechnology 26 (2017) 1325–1332. Available from: https://doi.org/10.1007/s10068-017-0160-9.

[139] F. Giacometti Cavalheiro, D. Parra Baptista, B. Domingues Galli, F. Negrão, M. Nogueira Eberlin, M. Lúcia Gigante, High protein yogurt with addition of *Lactobacillus helveticus*: peptide profile and angiotensin-converting enzyme ACE-inhibitory activity, Food Chemistry 333 (2020). Available from: https://doi.org/10.1016/j.foodchem.2020.127482.

[140] D. Solanki, S. Hati, Considering the potential of *Lactobacillus rhamnosus* for producing angiotensin I-converting enzyme (ACE) inhibitory peptides in fermented camel milk (Indian breed), Food Bioscience 23 (2018) 16–22. Available from: https://doi.org/10.1016/j.fbio.2018.03.004.

[141] L. Solieri, G.S. Rutella, D. Tagliazucchi, Impact of non-starter lactobacilli on release of peptides with angiotensin-converting enzyme inhibitory and antioxidant activities during bovine milk fermentation, Food Microbiology 51 (2015) 108–116. Available from: https://doi.org/10.1016/j.fm.2015.05.012.

[142] S.H. Kim, J.Y. Lee, M.P. Balolong, J.E. Kim, H.D. Paik, D.K. Kang, Identification and characterization of a novel antioxidant peptide from bovine skim milk fermented by *Lactococcus lactis* SL6, Korean Journal for Food Science of Animal Resources 37 (2017) 402–409. Available from: https://doi.org/10.5851/kosfa.2017.37.3.402.

[143] C. Gonzalez-Gonzalez, T. Gibson, P. Jauregi, Novel probiotic-fermented milk with angiotensin I-converting enzyme inhibitory peptides produced by *Bifidobacterium bifidum* MF 20/5, International Journal of Food Microbiology 167 (2013) 131–137. Available from: https://doi.org/10.1016/j.ijfoodmicro.2013.09.002.

[144] S. Hati, A. Sakure, S. Mandal, Impact of proteolytic *Lactobacillus helveticus* MTCC5463 on production of bioactive peptides derived from honey based fermented milk, International Journal of Peptide Research and Therapeutics 23 (2017) 297–303. Available from: https://doi.org/10.1007/s10989-016-9561-5.

[145] F.G. Amorim, L.B. Coitinho, A.T. Dias, A.G.F. Friques, B.L. Monteiro, L.C.D. de Rezende, et al., Identification of new bioactive peptides from Kefir milk through proteopeptidomics: bioprospection of antihypertensive molecules, Food Chemistry 282 (2019) 109–119. Available from: https://doi.org/10.1016/j.foodchem.2019.01.010.

[146] P. Worsztynowicz, W. Białas, W. Grajek, Integrated approach for obtaining bioactive peptides from whey proteins hydrolysed using a new proteolytic lactic acid bacteria, Food Chemistry 312 (2020). Available from: https://doi.org/10.1016/j.foodchem.2019.126035.

[147] B.P. Singh, S. Vij, Growth and bioactive peptides production potential of *Lactobacillus plantarum* strain C2 in soy milk: a LC-MS/MS based revelation for peptides biofunctionality, LWT 86 (2017). Available from: https://doi.org/10.1016/j.lwt.2017.08.013.

[148] B.P. Singh, S. Vij, In vitro stability of bioactive peptides derived from fermented soy milk against heat treatment, pH and gastrointestinal enzymes, LWT—Food Science and Technology 91 (2018) 303–307. Available from: https://doi.org/10.1016/j.lwt.2018.01.066.

[149] X.L. Zhu, K. Watanabe, K. Shiraishi, T. Ueki, Y. Noda, T. Matsui, et al., Identification of ACE-inhibitory peptides in salt-free soy sauce that are transportable across caco-2 cell monolayers, Peptides 29 (2008) 338–344. Available from: https://doi.org/10.1016/j.peptides.2007.11.006.

[150] M. Taniguchi, R. Aida, K. Saito, A. Ochiai, S. Takesono, E. Saitoh, et al., Identification and characterization of multifunctional cationic peptides from traditional Japanese fermented soybean Natto extracts, Journal of Bioscience and Bioengineering 127 (2019) 472–478. Available from: https://doi.org/10.1016/j.jbiosc.2018.09.016.

[151] H.J. Yang, D.Y. Kwon, N.R. Moon, M.J. Kim, H.J. Kang, D.Y. Jung, et al., Soybean fermentation with *Bacillus licheniformis* increases insulin sensitizing and insulinotropic activity, Food and Function 4 (2013) 1675–1684. Available from: https://doi.org/10.1039/C3FO60198F.

[152] B.J. Muhialdin, N.F. Abdul, Rani, A.S. Meor Hussin, Identification of antioxidant and antibacterial activities for the bioactive peptides generated from bitter beans (*Parkia speciosa*) via boiling and fermentation processes, LWT 131 (2020). Available from: https://doi.org/10.1016/j.lwt.2020.109776.

[153] B. Arulrajah, B.J. Muhialdin, M. Zarei, H. Hasan, N. Saari, Lacto-fermented Kenaf (*Hibiscus cannabinus* L.) seed protein as a source of bioactive peptides and their applications as natural preservatives, Food Control 110 (2020). Available from: https://doi.org/10.1016/j.foodcont.2019.106969.

[154] Y. Hu, A. Stromeck, J. Loponen, D. Lopes-Lutz, A. Schieber, M.G. Gänzle, LC-MS/MS quantification of bioactive angiotensin I-converting enzyme inhibitory peptides in rye malt sourdoughs, Journal of Agricultural and Food Chemistry 59 (2011) 11983–11989. Available from: https://doi.org/10.1021/jf2033329.

[155] C.G. Rizzello, A. Cassone, R. Di Cagno, M. Gobbetti, Synthesis of angiotensin I-converting enzyme (ACE)-inhibitory peptides and γ-aminobutyric acid (GABA) during sourdough fermentation by selected lactic acid bacteria, Journal of Agricultural and Food Chemistry 56 (2008) 6936–6943. Available from: https://doi.org/10.1021/jf800512u.

[156] P. Branco, D. Francisco, C. Chambon, M. Hébraud, N. Arneborg, M.G. Almeida, et al., Identification of novel GAPDH-derived antimicrobial peptides secreted by *Saccharomyces cerevisiae* and involved in wine microbial interactions, Applied Microbiology and Biotechnology 98 (2014) 843–853. Available from: https://doi.org/10.1007/s00253-013-5411-y.

[157] H. Korhonen, A. Pihlanto, Bioactive peptides: production and functionality, International Dairy Journal 16 (2006) 945–960. Available from: https://doi.org/10.1016/j.idairyj.2005.10.012.

[158] Y. Hou, Z. Wu, Z. Dai, G. Wang, G. Wu, Protein hydrolysates in animal nutrition: industrial production, bioactive peptides, and functional significance, Journal of Animal Science and Biotechnology 8 (2017) 1–13. Available from: https://doi.org/10.1186/s40104-017-0153-9.

[159] C. Raveschot, B. Cudennec, F. Coutte, C. Flahaut, M. Fremont, D. Drider, et al., Production of bioactive peptides by *Lactobacillus* species: from gene to application, Frontiers in Microbiology 9 (2018). Available from: https://doi.org/10.3389/fmicb.2018.02354.

[160] R. Chourasia, S. Padhi, L. Chiring Phukon, M.M. Abedin, S.P. Singh, A.K. Rai, A potential peptide from soy cheese produced using *Lactobacillus delbrueckii* WS4 for effective inhibition of SARS-CoV-2 main protease and S1 glycoprotein, Frontiers in Molecular Biosciences 7 (2020). Available from: https://doi.org/10.3389/fmolb.2020.601753.

[161] C. Adams, F. Sawh, J.M. Green-Johnson, et al., Characterization of casein-derived peptide bioactivity: differential effects on angiotensin-converting enzyme inhibition, cytokine, and nitric oxide production, Journal of Dairy Science 103 (2020). Available from: https://doi.org/10.3168/jds.2019-17976.

[162] C. Chaves-López, A. Serio, A. Paparella, M. Martuscelli, A. Corsetti, R. Tofalo, et al., Impact of microbial cultures on proteolysis and release of bioactive peptides in fermented milk, Food Microbiology 42 (2014) 117–121. Available from: https://doi.org/10.1016/j.fm.2014.03.005.

[163] S. Martini, A. Conte, D. Tagliazucchi, Effect of ripening and in vitro digestion on the evolution and fate of bioactive peptides in Parmigiano-Reggiano cheese, International Dairy Journal 105 (2020). Available from: https://doi.org/10.1016/j.idairyj.2020.104668.

[164] M. Garbowska, A. Pluta, A. Berthold-Pluta, Proteolytic and ACE-inhibitory activities of Dutch-type cheese models prepared with different strains of *Lactococcus lactis*, Food Bioscience 35 (2020). Available from: https://doi.org/10.1016/j.fbio.2020.100604.

[165] L. Abadía-García, A. Cardador, S.T. Martín del Campo, S.M. Arvízu, E. Castaño-Tostado, C. Regalado-González, et al., Influence of probiotic strains added to cottage cheese on generation of potentially antioxidant peptides, anti-listerial activity, and survival of probiotic microorganisms in simulated gastrointestinal conditions, International Dairy Journal 33 (2013) 191–197. Available from: https://doi.org/10.1016/j.idairyj.2013.04.005.

[166] A. Kocak, T. Sanli, E.A. Anli, A.A. Hayaloglu, Role of using adjunct cultures in release of bioactive peptides in white-brined goat-milk cheese, LWT 123 (2020). Available from: https://doi.org/10.1016/j.lwt.2020.109127.

[167] C. Chatterjee, S. Gleddie, C.W. Xiao, Soybean bioactive peptides and their functional properties, Nutrients 10 (2018) 8–11. Available from: https://doi.org/10.3390/nu10091211.

[168] S. Miri, R. Hajihosseini, H. Saedi, M. Vaseghi, A. Rasooli, Fermented soybean meal extract improves oxidative stress factors in the lung of inflammation/infection animal model, Annals of Microbiology 69 (2019) 1507–1515. Available from: https://doi.org/10.1007/s13213-019-01534-y.

[169] B. Yu, Z.X. Lu, X.M. Bie, F.X. Lu, X.Q. Huang, Scavenging and anti-fatigue activity of fermented defatted soybean peptides, European Food Research and Technology 226 (2008) 415–421. Available from: https://doi.org/10.1007/s00217-006-0552-1.

[170] Y. Zhang, S. Chen, X. Zong, C. Wang, C. Shi, F. Wang, et al., Peptides derived from fermented soybean meal suppresses intestinal inflammation and enhances epithelial barrier function in piglets, Food and Agricultural Immunology 31 (2020) 120–135. Available from: https://doi.org/10.1080/09540105.2019.1705766.

[171] T. Nakahara, A. Sano, H. Yamaguchi, K. Sugimoto, H. Chikata, E. Kinoshita, et al., Antihypertensive effect of peptide-enriched soy sauce-like seasoning and identification of its angiotensin I-converting enzyme inhibitory substances, Journal of Agricultural and Food Chemistry 58 (2010) 821–827. Available from: https://doi.org/10.1021/jf903261h.

[172] T. Nakahara, K. Sugimoto, A. Sano, H. Yamaguchi, H. Katayama, R. Uchida, Antihypertensive mechanism of a peptide-enriched soy sauce-like seasoning: the active constituents and its suppressive effect on renin-angiotensin-aldosterone system, Journal of Food Science 76 (2011) 1–6. Available from: https://doi.org/10.1111/j.1750-3841.2011.02362.x.

[173] Z. Zhu, J. Zhang, Y. Wu, W. Ran, Q. Shen, Comparative study on the properties of lipopeptide products and expression of biosynthetic genes from *Bacillus amyloliquefaciens* XZ-173 in liquid fermentation and solid-state fermentation, World Journal of Microbiology and Biotechnology 29 (2013) 2105–2114. Available from: https://doi.org/10.1007/s11274-013-1375-4.

[174] M. Asadpoor, C. Peeters, P.A.J. Henricks, S. Varasteh, R.J. Pieters, G. Folkerts, et al., Anti-pathogenic functions of non-digestible oligosaccharides in vitro, Nutrients 12 (2020) 1–30. Available from: https://doi.org/10.3390/nu12061789.

[175] K.L. Cheong, H.M. Qiu, H. Du, Y. Liu, B.M. Khan, Oligosaccharides derived from red seaweed: Production, properties, and potential health and cosmetic applications, Molecules 23 (2018) 1–18. Available from: https://doi.org/10.3390/molecules23102451.

[176] B. Gómez, B. Míguez, R. Yáñez, J.L. Alonso, Extraction of oligosaccharides with prebiotic properties from agro-industrial wastes, Water Extraction of Bioactive Compounds, Elsevier, 2017, pp. 131–161. Available from: https://doi.org/10.1016/B978-0-12-809380-1.00005-X.

[177] S.I. Mussatto, I.M. Mancilha, Non-digestible oligosaccharides: a review, Carbohydrate Polymers 68 (2007) 587–597. Available from: https://doi.org/10.1016/j.carbpol.2006.12.011.

[178] C. Zhao, Y. Wu, X. Liu, B. Liu, H. Cao, H. Yu, et al., Functional properties, structural studies and chemo-enzymatic synthesis of oligosaccharides, Trends in Food Science and Technology 66 (2017) 135–145. Available from: https://doi.org/10.1016/j.tifs.2017.06.008.

[179] S.G. Prapulla, V. Subhaprada, N.G. Karanth, Microbial production of oligosaccharides: a review, Advances in Applied Microbiology 47 (2000) 299–343. Available from: https://doi.org/10.1016/s0065-2164(00)47008-5.

[180] S.A. Belorkar, A.K. Gupta, Oligosaccharides: a boon from nature's desk, AMB Express 6 (2016). Available from: https://doi.org/10.1186/s13568-016-0253-5.

[181] C. Nobre, C.C. Castro, A.L. Hantson, et al., Strategies for the production of high-content fructo-oligo-saccharides through the removal of small saccharides by co-culture or successive fermentation with yeast, Carbohydrate Polymers 136 (2016) 274–281. Available from: https://doi.org/10.1016/j.carbpol.2015.08.088.

[182] C. Nobre, E.G. Alves Filho, F.A.N. Fernandes, E.S. Brito, S. Rodrigues, J.A. Teixeira, et al., Production of fructo-oligosaccharides by *Aspergillus ibericus* and their chemical characterization, LWT—Food Science and Technology 89 (2018) 58–64. Available from: https://doi.org/10.1016/j.lwt.2017.10.015.

[183] O. de la Rosa, D.B. Múñiz-Marquez, J.C. Contreras-Esquivel, J.E. Wong-Paz, R. Rodríguez-Herrera, C.N. Aguilar, Improving the fructooligosaccharides production by solid-state fermentation, Biocatalysis and Agricultural Biotechnology 27 (2020). Available from: https://doi.org/10.1016/j.bcab.2020.101704.

[184] A. Nasir, F. Sattar, I. Ashfaq, et al., Production and characterization of a high molecular weight levan and fructooligosaccharides from a rhizospheric isolate of *Bacillus aryabhattai*, LWT 123 (2020). Available from: https://doi.org/10.1016/j.lwt.2020.109093.

[185] P. Chanalia, D. Gandhi, P. Attri, S. Dhanda, Purification and characterization of β-galactosidase from probiotic *Pediococcus acidilactici* and its use in milk lactose hydrolysis and galactooligosaccharide synthesis, Bioorganic Chemistry 77 (2018) 176–189. Available from: https://doi.org/10.1016/j.bioorg.2018.01.006.

[186] T. Chand Bhalla, N. Bansuli, Thakur, Savitri, N. Thakur, Invertase of *Saccharomyces cerevisiae* SAA-612: production, characterization and application in synthesis of fructo-oligosaccharides, LWT—Food Science and Technology 77 (2017) 178–185. Available from: https://doi.org/10.1016/j.lwt.2016.11.034.

[187] A.A.I. Ratnadewi, M.H. Amaliyah Zain, A.A.N. Nara Kusuma, W. Handayani, A.S. Nugraha, T.A. Siswoyo, *Lactobacillus casei* fermentation towards xylooligosaccharide (XOS) obtained from coffee peel enzymatic hydrolysate, Biocatalysis and Agricultural Biotechnology 23 (2020). Available from: https://doi.org/10.1016/j.bcab.2019.101446.

[188] S.S. Reddy, C. Krishnan, Production of high-pure xylooligosaccharides from sugarcane bagasse using crude β-xylosidase-free xylanase of *Bacillus subtilis* KCX006 and their bifidogenic function, LWT—Food Science and Technology 65 (2016) 237–245. Available from: https://doi.org/10.1016/j.lwt.2015.08.013.

[189] S.F. Yuan, H.S. Alper, Metabolic engineering of microbial cell factories for production of nutraceuticals, Microbial Cell Factories 18 (2019) 1–11. Available from: https://doi.org/10.1186/s12934-019-1096-y.

[190] K.P. Stahmann, Vitamins and vitamin-like compounds: microbial production, Encyclopedia of Microbiology, 2019, pp. 569–580. Available from: https://doi.org/10.1016/B978-0-12-809633-8.13017-1.

[191] R. Chourasia, M.M. Abedin, L. Chiring Phukon, D. Sahoo, S.P. Singh, A.K. Rai, Biotechnological approaches for the production of designer cheese with improved functionality, Comprehensive Reviews in Food Science and Food Safety 20 (2021) 960–979. Available from: https://doi.org/10.1111/1541-4337.12680.

[192] L. Das, E. Bhaumik, U. Raychaudhuri, R. Chakraborty, Role of nutraceuticals in human health, Journal of Food Science and Technology 49 (2012) 173–183. Available from: https://doi.org/10.1007/s13197-011-0269-4.

[193] C. Ballabio, P. Restani, Lipids in functional foods, nutraceuticals and supplements, European Journal of Lipid Science and Technology 114 (2012) 369–371. Available from: https://doi.org/10.1002/ejlt.201200111.

[194] S. Kumar, B. Sharma, P. Bhadwal, P. Sharma, N. Agnihotri, Lipids as nutraceuticals: a shift in paradigm, handbook of food bioengineering, Therapeutic Foods, Academic Press, 2018, pp. 51–98. Available from: https://doi.org/10.1016/B978-0-12-811517-6.00003-9.

[195] A.K. Rai, N. Bhaskar, V. Baskaran, Bioefficacy of EPA-DHA from lipids recovered from fish processing wastes through biotechnological approaches, Food Chemistry 136 (2013) 80–86. Available from: https://doi.org/10.1016/j.foodchem.2012.07.103.

[196] T.O. Akanbi, C.J. Barrow, *Candida Antarctica* lipase A effectively concentrates DHA from fish and thraustochytrid oils, Food Chemistry 229 (2017) 509–516. Available from: https://doi.org/10.1016/j.foodchem.2017.02.099.

[197] Z. Xue, P.L. Sharpe, S.P. Hong, N.S. Yadav, D. Xie, D.R. Short, et al., Production of omega-3 eicosapentaenoic acid by metabolic engineering of *Yarrowia lipolytica*, Nature Biotechnology 31 (2013) 734–740. Available from: https://doi.org/10.1038/nbt.2622.

[198] L.T. Cordova, H.S. Alper, Production of α-linolenic acid in *Yarrowia lipolytica* using low-temperature fermentation, Applied Microbiology and Biotechnology 102 (2018) 8809–8816. Available from: https://doi.org/10.1007/s00253-018-9349-y.

[199] A. Patel, L. Matsakas, K. Hrůzová, U. Rova, P. Christakopoulos, Biosynthesis of nutraceutical fatty acids by the oleaginous marine microalgae *Phaeodactylum tricornutum* utilizing hydrolysates from organosolv-pretreated Birch and Spruce biomass, Marine Drugs 17 (2019) 1–17. Available from: https://doi.org/10.3390/md17020119.

Microbial transformation for improving food functionality

Antonia Terpou[1], Amit Kumar Rai[2]

[1]DEPARTMENT OF AGRICULTURAL DEVELOPMENT, AGRI-FOOD, AND NATURAL RESOURCES MANAGEMENT, SCHOOL OF AGRICULTURAL DEVELOPMENT, NUTRITION & SUSTAINABILITY, NATIONAL & KAPODISTRIAN UNIVERSITY OF ATHENS, PSACHNA, GREECE [2]INSTITUTE OF BIORESOURCES AND SUSTAINABLE DEVELOPMENT, REGIONAL CENTRE, GANGTOK, INDIA

2.1 Microbes and food production over the centuries

Microbes have been applied for food production since the beginning of human civilization as one of the most economical food processing and storage methods [1]. It has been well established that microbial fermentation can increase the shelf-life of food raw materials by converting their ingredients, forming inhibitory metabolites, while the beneficial microbial population acts antagonistically against other unwanted microorganisms during storage of fermented foods [2–4]. Back in the old times, during the Neolithic period, wild microbes were used to play important role in natural food fermentation. The scientific announcement of the important role of microorganisms in food fermentation came more recently by Van Leeuwenhoek and Hooke in 1665 [5]. As a result, microorganisms are being studied very thoroughly until nowadays, as selected strains can be applied in food fermentation as natural biopreservatives targeting improved nutritional value and products' exquisite organoleptic characteristics [6].

Food fermentation can be achieved by applying specific microbes under controlled conditions, which can achieve bioconversion of food substrates through their growth and metabolic activities. The microorganisms involved in food fermentation are mainly lactic acid bacteria, yeasts, bacilli, and filamentous molds [3,7]. Many different food substrates can be applied for fermentation, including milk, meat, seafoods, cereal grains, fruits, vegetables, root crops, or other miscellaneous food compounds. In each different food substrate, a selected microbial group can be applied depending on the different sugar substrate of fermentation as well as the fermentation capacity of each strain [6,8]. Specifically, food substrates containing monosaccharides and disaccharides are usually fermented by yeasts or lactic acid bacteria. In contrast, food substrates containing high amounts of polysaccharides are optimally fermented through combined microbial populations containing molds, yeasts, or bacilli targeting primary saccharification or proteolysis [8]. As a result of sugar fermentation, various metabolites can be produced, including organic acids, ethanol, and bacteriocins, which may impose an inhibitory effect against other unwanted microbes increasing shelf-life and microbial stability of the final products [6,9]. Likewise, during fermentation of food raw material, a wide variety of bioactive compounds can be synthesized either as metabolites or as substrates released by hydrolysis of organic compounds. These health-promoting bioactive compounds are mainly attributed to the metabolic activities of the microbiota that develops during fermentation. Their presence in the human diet is linked with various health-promoting effects that mainly depend on the compounds' biochemical state while reaching the bloodstream [6,9,10]. These biologically active compounds have been established to confer antimicrobial, antihypertensive, antidiabetic, antitumor, antioxidative, antiinflammatory, and antihyperlipidemic effects in addition to their regular nutritional value (Fig. 2–1).

The importance of food in health had also been acknowledged in the ancient times and especially by the philosopher Hippocrates establishing the phrase:

Ldet the food be your medicine and the medicine be your food

Nowadays, consumer's interest is linked toward high-quality and enhanced beneficial value safe food products. Likewise, a global trend is constantly rising toward the consumption of foods posing health-promoting attributes beyond the basic nutritional functions. Notably, fermented foods and their beneficial microflora have attracted consumers' and scientists' attention and the global industry as their microbially transformed metabolites can pose therapeutic activities. The main target is to sustain these metabolites within an active form [11,12].

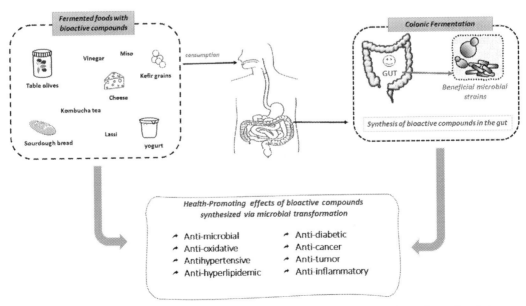

FIGURE 2–1 Health-promoting effects of bioactive compounds synthesized via microbial transformation during food fermentation or colonic fermentation.

Another worth mentioning factor is that fermentation process may also enhance digestibility of proteins and carbohydrates while living beneficial microbes may provide a similar impact after successful colonization in human intestinal tract [3,13]. The fermentation process of food raw material resulting may also occur in the gut, also known as colonic fermentation. A significant effect of colonic fermentation is the enhanced bioavailability of different bioactive compounds that could benefit the consumers, depending on the initial food substrate. Notably, the released bioactives are more potent to provide therapeutic effects as microbially transformed metabolites compared to their parent compounds (e.g., flavonoids, polyphenols, or terpenoids) [14,15]. As a result, the gut microbiota has become an important frontier nowadays, while recent studies combine beneficial microbes in the gut with the transformation of food compounds into several active metabolites. Notably, the fermentation of food components via microbial transformation may occur both during food production processes and within consumers' gastrointestinal (GI) tract providing in both cases metabolites with therapeutic activities (Fig. 2–1). In both cases, the produced bioactive compounds originate either from biosynthesis or biotransformation reactions, enhancing food products' nutritional value [16].

2.2 Bioactive compounds as functional food ingredients

Bioactive compounds are chemical compounds included in food products (e.g., lycopene, resveratrol, tannins, and indoles) that can promote good health. These compounds are usually found in small amounts within a food matrix. Bioactive compounds can often be mentioned as functional ingredients, but only when present in adequate amounts functionality can be claimed [17]. Bioactive compounds can be extracted from a food source (fruits, herbs, vegetables, etc.) or more rarely be manufactured in a laboratory or isolated from food processing residues [17,18]. In all cases, these functional ingredients should preserve their characteristics after extraction [19]. Notably, bioactive compounds are not food additives, but they naturally occur in food raw material, while sufficient scientific studies need to be conducted prior to public health claims.

The presence of bioactive ingredients in the diet can be linked with a great variety of health-promoting effects, as mentioned previously (Fig. 2–1). However, their beneficial claims depend on bioactive compounds' biochemical state reaching the bloodstream [10]. These ingredients are naturally present in the plant or animal foods, but they impose a great drawback as they lack bioavailability, and in order to impose their health-promoting and antidisease activities they must be released from the food matrix and amend in the GI tract of the consumer [20]. Bioactive compounds can usually be found in small amounts in natural food products such as fruits and vegetables, grains, oils and nuts, or various fermented food products as a result of microbial biotransformation. The amounts of bioactive compounds needed to confer health beneficial effects have been found to be greater than consumers' daily nutritional habits. Moreover, their unstable chemical state can also occur as a significant drawback [20]. Another important notice is that the application of bioactive compounds in the food industry is limited as they can easily be deactivated when exposed to harsh food manufacturing or storage conditions. As a result, the design of novel foods is consequently being explored with products reinforced with bioactive compounds such as antioxidants, living cells as are probiotic bacteria, prebiotics, structured lipids, fatty acids, carotenoids, vitamins, polysaccharides, proteins, phenols, peptides, sterols, alkaloids, mineral elements, or volatile oils [21]. These novel foods can be enhanced with bioactive ingredients either by supplementing the initial food product with an already existing compound or by fortifying the food product with a novel bioactive ingredient [19,22,23]. Moreover, the main concern regarding novel foods with enhanced bioactive ingredients is sustaining their bioactivity and increasing their concentration and bioavailability [19]. As a solution, many different techniques have been developed to be applied in food delivery systems targeting the sustained release of bioactive compounds within consumer's GI tract. Among these systems, the most successful ones have been immobilization and encapsulation methods and more specifically complex coacervates composed of proteins and polysaccharides to produce sustainable microcapsules which contain and protect the more sensitive bioactive compounds [19,24,25]. The most commonly applied nutraceutical food components which have been applied for encapsulation are lipids, proteins, carbohydrates, and probiotic bacteria [23].

Another major challenge is the successful delivery and bioavailability of bioactive compounds in foods. To overcome this challenge, appropriate food processing and storage conditions aim to be developed [19]. In this same view, bioactive compounds aim to maintain intact until consumption while biopreservation methods should achieve sustainable release and overcome the harsh conditions that meet within consumers' GI tract (pH, digestive enzymes, and long transit time). Many factors need to be taken into consideration when designing methods targeting bioactive compounds' sustainability and bioavailability. Initially, consumers' mechanisms of absorption and digestion, as well as the sensitivity of bioactive compounds to digestion, need to be considered. A representative and well-studied example is probiotics viability, which poses high sensitivity when inserted in the highly acidic environment and bile salts of the human stomach [3]. Probiotics bacteria, a well representative bioactive, need to be adequate to achieve colonization to the intestinal tract. Immobilization or encapsulation, especially in dietary fiber, which impose prebiotic characteristics, can beneficially affect their bioavailability in the GI tract [26–28]. Therefore, it is crucial to account whether or how bioactive compounds' chemical structure can alternate during food processing and storage conditions considering in parallel their bioavailability. Notably, the factors influencing the stability and availability of bioactive compounds must be evaluated in order to provide mechanisms that will benefit each type [19].

2.3 Health-promoting bioactive compounds synthesized *via* microbial transformation

Microbes are involved in producing fermented foods and render numerous health benefits to humans [8]. A great variety of biologically active compounds can be synthesized during raw food microbial transformation depending on the initial food substrate and the microbial strain to be applied for fermentation and sometimes on the conditions of fermentation (temperature, pH, oxygen) [3,23]. Wild-type microbial strains and selected starter cultures can enhance fermented food with important bioactive compounds released either as primary fermentation metabolites or as substances deriving from hydrolysis of foods organic compounds [29]. During food fermentation, the production of component or secondary metabolites such as organic acids, antimicrobial peptides, CO_2, and bacteriocins has been proved to suppress the growth of spoilage and pathogenic microorganisms act as a natural preservative during food storage [12]. The beneficial microbiota applied for food fermentation along with their health-promoting metabolites can reach the lower intestinal tract of the human through consumption, while colon fermentation may also take place, providing enhance beneficial metabolites after consumption [17]. The produced via microbial transformation bioactive metabolites may impose various health-promoting effects on the consumer, including antimicrobial, antioxidant, antidiabetic, anticancer, antihypertensive, antimutagenic, antitumor, and fibrinolytic activities [17,30].

2.3.1 Production of antimicrobial compounds

Foodborne disease remains, up until nowadays, a significant source of infection and mortality worldwide despite the significant upgraded in food production processes [9]. The human gut is colonized right after birth by gut microbiota, as a necessary event for the development of a functional intestine. Many pathogenic microorganisms may cause infection and disease or colonize consumers' GI tract causing in all cases unwanted health issues [3]. Despite the progress in understanding pathogenic microorganisms' mechanisms, they still pose a great risk on human health as antibiotics, improved control measures, and strict food production processes have not yet achieved to eliminate pathogens' undesirable effects [31]. Fermented food composed of bioactive compounds can act as antimicrobial agents and can provide great support in such cases as they shield the human immune system against pathogenic infections [12]. So, consumers' main target should be the foundation of a strong immune system achieved through daily basis consumption of healthy foods containing antimicrobial agents. For instance, lactic acid bacteria are known to metabolize sugars, especially lactose, producing adequate amounts of lactic acid [32]. The produced lactic acid lowers the pH (4.6–3.0) of fermented foods and therefore enhancing their self-life acting in parallel as an antimicrobial agent against various pathogenic or spoilage microorganisms which cannot survive these low pH values [3,12,27].

Another important antimicrobial group is bacteriocins, which act against other microbial strains limiting their presence [12]. Bacteriocins are known to be produced by several species from bacteria, with lactic acid bacteria being their most popular health-promoting indicator. They are molecules consisting of amino acids (12–100), which are synthesized by microorganism within the cell and released to act extracellularly against treating microbial strains. More than 1700 different kinds of bacteriocins have been reported to act against (kill or inhibit) other microbes' growth [12]. The implication of bacteriocins against other species may be achieved via various different biological mechanisms such as infection and destabilization of viral capsules, damage of bacterial or fungal membranes, or even recruitment of hosts immune cells in inflammatory sites to act against pathogenic infections [12,33]. An important representative example was reported when a bacteriocin was used against the sporous intestinal pathogen *Enterocytozoon bieneusi* that causes diarrhea. Subsequently, the encapsulation of bacteriocin produced by *Lactobacillus acidophilus* and applied against *E. bieneusi* resulted in a significant reduction in the number of intestinal spores [34]. Likewise, probiotic strains and especially ones that can produce antimicrobial peptides are considered most valuable within fermented foods as they can provide a competitive survival advantage when consumed at adequate amounts compared to other prokaryotes in the same niche [12]. Notably, lactic acid fermented foods containing viable beneficial microbes along with their precious metabolites can pose significant health benefits to the consumer [8].

2.3.2 Production of antioxidant compounds

According to recent literature, free radicals are able to affect the regular biological process of the human body and can become pathogenic posing a series of diseases like Alzheimer's, Parkinson's disease, cancer, diabetes, atherosclerosis, and arthritis [35,36]. Free radicals are molecules, which contain at least one unpaired electron and thus being highly reactive. They can be frequently generated in the body during metabolic processes and stress conditions. Also, free radicals have a controversial impact on the immune system as they are found to eliminate living microbial cells beneficial or pathogenic. The free radicals are naturally occurring in the body as they take part in cell signaling, expression of genes and ion transportation, but the main consideration refers to free radicals occurring from external sources such as stressful environment, smoking, and pollution as these factors can easily cause oxidative stress and damage the human cells [35,36]. Oxidative factors can be prevented when adequate amounts of antioxidant compounds are found in the body. In specific, food antioxidants can be classified into different categories regarding their properties as water-soluble compounds like citrates, flavonoids, anthocyanins, and lipid-soluble compounds like terpenoids, carotenoids, tocopherols, and vitamins [37].

Fermented food products containing antioxidant compounds such as hydrolyzed antioxidative peptides, free polyphenols, isoflavones, and flavonoids can inactivate free radicals alternating the caused oxidative stress and thus be more effective against radical oxidation [35]. For example, probiotic bacteria-based fermented foods can act against allergies, cancer, and pathogenic microbes while being rich in antioxidant compounds [3]. Likewise, the plant-based beverage "Kombucha tea," which is made of black tea and sugar fermented with a mixed culture of yeasts and acetic acid bacteria, states antioxidant activity mediated via microbial metabolism of tea polyphenols [38]. Having a long history as one to provide benefits on cardiovascular health, red wine, has evidently been related to one of the main nutraceuticals' polyphenols [39]. Another important category providing a plethora of polyphenols as a source of antioxidant activity are fermented foods, and beverages as more than 500 different polyphenols have been detected over the years [18,40,41]. Among wide compounds, flavonoids and phenolic acids are posed as the most significant antioxidant substrates providing antimicrobial, antiinflammatory, antineoplastic, hepatoprotective, and immune modulating activities to the consumer [39].

2.3.3 Production of oligosaccharides with prebiotic potential

Carbohydrates are essential and most abundant biomolecules most significant in human and animal nutrition [42]. Functional carbohydrates include nondigestible dietary fibers and other ingredients providing a dietary function [43]. Carbohydrates can be classified into three main groups according to their carbon chain, including monosaccharides, disaccharides, oligosaccharides, and polysaccharides. In general, carbohydrates fermentation, mainly posed as saccharolytic fermentation, may occur during industrial fermentation and subsequently after food intake can take place in the proximal colon. Microbes associated with fermentation have the ability to hydrolyze a broad range of carbohydrates. As a result, oligosaccharides or monosaccharides can occur after fermentation of food products or decomposed enzymatically via microbial fermentation within the GI tract [44].

Oligosaccharides can provide a functional effect on food products such as prebiotic effects, antimicrobial properties, cytotoxic effects, cancer prevention, or even therapy of chronic diseases [43]. Prebiotics are dietary ingredients, which can be found in a great variety of sources, such as fruits, vegetables, milk, honey, or marine algae and are very popular as they can selectively enhance beneficial gut microbiota [3]. Prebiotics can be categorized into different types including galactooligosaccharides as inulin of different chain lengths and fructooligosaccharides as starch and lactulose. For instance, lactulose (4-O-β-D-galactopyranosyl-D-fructofuranose) which is composed of fructose and galactose is known to be highly tolerant in human digestive enzymes and can reach the intestine where it is selectively fermented by Bifidobacteria and Lactic acid bacteria [45]. Lactulose can be fermented within the intestine, providing a new group of lactulose-derived oligosaccharides with potential beneficial health effects [43,45].

2.3.4 Production of exopolysaccharides with associated health benefits

Towards the last decade, another important group that provides foods with functionality has attracted attention. This group parts from complex polysaccharide metabolites that can be synthesized via microbial transformation of food substrates. These valuable polysaccharide metabolites, deriving from different biogenetic resources, may be found either attached to the microbial cell wall (capsular polysaccharides) or released from the cells from the cell surface [exopolysaccharides (EPSs)] [42]. EPSs can be found either slightly bound to the cell surface or totally independent of the external environment forming mucus, ropiness, and slimy materials. Naturally formed EPSs are nontoxic biodegradable polymers that play an important role in various biological mechanisms and are thus considered bioactive compounds [46]. The main biofunctional characteristics of EPSs include protection from adverse environmental effects, immediate response to the immune system, and increased microbial resistance to stressful environmental conditions [47]. In addition, EPSs occurring in fermented foods produced mainly by Lactobacilli have been reported to favor colonization of beneficial microbes (e.g., probiotics) and prebiotic activity while acting as a physical barrier against pathogenic microorganisms by intercellular signal transduction, which "informs" the cell to initiate its protection mechanism [47]. As a result, EPSs can provide the consumer a number of health-promoting activities, including cholesterol reduction, antioxidant activity, immunomodulation, and antimicrobial characteristics that act against pathogenic microorganisms prebiotic effect, which can enhance the viability of beneficial microorganisms, antiproliferative activities, and anticancer effect among others [46,48]. The production of EPSs can occur during fermented foods production as well as within consumer's GI tract as a metabolite of colon fermentation. Especially during microbial transformation within the GI tract fermentation of EPSs is involved in short-chain fatty generation [46–48]. As a result, consumption of functional fermented foods containing viable beneficial microbes and bioactive compounds is of major importance to promote health, even after consumption, targeting colon fermentation and enhanced nutraceuticals bioavailability [42].

2.4 Trends in the delivery of bioactive compounds targeting enhanced food functionality

The ingestion of foods with incorporated live probiotic strains cannot be efficient if adequate protection is not provided against occurring adverse conditions (e.g., harsh food processing and storage conditions, harsh GI conditions). Likewise, to exert their functional properties, probiotic strains must maintain high survival rates during food processing and storage conditions and survive through the harsh GI conditions initiated after consumption. Encapsulation of probiotic cells using wall material like resins, maltodextrin, cellulose, and proteins is a very promising approach to address these challenges achieving high survival rates [3]. The protective effect of different probiotic encapsulation vehicles is tested using various different strains resulting in many cases in tailor-made encapsulation vehicles for each food category. For instance, encapsulated probiotic cells in mastic and other resins applied in dairy products have shown enhanced culture growth and viability during storage when compared to free probiotic cultures [2,27,49]. However, after a more thorough investigation recent studies reported that the combination of different wall encapsulation materials with different properties was proved to confer not only improved protection but also controlled release of probiotics when tested under gastric conditions [27,50,51]. As a result, the application of multilayer structured vehicles is now preferred as a more effective technique in improving the survivability as well as controlled release of encapsulated probiotic cells [50].

Encapsulation is the technique, which involves the introduction of bioactive compounds into a matrix. More specifically, with encapsulation the compound of interest (bioactive compound) can be entrapped within or coated with another material or system (encapsulating agent) targeting the protection of the active compounds from external conditions, or restraining their contact with the environment and/or controlling their release [52]. Encapsulation has been successfully applied to protect the bioactive compounds against adverse environmental conditions, improve their stability during processing and storage, and allow their controlled release after consumption [24,53]. Another significant application of encapsulation is that it can be used to mask undesirable odors or bitter tastes contained in food products either via the odor effect of bioactive compounds or through the aromatic characteristics provided by the encapsulating agent [24,52,53]. Encapsulation technology is nowadays successfully applied in pharmaceuticals, cosmetics, and food industries utilizing many different methods such as spray-drying, entrapment in viscus matrix, coating, complexation and coacervation, emulsification, spray chilling, and cooling or even combination of different technologies [21,54].

Bioavailability is a key step in ensuring bioefficacy of bioactive compounds after consumption [55]. Most of the bioactive compounds show low bioavailability, which is the major concern regarding the impact of their beneficial effects. Numerous encapsulation techniques that have been enlisted target to enhance the bioavailability of bioactive compounds [25,30,56,57]. Bioactive compounds can be unstable during food processing and storage conditions. On the other hand, food-grade carrier agents can be applied for encapsulation of bioactive components verifying their stability as they target to be released after consumption facilitating bioactive uptake [21,25,53,58,59]. Finally, regarding the food sector two main aspects for applying these insights need to be kept under consideration. The procedure and the encapsulation material must be of low cost, and more importantly, the encapsulation carriers must be of natural origin, including food-grade characteristics.

2.4.1 Encapsulation of living cells

The potential for bringing encapsulated beneficial microorganisms into commercial food products is great. Concepts, which are extensively studied in pharmaceutics, may revolutionize the food sector as many techniques are characterized as food grades and low cost. For example, the design to selectively target probiotic bacteria and selectively enhance their growth factors within specific encapsulation material (e.g., whey protein, prebiotics) can finally promote their growth in the highly competitive gut environment [3]. Viability is the most important parameter in selecting an appropriate encapsulation technology. Several encapsulations have been applied to probiotics that include spray-drying, coating, entrapment, coacervation, and gel-particle technologies [3,56]. Among them, spray-drying techniques are the most promising as the cells can be partially adapted to the high processing temperatures. Probiotics present two sets of problems when considering microencapsulation; their size (diameter: 1–5 μm), which immediately excludes nanotechnologies and cell viability [60]. Complex coacervation shows many promising characteristics, which may result in a leading encapsulation method for probiotics [56].

2.4.2 Encapsulation of antioxidants

Antioxidants have gained exquisite attention the recent years as they may prevent food deterioration in parallel with preserving the value of their nutritional food. In order for antioxidant compounds to be characterized as efficient, they need to confer high reactivity, biological availability, ubiquity, versatility, and the ability to cross physiologic barriers. More importantly, bioactivity of antioxidants may be influenced by oxygen, light, temperature, and moisture [53]. Thus, encapsulation techniques have been extensively studied targeting their efficient stabilization thought-out food processing and storage [53]. Apart from oxygen and free radicals, another worth mentioning factor is acid. Acids are contained in food products and may cause a negative impact in combination with other food ingredients. For instance, the conjunction of acids with different food ingredients may alter food flavor, provide undesirable odors or even alter the food pH especially when occurring in significant amounts [61]. As a result, natural or synthetic antioxidant constituents tend to be applied to restore or delay any oxidation or undesirable effects [52]. For instance, ascorbic acid, a water-soluble vitamin, is considered one of the most important antioxidants. However, it is considered as a very unstable antioxidants as various environmental factors, such as oxygen, pH, storage temperature, and radiation can affect its stability. The stability of ascorbic acid can be enforced via microencapsulation eliminating the effects harsh environmental conditions [62]. Likewise, instability and low bioavailability of polyphenols could greatly limit their inventions health benefits. As a result, emulsion-based encapsulation has been applied for its successful delivery [63]. The incorporation of novel technologies seems to be very promising for revealing the incentive quality of processed substances, and tailor-made microcapsules need to be considered for their specific application [53].

2.5 Fermented food as potential for antiviral therapy

Microorganisms and bioactive metabolites produced in fermented foods can be responsible for immunomodulatory and antiviral effects [64–66]. Bioactive peptide from different types of fermented food products has been reported to boost immunity as well as antiviral therapeutics [65–68]. Food-derived bioactive peptides as well as bacteriocins produced by lactic acid bacteria produced during food fermentation can be explored as inhibitors of key targets identified for screening metabolites for anti-COVID therapeutics [65,66].

In a recent computational study, nisin (a bacteriocin) produced by strains of *Lactococcus lactis* showed effective interaction with the human angiotensin-converting enzyme 2 (hACE2) that can help in prevention of binding with SARS-CoV-2 spike protein receptor [66,69]. Among nisin variants studied on computational analysis, nisin H showed most effective binding with hACE2 receptor [66]. A multitarget peptide "KFVPKQPNMIL" that can inhibit the target proteins (SARS-CoV-2 main protease and S1 glycoprotein) has been identified in the proteome of soybean cheese fermented using *Lactobacillus delbrueckii* WS4 [67]. Proteomics and computational analysis of fermented soybean using different *Bacillus* species resulted in identification of a multifunctional peptide that can have antiviral as well as immunomodulatory effects [68]. The peptide "ALPEEVIQHTFNLKSQ" from fermented soybean can be explored further for Toll-like receptor 4 modulation and SARS-CoV-2 S1 receptor-binding domain inhibition [68]. Bioactive peptides from fermented foods can be exhibit antiviral effect, which depends on the amino acid composition and their sequence in the peptide [67,68].

Similarly, a small dsRNA produced by *Bacillus subtilis* isolated from alkaline fermented soybean product of North East India showed antiviral effect against HIV-1 [70]. Probiotics has been proposed as effective agents to contribute to flatten the curve of COVID-19 pandemic [64]. Bioactive metabolites and probiotics from traditionally fermented food products have shown to exhibit immunomodulatory and antiviral properties by different mechanisms [71]. Some of the mechanisms include (1) enhancement of the cytotoxic of T lymphocytes, (2) elevating natural killers cell toxicity, and (3) proinflammatory cytokines production [71]. Bioactive compounds in fermented food have immense potential, which need to be further explored for immune boosting and antiviral effects. There are possibilities of finding novel biomolecules having antiviral effect formed during microbial transformation of food components.

2.6 Conclusions and perspectives

Bioactive compounds can be found naturally in food and plants. Nowadays, novel foods of enriched bioactive compounds are introduced into the market, providing many health claims. As a result of this trend, recent research interest is linked toward the microbial transformation of foods and food components during industrial and colon fermentation due to the enhancement in bioactive compounds providing health claims. The bioavailability of bioactive compounds mainly depends on the environment meet within the consumers' GI tract and other food ingredients consumed along with the enriched product. Moreover, a consumed bioactive compound's stability may alter between different individuals depending on the consumer's health and the varieties in microflora of each individual's GI tract. Hence, proper experimental designs regarding the successful delivery of bioactive compounds are of paramount importance. For the enhancement of their delivery, encapsulation methods can be applied. A more ambitious perspective of bioactive delivery includes smart, selective targeted delivery systems designed to release bioactive compounds to particular regions or cells of the body, providing personalized actions based on each individual's health needs.

References

[1] F. Bourdichon, S. Casaregola, C. Farrokh, J.C. Frisvad, M.L. Gerds, W.P. Hammes, et al., Food fermentations: microorganisms with technological beneficial use, International Journal of Food Microbiology 154 (2012) 87−97.

[2] A. Terpou, P.S. Nigam, L. Bosnea, M. Kanellaki, Evaluation of chios mastic gum as antimicrobial agent and matrix forming material targeting probiotic cell encapsulation for functional fermented milk production, LWT Food Science and Technology 97 (2018) 109−116.

[3] A. Terpou, A. Papadaki, I.K. Lappa, V. Kachrimanidou, L.A. Bosnea, N. Kopsahelis, Probiotics in food systems: significance and emerging strategies towards improved viability and delivery of enhanced beneficial value, Nutrients 11 (2019) 1591.

[4] H.T. Mouafo, A. Mbawala, K. Tanaji, D. Somashekar, R. Ndjouenkeu, Improvement of the shelf life of raw ground goat meat by using biosurfactants produced by lactobacilli strains as biopreservatives, LWT Food Science and Technology 133 (2020) 110071.

[5] H. Gest, The discovery of microorganisms by Robert Hooke and Antoni van Leeuwenhoek, fellows of the Royal Society, Notes and Records of the Royal Society 58 (2004) 187−201.

[6] P.M. O'Connor, T.M. Kuniyoshi, R.P.S. Oliveira, C. Hill, R.P. Ross, P.D. Cotter, Antimicrobials for food and feed; a bacteriocin perspective, Current Opinion in Biotechnology 61 (2020) 160–167.

[7] M. Venturini, Copetti, yeasts and molds in fermented food production: an ancient bioprocess, Current Opinion in Food Science 25 (2019) 57–61.

[8] M.L. Marco, D. Heeney, S. Binda, C.J. Cifelli, P.D. Cotter, B. Foligné, et al., Health benefits of fermented foods: microbiota and beyond, Current Opinion in Biotechnology 44 (2017) 94–102.

[9] A.G. Abdelhamid, N.K. El-Dougdoug, Controlling foodborne pathogens with natural antimicrobials by biological control and antivirulence strategies, Heliyon 6 (2020) e05020.

[10] S.M. Martínez-Sánchez, J.A. Gabaldón-Hernández, S. Montoro-García, Unravelling the molecular mechanisms associated with the role of food-derived bioactive peptides in promoting cardiovascular health, Journal of Functional Foods 103645 (2019).

[11] R. Chourasia, M.M. Abedin, L.C. Phukon, D. Sahoo, S.P. Singh, A.K. Rai, Biotechnological approaches for the production of designer cheese with improved functionality, Comprehensive Reviews in Food Science and Food Safety 20 (2021) 960–979.

[12] G.M. Daba, W.A. Elkhateeb, Bacteriocins of lactic acid bacteria as biotechnological tools in food and pharmaceuticals: current applications and future prospects, Biocatalysis and Agricultural, Biotechnology 28 (2020) 101750.

[13] W. Morovic, C.R. Budinoff, Epigenetics: a new frontier in probiotic research, Trends in Microbiology 29 (2020) 117–126.

[14] Y.A. Chait, A. Gunenc, F. Bendali, F. Hosseinian, Simulated gastrointestinal digestion and in vitro colonic fermentation of carob polyphenols: bioaccessibility and bioactivity, LWT Food Science and Technology 117 (2020) 108623.

[15] S. Wu, A.E.-D.A. Bekhit, Q. Wu, M. Chen, X. Liao, J. Wang, et al., Bioactive peptides and gut microbiota: candidates for a novel strategy for reduction and control of neurodegenerative diseases, Trends in Food Science and Technology 108 (2021) 164–176.

[16] J.E.T. van Hylckama Vlieg, P. Veiga, C. Zhang, M. Derrien, L. Zhao, Impact of microbial transformation of food on health—from fermented foods to fermentation in the gastro-intestinal tract, Current Opinion in Biotechnology 22 (2011) 211–219.

[17] S. Misra, P. Pandey, H.N. Mishra, Novel approaches for co-encapsulation of probiotic bacteria with bioactive compounds, their health benefits and functional food product development: a review, Trends in Food Science and Technology 109 (2021) 340–351.

[18] I. Mantzourani, A. Terpou, A. Bekatorou, A. Mallouchos, A. Alexopoulos, A. Kimbaris, et al., Functional pomegranate beverage production by fermentation with a novel synbiotic L. paracasei biocatalyst, Food Chemistry 308 (2020) 125658.

[19] C. Bao, P. Jiang, J. Chai, Y. Jiang, D. Li, W. Bao, et al., The delivery of sensitive food bioactive ingredients: absorption mechanisms, influencing factors, encapsulation techniques and evaluation models, Food Research International 120 (2019) 130–140.

[20] M. Nooshkam, M. Varidi, Maillard conjugate-based delivery systems for the encapsulation, protection, and controlled release of nutraceuticals and food bioactive ingredients: a review, Food Hydrocolloids 100 (2020) 105389.

[21] B. Niu, P. Shao, Y. Luo, P. Sun, Recent advances of electrosprayed particles as encapsulation systems of bioactives for food application, Food Hydrocolloids 99 (2020) 105376.

[22] M.P. Caporgno, A. Mathys, Trends in microalgae incorporation into innovative food products with potential health benefits, Frontiers in Nutrition 5 (2018) 58.

[23] J. Chai, P. Jiang, P. Wang, Y. Jiang, D. Li, W. Bao, et al., The intelligent delivery systems for bioactive compounds in foods: physicochemical and physiological conditions, absorption mechanisms, obstacles and responsive strategies, Trends in Food Science and Technology 78 (2018) 144–154.

[24] C.P. Champagne, P. Fustier, Microencapsulation for the improved delivery of bioactive compounds into foods, Current Opinion in Biotechnology 18 (2007) 184–190.

[25] M.A. Colín-Cruz, D.J. Pimentel-González, H. Carrillo-Navas, J. Alvarez-Ramírez, A.Y. Guadarrama-Lezama, Co-encapsulation of bioactive compounds from blackberry juice and probiotic bacteria in biopolymeric matrices, LWT Food Science and Technology 110 (2019) 94–101.

[26] L.A. Bosnea, N. Kopsahelis, V. Kokkali, A. Terpou, M. Kanellaki, Production of a novel probiotic yogurt by incorporation of L. casei enriched fresh apple pieces, dried raisins and wheat grains, Food and Bioproducts Processing 102 (2017) 62–71.

[27] V. Schoina, A. Terpou, L. Bosnea, M. Kanellaki, P.S. Nigam, Entrapment of Lactobacillus casei ATCC393 in the viscus matrix of Pistacia terebinthus resin for functional myzithra cheese manufacture, LWT - Food Science and Technology 89 (2018) 441–448.

[28] A. Terpou, A. Bekatorou, L. Bosnea, M. Kanellaki, V. Ganatsios, A.A. Koutinas, Wheat bran as prebiotic cell immobilisation carrier for industrial functional Feta-type cheese making: chemical, microbial and sensory evaluation, Biocatalysis and Agricultural Biotechnology 13 (2018) 75–83.

[29] A. Valdés, A. Cifuentes, C. León, Foodomics evaluation of bioactive compounds in foods, TrAC Trends in Analytical Chemistry 96 (2017) 2–13.

[30] M. Mohammadian, M.I. Waly, M. Moghadam, Z. Emam-Djomeh, M. Salami, A.A. Moosavi-Movahedi, Nanostructured food proteins as efficient systems for the encapsulation of bioactive compounds, Food Science and Human Wellness (2020). Available from: https://doi.org/10.1016/j.fshw.2020.04.009.

[31] C.M. Galanakis, The food systems in the era of the Coronavirus (COVID-19) pandemic crisis, Foods 9 (2020) 523.

[32] I.K. Lappa, A. Papadaki, V. Kachrimanidou, A. Terpou, D. Kourougliotis, E. Eriotou, et al., Cheese whey processing: integrated biorefinery concepts and emerging food applications, Foods 8 (2019) 347.

[33] A. Radaic, M.B. de Jesus, Y.L. Kapila, Bacterial anti-microbial peptides and nano-sized drug delivery systems: the state of the art toward improved bacteriocins, Journal of Controlled Release 321 (2020) 100–118.

[34] S.F. Mossallam, E.I. Amer, R.G. Diab, Potentiated anti-microsporidial activity of Lactobacillus acidophilus CH1 bacteriocin using gold nanoparticles, Experimental Parasitology 144 (2014) 14–21.

[35] Z.Q. Liu, Bridging free radical chemistry with drug discovery: a promising way for finding novel drugs efficiently, European Journal of Medicinal Chemistry 189 (2020) 112020.

[36] C. Peña-Bautista, M. Baquero, M. Vento, C. Cháfer-Pericás, Free radicals in Alzheimer's disease: lipid peroxidation biomarkers, Clinica Chimica Acta 491 (2019) 85–90.

[37] G. Ozkan, P. Franco, I. De Marco, J. Xiao, E. Capanoglu, A review of microencapsulation methods for food antioxidants: principles, advantages, drawbacks and applications, Food Chemistry 272 (2019) 494–506.

[38] A. Chandrasekara, F. Shahidi, Herbal beverages: bioactive compounds and their role in disease risk reduction—a review, Journal of Traditional and Complementary Medicine 8 (2018) 451–458.

[39] R. Rodrigo, A. Miranda, L. Vergara, Modulation of endogenous antioxidant system by wine polyphenols in human disease, Clinica Chimica Acta 412 (2011) 410–424.

[40] I. Mantzourani, C. Nouska, A. Terpou, A. Alexopoulos, E. Bezirtzoglou, M.I. Panayiotidis, et al., Production of a novel functional fruit beverage consisting of Cornelian cherry juice and probiotic bacteria, Antioxidants 7 (2018) 163.

[41] X. Du, A.D. Myracle, Fermentation alters the bioaccessible phenolic compounds and increases the alpha-glucosidase inhibitory effects of aronia juice in a dairy matrix following in vitro digestion, Food and Function 9 (2018) 2998–3007.

[42] W. Chaisuwan, K. Jantanasakulwong, S. Wangtueai, Y. Phimolsiripol, T. Chaiyaso, C. Techapun, et al., Microbial exopolysaccharides for immune enhancement: Fermentation, modifications and bioactivities, Food Bioscience 35 (2020) 100564.

[43] A. Kruschitz, B. Nidetzky, Downstream processing technologies in the biocatalytic production of oligosaccharides, Biotechnology Advances 43 (2020) 107568.

[44] U.P. Tiwari, A.K. Singh, R. Jha, Fermentation characteristics of resistant starch, arabinoxylan, and β-glucan and their effects on the gut microbial ecology of pigs: a review, Animal Nutrition 5 (2019) 217–226.

[45] X. Zhang, J. Zheng, N. Jiang, G. Sun, X. Bao, M. Kong, et al., Modulation of gut microbiota and intestinal metabolites by lactulose improves loperamide-induced constipation in mice, European Journal of Pharmaceutical Sciences 158 (2021) 105676.

[46] M. Oleksy-Sobczak, E. Klewicka, L. Piekarska-Radzik, Exopolysaccharides production by Lactobacillus rhamnosus strains—optimization of synthesis and extraction conditions, LWT Food Science and Technology 122 (2020) 109055.

[47] J. Wang, D.R. Salem, R.K. Sani, Extremophilic exopolysaccharides: a review and new perspectives on engineering strategies and applications, Carbohydrate Polymers 205 (2019) 8–26.

[48] Y. Rahbar, Saadat, A. Yari Khosroushahi, B. Pourghassem Gargari, A comprehensive review of anticancer, immunomodulatory and health beneficial effects of the lactic acid bacteria exopolysaccharides, Carbohydrate Polymers 217 (2019) 79–89.

[49] V. Schoina, A. Terpou, G. Angelika-Ioanna, A. Koutinas, M. Kanellaki, L. Bosnea, Use of Pistacia terebinthus resin as immobilization support for Lactobacillus casei cells and application in selected dairy products, Journal of Food Science and Technology 52 (2015) 5700–5708.

[50] M. de Araújo Etchepare, G.L. Nunes, B.R. Nicoloso, J.S. Barin, E.M. Moraes Flores, R. de Oliveira Mello, et al., Improvement of the viability of encapsulated probiotics using whey proteins, LWT Food Science and Technology 117 (2020).

[51] A. Terpou, A. Papadaki, L. Bosnea, M. Kanellaki, N. Kopsahelis, Novel frozen yogurt production fortified with sea buckthorn berries and probiotics, LWT Food Science and Technology 105 (2019) 242–249.

[52] S. Sharma, S.F. Cheng, B. Bhattacharya, S. Chakkaravarthi, Efficacy of free and encapsulated natural antioxidants in oxidative stability of edible oil: special emphasis on nanoemulsion-based encapsulation, Trends in Food Science and Technology 91 (2019) 305–318.

[53] J. Aguiar, B.N. Estevinho, L. Santos, Microencapsulation of natural antioxidants for food application—the specific case of coffee antioxidants—a review, Trends in Food Science and Technology 58 (2016) 21–39.

[54] M. Zanetti, T.K. Carniel, F. Dalcanton, R.S. dos Anjos, H. Gracher Riella, P.H.H. de Araújo, et al., Fiori, use of encapsulated natural compounds as antimicrobial additives in food packaging: a brief review, Trends in Food Science and Technology 81 (2018) 51–60.

[55] E. Nowak, Y. DLivney, Z. Niu, H. Singh, Delivery of bioactives in food for optimal efficacy: what inspirations and insights can be gained from pharmaceutics? Trends in Food Science and Technology 91 (2019) 557–573.

[56] L.A. Bosnea, T. Moschakis, C.G. Biliaderis, Complex coacervation as a novel microencapsulation technique to improve viability of probiotics under different stresses, Food and Bioprocess Technology 7 (2014) 2767–2781.

[57] S.B. Doherty, M.A. Auty, C. Stanton, R.P. Ross, G.F. Fitzgerald, A. Brodkorb, Application of whey protein micro-bead coatings for enhanced strength and probiotic protection during fruit juice storage and gastric incubation, Journal of microencapsulation 29 (2012) 713–728.

[58] Y. Alvarado, C. Muro, J. Illescas, M.d.C. Díaz, F. Riera, Encapsulation of antihypertensive peptides from whey proteins and their releasing in gastrointestinal conditions, Biomolecules 9 (2019) 164.

[59] G.B. Brinques, M.A.Z. Ayub, Effect of microencapsulation on survival of *Lactobacillus plantarum* in simulated gastrointestinal conditions, refrigeration, and yogurt, Journal of Food Engineering 103 (2011) 123–128.

[60] P. de Vos, M.M. Faas, M. Spasojevic, J. Sikkema, Encapsulation for preservation of functionality and targeted delivery of bioactive food components, International Dairy Journal 20 (2010) 292–302.

[61] M. Trindade, C. Grosso, The stability of ascorbic acid microencapsulated in granules of rice starch and in gum Arabic, Journal of Microencapsulation 17 (2000) 169–176.

[62] K.G.H. Desai, H.J. Park, Recent developments in microencapsulation of food ingredients, Drying Technology 23 (2005) 1361–1394.

[63] W. Lu, A.L. Kelly, S. Miao, Emulsion-based encapsulation and delivery systems for polyphenols, Trends in Food Science and Technology 47 (2016) 1–9.

[64] D. Baud, V. Dimopoulou, G.R. Agri, G. Gibson, E. Reid, Giannoni, Using probiotics to flatten the curve of coronavirus disease COVID-2019 pandemic, Frontiers in Public Health 8 (2020) 186.

[65] Z. Yu, R. Kan, H. Ji, S. Wu, W. Zhao, et al., Identification of tuna protein-derived peptides as potent SARS-CoV-2 inhibitors via molecular docking and molecular dynamic simulation, Food Chemistry 128366 (2020).

[66] R. Bhattacharya, A.M. Gupta, S. Mitra, S. Mandal, S.R. Biswas, A natural food preservative peptide nisin can interact with the SARS-CoV-2 spike protein receptor human ACE2, Virology 552 (2021) 107–111.

[67] R. Chourasia, S. Padhi, L.C. Phukon, M.M. Abedin, S.P. Singh, A.K. Rai, A potential peptide from soy cheese produced using Lactobacillus delbrueckii WS4 for effective inhibition of SARS-CoV-2 main protease and S1 glycoprotein, Frontiers in Molecular Biosciences 7 (2020) 601753.

[68] S. Padhi, S. Sanjukta, R. Chourasia, R.K. Labala, S.P. Singh, A.K. Rai, A multifunctional peptide from *Bacillus* fermented soybean for effective inhibition of SARS-CoV-2 S1 receptor binding domain and modulation of toll like receptor 4: a molecular docking study, Frontiers in Molecular Biosciences 8 (2021) 636647.

[69] N. Kitagawa, T. Otani, T. Inai, Nisin, a food preservative produced by Lactococcus lactis, affects the localization pattern of intermediate filament protein in HaCaT cells, Anatomical Science International 94 (2) (2019) 163–171.

[70] R.K. Imrat, S. Labala, S. Velhal, P. Bhagat, K. Vainav, Jeyaram, Small double-stranded RNA with anti-HIV activity abundantly produced by *Bacillus subtilis* MTCC5480 isolated from fermented soybean, International Journal of Biological Macromolecules 161 (2020) 828–835.

[71] B.J. Muhialdin, N. Zawawi, A.F.A. Razis, J. Bakar, M. Zarei, Antiviral activity of fermented foods and their probiotics bacteria towards respiratory and alimentary tracts viruses, Food Control 127 (2021) 108140.

Bioactive peptide production in fermented foods

Anja Dullius[1,2], Gabriela Rabaioli Rama[1,2], Maiara Giroldi[1,2], Márcia Inês Goettert[2,3], Daniel Neutzling Lehn[1], Claucia Fernanda Volken de Souza[1,2]

[1]FOOD BIOTECHNOLOGY LABORATORY, UNIVERSITY OF VALE DO TAQUARI – UNIVATES, LAJEADO, RS, BRAZIL [2]BIOTECHNOLOGY GRADUATE PROGRAM, UNIVERSITY OF VALE DO TAQUARI – UNIVATES, LAJEADO, RS, BRAZIL [3]CELL CULTURE LABORATORY, UNIVERSITY OF VALE DO TAQUARI – UNIVATES, LAJEADO, RS, BRAZIL

3.1 Introduction

Many dietary proteins have specific biological and functional properties with positive effects on humans due to health-promoting properties. These properties are attributed to the presence of active peptides [1,2]. These peptides are released from precursor proteins of animal or vegetable origin, during the manufacture of fermented foods [3,4]. Bioactive peptides (BAPs) are the most frequently studied and are present in fermented products such as dairy products, legumes, cereals, and meat-derived products. BAPs are low molecular weight protein fragments majoritarily comprised 2–20 amino acid residues [3,5]. BAP production enables the use of these peptides in the development of functional food, due to their several interesting health-promoting bioactivities, for example, antidiabetic, antihypertensive, mucin-stimulating, antimicrobial, antioxidant, or immunomodulatory properties, shown in Fig. 3–1 [6–8]. Importantly, the human digestive system releases a particularly large number of BAPs from dietary proteins, thus naturally providing health benefits [9,10]. These peptides are inactive within the parent protein sequence and can be released during gastrointestinal digestion or food processing; depending on the amino acid sequence, these peptides may perform several activities. However, since the consumption of protein-rich foods is often not enough to achieve the positive physiological effects, the use of peptide additives is recommended [1,11]. This wide array of activities broadens their use in the biopreservation of food products, due to antimicrobial and antioxidant activities, and several peptide sequences have either bifunctional or multifunctional activities [12–14]. Nowadays, there is a growing trend and interest in the use of food proteins of different origins such as milk, eggs, soy, fish, and meat as a source of biologically active peptides with health-promoting properties [15]. As a result, much research has been recently focused on the generation and processing of BAPs,

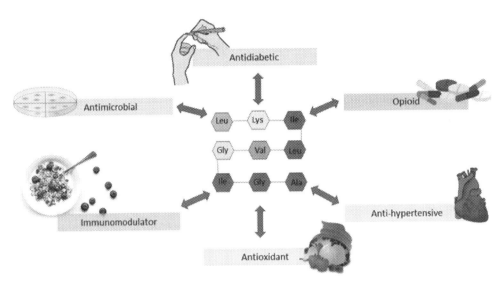

FIGURE 3–1 Major biological activities resulting from the intake of bioactive peptides derived from fermented foods.

and the previously underused protein-rich by-products of the food industries. Considering that the functional foods business has been thriving, BAP production boosts the commercial exploitation of fermented foods.

However, both BAPs identification and purification steps are time-consuming, and there are still no application systematic methodologies. The main difficulties of these steps are protein and enzyme selection, peptides detection, and the selection of the appropriate enrichment method, such as quantitative structure–activity relationship and in silico-based bioinformatics [2,16,17]. Therefore this chapter describes the most recent investigations that support the potential health benefits of BAPs, and BAP production, identification, and functional activities in fermented dairy, meat, and plant products. Finally, it provides an overview of the biotechnological progress of industrial BAP production.

3.2 Bioactive peptides in fermented dairy products

Milk composition has basically two groups of proteins: caseins and whey proteins. Caseins occur in higher concentration (80% of the total protein in dry base) and are divided into the following types: α_{s1}-casein (α_{s1}-CN), α_{s2}-CN, β-CN, and κ-CN. The remaining total protein content is divided in α-lactoalbumin (α-LA), β-lactoglobulin (β-LG), immunoglobulins, bovine serum albumin, and others [18]. Aside from playing an important role in food manufacture, dairy proteins (DPs) are known for their nutritional value, due to the presence of essential amino acids. That is why milk is considered as one of the major sources of nutrients for newborns and adults [19,20]. DPs have a series of bioactive fragments (BAPs) in their native sequences that can play different roles in pattern protein [21].

Therefore an array of activities related to milk-derived BAPs has been described in literature since the 90s. They include mineral-binding, cardioprotector properties, antihypertensive, antidiabetes, agonistic orantagonistic opioid, antimicrobial, immunomodulatory, antioxidant, and anticancer properties [22]. Overall, these BAPs can be obtained using the following methods: (1) conventional enzyme hydrolysis; (2) microbial fermentation; (3) a combination of methods (1) and (2); (4) food manufacture processes with high temperatures and pressures; and (5) in vivo digestion via gastrointestinal enzymes [16,23].

Thus fermented dairy products, aside from being a source of energy and nutrients, can be considered an important source of BAPs that can provide benefits to consumer health. This is due to the fact that many of the microorganisms used industrially (as starters or nonstarters) are highly proteolytic, metabolizing DPs as a primary source to obtain amino acids essential for their growth [24]. Lactic acid bacteria (LAB) are the most frequently employed microorganisms for this purpose, although some yeast species are also used [25]. LAB are oxygen-sensitive microorganisms, and that is why most of their genes have developed tolerance to oxidative stress [26]. LAB biosynthetic activities for amino acid synthesis are limited. Therefore these microorganisms have a complex proteolytic system that allows the obtaining of amino acids from protein sources by enzymatic reactions. Therefore characteristics of fermentation-generated BAPs vary according to the species and genus of the bacteria employed, especially due to the differences between proteolytic enzymes found in different bacteria [26,27]. It is also worth noting that peptides are released extracellularly, since they are considered protein residues and are therefore not used for nitrogen assimilation by LAB [28].

Therefore the LAB proteolytic system, for example, *Lactobacillus helveticus*, *Lactobacillus delbrueckii* subsp. *bulgaricus*, *Lactococcus lactis* subsp. *diacetylactis*, *Lactococcus lactis* subsp. *cremoris*, *Streptococcus salivarius* subsp. *thermophilus*, and *Enterococcus faecalis*, proved capable of generating BAPs with several activities, including immunomodulatory, cytomodulatory, hypocholesterolemic, antioxidant, antimicrobial, mineral-binding, opioid-binding, and bone formation properties [29]. In this regard, peptides with different bioactivities are found in several dairy products, such as cheese, yogurts, and fermented milk [30]. Table 3−1 shows the description of BAPs present in fermented dairy products, as well as their biological activities and proteins of origin.

3.2.1 Health-promoting activities of peptides in fermented dairy products

BAP microbial activity is related to the inhibition of growth of a series of Gram-positive and Gram-negative pathogen bacteria, such as *Escherichia coli*, *Aeromonas hydrophila*, *Salmonella typhi*, *Bacillus cereus*, *Salmonella typhimurium*, *Salmonella enteritidis*, *Staphylococcus aureus* and others, as well as some molds and yeasts [58]. Therefore the use of these BAPs as biopreservatives or diet supplements with similar effects as antibiotics has already been proposed [22]. Peptides with antimicrobial activity (antimicrobial peptides, AMPs) contain from 12 to 100 amino acids and their mechanism of action can be triggered by infectious stimuli by, for example, bacteria, bacterial metabolism-derived molecules, or pro-inflammatory cytokines that induce innate immunity. Due to their small size, AMPs are diffused and secreted by cells, triggering the defense against pathogen microorganisms, through damages to cell membranes [59].

Table 3–1 Bioactive peptides in fermented foods of dairy origin.

Activity	Food of origin	Microorganism	Source
Antimicrobial	Fermented camel milk	*Lactobacillus acidophilus* or *Lactobacillus helveticus* + *Streptococcus thermophilus*	[31]
	Fermented milk	*L. acidophilus*	[32]
	Kefir	*Lactococcus* sp. + *Lactobacillus* sp.	[33][a]
	Fermented donkey milk	*Enterococcus* sp., *Lactobacillus* sp., and *Leuconostoc mesenteroides*	[34][a]
	Yogurt	*Lactobacillus* sp.	[35]
	Yogurt	*L. helveticus* and *L. acidophilus*	[36][a]
Antihypertensive	Fermented camel milk	*L. acidophilus* or *L. helveticus* + *S. thermophilus*	[31]
	Fermented donkey milk	*Enterococcus faecalis* and *L. helveticus*	[34][a]
	Commercial fermented milk	Data not shown	[37][a]
	Fermented whey	*Pediococcus acidilactici*	[38][a]
	Fermented whey with kombucha	Kombucha culture[b]	[39]
	Lassi[c]	*L. acidophilus*	[40][a]
	Yogurt and fermented milk	*L. helveticus* and *Lactobacillus delbrueckii* subsp. *bulgaricus*	[41][a]
	Fermented milk	*L. helveticus*	[42]
Immunomodulator	Yogurt (commercial)	Data not shown	[43][a]
	Fermented milk	*L. helveticus*	[42]
	Fermented milk	*S. thermophilus*, *Bifidobacterium longum*, and *L. helveticus*	[44]
	Fermented milk	*L. delbrueckii* subsp. *bulgaricus*	[45]
	Fermented milk	*L. delbrueckii* subsp. *bulgaricus*	[46]
Antioxidant	Kefir	*Lactococcus* sp. + *Lactobacillus* sp.	[33][a]
	Fermented donkey milk	*Enterococcus* sp., *Lactobacillus* sp., and *Leuconostoc mesenteroides*	[34][a]
	Fermented milk (commercial)	Data not shown	[37][a]
	Fermented milk	*L. helveticus*	[42]
	Yogurt	*L. helveticus* and *L. acidophilus*	[36][a]
	Fermented milk	*Lactobacillus casei* and *Lactobacillus rhamnosus*	[47]
	Lassi[c]	*L. acidophilus* and *Lactobacillus paracasei*	[48][a]
	Fermented milk	*Lactobacillus* spp.	[49]
	Fermented milk	*L. delbrueckii* subsp. *bulgaricus*	[46]
	Fermented goat milk	*L. casei*	[50]
	Fermented camel milk	*Lactobacillus rhamnosus*	[51]
	Fermented milk	*Lactobacillus* spp.	[52]
Opioid	Yogurt (commercial)	Data not shown	[43][a]
	Fermented milk	*L. helveticus*	[53]
	Matured cheese	Data not shown	[54]
	Yogurt	*S. thermophilus* and *L. delbrueckii* subsp. *bulgaricus*	[55][a]
	Yogurt	*S. thermophilus* and *L. delbrueckii* subsp. *bulgaricus*	[56]

[a]These authors show the peptide sequences obtained during fermentation.
[b]Hrnjez et al. [39] indicate that, according to Malbaša et al. [57], aside from bacteria of the *Acetobacter* genus, the kombucha culture in question contains the following microorganisms: *Saccharomycodes ludwigii*, *Saccharomyces cerevisiae*, *Saccharomyces bisporus*, *Torulopsis* sp., and *Zygosaccharomyces* sp.
[c]*Lassi* is fermented milk of Indian origin, with addition of water, spices, and sometimes, fruits.

Under this perspective, de Lima et al. [33] observed that lactobacillus and lactococcus species were capable of releasing several peptides during the fermentation of kefir using sheep milk. The authors showed that the product had antimicrobial activity against some of the major dietary pathogens (*E. coli, Pseudomonas aeruginosa, Klebsiella pneumoniae, B. cereus, Bacillus subtilis*, and *S. aureus*). The ability to reduce blood pressure is one of the most relevant bioactivities of BAPs [60]. Since the late 90s, there have been reports of BAPs from DPs with antihypertensive activity [6,19,61,62]. More recently, some authors such as Padghan et al. [40] and Elkhtab et al. [63] have also obtained BAPs with antihypertensive activity. Hernández-Ledesma et al. [37] showed that the activity of angiotensin-converting enzyme (ACE) inhibitory peptides is higher in fermented milk than in nonfermented milk. Additionally, Daliri et al. [38] obtained good results in ACE inhibition with cheese whey fermented with *Pediococcus acidilactici*. In this case, the sample inhibited $84.70 \pm 0.67\%$ of enzyme activity and obtained IC_{50} four times the value obtained by the positive control of captopril, with inhibition of $90.82 \pm 10.30\%$.

Peptides with immunomodulating activity are able to improve in vivo immune function by stimulating or suppressing the immune system [22]. Plaisancié et al. [43] studied the stimulation of intestinal mucin release from peptides present in yogurts. Among the BAPs tested, the authors observed that the ingestion of yogurt containing the 92–123 fragment of β-CN was associated with a higher expression of mucins Muc2 and Muc4, indicating that this BAP might be responsible for the maintenance or restructuring of homeostasis and for protecting the intestinal lumen against degrading agents.

BAPs with antioxidant activity can prevent the formation of oxygen-reactive species or eliminate free radicals and peroxides involved in the oxidation of membranes, cell proteins, DNA, and enzymes and are employed to promote human health. However, antioxidant BAPs can also be used in food processing to prevent oxidation of compounds such as fats [64]. Dairy products are considered an important source of antioxidant compounds. An extensive revision conducted by Fardet and Rock [65] indicated that the antioxidant potential is an essential nutritional property of functional fermented dairy foods.

Taha et al. [36] demonstrated that *L. helveticus* and *Lactobacillus acidophilus* species were able to release BAPs with antioxidant activity during fermentation and storage of yogurt manufactured from buffalo's milk. The quantification of 1,1-diphenyl-2-picrylhydrazyl scavenging activity showed results ranging from 70.14% to 80.62% in yogurt fermented with *L. acidophilus* and from 79.45% to 81.62% in yogurt fermented with *L. helveticus*, thus indicating the potential of these strains to produce antioxidant BAPs. Similarly, Solieri et al. [47] demonstrated that two endogenous strains of *Lactobacillus casei* and *Lactobacillus rhamnosus* isolated from Parmigiano Reggiano and Pecorino Siciliano cheeses were capable of releasing BAPs with antioxidant activity during both cheese fermentation and ripening.

Peptides with opioid properties bind to specific receptors in the human organism and have a similar activity as substances that are usually employed. When they derive from casein, opioid BAPs are called casomorphins [22]. Similar characteristics among opioid peptides are the presence of a N-terminal tyrosine residue (except when they derive from α-CN) and the presence of another aromatic residue, for example, phenylalanine or tyrosine, at positions three or four. Proline residue in the second position seems to be essential for the maintenance of the opioid activity of BAPs [66].

As for the existence of opioid BAPs in fermented dairy, De Noni and Cattaneo [54] reported the presence of β-casomorphin-7 in Gorgonzola, Brie, Gouda, Fontina, and Cheddar cheeses, at concentrations ranging from 0.01 to 0.15 mg/kg; Brie cheese has the highest concentration among the others. After simulated gastrointestinal digestion, this BAP was found in all tested samples (cheeses, yogurts, and fermented milk). Thus the authors [54] observed that β-casomorphin-7 release from β-CN in dairy products is only possible upon the action of the LAB proteolytic system during cheese maturation, as this peptide was not found in the other products prior to gastrointestinal digestion.

3.3 Bioactive peptides in fermented meat products

Aside from fermented dairy foods, fermented meat products are also sources of naturally produced BAPs. Therefore the consumption of these products might prevent diseases by modulating physiological systems, such as the cardiovascular, immune, endocrine, digestive, and nervous systems, which emphasizes the health-promoting potential of these meat products [67,68]. Peptides with biological activity can be generated in fermented meat products from their parent proteins by two pathways: (1) due to the action of endogenous muscle enzymes and (2) through the action of enzymes derived from proteolytic microorganisms present in the autochthonous microbiota or added like starter or nonstarter cultures [69]. These processes catalyzed by both muscle and microbial peptidases, for example, endopeptidases and exopeptidases, release physiologically active small-sized peptides that are encrypted in meat native protein molecules and are thus efficient strategies for designing functional or health-promoting meat foods [70]. This potential is already well demonstrated by the presence of BAPs in fermented dairy foods [22]. Furthermore, the effect of addition of sodium caseinate to fermented meat products results on naturally produced BAPs that have been previously described in dairy products [69].

There is broad knowledge about both meat protein-derived BAPs and bioactive compounds from fermented meat products [71–74]. However, there are few studies regarding the BAPs released from meat products during fermentation. Only in recent years, there have been scientific findings regarding the identification [75,76] and quantification [76] of BAPs derived from fermented meat products, and the roles of such foods in the prevention of diseases. Moreover, the release of BAPs by microbial fermentation from meat proteins has been reported in some studies [75–78]. So far, fermented meat protein-derived BAPs have been identified in products such as chorizo [79], dry-fermented camel sausages [76], and three typical European dry-fermented sausages from Spain, Italy, and Belgium [75].

3.3.1 The generation of bioactive peptides in fermented meat products

The processing of fermented meat products has essentially three phases: formulation/mixture, fermentation, and ripening/drying. The improvement in sensory characteristics (e.g., texture, taste, and flavor) of fermented meat products due to protein enzyme hydrolysis has been studied extensively [80]. However, little is known about BAPs generated during these fermentative processes and the effects of these peptides on human health. In fact, the first study using peptidomic approach for fermented meat products was reported by López et al. [81].

BAPs of different sizes can be generated, primarily, both during fermentation and during ripening/drying of fermented meat products, due to the combined action of endogenous muscles and microbial peptidases. Protein extracted from a naturally dried fermented sausage from Serbia (Petrovac Sausage) exhibited antioxidant and ACE-inhibitory properties, respectively, approximately 2 and threefold higher in the end product than in the initial sausage mixture [82]. This indicates that biologically active peptides derived from sarcoplasmic and myofibrillar proteins can be produced during sausage ripening, especially at the end of the ripening process due to the production of small-sized BAPs [75].

BAP production during processing of fermented meat products occurs due to the combined action of both muscle and microbial proteases. Sarcoplasmic and myofibrillar proteins are hydrolyzed by endogenous endopeptidases from the skeletal muscle, particularly calpains and cathepsins, followed by endogenous exopeptidases, such astripeptidyl peptidases, dipeptidyl peptidases, peptidyldipeptidases, tripeptidases, dipeptidases, aminopeptidases, and carboxypeptidases. The action of microbial peptidases, especially tripeptidyl peptidases and dipeptidyl peptidases, has also been reported as responsible for meat protein proteolysis [67,69,81,83]. Microorganisms such as LAB, *Staphylococcus* spp., *Micrococcus* spp., *Kocuria* spp., *Debaryomyces* spp., and *Penicillium* spp. have been widely reported in the fermentation and ripening of meat products. The proteolytic activity of these microorganisms is responsible for hydrolyzes of sarcoplasmic and myofibrillar proteins releasing BAPs. Furthermore, LAB affect meat protein degradation by causing a decrease in pH, which results in increased muscle protease activity [73,84]. Accordingly, the presence of naturally produced BAPs in fermented meat products depends on the equilibrium between their formation and the degradation performed by the proteolytic systems involved in meat fermentation.

3.3.2 Identification and functional activities of bioactive peptides in fermented meat products

So far, the main bioactivities identified in naturally produced peptides from fermented meat products are antimicrobial, antihypertensive, and antioxidant. One study has shown the characterization and partial purification of a bacteriocin produced by *Lactobacillus plantarum* LP31, a strain isolated from Argentinian dry-fermented sausage. The antimicrobial compound consisted of 14 amino acid residues and exhibited bactericidal effect against the foodborne pathogenic bacteria *S. aureus*, *Listeria monocytogenes*, *Pseudomonas* sp., and *B. cereus* [85]. In addition, autochthonous *Lactobacillus curvatus* strains isolated from Italian salamis were able to produce bacteriocins with activities against *L. monocytogenes* [86].

In addition to *Lactobacillus* genus, other starter cultures used in the fermentation of meat products have been shown to produce BAPs with antimicrobial activity. In fact, different nitrogen compounds with antimicrobial properties, including small peptides, have been detected throughout the ripening process of Iberian dry-fermented pork sausage "salchichón" elaborated with the autochthonous starter cultures *Pediococcus acidilactici* Ms200 and *Staphylococcus vitulus* RS34. Fermented sausage extracts obtained after 90 days of ripening exhibited inhibition activity, especially against *B. cereus* CECT131 [87].

Moreover, one of the first studies about naturally produced BAPs derived from fermented meat products reported that the action of LAB on meat protein molecules generates bioactivities beneficial to human health [88]. The authors observed ACE-inhibitory and antihypertensive activities generated by LAB from porcine skeletal muscle proteins. According to Arihara [70], these same activities were measured on fermented meat extracts, and ACE-inhibitory activity levels of European fermented sausage extracts were higher than those of extracts obtained from nonfermented pork products. In this regard, the effect of microbial exopeptidases derived from *Lactobacillus sakei* CRL1862 and *Lactobacillus curvatus* CRL705 on sarcoplasmic and myofibrillar porcine proteins released different peptides with ACE-inhibitory activity [77].

Regarding antioxidant peptide production during fermented meat processing, one study showed that both carnosine and L-carnitine dipeptides together with Phe-Gly-Gly are the major compounds responsible for antioxidant activity during the ripening of fermented "chorizo" sausages [79]. Moreover, Glu, His, Ala, Phe, Val, Pro, Gly, Leu, and Ile have been reported in antioxidant peptides released in the proteolysis process of ripening of typical European dry-fermented sausages from Spain, Italy, and Belgium [75].

In addition, both ACE-inhibitory and antioxidant properties were reported for protein hydrolysates of Petrovac Sausage (traditional dry-fermented sausage) throughout the ripening period of 90 days [82]. In a more recent study, sausages fermented by either *Lactobacillus sakei* strain no. 23 or *Lactobacillus curvatus* strain no. 28 exhibited both antioxidant and ACE-inhibitory activities higher than the sausages manufactured without these LAB on the twenty-first day of preparation [78]. According to the authors, these bioactivities were probably originated by peptides generated during the meat fermentation process by *Lactobacillus* spp. Moreover, low molecular weight peptides (below 3 kDa) that are naturally produced during the ripening of fermented meat products can both aid in the development of flavor and perform antihypertensive activity [69].

Both bioactivities generated by peptides released during ripening also were studied in fermented sausages formulated using camel meat. Four dry-fermented camel meat sausages were prepared; different mixtures of starter cultures were used in three formulations and one formulation (named control) was elaborated without adding any starter microbial culture. Both inoculated and noninoculated dry-fermented camel sausages had increased ACE-inhibitory and antioxidant activities during ripening, and the extracts containing peptides below 3 kDa had the highest antihypertensive and antioxidant activities. The BAP production with ACE-inhibitory activity was higher in the dry-fermented sausage inoculated with *Staphylococcus xylosus* and *Lactobacillus plantarum*, indicating that the production of small-sized ACE-inhibitory BAPs depends on the starter culture. Naturally produced peptides with the sequences EDDEVEH and AGDDAPR were observed in the extracts containing peptides below 3 kDa of the four dry-fermented camel sausage batches. In addition, Pro, Arg, and Lys were described as residues present in antihypertensive peptides, which were released in the proteolysis process that occurred during ripening of dry-fermented sausages [76].

More recently, Spanish, Italian, and Belgian dry-fermented sausages have been investigated as natural sources of antihypertensive and antioxidant peptides by Gallego et al. [75].

Typical European dry-fermented sausages from Spain, Italy, and Belgium were inoculated, respectively, with a starter culture containing *Lactobacillus sakei, Pediococcus pentosaceus, Staphylococcus xylosus*, and *Staphylococcus carnosus*; *Lactobacillus sakei* and a mixture of *Staphylococcus xylosus* strains; and *Lactobacillus sakei, Staphylococcus carnosus*, and *Staphylococcus xylosus*. The water-soluble peptide extracts from Spanish and Belgian fermented sausages had the highest ACE-inhibitory activities, while the Belgian sausage had the highest antioxidant activity. According to the authors [75], further studies are required to identify the peptide sequences responsible for each bioactivity observed.

ACE-inhibitory BAPs are among the most widely studied BAPs derived from fermented meat proteins. Therefore fermented meat products are a natural source of ACE-inhibitory activity peptides that provide benefits to human health against the high amounts of sodium found in these products and might thus mitigate the risk of hypertension [73]. Furthermore, these previous studies show that microbial fermentation, particularly performed by LAB, seems to be a potential natural technology applicable to BAP production from meat proteins, benefiting the human health against lifestyle diseases.

Few studies have been published about naturally produced BAPs from fermented meat products and even fewer studies have been conducted on the isolation of these substances from these processed meats. Although it is known that the action of endogenous and microbial enzymes aids in the release of biologically active peptides from proteins during fermented meat processing, further studies are required to characterize, identify, and quantify these compounds with focus on fermented meat products.

3.4 Bioactive peptides in fermented vegetables

Although there are many studies on animal protein-derived BAPs, studies related to vegetable protein-derived BAPs have increased due to their sustainable appeal, along with the growing demand for healthy and balanced diets. Investigating and understanding new functional foods, developed from vegetable sources, is a promising strategy in order to improve and implement their application as ingredients in the formulation of products and the direct uptake of these nutrients [89,90]. Vegetable foods comprise an alternative and low-cost source of proteins, which indicates their importance in BAP production. One example is cereals with a high content of essential amino acids, and which release BAPs via fermentation or enzyme hydrolysis. Bioinformatics has shown that wheat gliadin is the most susceptible vegetable protein for BAP release [90–93].

Considering the latent state in which BAPs occur, as part of a protein sequence, they need to be released, and fermentation is one possible method to obtain these compounds. Bacteria and fungi involved in this process secrete their proteolytic enzymes in the growth medium to release peptides from the original proteins (Fig. 3–2) [90,94,95]. Microbial fermentation is widely applied in the food industry to attribute distinct sensory characteristics. These fermentative processes also release BAPs, representing an economical method for their production [38,96,97].

Specific culture media for microbial fermentation to generate peptides have costly formulations, accounting for up to 80% of production costs [98]. Therefore the use of agroindustrial

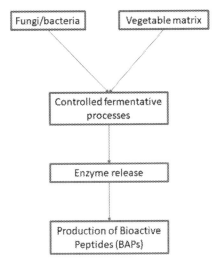

FIGURE 3–2 Schematics of the production of peptides via fermentation using bacteria or fungi.

by-products in the production of fermented foods with Generally Recognized As Safe (microorganisms is relevant considering the economic, environmental, and food safety aspects [25]. In this context, technologies for industrial expansion are required to ensure that the production cost of these compounds remains feasible. There is an interest in obtaining higher added value in products added to residues from agroindustrial processes, which is in agreement with concepts of circular economy and sustainability [99–101].

Fermentation of vegetable products, via LABs, might lead to improved sensory characteristics of the end product, eliminate antinutritional factors, and generate BAPs. However, beneficial properties depend mostly on the vegetable and microbial material [102–105]. In the case of fermented legumes, studies have reported that several vegetable-derived BAPs, for example, soybean seeds [106] and chickpea [107], which are extracted via fermentation, had a high antioxidant activity [108–110]. Singh and Vij [111] used the *L. plantarum* C2 strain in soymilk fermentation, sequenced the peptides from the 10 kDa fraction obtained in the fermentative process and evaluated it using a specific database of soybean BAPs. Among all soybean peptides identified, 18 had antioxidant activities, 16 were ACE inhibitors, and 17 remaining peptides exhibited both activities.

The fermentation process using *L. plantarum 299 v* in pea seeds conducted by Jakubczyk et al. [112] showed increased ACE-inhibitory peptides released during in vitro digestion under gastrointestinal conditions. The fermented seeds were the most effective source of potentially antihypertensive activity after 7 days of process at 22°C. The sequence of the pea protein-derived peptide was KEDDEEEEQGEEE. These results suggest that the use of fermented seeds for the production of products, such as pasta and cookies, might provide countless benefits to health. In 2017 another study conducted by Jakubczyk et al. [113] with beans fermented by the same strain of *L. plantarum*, similar bioactivities as the previous study were observed due to the release of BAPs after going through the simulated gastrointestinal tract. The condition that favored BAP release with the highest ACE-inhibitory activity (CI_{50} = 0.28 mg/mL) was fermentation of 3 days at 30°C.

Regarding cereals, fermented quinoa with autochthonous LAB [94] and *Rhizopus oligosporus* [114] has released BAPs with antioxidant activities. Rizzello et al. [115] investigated ACE-inhibitory BAP production during the fermentation of white wheat, whole-wheat, and rye flours. The highest ACE-inhibitory activity was observed in fermented flour under semiliquid conditions, and particularly with whole-wheat flour. Moreover, 14 ACE-inhibitory peptides were identified and their bioactivities were quantified (CI_{50} ranging from 0.19 to 0.54 mg/mL).

Coda et al. [116] identified 25 antioxidant peptides comprised 8–57 amino acid residues in the active fractions of fermented cereal flours. Additionally, several ACE-inhibitory sequences were identified in whole-wheat (DPVAPLQRSGPEI, PVAPQLSRGLL, ELEIVMASPP, QILLPRPGQAA, and VPFGVG) and rye malt-based products (LQP, IPP, LLP, and VPP) that were fermented using different LAB strains [115,117,118]. Zhang et al. [119] studied peanut flour fermentation using *B. subtilis* and suggested that it is possible to produce a natural antioxidant using peanut flour. Similarly, Wu et al. [120] performed nut meal protein fermentation via *B. subtilis*. An analysis of the composition of amino acids in the nut peptide fraction with molecular mass lower than 5 kDa showed that Asp, Glu, and Arg were the most abundant amino acids, proving their antioxidant activity.

Regular uptake of vegetable matrices, whether they are fruits, legumes, leaves, or stems, might prevent chronic diseases due to the presence of BAPs, as well as fibers and micronutrients, such as vitamins, minerals, and phytochemicals [102]. Studies to obtain BAPS via fermentation, using legumes, cereals or seeds, have constantly advanced with promising results. There are many studies related to the natural release of bioactive compounds, such as polyphenols, flavonoids, anthocyanins, phenolic compounds, among others [121–125]. However, studies on how to obtain BAPs from proteins of fermented fruits are scarce. Three ACE-inhibitory peptides, IPP (0.42–0.49 mg/kg), LPP (0.30–0.33 mg/kg), and VPP (0.32–0.35 mg/kg), have been identified in cucumber fermentation. A fourth ACE-inhibitory peptide (0.93–1.5 mg/kg) occurred with higher concentration in fermented cucumbers than in acidified cucumbers. With this study, Fideler et al. [126] showed that the fermentation of vegetable foods via LAB can increase BAP release.

The food industry is constantly searching for alternative sources of vegetable proteins, considering the growing interest in a healthy diet by consumers. This indicates a promising scenario for the use of vegetable matrices in BAP production employing controlled fermentation processes. Vegetable fermentation to obtain these peptides at an industrial scale is not yet disseminated, there are several issues to be studied, such as the required fermentation time to obtain the desired activities in a satisfactory active concentration, and the behavior of BAPs when they undergo in vivo digestion.

3.5 Biotechnological progress for industrial production of bioactive peptides

BAPs are largely found in animals and their products, in vegetables, and microorganisms in nature, where they are synthesized by the organism, or can be found encrypted in proteins

[4]. Furthermore, protein-rich by-products derived from food industry processes, once regarded as useless effluents, are nowadays considered as substrates of commercial value, as they can represent high-quality sources for BAPs production [127]. One extensively investigated food protein source fulfilling these criteria are whey proteins derived from cheese production, due to their easy availability, low production costs, and high nutritious value [128,129]. As native whey proteins are well known for having different bioactivities, they are considered a valuable source, rich in BAPs [130].

Within the principles of the rapidly developing bioeconomy, the search for new protein sources usable for BAP production are still a topic of great interest in an array of studies [127]. With approximately 1.6 billion tons of food waste generated worldwide, which means an economic loss of approximately US$1 trillion, maximizing industrial food production is required in order to prevent loss of raw materials with biofunctional and nutritional value [131]. Several protein sources from the food industry are currently under investigation, for example, meat processing wastes [132], seafood processing wastes [133,134], and biowastes of plant origin [135]. For large-scale recovery of proteins from industrial waste, processes such as membrane technologies, precipitation, and flocculation must be improved and simplified [136]. It is worth highlighting that membrane processes have been considered "green technologies," as they have no negative impact on the environment; they do not require the use of solvents or other process compounds [137].

The evaluation of novel protein sources has been performed so far according to nutritional criteria, such as essential amino acid content or the Protein Digestibility Corrected Amino Acid Score [129]. Rutherford-Markwick [138] suggested including the potential to generate BAPs in classifying dietary protein quality. In general, the branched-chain amino acid (BCAA) content of proteins, represented by amino acids valine, leucine, and isoleucine, seems to be a crucial factor for protein selection, as peptides that exhibit a wide range of physiologic activities had high BCAA content or harbor a BCAA in a specific position [139,140]. Tryptophan was found to be a driving amino acid residue in peptides with ACE inhibition, plays antioxidant, antidiabetic, satiating, and cognitive roles, and is present in high percentages in egg, milk, soy, beans, seafood, and poultry proteins [20].

Regarding BAPs production generally, they can be obtained during microbial fermentation processes or via in vitro enzyme hydrolysis by using specific proteolytic enzymes [60]. Despite extensive studies about the effectiveness of BAPs over the last four decades, in which countless new active sequences have been discovered, availability of peptide-based food additives in the global market is still scarce (Table 3–2), as the development of production processes has posed several challenges [17,141].

Natural BAP production via microbial fermentation processes can be observed particularly in traditional fermented foods, such as in cheese, fermented milk products, yogurt, kefir, or meat products, such as ham, where proteolytic LAB break down proteins [142]. Aside from animal- and vegetable-derived proteases, microorganisms represent an excellent source of proteases, due to their advanced technical stage for production, as well as their biochemical diversity and potential for genetic manipulation [143]. Moreover, fermented products, which are a part of the tradition in some ethnic groups, are a valuable source for previously

Table 3–2 Food ingredients available in the global market based on bioactive peptides derived from whey proteins.

Product name	Manufacturer	Product type	Health claim
BioZate product line	Davisco, United States	Fragments from hydrolyzed whey protein β-lactoglobulin	Reduction of blood pressure
BioPureGMP	Davisco, United States	Whey-derived GMP f(106–169)	Prevention of dental caries, blood clotting, antibacterial and antivirus activity
Vivinal ALPHA	Borculo Domo Ingredients (BDI), the Netherlands	Ingredient/peptide from whey protein hydrolysate	Relaxation and sleep
Praventin	DMV International, the Netherlands	Dietary supplement/lactoferrin-enriched whey protein hydrolysate	Acne reduction
Dermylex	Advitec, Inc., Canada	Dietary supplement/whey protein extract XP-828L	Reduction of psoriasis symptoms
Hilmar 8390	Hilmar Ingredients, United States	Ingredient/whey protein hydrolysate	ACE inhibition, DPP-IV enzyme inhibition
NOP-47	Glanbia Nutritionals, United States	Ingredient/nutraceutical/peptide from whey protein hydrolysate	Antiinflammatory properties
AmealPeptide	Maypro Industries LLC, United States	Ingredient/milk protein-derived peptides VPP and IPP obtained via fermentation and enzyme hydrolysis	ACE inhibition
KEFPEP	PHERMPEP, Taiwan	Ingredient/enriched peptide mix obtained via fermentation of milk proteins	ACE inhibition; antiobesity; antiosteoporotic
Egg white peptide EP-1	Kewpie, Japan	Ingredient/egg white protein hydrolysate	Anticholesterolemic and antioxidant
NWT-03	Newtricious, the Netherlands	Ingredient and pharmaceutical applications/egg white protein hydrolysate	ACE inhibition, DPP-IV enzyme inhibition
ProwLiz	Cargill, United States	Ingredient/wheat protein hydrolysate	Satiety benefits
PeptACE	Natural Factors Inc., United States	Dietary supplement/mixture of seven highly purified small peptides from hydrolyzed bonito fish protein	ACE inhibition
Garum Armoricum	Research Allergy Group, France	Dietary supplement/mix of oligopeptides from blue ling fish protein hydrolysate	Anxiolytic
Seacur	Proper Nutrition Inc., United States	Dietary supplement/hydrolysate preparation containing pacific whiting fish protein fermented by *Hansenula* yeast culture	Intestinal health and antiinflammatory

ACE, Angiotensin-converting enzyme; *DPP-IV*, dipeptidylpeptidase IV; *GMP*, glycomacropeptide; *IPP*, isoleucyl-prolyl-proline; *VPP*, valyl-prolyl-proline.
Source: Data updated from A. Dullius, M.I. Goettert, C.F.V. de Souza, Whey protein hydrolysates as a source of bioactive peptides for functional foods—biotechnological facilitation of industrial scale-up, Journal of Functional Foods, 42 (2018) 58–74 [21].

unknown microorganisms, and therefore, of new proteases, as well as new BAPs sequences [144,145].

The bottleneck of microbial fermentation is to reach a concentration of BAPs that is sufficient to ensure the desired physiologic efficiency of the product [141]. Based on the cheese ripening process, it is well known that peptides and bioactivity pattern depend strongly on pH, type of enzymes, salt-to-moisture ratio, humidity, storage time, and temperature, and therefore, varies according to different cheese types manufactured using different methods [146]. Thus to maximize BAPs production from conventional batch fermentations, Mechmeche et al. [147] used response surface methodology and central composite rotatable design to obtain high cell growth and protease production. Ruan et al. [148] used an ultrasound-assisted fermentation to enhance proliferation of *B. subtilis* cells, which in turn increased yield, purity, and ACE-inhibitory activity of peptides. As a low-cost BAP production alternative, Pérez et al. [149] used encapsulated *B. subtilis* cells for the production of ACE-inhibitory and antioxidant peptides from whey protein. It was possible to maintain the continuous enzyme production of cells for 5 weeks by keeping a stable hydrolysis process.

The development of peptide production as a food additive outside food formulations is still limited by the lack of a systematic development methodology, beginning with protein and enzyme selection, which goes hand in hand with peptide detection, and ending with the selection of an appropriate enrichment method [16,17]. The main challenge thereby is peptide enrichment after enzyme hydrolysis, as this step usually accounts for up to 90% of the downstream process costs, and therefore is essential in determining the economic viability of the production of food additives at an industrial scale [150–153]. Furthermore, scaling-up is rendered even more difficult by the fact that activity profiles and peptide fingerprints can deviate quickly if different methods are used during the scale-up [154].

To overcome these hindrances, Dullius et al. [21] suggested integrating hybrid methods for process design, thus applying bioinformatics to select the protein source and proteases used for enzyme hydrolysis, as well as for the selection of the appropriate enrichment process. In silico methods are complemented with knowledge-based information, as well as mechanistic models, as the latter are commonly used in biopharmaceutical industry (Fig. 3–3). An increasing number of computational tools have enabled a more focused approach to solve systematically the investigation of a protein source, the selection of appropriate proteases, and the subsequent hydrolysis and downstream process [140].

The BIOPEP database provides a list of databases or prediction tools on BAPs that is constantly updated (http://www.uwm.edu.pl/biochemia/index.php/pl/biopep/32-bioactive-peptide-databases), and that currently has 68 entries (accessed April 08.04.20) [155]. Similar to the BIOPEP database, but specifically for food-derived BAPs, the available online database BioPepDB enables in silico protein hydrolysis and bioactivity prediction using motif search or alignment against over 4000 peptide sequences, which are described in detail [156]. Furthermore, BIOPEP provides investigation of peptide bitterness and toxicity and provides an enzyme selection tool to release the desired peptide sequences [157,158].

Information in the literature about systematic in silico enzyme selection is quite rare. There are different approaches for protein selections, which means that this development

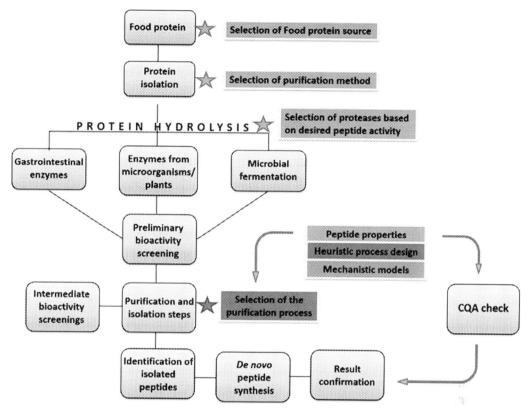

FIGURE 3–3 The development of process design involves applying hybrid methods, such as algorithms (marked with *stars*), knowledge-based approaches, and mechanistic models. *CQA*, Critical quality attribute, for instance, measurable product properties within product quality requirements. *From: A. Dullius, M.I. Goettert, C.F.V. de Souza, Whey protein hydrolysates as a source of bioactive peptides for functional foods—biotechnological facilitation of industrial scale-up, Journal of Functional Foods, 42 (2018) 58–74 [21].*

step is still in the beginning. Similarity searches to find BAPs sequences, with subsequent evaluation of the frequency of occurrence of the peptides, as well as a ranking with the probability that the fragment has bioactivity, have been successfully applied for protein and enzyme selection [159–161]. Nongonierma et al. [162] developed an in silico method as another evaluation approach to investigate dietary proteins for production of DPP-IV inhibitory peptides, based on a peptide inhibitory index and a corrected protein coverage value to avoid overlapping of peptides within the protein sequence.

The application of fold assignment techniques was proposed as an alternative to sequence similarity searches, as a technical alternative to widen the window for peptide detection [163]. The result of this research was that BAPs with different activities could overlap in some regions (hotspots) in a superfamily template when the protein sequence harbors more than 5 BAP sequences. These results could be linked to the development of enzyme digestion mechanisms, which in practice are driven by enzyme specificity, meaning that under determined conditions an enzyme either has access to cleavage sites within the protein structure or not [164,165]. Leeb et al. [166] mentioned that to predict hydrolysate composition detailed knowledge about the reaction dynamics of protein breakdown and peptide release is required. They investigated the cleavage behavior of trypsin during hydrolysis of β-lactoglobulin. The authors developed a mathematical model, considering the observation according to which released peptides can be grouped into those released from the cleavage of the intact protein or the released by intermediate hydrolysis. A consequence of such two-step reaction is that the reaction rate decreases after a certain period, due to competition of the peptide intermediates with protein residues for free enzyme-binding sites. Adding steric accessibility to the model in the form of enzymatic accessibility constant, which depends on substrate chain length, led to a rough prediction of peptide release, for which feeding experimental data about substrate degradation is required.

However, regarding enzyme specificity, the complexity of enzyme protein degradation becomes clear as specificity also depends on the type reactor models used. Suwal et al. [167] observed different peptide patterns, by comparing those obtained in ex situ digestion with those obtained in situ (simultaneous) and by separation using electrodialysis with ultrafiltration membrane. For monolith based immobilized trypsin reactors, Mao et al. [168] showed, using immobilized trypsin, which β-LG protein cleavage resulted in higher specificity than free trypsin. They also showed that adding ethanol up to 20% as a solvent also increased the release of those peptides, which are located in the core region of β-lactoglobulin. Antink et al. [169] revealed that the immobilization material affects enzyme orientation, which results in different protein cleavage specificities.

The examples mentioned above have clearly shown that the production of a protein hydrolysate with certain bioactivity is a question of fine-tuned parameters, and finding them goes hand in hand with in silico models and evaluation methods, which in a near future might allow us to have rough estimations, from which developing economic viable BAPs production processes could be possible.

3.6 Conclusions and perspectives

Health-promoting functional foods, pharmaceutical preparations, and supplements containing BAPs shall be increasingly demanded, given the search for a healthy diet by consumers. BAP production from food matrices via controlled fermentation processes has raised the interest of the scientific community. Regarding BAP generation, there is a higher number of studies employing DPs, there has been a recent development of scientific studies with meat fermented, and a promising beginning of studies on BAP production using vegetable matrices. Fermented products such as cheese and sausages have shown several properties with nutraceutical interest, while fermented vegetables can provide sensory improvements, as well as BAP production. BAPs have been produced in food matrices studied as sources of proteins, and have showing particularly antihypertensive activity, as well as antioxidant, antimicrobial, and immunomodulating properties.

A relevant piece of information is that amino acid sequences that have the same bioactivity can be obtained from native animal and vegetable proteins. A general analysis of the study peptides showed excellent potential for providing health, as they had the properties to prevent several chronic diseases. However, clinical studies are required to observe gastrointestinal stability, bioavailability, and safety in the use of these peptides as supplements, nutraceuticals, or even in functional foods.

A factor that drives the industrial production of BAPs is the potential for using by-products/waste from the food processing industry as raw material. Therefore a highlight is the study of the application of membrane separation techniques to extract protein of this origin. Another important point is the interest in biotechnological development for the production of proteases. Systematic methodologies are required for the development of industrial scale processes, in which innovations are required for method hybridization when designing the process, with the help of tools such as bioinformatics.

References

[1] H. Korhonen, A. Pihlanto, Food-derived bioactive peptides-opportunities for designing future foods, Current Pharmaceutical Design 9 (2003) 1297−1308.

[2] S. Chakrabarti, S. Guha, K. Majumder, Food-derived bioactive peptides in human health: Challenges and opportunities, Nutrients 10 (2018) 1738.

[3] L. Mora, M.-C. Aristoy, F. Toldrá, Bioactive peptides, Reference Module in Food Science (eBook), 2018.

[4] A.R. Madureira, T. Tavares, A.M.P. Gomes, M.E. Pintado, F.X. Malcata, Invited review: physiological properties of bioactive peptides obtained from whey proteins, Journal of Dairy Science 93 (2010) 437−455.

[5] L.Y. Chew, G.T. Toh, A. Ismail, Application of proteases for the production of bioactive peptides, In: M. Kuddus (Eds.), Enzymes in Food Biotechnology, Academic Press, 2019, pp. 247−261.

[6] A. Pihlanto-Leppälä, Bioactive peptides derived from bovine whey proteins: opioid and ace-inhibitory peptides, Trends in Food Science & Technology 11 (2000) 347−356.

[7] L. Das, E. Bhaumik, U. Raychaudhuri, R. Chakraborty, Role of nutraceuticals in human health, Journal of Food Science and Technology 49 (2012) 173−183.

[8] A. Dullius, C.M. Rocha, S. Laufer, C.F.V. de Souza, M.I. Goettert, Are peptides a solution for the treatment of hyperactivated JAK3 pathways? Inflammopharmacology (2019) 1−20.

[9] D. Dupont, Peptidomic as a tool for assessing protein digestion, Current Opinion in Food Science 16 (2017) 53−58.

[10] J. Sanchón, S. Fernández-Tomé, B. Miralles, B. Hernández-Ledesma, D. Tomé, C. Gaudichon, et al., Protein degradation and peptide release from milk proteins in human jejunum: comparison with in vitro gastrointestinal simulation, Food Chemistry 239 (2018) 486−494.

[11] S.F. Gauthier, Y. Pouliot, D. Saint-Sauveur, Immunomodulatory peptides obtained by the enzymatic hydrolysis of whey proteins, International Dairy Journal 16 (2006) 1315−1323.

[12] M. Chalamaiah, W. Yu, J. Wu, Immunomodulatory and anticancer protein hydrolysates (peptides) from food proteins: a review, Food Chemistry 245 (2018) 205−222.

[13] C. Lammi, G. Aiello, G. Boschin, A. Arnoldi, Multifunctional peptides for the prevention of cardiovascular disease: a new concept in the area of bioactive food-derived peptides, Journal of Functional Foods 55 (2019) 135−145.

[14] R. Vilcacundo, C. Martínez-Villaluenga, B. Miralles, B. Hernández-Ledesma, Release of multifunctional peptides from kiwicha (*Amaranthus caudatus*) protein under in vitro gastrointestinal digestion, Journal of the Science of Food and Agriculture 99 (2019) 1225–1232.

[15] M. Mohammadian, M. Salami, Z. Emam-Djomeh, F. Alavi, Nutraceutical properties of dairy bioactive peptides, Dairy in Human Health and Disease Across the Lifespan, Academic Press, 2017, pp. 325–342.

[16] D. Agyei, C.M. Ongkudon, C.Y. Wei, A.S. Chan, M.K. Danquah, Bioprocess challenges to the isolation and purification of bioactive peptides, Food and Bioproducts Processing 98 (2016) 244–256.

[17] K. Anekthanakul, A. Hongsthong, J. Senachak, M. Ruengjitchatchawalya, SpirPep: an in silico digestion-based platform to assist bioactive peptides discovery from a genome-wide database, BMC Bioinformatics 19 (2018) 149.

[18] G. Bylund, Dairy Processing Handbook, Tetra Pak Processing Systems, 1995.

[19] D.A. Clare, H.E. Swaisgood, Bioactive milk peptides: a prospectus, Journal of Dairy Science 83 (2000) 1187–1195.

[20] A.B. Nongonierma, R.J. FitzGerald, The scientific evidence for the role of milk protein-derived bioactive peptides in humans: a review, Journal of Functional Foods 17 (2015) 640–656.

[21] A. Dullius, M.I. Goettert, C.F.V. de Souza, Whey protein hydrolysates as a source of bioactive peptides for functional foods–biotechnological facilitation of industrial scale-up, Journal of Functional Foods 42 (2018) 58–74.

[22] A.B. Nongonierma, M.B. O'keeffe, R.J. FitzGerald, Milk protein hydrolysates and bioactive peptides, Advanced Dairy Chemistry, Springer, 2016, pp. 417–482.

[23] M. Tu, S. Cheng, W. Lu, M. Du, Advancement and prospects of bioinformatics analysis for studying bioactive peptides from food-derived protein: sequence, structure, and functions, Trends in Analytical Chemistry 105 (2018) 7–17.

[24] J. Choi, L. Sabikhi, A. Hassan, S. Anand, Bioactive peptides in dairy products, International Journal of Dairy Technology 65 (2012) 1–12.

[25] M.G. Venegas-Ortega, A.C. Flores-Gallegos, J.L. Martínez-Hernández, C.N. Aguilar, G.V. Nevárez-Moorillón, Production of bioactive peptides from lactic acid bacteria: a sustainable approach for healthier foods, Comprehensive Reviews in Food Science and Food Safety 18 (2019) 1039–1051.

[26] E. Pessione, Lactic acid bacteria contribution to gut microbiota complexity: lights and shadows, Frontiers in Cellular and Infection Microbiology 2 (2012) 86.

[27] E. Pessione, S. Cirrincione, Bioactive molecules released in food by lactic acid bacteria: encrypted peptides and biogenic amines, Frontiers in Microbiology 7 (2016) 876.

[28] K. Savijoki, H. Ingmer, P. Varmanen, Proteolytic systems of lactic acid bacteria, Applied Microbiology and Biotechnology 71 (2006) 394–406.

[29] A. Pihlanto, H. Korhonen, Bioactive peptides from fermented foods and health promotion, advances in fermented foods and beverages: improving quality, Technologies and Health Benefits, Elsevier Ltd., 2014.

[30] H. Korhonen, Milk-derived bioactive peptides: from science to applications, Journal of Functional Foods 1 (2009) 177–187.

[31] O.A. Alhaj, A.A. Metwalli, E.A. Ismail, H.S. Ali, A.S. Al-Khalifa, A.D. Kanekanian, Angiotensin converting enzyme-inhibitory activity and antimicrobial effect of fermented camel milk (*Camelus dromedarius*), International Journal of Dairy Technology 71 (2018) 27–35.

[32] A. Shafi, H. Naeem Raja, U. Farooq, K. Akram, Z. Hayat, A. Naz, et al., Antimicrobial and antidiabetic potential of synbiotic fermented milk: a functional dairy product, International Journal of Dairy Technology 72 (2019) 15–22.

[33] M.D.S.F. de Lima, R.A. da Silva, M.F. da Silva, P.A.B. da Silva, R.M.P.B. Costa, J.A.C. Teixeira, et al., Brazilian kefir-fermented sheep's milk, a source of antimicrobial and antioxidant peptides, Probiotics and Antimicrobial Proteins 10 (2018) 446–455.

[34] M. Aspri, G. Leni, G. Galaverna, P. Papademas, Bioactive properties of fermented donkey milk, before and after in vitro simulated gastrointestinal digestion, Food Chemistry 268 (2018) 476–484.

[35] I. Tzvetkova, M. Dalgalarrondo, S. Danova, I. Iliev, I. Ivanova, J.M. Chobert, et al., Hydrolysis of major dairy proteins by lactic acid bacteria from Bulgarian yogurts, Journal of Food Biochemistry 31 (2007) 680–702.

[36] S. Taha, M. El Abd, C. De Gobba, M. Abdel-Hamid, E. Khalil, D. Hassan, Antioxidant and antibacterial activities of bioactive peptides in buffalo's yoghurt fermented with different starter cultures, Food Science and Biotechnology 26 (2017) 1325–1332.

[37] B. Hernández-Ledesma, B. Miralles, L. Amigo, M. Ramos, I. Recio, Identification of antioxidant and ACE-inhibitory peptides in fermented milk, Journal of the Science of Food and Agriculture 85 (2005) 1041–1048.

[38] E.B.M. Daliri, B.H. Lee, B.J. Park, S.H. Kim, D.H. Oh, Antihypertensive peptides from whey proteins fermented by lactic acid bacteria, Food Science and Biotechnology 27 (2018) 1781–1789.

[39] D. Hrnjez, Ž. Vaštag, S. Milanović, V. Vukić, M. Iličić, L. Popović, et al., The biological activity of fermented dairy products obtained by kombucha and conventional starter cultures during storage, Journal of Functional Foods 10 (2014) 336–345.

[40] P.V. Padghan, B. Mann, R. Sharma, R. Bajaj, P. Saini, Production of angiotensin-i-converting-enzyme-inhibitory peptides in fermented milks (*lassi*) fermented by *Lactobacillus acidophillus* with consideration of incubation period and simmering treatment, International Journal of Peptide Research and Therapeutics 23 (2017) 69–79.

[41] C. Raveschot, B. Cudennec, B. Deracinois, M. Frémont, M. Vaeremans, J. Dugersuren, et al., Proteolytic activity of *Lactobacillus* strains isolated from Mongolian traditional dairy products: a multiparametric analysis, Food Chemistry 304 (2020) 125415.

[42] K.R. Elfahri, O.N. Donkor, T. Vasiljevic, Potential of novel *Lactobacillus helveticus* strains and their cell wall bound proteases to release physiologically active peptides from milk proteins, International Dairy Journal 38 (2014) 37–46.

[43] P. Plaisancie, J. Claustre, M. Estienne, G. Henry, R. Boutrou, A. Paquet, et al., A novel bioactive peptide from yoghurts modulates expression of the gel-forming MUC2 mucin as well as population of goblet cells and Paneth cells along the small intestine, The Journal of Nutritional Biochemistry 24 (2013) 213–221.

[44] L.E. Wagar, C.P. Champagne, N.D. Buckley, Y. Raymond, J.M. Green-Johnson, Immunomodulatory properties of fermented soy and dairy milks prepared with lactic acid bacteria, Journal of Food Science 74 (2009) M423–M430.

[45] D. Regazzo, L. Da Dalt, A. Lombardi, C. Andrighetto, A. Negro, G. Gabai, Fermented milks from *Enterococcus faecalis* TH563 and *Lactobacillus delbrueckii* subsp. *bulgaricus* LA2 manifest different degrees of ACE-inhibitory and immunomodulatory activities, Dairy Science & Technology 90 (2010) 469–476.

[46] B. Qian, M. Xing, L. Cui, Y. Deng, Y. Xu, M. Huang, et al., Antioxidant, antihypertensive, and immunomodulatory activities of peptide fractions from fermented skim milk with *Lactobacillus delbrueckii* ssp. *bulgaricus* LB340, Journal of Dairy Research 78 (2011) 72–79.

[47] L. Solieri, G.S. Rutella, D. Tagliazucchi, Impact of non-starter lactobacilli on release of peptides with angiotensin-converting enzyme inhibitory and antioxidant activities during bovine milk fermentation, Food Microbiology 51 (2015) 108–116.

[48] P.V. Padghan, B. Mann, S. Hati, Purification and characterization of antioxidative peptides derived from fermented milk (*lassi*) by lactic cultures, International Journal of Peptide Research and Therapeutics 24 (2018) 235–249.

[49] A. Osuntoki, I. Korie, Antioxidant activity of whey from milk fermented with *Lactobacillus* species isolated from Nigerian fermented foods, Food Technology and Biotechnology 48 (2010) 505–511.

[50] G. Shu, X. Shi, L. Chen, J. Kou, J. Meng, H. Chen, Antioxidant peptides from goat milk fermented by *Lactobacillus casei* L61: preparation, optimization, and stability evaluation in simulated gastrointestinal fluid, Nutrients 10 (2018) 797.

[51] M. Moslehishad, M.R. Ehsani, M. Salami, S. Mirdamadi, H. Ezzatpanah, A.N. Naslaji, et al., The comparative assessment of ACE-inhibitory and antioxidant activities of peptide fractions obtained from fermented camel and bovine milk by Lactobacillus rhamnosus PTCC 1637, International Dairy Journal 29 (2013) 82–87.

[52] V. Ramesh, R. Kumar, R.R.B. Singh, J.K. Kaushik, B. Mann, Comparative evaluation of selected strains of lactobacilli for the development of antioxidant activity in milk, Dairy Science & Technology 92 (2012) 179–188.

[53] C. Matar, J. Goulet, β-casomorphin 4 from milk fermented by a mutant of *Lactobacillus helveticus*, International Dairy Journal 6 (1996) 383–397.

[54] I. De Noni, S. Attaneo, Occurrence of β-casomorphins 5 and 7 in commercial dairy products and in their digests following in vitro simulated gastro-intestinal digestion, Food Chemistry 119 (2010) 560–566.

[55] P.B. Kunda, F. Benavente, S. Catalá-Clariana, E. Giménez, J. Barbosa, V. Sanz-Nebot, Identification of bioactive peptides in a functional yogurt by micro liquid chromatography time-of-flight mass spectrometry assisted by retention time prediction, Journal of Chromatography A 1229 (2012) 121–128.

[56] A. Schieber, H. Brückner, Characterization of oligo-and polypeptides isolated from yoghurt, European Food Research and Technology 210 (2000) 310–313.

[57] R.V. Malbaša, S.D. Milanović, E.S. Lončar, M.S. Djurić, M.Ð. Carić, M.D. Iličić, L. Kolarov, Milk-based beverages obtained by Kombucha application, Food Chemistry 112 (2009) 178–184.

[58] D.P. Mohanty, S. Mohapatra, S. Misra, P.S. Sahu, Milk derived bioactive peptides and their impact on human health—a review, Saudi, Journal of Biological Sciences 23 (2016) 577–583.

[59] D. Mohanty, R. Jena, P.K. Choudhury, R. Pattnaik, S. Mohapatra, M.R. Saini, Milk derived antimicrobial bioactive peptides: a review, International Journal of Food Properties 19 (2016) 837–846.

[60] A. Brandelli, D.J. Daroit, A.P.F. Corrêa, Whey as a source of peptides with remarkable biological activities, Food Research International 73 (2015) 149–161.

[61] M.M. Mullally, H. Meisel, R.J. FitzGerald, Identification of a novel angiotensin-I-converting enzyme inhibitory peptide corresponding to a tryptic fragment of bovine β-lactoglobulin, FEBS Letters 402 (1997) 99–101.

[62] A. Pihlanto-Leppälä, P. Koskinen, K. Piilola, T. Tupasela, H. Korhonen, Angiotensin I-converting enzyme inhibitory properties of whey protein digests: concentration and characterization of active peptides, Journal of Dairy Research 67 (2000) 53–64.

[63] E. Elkhtab, M. El-Alfy, M. Shenana, A. Mohamed, A.E. Yousef, New potentially antihypertensive peptides liberated in milk during fermentation with selected lactic acid bacteria and kombucha cultures, Journal of Dairy Science 100 (2017) 9508–9520.

[64] S.D. Nielsen, R.L. Beverly, Y. Qu, D.C. Dallas, Milk bioactive peptide database: a comprehensive database of milk protein-derived bioactive peptides and novel visualization, Food Chemistry 232 (2017) 673–682.

[65] A. Fardet, E. Rock, In vitro and in vivo antioxidant potential of milks, yoghurts, fermented milks and cheeses: a narrative review of evidence, Nutrition Research Reviews 31 (2018) 52–70.

[66] H. Meisel, R.J. Fitzgerald, Opioid peptides encrypted in intact milk protein sequences, British Journal of Nutrition 84 (2000) 27–31.

[67] L. Mora, M. Gallego, M. Reig, F. Toldrá, Challenges in the quantitation of naturally generated bioactive peptides in processed meats, Trends in Food Science & Technology 69 (2017) 306–314.

[68] J. Stadnik, P. Kęska, Meat and fermented meat products as a source of bioactive peptides, Acta Scientiarum Polonorum Technologia Alimentaria 14 (2015) 181–190.

[69] L. Mora, E. Escudero, M.C. Aristoy, F. Toldrá, A peptidomic approach to study the contribution of added casein proteins to the peptide profile in Spanish dry-fermented sausages, International Journal of Food Microbiology 212 (2015) 41–48.

[70] K. Arihara, Strategies for designing novel functional meat products, Meat Science 74 (2006) 219–229.

[71] F. Melini, V. Melini, F. Luziatelli, A.G. Ficca, M. Ruzzi, Health-promoting components in fermented foods: an up-to-date systematic review, Nutrients 11 (2019) 1189.

[72] L. Xing, R. Liu, S. Cao, W. Zhang, Z. Guanghong, Meat protein based bioactive peptides and their potential functional activity: a review, International Journal of Food Science & Technology 54 (2019) 1956–1966.

[73] R. Bou, S. Cofrades, F. Jiménez-Colmenero, Fermented meat sausages, Fermented Foods in Health and Disease Prevention, Academic Press, 2017, pp. 203–235.

[74] M. Hayes, M. García-Vaquero, Bioactive compounds from fermented food products, Novel Food Fermentation Technologies, Springer, 2016, pp. 293–310.

[75] M. Gallego, L. Mora, E. Escudero, F. Toldrá, Bioactive peptides and free amino acids profiles in different types of European dry-fermented sausages, International Journal of Food Microbiology 276 (2018) 71–78.

[76] L. Mejri, R. Vásquez-Villanueva, M. Hassouna, M.L. Marina, M.C. García, Identification of peptides with antioxidant and antihypertensive capacities by RP-HPLC-Q-TOF-MS in dry fermented camel sausages inoculated with different starter cultures and ripening times, Food Research International 100 (2017) 708–716.

[77] P. Castellano, M.C. Aristoy, M.Á. Sentandreu, G. Vignolo, F. Toldrá, Peptides with angiotensin I converting enzyme (ACE) inhibitory activity generated from porcine skeletal muscle proteins by the action of meat-borne *Lactobacillus*, Journal of Proteomics 89 (2013) 183–190.

[78] S. Takeda, H. Matsufuji, K. Nakade, S.I. Takenoyama, A. Ahmed, R. Sakata, et al., Investigation of lactic acid bacterial strains for meat fermentation and the product's antioxidant and angiotensin-I-converting-enzyme inhibitory activities, Animal Science Journal 88 (2017) 507–516.

[79] J.M. Broncano, J. Otte, M.J. Petrón, V. Parra, M.L. Timón, Isolation and identification of low molecular weight antioxidant compounds from fermented "chorizo" sausages, Meat Science 90 (2012) 494–501.

[80] P. Kumar, M.K. Chatli, A.K. Verma, N. Mehta, O.P. Malav, D. Kumar, et al., Quality, functionality, and shelf life of fermented meat and meat products: a review, Critical Reviews in Food Science and Nutrition 57 (2017) 2844–2856.

[81] C.M. López, E. Bru, G.M. Vignolo, S.G. Fadda, Identification of small peptides arising from hydrolysis of meat proteins in dry fermented sausages, Meat Science 104 (2015) 20–29.

[82] Ž. Vaštag, L. Popović, S. Popović, L. Petrović, D. Peričin, Antioxidant and angiotensin-I converting enzyme inhibitory activity in the water-soluble protein extract from Petrovac Sausage (Petrovská Kolbása), Food Control 21 (2010) 1298–1302.

[83] L. Mora Soler, M. Gallego Ibáñez, E. Escudero, M. Reig Riera, M.C. Aristoy, F. Toldrá, Vilardell, Small peptides hydrolysis in dry-cured meat, International Journal of Food Microbiology 212 (2015) 9–15.

[84] M. Laranjo, M. Elias, M.J. Fraqueza, The use of starter cultures in traditional meat products, Journal of Food Quality 2017 (2017) 1–18.

[85] D.M. Müller, M.S. Carrasco, G.G. Tonarelli, A.C. Simonetta, Characterization and purification of a new bacteriocin with a broad inhibitory spectrum produced by *Lactobacillus plantarum* LP 31 strain isolated from dry-fermented sausage, Journal of Applied Microbiology 106 (2009) 2031–2040.

[86] M. de Souza Barbosa, S.D. Todorov, I. Ivanova, J.M. Chobert, T. Haertlé, B.D.G. de Melo, Franco, Improving safety of salami by application of bacteriocins produced by an autochthonous Lactobacillus curvatus isolate, Food Microbiology 46 (2015) 254–262.

[87] M. Fernández, S. Ruiz-Moyano, M.J. Benito, A. Martín, A. Hernández, M. de Guía, Córdoba, Potential antimicrobial and antiproliferative activities of autochthonous starter cultures and protease EPg222 in dry-fermented sausages, Food & Function 7 (2016) 2320–2330.

[88] K. Arihara, Y. Nakashima, S. Ishikawa, M. Itoh, Antihypertensive activities generated from porcine skeletal muscle proteins by lactic acid bacteria, Abstracts of 50th International Congress of Meat Science and Technology, Helsinki, Finland, 2004, p. 236.

[89] C. Ruiz, M. Pla, J. Riudavets, A. Nadal, High CO_2 concentration as an inductor agent to drive production of recombinant phytotoxic antimicrobial peptides in plant biofactories, Plant Molecular Biology 90 (2016) 329–334.

[90] M.C. García, P. Puchalska, C. Esteve, M.L. Marina, Vegetable foods: a cheap source of proteins and peptides with antihypertensive, antioxidant, and other less occurrence bioactivities, Talanta 106 (2013) 328–349.

[91] M. Hayes, S. Bleakley, Peptides from plants and their applications, Peptide Applications in Biomedicine, Biotechnology and Bioengineering (2018) 603–622.

[92] N. Zhang, C. Zhang, Y. Chen, B. Zheng, Purification and characterization of antioxidant peptides of Pseudosciaena crocea protein hydrolysates, Molecules 22 (2016) 57.

[93] C.E. Salas, J.A. Badillo-Corona, G. Ramírez-Sotelo, C. Oliver-Salvador, Biologically active and antimicrobial peptides from plants, BioMed Research International 2015 (2015) 102–129.

[94] C.G. Rizzello, A. Lorusso, V. Russo, D. Pinto, B. Marzani, M. Gobbetti, Improving the antioxidant properties of quinoa flour through fermentation with selected autochthonous lactic acid bacteria, International Journal of Food Microbiology 241 (2017) 252–261.

[95] C.G. Rizzello, A. Lorusso, M. Montemurro, M. Gobbetti, Use of sourdough made with quinoa (*Chenopodium quinoa*) flour and autochthonous selected lactic acid bacteria for enhancing the nutritional, textural and sensory features of white bread, Food Microbiology 56 (2016) 1–13.

[96] Z. Hafeez, C. Cakir-Kiefer, E. Roux, C. Perrin, L. Miclo, A. Dary-Mourot, Strategies of producing bioactive peptides from milk proteins to functionalize fermented milk products, Food Research International 63 (2014) 71–80.

[97] A. Pihlanto, Lactic fermentation and bioactive peptides, In: Lactic Acid Bacteria—R & D for Food, Health and Livestock Purposes (2013) 309–332.

[98] J.A. Vázquez, M.I. Montemayor, J. Fraguas, M.A. Murado, Hyaluronic acid production by *Streptococcus zooepidemicus* in marine by-products media from mussel processing wastewaters and tuna peptone viscera, Microbial Cell Factories 9 (2010) 46.

[99] D. Montesano, M. Gallo, F. Blasi, L. Cossignani, Biopeptides from vegetable proteins: new scientific evidences, Current Opinion in Food Science 31 (2020) 31–37.

[100] R.O. Arise, J.J. Idi, I.M. Mic-Braimoh, E. Korode, R.N. Ahmed, O. Osemwegie, In vitro angiotesin-1-converting enzyme, α-amylase and α-glucosidase inhibitory and antioxidant activities of *Luffa cylindrical* (L.) M. Roem seed protein hydrolysate, Heliyon 5 (2019) 1634.

[101] M.E. Valverde, D. Orona-Tamayo, B. Nieto-Rendón, O. Paredes-López, Antioxidant and antihypertensive potential of protein fractions from flour and milk substitutes from canary seeds (*Phalaris canariensis* L.), Plant Foods for Human Nutrition 72 (2017) 20–25.

[102] A. Septembre-Malaterre, F. Remize, P. Poucheret, Fruits and vegetables, as a source of nutritional compounds and phytochemicals: changes in bioactive compounds during lactic fermentation, Food Research International 104 (2018) 86–99.

[103] M.A. Oyarekua, Biochemical and microbiological changes during the production of fermented pigeon pea (*Cajanus cajan*) flour, African Journal of Food Science and Technolology 2 (2011) 223–231.

[104] G. Oboh, K.B. Alabi, A.A. Akindahunsi, Fermentation changes the nutritive value, polyphenol distribution, and antioxidant properties of *Parkia biglobosa* seeds (African locust beans), Food Biotechnology 22 (2008) 363–376.

[105] D.W. Schaffner, L.R. Beuchat, Fermentation of aqueous plant seed extracts by lactic acid bacteria, Applied and Environmental Microbiology 51 (1986) 1072–1076.

[106] A.L. Capriotti, G. Caruso, C. Cavaliere, R. Samperi, S. Ventura, R.Z. Chiozzi, et al., Identification of potential bioactive peptides generated by simulated gastrointestinal digestion of soybean seeds and soy milk proteins, Journal of Food Composition and Analysis 44 (2015) 205–213.

[107] Z. Xue, H. Wen, L. Zhai, Y. Yu, Y. Li, W. Yu, et al., Antioxidant activity and anti-proliferative effect of a bioactive peptide from chickpea (*Cicer arietinum* L.), Food Research International 77 (2015) 75–81.

[108] C. Beermann, M. Euler, J. Herzberg, B. Stahl, Anti-oxidative capacity of enzymatically released peptides from soybean protein isolate, European Food Research and Technology 229 (2009) 637–644.

[109] V.P. Dia, W. Wang, V.L. Oh, B.O. De Lumen, E.G. De, Mejia, Isolation, purification and characterisation of lunasin from defatted soybean flour and in vitro evaluation of its anti-inflammatory activity, Food Chemistry 114 (2009) 108–115.

[110] S. Wang, P. Rao, X. Ye, Isolation and biochemical characterization of a novel leguminous defense peptide with antifungal and antiproliferative potency, Applied Microbiology and Biotechnology 82 (2009) 79–86.

[111] B.P. Singh, S. Vij, Growth and bioactive peptides production potential of *Lactobacillus plantarum* strain C2 in soy milk: A LC-MS/MS based revelation for peptides biofunctionality, LWT- Food Science and Technology 86 (2017) 293–301.

[112] A. Jakubczyk, M. Karaś, B. Baraniak, M. Pietrzak, The impact of fermentation and *in vitro* digestion on formation angiotensin converting enzyme (ACE) inhibitory peptides from pea proteins, Food Chemistry 141 (2013) 3774–3780.

[113] A. Jakubczyk, M. Karaś, U. Złotek, U. Szymanowska, Identification of potential inhibitory peptides of enzymes involved in the metabolic syndrome obtained by simulated gastrointestinal digestion of fermented bean (*Phaseolus vulgaris* L.) seeds, Food Research International 100 (2017) 489–496.

[114] J. Hur, T.T.H. Nguyen, N. Park, J. Kim, D. Kim, Characterization of quinoa (*Chenopodium quinoa*) fermented by *Rhizopus oligosporus* and its bioactive properties, AMB Express 8 (2018) 143.

[115] C.G. Rizzello, A. Cassone, R. Di Cagno, M. Gobbetti, Synthesis of angiotensin I-converting enzyme (ACE)-inhibitory peptides and γ-aminobutyric acid (GABA) during sourdough fermentation by selected lactic acid bacteria, Journal of Agricultural and Food Chemistry 56 (2008) 6936–6943.

[116] R. Coda, C.G. Rizzello, D. Pinto, M. Gobbetti, Selected lactic acid bacteria synthesize antioxidant peptides during sourdough fermentation of cereal flours, Applied and Environmental Microbiology 78 (2012) 1087–1096.

[117] Y. Hu, A. Stromeck, J. Loponen, D. Lopes-Lutz, A. Schieber, M.G. Gänzle, LC-MS/MS quantification of bioactive angiotensin I-converting enzyme inhibitory peptides in rye malt sourdoughs, Journal of Agricultural and Food Chemistry 59 (2011) 11983–11989.

[118] T. Nakamura, T. Hirota, K. Mizushima, K. Ohki, Y. Naito, N. Yamamoto, et al., Milk-derived peptides, val-pro-pro and ile-pro-pro, attenuate atherosclerosis development in apolipoprotein E–deficient mice: a preliminary study, Journal of Medicinal Food 16 (2013) 396–403.

[119] Y. Zhang, J. Liu, X. Lu, H. Zhang, L. Wang, X. Guo, et al., Isolation and identification of an antioxidant peptide prepared from fermented peanut meal using *Bacillus subtilis* fermentation, International Journal of Food Properties 17 (2014) 1237–1253.

[120] W. Wu, S. Zhao, C. Chen, F. Ge, D. Liu, X. He, Optimization of production conditions for antioxidant peptides from walnut protein meal using solid-state fermentation, Food Science and Biotechnology 23 (2014) 1941–1949.

[121] I. Mantzourani, A. Terpou, A. Alexopoulos, A. Kimbaris, E. Bezirtzoglou, A.A. Koutinas, et al., Production of a potentially synbiotic pomegranate beverage by fermentation with *Lactobacillus plantarum* ATCC 14917 adsorbed on a prebiotic carrier, Applied Biochemistry and Biotechnology 188 (2019) 1096–1107.

[122] Z.P. Zhang, J. Ma, Y.Y. He, J. Lu, D.F. Ren, Antioxidant and hypoglycemic effects of *Diospyros lotus* fruit fermented with *Microbacterium flavum* and *Lactobacillus plantarum*, Journal of Bioscience and Bioengineering 125 (2018) 682–687.

[123] R. Di Cagno, G. Minervini, C.G. Rizzello, M. De Angelis, M. Gobbetti, Effect of lactic acid fermentation on antioxidant, texture, color and sensory properties of red and green smoothies, Food Microbiology 28 (2011) 1062–1071.

[124] A. Fessard, E. Bourdon, B. Payet, F. Remize, Identification, stress tolerance, and antioxidant activity of lactic acid bacteria isolated from tropically grown fruits and leaves, Canadian Journal of Microbiology 62 (2016) 550–561.

[125] W. Randazzo, O. Corona, R. Guarcello, N. Francesca, M.A. Germana, H. Erten, et al., Development of new non-dairy beverages from Mediterranean fruit juices fermented with water kefir microorganisms, Food Microbiology 54 (2016) 40–51.

[126] J. Fideler, S.D. Johanningsmeier, M. Ekelöf, D.C. Muddiman, Discovery and quantification of bioactive peptides in fermented cucumber by direct analysis IR-MALDESI mass spectrometry and LC-QQQ-MS, Food Chemistry 271 (2019) 715–723.

[127] M. Hayes, Food proteins and bioactive peptides: new and novel sources, characterisation strategies and applications, Foods 7 (2018) 38.

[128] R. Kankanamge, C. Jeewanthi, N.K. Lee, S.K. Lee, Y.C. Yoon, H.D. Paik, Physicochemical characterization of hydrolysates of whey protein concentrates for their use in nutritional beverages, Food Science and Biotechnology 24 (2015) 1335–1340.

[129] G.W. Smithers, Whey-ing up the options–yesterday, today and tomorrow, International Dairy Journal 48 (2015) 2–14.

[130] O. Kareb, M. Aïder, Whey and its derivatives for probiotics, prebiotics, synbiotics, and functional foods: a critical review, Probiotics and Antimicrobial Proteins 11 (2019) 348–369.

[131] F.A.O., Food Wastage footprint. Full-cost accounting, Final Report, 2014.

[132] L. Drummond, C. Álvarez, A.M. Mullen, Proteins recovery from meat processing coproducts, Sustainable Meat Production and Processing (2019) 69–83.

[133] I.R. Amado, M.P. González, M.A. Murado, J.A. Vázquez, Shrimp wastewater as a source of astaxanthin and bioactive peptides, Journal of Chemical Technology & Biotechnology 91 (2016) 793–805.

[134] S. Ghalamara, S. Silva, C. Brazinha, M. Pintado, Valorization of fish by-products: purification of bioactive peptides from codfish blood and sardine cooking wastewaters by membrane processing, Membranes 10 (2020) 44.

[135] J.K. Lee, S.K.S. Patel, B.H. Sung, V.C. Kalia, Biomolecules from municipal and food industry wastes: an overview, Bioresource Technology 298 (2019) 122346.

[136] M. Gong, A.M. Aguirre, A. Bassi, Technical issues related to characterization, extraction, recovery, and purification of proteins from different waste sources, Protein Byproducts (2016) 89–106.

[137] S. Álvarez-Blanco, J.A. Mendoza-Roca, M.J. Corbatón-Báguena, M.C. Vincent-Vela, Valuable products recovery from wastewater in agrofood by membrane processes, Sustainable Membrane Technology for Water and Wastewater Treatment (2017) 295–318.

[138] K.J. Rutherfurd-Markwick, Food proteins as a source of bioactive peptides with diverse functions, British Journal of Nutrition 108 (2012) S149–S157.

[139] M. Cermeño, A. Connolly, M.B. O'Keeffe, C. Flynn, A.M. Alashi, R.E. Aluko, et al., Identification of bioactive peptides from brewers' spent grain and contribution of Leu/Ile to bioactive potency, Journal of Functional Foods 60 (2019) 103455.

[140] A. Dullius, P. Fassina, M. Giroldi, M.I. Goettert, C.F.V. de Souza, A biotechnological approach for the production of branched chain amino acid containing bioactive peptides to improve human health: a review, Food Research International 109002 (2020).

[141] M.R. De Oliveira, T.J. Silva, E. Barros, V.M. Guimarães, M.C. Baracat-Pereira, M.R. Eller, et al., Antihypertensive peptides derived from caseins: mechanism of physiological action, production bioprocesses, and challenges for food applications, Applied Biochemistry and Biotechnology 185 (2018) 884–908.

[142] M.L. Marco, D. Heeney, S. Binda, C.J. Cifelli, P.D. Cotter, B. Foligné, et al., Health benefits of fermented foods: microbiota and beyond, Current Opinion in Biotechnology 44 (2017) 94–102.

[143] J.G. Dos Santos Aguilar, H.H. Sato, Microbial proteases: production and application in obtaining protein hydrolysates, Food Research International 103 (2017) 253–262.

[144] C. Chaves-López, A. Serio, A. Paparella, M. Martuscelli, A. Corsetti, R. Tofalo, et al., Impact of microbial cultures on proteolysis and release of bioactive peptides in fermented milk, Food Microbiology 42 (2014) 117–121.

[145] M. Pourjoula, G. Picariello, G. Garro, G. D'Auria, C. Nitride, A.R. Ghaisari, et al., The protein and peptide fractions of kashk, a traditional Middle East fermented dairy product, Food Research International 132 (2020) 109107.

[146] I. López-Expósito, B. Miralles, L. Amigo, B. Hernández-Ledesma, Health effects of cheese components with a focus on bioactive peptides, Fermented foods in health and disease prevention (2017) 239–273. Academic Press.

[147] M. Mechmeche, F. Kachouri, H. Ksontini, M. Hamdi, Production of bioactive peptides from tomato seed isolate by *Lactobacillus plantarum* fermentation and enhancement of antioxidant activity, Food Biotechnology 31 (2017) 94–113.

[148] S. Ruan, J. Luo, Y. Li, Y. Wang, S. Huang, F. Lu, et al., Ultrasound-assisted liquid-state fermentation of soybean meal with *Bacillus subtilis*: effects on peptides content, ACE inhibitory activity and biomass, Process Biochemistry 91 (2020) 73–82.

[149] Y.A. Pérez, C.M. Urista, A.M. Cerda, J.Á. Sánchez, F.R. Rodríguez, Antihypertensive and antioxidant properties from whey protein hydrolysates produced by encapsulated *Bacillus subtilis* cells, International Journal of Peptide Research and Therapeutics 25 (2019) 681–689.

[150] H. Chmiel, Bioprozesstechnik, Springer-Verlag, 2011.

[151] D. Agyei, M.K. Danquah, Industrial-scale manufacturing of pharmaceutical-grade bioactive peptides, Biotechnology Advances 29 (2011) 272–277.

[152] B.N.P. Sah, T. Vasiljevic, S. McKechnie, O.N. Donkor, Antioxidative and antibacterial peptides derived from bovine milk proteins, Critical Reviews in Food Science and Nutrition 58 (2018) 726–740.

[153] T. Lafarga, M. Hayes, Bioactive protein hydrolysates in the functional food ingredient industry: Overcoming current challenges, Food Reviews International 33 (2017) 217–246.

[154] E.C. Li-Chan, Bioactive peptides and protein hydrolysates: research trends and challenges for application as nutraceuticals and functional food ingredients, Current Opinion in Food Science 1 (2015) 28–37.

[155] P. Minkiewicz, J. Dziuba, A. Iwaniak, M. Dziuba, M. Darewicz, BIOPEP database and other programs for processing bioactive peptide sequences, Journal of AOAC International 91 (2008) 965–980.

[156] Q. Li, C. Zhang, H. Chen, J. Xue, X. Guo, M. Liang, et al., BioPepDB: an integrated data platform for food-derived bioactive peptides, International Journal of Food Sciences and Nutrition 69 (2018) 963–968.

[157] A. Iwaniak, P. Minkiewicz, M. Darewicz, K. Sieniawski, P. Starowicz, BIOPEP database of sensory peptides and amino acids, Food Research International 85 (2016) 155–161.

[158] P. Minkiewicz, A. Iwaniak, M. Darewicz, BIOPEP-UWM Database of bioactive peptides: current opportunities, International Journal of Molecular Sciences 20 (2019) 5978.

[159] Z. Agirbasli, L. Cavas, In silico evaluation of bioactive peptides from the green algae Caulerpa, Journal of Applied Phycology 29 (2017) 1635–1646.

[160] R. Han, J. Maycock, B.S. Murray, C. Boesch, Identification of angiotensin converting enzyme and dipeptidyl peptidase-IV inhibitory peptides derived from oilseed proteins using two integrated bioinformatic approaches, Food Research International 115 (2019) 283–291.

[161] Y. Kumagai, Y. Miyabe, T. Takeda, K. Adachi, H. Yasui, H. Kishimura, In silico analysis of relationship between proteins from plastid genome of red alga *Palmaria* sp. (Japan) and angiotensin I converting enzyme inhibitory peptides, Marine Drugs 17 (2019) 190.

[162] A.B. Nongonierma, R.J. FitzGerald, An in silico model to predict the potential of dietary proteins as sources of dipeptidyl peptidase IV (DPP-IV) inhibitory peptides, Food Chemistry 165 (2014) 489–498.

[163] A.E. Nardo, M.C. Añón, G. Parisi, Large-scale mapping of bioactive peptides in structural and sequence space, PloS One 13 (2018) 0191063.

[164] E. Leeb, A. Götz, T. Letzel, S.C. Cheison, U. Kulozik, Influence of denaturation and aggregation of β-lactoglobulin on its tryptic hydrolysis and the release of functional peptides, Food Chemistry 187 (2015) 545–554.

[165] Y. Deng, F. van der Veer, S. Sforza, H. Gruppen, P.A. Wierenga, Towards predicting protein hydrolysis by bovine trypsin, Process Biochemistry 65 (2018) 81–92.

[166] E. Leeb, T. Stefan, T. Letzel, J. Hinrichs, U. Kulozik, Tryptic hydrolysis of β-lactoglobulin: a generic approach to describe the hydrolysis kinetic and release of peptides, International Dairy Journal 105 (2020) 104666.

[167] S. Suwal, E. Rozoy, M. Manenda, A. Doyen, L. Bazinet, Comparative study of in situ and ex situ enzymatic hydrolysis of milk protein and separation of bioactive peptides in an electromembrane reactor, ACS Sustainable Chemistry & Engineering 5 (2017) 5330–5340.

[168] Y. Mao, M. Krischke, U. Kulozik, β-Lactoglobulin hydrolysis by a flow-through monolithic immobilized trypsin reactor in ethanol/aqueous solvents, Process Biochemistry 82 (2019) 84–93.

[169] M.M.H. Antink, T. Sewczyk, S. Kroll, P. Árki, S. Beutel, K. Rezwan, et al., Proteolytic ceramic capillary membranes for the production of peptides under flow, Biochemical Engineering Journal 147 (2019) 89–99.

4

Probiotics in fermented products and supplements

Neha Baliyan[1,2], Madhu Kumari[2,3], Poonam Kumari[1], Kiran Dindhoria[1,2], Srijana Mukhia[1,4], Sanjeet Kumar[1], Mahesh Gupta[2,3], Rakshak Kumar[1,2]

[1]BIOTECHNOLOGY DIVISION, CSIR-INSTITUTE OF HIMALAYAN BIORESOURCE TECHNOLOGY, PALAMPUR, INDIA [2]ACADEMY OF SCIENTIFIC AND INNOVATIVE RESEARCH (ACSIR), CSIR-HUMAN RESOURCE DEVELOPMENT CENTRE, GHAZIABAD, INDIA [3]FOOD AND NUTRACEUTICAL DIVISION, CSIR-INSTITUTE OF HIMALAYAN BIORESOURCE TECHNOLOGY, PALAMPUR, INDIA [4]DEPARTMENT OF MICROBIOLOGY, GURU NANAK DEV UNIVERSITY, AMRITSAR, INDIA

4.1 Introduction

Fermentation is a traditional method of food preservation that adds nutritional value to the food. Fermented foods are prepared through controlled microbial growth and enzymatic conversion of food components [1]. These foods constituted a significant part of the human diet across all continents and diversified based on cultural differences and geographical locations of different regions. Consequently, large varieties of fermented foods are consumed worldwide that differ based on the microbial culture, the substrate used for the production, and the product type [2]. The initial evidence of fermentation has been dated back to the Neolithic period around 10,000 BCE [2,3]. However, the beneficial role of probiotics in fermentation was unknown until 1907 when Nobel laureate Elie Metchnikoff found the link between longevity of Bulgarian peasants and ingestion of fermented milk and gave the probiotics concept [4]. After Metchnikoff's observation, many scientific progressions were made, and fermented foods containing probiotics are now recognized as a critical dietary supplement for improving human health [5]. Probiotics affect human health by maintaining gut microbiota homeostasis. At present, probiotics have become a multibillion-dollar global industry. The global market of probiotics is estimated to reach US$7 billion during 2020–25 (https://www.statista.com).

According to FAO/WHO, probiotics are defined as "live microorganisms that, when administered in adequate amounts, confer a health benefit on the host" [6]. Probiotic benefits can be categorized into nutritional and therapeutic. Former includes enhanced nutritional bioavailability, vitamin synthesis, short-chain fatty acids (SCFAs) production and increased digestibility of complex compounds [7]. The latter include growth inhibition of pathogenic microbes in the gastrointestinal tract (GIT), homeostasis of gut microbiota, and host immune system modulation [8,9]. Probiotics also help prevent/treat some health disorders, such as diarrhea, irritable bowel syndrome, lactose intolerance, anxiety, cardiovascular diseases, colon cancer, arthritis, and diabetes [10,11]. Probiotics include bacteria, yeast, and some molds. However, *Bifidobacteria* and *Lactobacillus* are the major groups of probiotics [12]. *Lactobacillus* genera are the most commonly used dairy-based probiotics. Also, there is an increase in demand for nondairy-based fermented products due to lactose intolerance in adults.

This chapter has introduced fermented food products (dairy and nondairy based) and supplements with their potential probiotic properties in human health and disease.

4.2 Probiotic bacteria

Bacteria, yeast, and some molds are used as probiotics; however, lactic acid bacteria (LAB) are the main groups used as probiotics because of their fermentative properties of proteolysis, acidification, and aromatic acid production [13]. Among LAB, *Lactobacillus* spp. are most commonly used as probiotics in humans since they belong to the human gut [14], and the genus consists of a high number of generally recognized as safe (GRAS) species [15,16]. In addition to them, certain species of *Streptococcus*, *Enterococcus*, *Bacillus*, *Escherichia*, and *Saccharomyces* are also used as probiotics. Probiotics could be used as single species or several bacteria/yeast species that become effective at different levels. Still, it cannot be possible to provide one count for all probiotics.

4.2.1 Lactic acid bacteria

LAB represent a phylogenetically diverse cluster of bacteria similar in morphological, physiological, and metabolic characteristics with a G + C content of less than $\sim$50 mol% [17]. LAB are Gram-positive, catalase-negative, nonsporulating, cocci/rods, cytochrome-negative, fastidious, and aerotolerant microorganisms that produce acid (lactic) as the end product of carbohydrate fermentation [17,18]. Although lacking catalase, they are protected from oxygen (O_2) by-products (e.g., H_2O_2) through peroxidase enzymes [17].

4.2.1.1 Metabolism
Carbohydrate fermentation is an essential feature of LAB metabolism. They are restricted to sugar-rich environments since sugars are the only source of energy for the LAB. LAB also has limited biosynthetic ability; hence, it requires nutrient-rich environments for growth [19]. Based on the sugar fermentation pattern, LAB are divided into two broad categories, that is, homofermentative and heterofermentative (Fig. 4−1).

4.2.1.1.1 Homofermentative metabolism
Homofermentative bacteria metabolize the sugars entirely into lactic acid using the Embden − Meyerhof pathway (glycolysis, E-M pathway). In homofermentative, each molecule of glucose produces two molecules of lactic acid. They are commercially exploited for

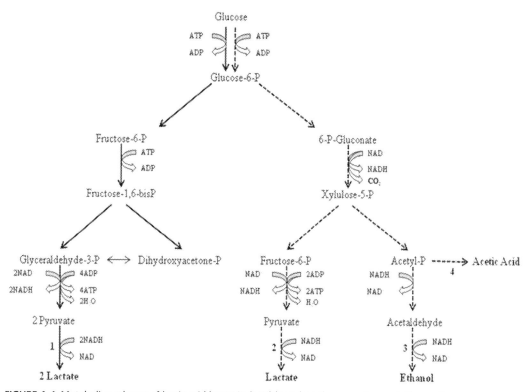

FIGURE 4–1 Metabolic pathway of lactic acid bacteria (LAB) based on the sugar fermentation pattern. Homofermentative LABs completely metabolize the sugars into lactic acid using the Embden–Meyerhof pathway. In homofermentation, each molecule of glucose produces two molecules of lactic acid. Heterofermentative LABs metabolizes sugars using the phosphorketolase pathway and produces lactate, ethanol, acetate, and CO_2.

the production of lactic acid. Homofermentative LAB includes species of *Enterococci, Tetragenococci, Lactococci, Vagococci, Lactobacilli, Pediococci,* and *Streptococci.*

4.2.1.1.2 Heterofermentative metabolism

Heterofermentative LAB metabolizes sugars using the phosphoketolase pathway (pentose phosphate pathway, PK). The difference in the enzyme level between homo- and heretofermentation is the absence of enzymes fructose 1, 6-diphosphate of the E-M pathway, and the phosphoketolase enzyme of the PK pathway. The final products of the PK pathway are lactate, ethanol, acetate, and CO_2. Obligate heterofermentative LAB includes *Oenococcus, Leuconostoc Weissella,* and some *Lactobacilli. Leuconostoc* prefers to produce lactate and ethanol in slightly aerated conditions, whereas in an oxygen-rich environment, they produce lactate and acetate.

4.2.1.2 Classification of LAB

Orla-Jensen gave the basis of the modern classification of LAB [20]. He divided LAB into four genera: *Lactobacillus, Pediococcus, Leuconostoc,* and *Streptococcus* based on cellular morphology, mode of glucose fermentation, temperature range, and sugar utilization pattern. Later discovering new taxonomic tools such as 16S rRNA gene sequencing, DNA–DNA hybridization, and the mol% G + C content, etc., led to many new genera in the LAB group. The major genera of the LAB are *Lactobacillus, Pediococcus, Lactococcus, Streptococcus, Carnobacterium, Enterococcus, Leuconostoc, Oenococcus, Tetragenococcus, Vagococcus,* and *Weissella* [21,22]. *Bifidobacteria* are sometimes referred to as LAB because of similar physiological and biochemical properties [23]. However, due to their different metabolic pathways for the degradation of hexoses, higher DNA G + C content (>50%), and poor phylogenetic relatedness to other LAB spp., they are not included in the LAB group [18]. Phylogenetic studies have shown that the core LAB species belong to the order *Lactobacillales* of the phylum *Firmicutes*. Among the genera of the LAB, *Lactobacillus* is the most dominant genus, followed by *Pediococcus*.

4.2.1.2.1 Lactobacillus

The genus *Lactobacillus* belongs to the family *Lactobacillaceae*, order *Lactobacillales*, and class *Bacilli* of the phylum *Firmicutes*. It includes Gram-positive, nonspore-forming, fermentative bacteria that can occur as rods or coccobacilli. *Lactobacilli* are overly fastidious organisms and require complex organic substrates to grow [24]. *Lactobacillus* genus contained 261 species (in March 2020) with a broad diversity of phenotypic, ecological, and genotypic properties. Due to the large variations within the genus, scientists reclassified the *Lactobacillus* genus into 25 genera [25]. They evaluated the taxonomy of *Lactobacillaceae* and *Leuconostocaceae* based on whole-genome sequences. The 25 genera include the emended genus *Lactobacillus* (*L. delbrueckii* group and *Paralactobacillus*) and 23 novel genera. Based on the fermentative characteristics, these genera have been divided into homofermentative and heterofermentative as follows: for homofermentative *Lactobacillus delbrueckii* group, *Paralactobacillus, Amylolactobacillus, Holzapfelia, Bombilactobacillus, Companilactobacillus, Lapidilactobacillus, Agrilactobacillus, Schleiferilactobacillus, Lacticaseibacillus, Latilactobacillus, Loigolactobacillus, Dellaglioa, Liquorilactobacillus, Ligilactobacillus, Lactiplantibacillus* and for heterofermentative *Furfurilactobacillus, Paucilactobacillus, Limosilactobacillus, Secundilactobacillus, Levilactobacillus, Fructilactobacillus, Acetilactobacillus, Apilalactobacillus, Lentiactobacillus, Lapidilactobacillus, Lacticaseibacillus, Loigolactobacillus,* and *Lactiplantibacillus* are reported.

Lactobacilli are well known for their health-promoting properties and most commonly used as probiotics. Several studies illustrated that *Lactobacilli* effectively prevents several diseases such as antibiotic-associated diarrhea, traveler's diarrhea, eczema, irritable bowel syndrome (IBS), bacterial vaginosis, etc. [26,27]. Table 4–1 shows some crucial probiotics new names according to the reclassification given by Zheng [25]. Whereas the genus names of some probiotics *Lactobacillus* species such as *L. acidophilus, L. delbrueckii* subsp. *bulgaricus, L. crispatus, L. gasseri, L. johnsonii,* and *L. helveticus* remained the same. In the present chapter, we used the old nomenclature of the genus *Lactobacillus*.

4.2.1.2.2 Pediococcus

Pediococcus strains are homofermentative bacteria that are spherical and exist in a tetrad or diploid forms. This genus belongs to the family *Lactobacillaceae* of the phylum *Firmicutes*. *Pediococci* are commonly found in fermented meat, fish products, and natural cheese, where they play a role in the development of cheese flavor. Genus consists of nine species, including *P. acidilactici, P. claussenii, P. cellicola, P. damnosus, P. dextrinicus, P. inopinatus,*

Table 4–1 New names of some common *Lactobacillus* probiotic species [25].

Current name	New name
Lactobacillus casei	*Lacticaseibacillus casei*
Lactobacillus paracasei	*Lacticaseibacillus paracasei*
Lactobacillus rhamnosus	*Lacticaseibacillus rhamnosus*
Lactobacillus plantarum	*Lactiplantibacillus plantarum*
Lactobacillus brevis	*Levilactobacillus brevis*
Lactobacillus salivarius	*Ligilactobacillus salivarius*
Lactobacillus fermentum	*Limosilactobacillus fermentum*
Lactobacillus reuteri	*Limosilactobacillus reuteri*

P. parvulu, *P. pentosaceus* subsp. *pentosaceus*, *P. pentosaceus* subsp. *intermedius*, and *P. stilesii*. Among them, *P. acidilactici* and *P. pentosaceus* have been reported to possess probiotics potential. *P. pentosaceus* is usually used as a starter culture to ferment various foods like meats, vegetables, and cheeses. It has been approved to be used as an animal feed additive in several countries [28], but now several studies have characterized its potential to be used in humans [29,30]. *P. pentosaceus* KID7 has been reported to exhibit cholesterol-lowering activity and essential probiotic attributes [31]. In addition to *P. pentosaceus*, *P. acidilactici* also reported as a potential probiotic. Balgir (2014) said that *P. acidilactici* MTCC5101 improved iron absorption in anemic young women [32]. A study conducted by Kaur [33] provided evidence of controlling peptic ulcer disease by pediocin-producing *P. acidilactici* BA28.

4.2.1.2.3 Enterococcus and Streptococcus

The genus *Enterococcus* belongs to the family *Enterococcaceae* of the phylum *Firmicutes*. Based on DNA–DNA and DNA–rDNA hybridization studies, this genus was separated from *Streptococcus* [34], which was later confirmed by 16S rRNA studies [35]. Phylogenetically, *Enterococcus* species are more closely related to the genus *Leuconostoc* [36]. Enterococci are homofermentative LAB having DNA G + C content ranges from 35.1 to 44.9 (mol%). Bacterial cells are ovoid, occur singly, in pairs, or short chains, usually nonmotile and nonpigmented. The best-studied probiotics strain of genus *Enterococcus* is *E. faecium* SF68. Although *E. faecium* has been reported as probiotics for preventing antibiotic-associated diarrhea, some strains contain antibiotic resistance and virulence genes. Few strains also act as pathogens associated with nosocomial infections, bacteremia, and urinary tract infections; therefore *Enterococci* is generally not treated as safe for humans use [37].

Streptococcus belongs to the LAB group; however, most of the species are associated with several health conditions. One exception in the genus is *S. thermophilus*, which is reported to possess probiotics attributes and used in the dairy industry to prepare yogurt. It is also the only species in the genus given GRAS status by Food and Drug Administration (FDA). *S. thermophilus* is a thermotolerant, Gram-positive bacterium having ovoid cells [38]. Some animal and human studies have shown that *S. thermophilus* helps prevent chronic gastritis, diarrhea, IBS symptoms, lactose intolerance, and necrotizing enterocolitis in preterm infants [39–41].

4.2.1.2.4 Leuconostoc

Leuconostoc is a diverse group of bacteria that share many characteristics with the genus *Lactobacillus*. It belongs to the family *Leuconostocaceae*, within the order *Lactobacillales* of phylum *Firmicutes*. Members of the *Leuconostoc* genus are heterofermentative cocci arranged in pairs or chains [42], lack cytochrome, and nonhemolytic [43]. The species are nonacidophilic, with an optimum temperature range between 20°C and 30°C; they can grow at 5°C but not above 40°C. *Leuconostoc* spp. are commonly present in fresh plants. The *Leuconostoc* genus comprises *L. mesenteroides*, *L. citreum*, *L. carnosum*, *L. durionis*, *L. fallax*, *L. ficulneum*, *L. pseudoficulneum*, *L. fructosum*, *L. gasicomitatum*, *L. gelidum*, *L. inhae*, *L. kimchii*, *L. lactis*, and *L. pseudomesenteroides* [44]. *Leuconostoc* species play a significant role in several industrial processes [45,46]. *Leuconostoc* are generally considered nonpathogenic to humans but can act as opportunistic pathogens [47]. Despite its high technological values, *Leuconostoc* has not gained much popularity as probiotics. Some strains of this genus are also used as probiotics. Agarwal [47,48] reported *L. mesenteroides*, combined with other probiotics strains, help control children's diarrhea. Also, they are said to possess immunomodulatory effects and antimicrobial activity [48,49,50].

4.2.1.2.5 Weissella

The genus *Weissella* belongs to the family *Leuconostocaceae* of the phylum *Firmicutes* with DNA G + C content ranging from 37% to 47% [51,52]. Bacteria belonging to this genus are heterofermentative with coccoid or rod-shaped morphology [52,53]. All species can grow at 15°C, whereas growth varies at 45°C. *Weissella* genus is differentiated into four main branches: first contains *W. hellenica*, *W. paramesenteroides*, *W. thailandensis*, and *W. confuse*, *W. cibaria* represents the second lineage. *W. minor*, *W. viridescens*, and *W. halotolerans* form the third line, and the fourth line includes *W. kandleri* [54]. *Weissella* is an industrially important genus that is used in several fermentation processes. Strains belonging to species *W. confusa* and *W. cibaria* have been widely studied for the production of exopolysaccharides (EPSs) and nondigestible oligosaccharides [55]. Besides, some strains of the *Weissella* genus act as probiotics [51]. *W. cibaria* has been used as probiotics for oral health [56]. *W. hellenica* DS-12 has been used as fish probiotics, owing to its inhibitory potential against fish pathogens [57]. Furthermore, *W. koreensis* OK1-6 has been reported to possess antiobesity effects in mice [58,59]. Despite this, the use of *Weissella* spp. as probiotics is controversial since some genus species act as opportunistic pathogens and contain antibiotic resistance genes [60].

4.2.1.2.6 Lactococci

Lactococci are homofermentative, coccoid-shaped bacteria that form short chains and produce L (+)-lactic acid. They are commonly called mesophilic lactic streptococci that can grow at temperature 10°C but not at 45°C. The DNA G + C content of *Lactococcus* ranges from 34 to 43 mol % [61,62]. *Lactococci* are important starter cultures used in the preparation of fermented dairy products for hundreds of years. Its technological properties, such as EPS and bacteriocin production, proteolytic activity, and lactose metabolism, are well identified. *Lactococcus* is an important genus of the LAB that is GRAS. Some species like *Lactococcus lactis* subsp. *lactis* has been documented to possess probiotics properties and prevent/treat antibiotic-associated diarrhea, intestinal inflammation, allergies, hypertension, cancer, and diabetes [61]. This strain has also been reported to produce gamma-aminobutyric acid, vitamins, and bioactive peptides [63].

4.2.1.2.7 Bacillus

In addition to LAB, other bacteria such as *Bacillus* spp. are also reported to possess probiotics potential. Genus *Bacillus* includes Gram-positive, rod-shaped, spore-forming, aerobic bacteria widespread in the environment [64]. This genus consists of 318 species of bacteria. *Bacillus* spp. *are* predominantly used in the preparation of fermented soya products. *B. subtilis* var. *natto* is used to produce *natto*, a popular soya fermented food in Japan [65]. Several in vitro, in vivo, and clinical studies, proved the probiotics candidature of *Bacillus* spp. They are documented to prevent or cure diarrhea, gingivitis, *H. pylori* infection, maintaining intestinal homeostasis, immunomodulation, etc. [66,67]. They are also used for the production of vitamins, carotenoids, and various food grade enzymes [64]. One species that have been extensively examined as probiotics is *B. coagulans*. It has characteristics of both genera *Lactobacillus* and *Bacillus*, and until 1974 it was classified as *Lactobacillus sporogenes* [68]. *B. coagulans* combined with other probiotics strains have proven to be most successful in treating antibiotic-associated diarrhea [69]. Other *Bacillus* spp. used as probiotics includes *B. cereus, B. clausii, B. subtilis*, and *B. licheniformis* [68]. The essential advantages of *Bacillus* probiotics are their high survivability in the GIT and stability of strains in the product for a longer duration. However, compared with LAB probiotics, *Bacillus* probiotics are not much popular in the functional food industry due to their relatedness to a few human pathogens.

4.2.2 *Bifidobacterium*

The genus *Bifidobacterium* belongs to the family *Bifidobacteriaceae* of the phylum *Actinobacteria* [70]. Bifidobacteria are Gram-positive, nonmotile, nonspore-forming, nongas-producing, anaerobic, heterofermentative, and catalase-negative bacteria with a high G + C DNA content (55% to 67%). They were first isolated in 1899 by Tissier from an infant's feces and named *Bacillus bifidus* [71]. Later, due to their metabolic similarity to LAB, they were included in the genus *Lactobacillus*. In 1974 *Bifidobacteria* were reclassified as a separate genus [72]. Bifidobacterium is the most commonly isolated from the intestinal tracts of animals. *Bifidobacterium bifidum, Bifidobacterium breve*, and *Bifidobacterium longum* are dominant bacteria in the gut of breastfed infants [71]. They have also been reported from other environmental niches such as an oral cavity and vagina of humans, sewage, blood, and fermented food. For example, *B. minimum* and *B. subtilis* were isolated from sewage, *B. mongoliense* from koumiss, *B. aquikefiry* from water kefir, *B. crudilactis*, and *B. mongoliense*, from raw milk cheeses [73].

4.2.3 Yeast

Some nonbacterial microorganisms, for example, yeasts, are also used as human probiotics [74]. *Saccharomyces* is a symbiotic yeast that inhabits the GIT, respiratory tract, and vaginal mucosa of humans. They belong to the family *Saccharomycetes* of the phylum *Ascomycota*. The genus *Saccharomyces* includes various species, including *S. cerevisiae*, and *S. boulardii* [75].

The taxonomic position of *S. boulardii* was redetermined using multilocus sequence analysis, and now it is considered a subspecies of *S. cerevisiae* [76]. The change *S. cerevisiae* var. *boulardii* has been done, but the designation *S. boulardii* is still primarily used in the scientific literature. This strain has been utilized worldwide as a preventive/therapeutic agent for GI disorders caused by antimicrobial agents [77]. It imparts gastrointestinal (GI) benefits by increasing the number of beneficial microbes like *Bifidobacteria* and *Lactobacilli* while decreasing disease-causing organisms [78]. *S. boulardii* also reported preventing/alleviate the symptoms of inflammatory bowel disease, acute gastroenteritis, and *Helicobacter pylori* infection [79,80]. In healthy individuals, these strains have an excellent reputation for safety; however, in immunocompromised individuals, it has been associated with fungemia [81].

4.3 Probiotics in fermented foods

4.3.1 History of fermented foods

Humans have been managing the fermentation process for thousands of years, especially in fermented beverages in the earliest days. Fermented beverages prepared from fruit, honey, and rice found in Neolithic China at around 7000–6600 BCE. Humans were fermenting dairy products even before fermented alcoholic beverages were generated. The milk of goats, sheep, cattle, and camels was fermented naturally dates back to 10,000 BCE. Fermentation is a process, which takes place spontaneously due to naturally occurring microflora in the milk [82]. The subtropical environment likely played a significant role in dairy fermentation, as the heat of this climate supports thermophilic lactic acid fermentation. It has been advocated that the earliest production of yogurts took place in goat bags draped over the backs of camels in the high temperature of North Africa, where weather prevails at around 110°F during the daytime, made an ideal condition for the fermentation process. Amidst 1800, the people understood what was happening to make their food ferment. In 1856 Louis Pasteur, a French chemist, linked yeast's role to the fermentation process. He defined fermentation as "respiration without air." It was not till 1910; these fermented foods were first taken as helpful to health. Elie Metchnikoff, a Russian bacteriologist, noticed that Bulgarians had an average age of 87 years, which was unique for the early 1900s. During the inspection of the Bulgarian lifestyle that has set them aside and bestowed with a longer lifespan, Metchnikoff observed a larger consumption of fermented milk than other cultures. He named the bacteria studied in these fermented milk *Bulgarian bacillus*. He advised or promoted the use of fermented milk as much as possible by elaborating on many health-promoting benefits and long lifespan due to this strain of bacteria. *B. bacillus*, was later designated as *Lactobacillus bulgaricus*, was unable to survive in humans' digestive system by Leo F. Rettger of Yale [83,84]. This finding brought a fall off of the fermented food concept. Rettger investigated different strains of *Lactobacillus* and he concluded in 1935 that some strains of *Lactobacillus acidophilus* not only could survive in the human gut even they were very active.

Extensive work has been done in the last 40 years to explore the health benefits of taking friendly bacteria. Relation appears between taking friendly bacteria in a diet and improved detoxification and digestion in different areas. Modern food habits show these findings in the popularization of "Probiotics" products. These products are simple food that consisted of friendly bacteria. According to Pasteur's study, fermented foods are naturally high in such types of friendly bacteria. Fermentation is a process that has long been exploited to enhance preserve the shelf-life, texture, flavor, and health-promoting properties of food [85]. Nowadays, the consumption of these fermented products enriched with live, friendly microbes has come out as an essential dietary alternative for enhancing human health Marco [5]. LAB from various genera, like *Streptococcus*, *Leuconostoc*, and *Lactobacillus*, are enormous in fermented foods, but other microbes like yeast and fungus contribute to the fermentation process. As probiotics hold importance in terms of the health of human beings, but there is still confusion about which fermented food products consist of live microbes and studying or understanding the important role of these microorganisms on the gut microbiota [86].

Nonetheless, cultured dairy items and yogurt are generally consumed by people as an essential source of live microbes, positively affecting human health [87]. Examining 335 adults was carried out; yogurt was the leading fermented food linked with probiotics microbes [88]. It is strongly believed that fermented products were probably found spontaneously. The history of these probiotics fermented foods with live microbes goes parallel with the evolution of human lives. Because of the sophisticated techniques, they can be traced back to ancient periods, around 10,000 years before.

4.3.2 Dairy fermented foods

Dairy fermented products have been part of the dietary culture of humans since ancient times. Even though dairy fermented foods are as of now the most well-known food transporter to convey probiotics. Among dairy products, the significant items are fermented milk, kefir, and yogurt [89]. All the more, as of late, the medical advantages of many fermented dairy items have been the subject of serious examination. A few models incorporate explicit investigations showing the effect of kefir utilization on the bone mineral thickness and bone digestion, proof of a diminished danger of type 2 diabetes related to yogurt utilization, and examinations uncovering that a fermented milk item containing an assortment of microorganisms influences the action of the mind. The formation of yogurt involves *Streptococcus salivarius* subsp. *thermophilus* and *Lactobacillus delbrueckii* subsp. *bulgaricus* [90]. Various examples of fermented dairy products are figured in Table 4–2.

4.3.3 Nondairy fermented foods

With an expansion in vegetarianism all through the developed nations, there is additional interest in nondairy products. Nondairy food products have some unique characteristics and preferences as an option in contrast to dairy probiotics. Nondairy fermented foods are an

Table 4–2 Probiotics in dairy fermented foods.

Product	Microorganisms	Region	Reference
Yogurt	Lactobacillus paracasei, Lactobacillus helveticus, Lactobacillus fermentum, Lactobacillus bulgaricus, Streptococcus thermophiles	Greece, Turkey	[88]
Cheeses	Lactobacillus spp., Streptococcus spp., Lactococcus, spp., Lactococcus lactis	Worldwide	[91]
Ayran	Streptococcus salivarius, Lactobacillus delbrueckii	Turkey	[92]
Matzoon	Streptococcus thermophilus, Lactobacillus bulgaricus	Armenia and Georgia	[92]
Nunu	Leuconostoc mesenteroides, Lactobacillus fermentum, Lactobacillus plantarum, Lactobacillus helveticus	West Africa	[92]
Skyr	Lactobacillus bulgaricus, Streptococcus thermophiles	Iceland	[92]
Viili	Geotrichum candidum, Pichia fermentans, Kluveromyces marxianus, Leuconostoc mesenteroides subsp. Cremoris, Lactococcus diacetylactis, Lactobacillus lactis subsp. cremoris	Scandinavia	[93]
Ymer	Lactobacillus lactis	Denmark	[92]
Gioddu	Lactobacillus casei, Streptococcus thermophilus, Leuconostoc mesenteroides subsp. mesenteroides, Lactobacillus bulgaricus	Italy	[94]
Kefir	Lactobacillus lactis, Saccharomyces cerevisiae, Kluyveromyces marxianus, Streptococcus thermophilus, Lactobacillus acidophilus, Bifidobacterium bifidum, Lactobacillus bulgaricus, Lactobacillus helveticus, Lactobacillus kefiranofaciens, Leuconostoc spp., Kluyveromyces lactis, Saccharomyces kefir	Russia	[92]
Koumiss	Enterococcus faecalis, Lactobacillus acidophilus, Enterococcus faecium	Russia	[95]
Shubat or Chal	Streptococcus thermophilus Lactobacillus paracasei, Lactobacillus helveticus	Turky	[92]
Gariss	Pichia kudriavzevii, Kluyveromyces marxianus, Lactobacillus fermentum, Candida kefyr, Streptococcus infantarius subsp.	West Africa	[92]
Dahi	Lactobacillus cremoris, Lactobacillus bulgaricus, Streptococcus cremoris, Lactobacillus helveticus, Streptococcus thermophilus, Lactobacillus acidophilus, Streptococcus lactis	India	[94]
Chhurpi	Leuconostoc mesenteroides, Lactobacillus plantarum, Lactobacillus fermentum, Enterococcus faecium, Lactobacillus paracasei subsp. pseudoplantarum, Lactobacillus alimentarius, Lactobacillus kefir, Lactobacillus curvatus	Arunachal Pradesh, India	[92]
Chhu	Lactococcus lactis subsp. cremoris, Lactobacillus farciminis, Lactobacillus Alimentarius, Lactobacillus brevis	Sikkim, India	[96]
Somar	Lactococcus lactis subsp. cremoris, Lactobacillus paracasei subsp. pseudoplantarum	Northeast, India	[97]
Philu	Enterococcus faecium, Lactobacillus bifermentans	Northeast, India	[97]
Mor kuzhambhu	Weissella paramesentroides	Tamil Nadu, India	[97]

acceptable option for dairy-based foods since some dairy products cause allergies. For example, the benefit of nondairy products, the nonappearance of cholesterol and lactose sugar causes cardiovascular diseases, obesity. Nondairy probiotics drinks are accessible in a showcase, consumed by individuals having lactose intolerance [98]. Cereal based foods, fruits, cabbages, meat, and fish are part of nondairy-based fermented products (Table 4–3). These products contain a wide variety of aromas, texture, and flavors. *Lactobacillus plantarum*,

Table 4-3 Probiotics in nondairy fermented foods.

Substrate	Products	Probiotic microorganisms	Region	Reference
Soybeans	Soy sauce (shoyu)	Lactobacillus delbrueckii, Aspergillus oryzae, Zygosaccharomyces rouxii, Aspergillus soyae	Japan	[99]
	Tempeh	Rhizopus oryzae, Rhizopus oligosporus	Indonesia, New Guinea, Surinam	[100]
	Tao-si	Aspergillus oryzae	Philippines	[101]
	Miso	Zygosaccharomyces rouxii, Aspergillus oryzae	Japan	[99]
	Kinema	Bacillus subtilis, Bacillus cereus, Bacillus licheniformis	Sikkim, India	[102]
	Hawaijar	Proteus mirabilis, Bacillus subtilis, Providencia rettgers, Bacillus amyloliquefaciens, Alkaligenes sp., Bacillus cereus, Bacillus licheniformis	Manipur, India	[103]
	Bekang	Lysinibacillus fusiformis, Bacillus subtilis, Bacillus licheniformis, Bacillus coagulans, Bacillus brevis	Mizoram, India	[104]
	Aakhone/ axone	Proteus mirabilis, Bacillus subtilis	Nagaland, India	[105]
	Peruyyan	Enterococcus faecalis, Bacillus subtilis, Vagococcus lutrae, Bacillus amyloliquefaciens	Arunachal Pradesh, India	[105]
Wheat based	Tungrymbai	Bacillus subtilis, B. licheniformis	Meghalaya, India	[104]
	Kishk	Bacillus subtilis, Lactobacillus plantarum, Lactobacillus casei, Lactobacillus brevis	Egypt, Syria, and many Arabic countries	[106]
	Tarhana	Lactobacillus lactis, Lactobacillus bulgaricus, Streptococcus thermophilus	Turkey	[107]
	Jilebi	Streptococcus lactis, Enterococcus faecalis, Lactobacillus buchneri, Lactobacillus fermentum	South India	[108]
	Bhaturu (indigenous bread)	Lactobacillus acidophilus, Lactobacillus plantarum, Lactobacillus lactis, Lactobacillus mesenteroides	Himachal Pradesh, India	[109]
	Aet, aktori, baari, babroo, bhatooru	Lactobacillus plantarum, Pediococcus acidilactici, Lactobacillus fermentum	Himachal Pradesh, India	[110]
Corn based	Ogi	Saccharomyces cerevisiae, Lactobacillus brevis, Candida mycoderma, Lactobacillus plantarum, Lactobacillus fermentum	Nigeria	[99]
	Kenkey	Candida krusei, Saccharomyces cerevisiae, Lactobacillus fermentum, Lactobacillus reuteri	Ghana	[99]
	Pozol	Lactobacillus delbrueckii, Lactobacillus plantarum, Lactobacillus alimentarium, Lactococcus lactis, Lactobacillus casei, Streptococcus suis	South-eastern Mexico	[111]
Sorghum-based fermented food	Injera	Hormodendrum, Pullaria sp., Rhodotorula sp., Aspergillus sp.	Ethiopia and Eritrea	[112]
	Kisra	Lactobacillus coprophilus, Pediococcus pentosaceus	Sudan	[110]
Cereal legume-based fermented foods	Idli	Lactobacillus fermentum, Lactobacillus delbrueckii, Leuconostoc mesenteroides	India and Sri Lanka	[99]
	Dosa	Saccharomyces cerevisiae, Leuconostoc mesenteroides, Bacillus amyloliquefaciens, Lactobacillus fermentum, Streptococcus faecalis	India and Sri Lanka	[99]
	Dhokla	Lactobacillus fermentum, Leuconostoc mesenteroides	India	[99]
Rice based	Rabdi (rabadi)	Micrococcus sp., Bacillus sp.	Rajasthan India	[99]
	Pazhaiya soru	Enterococcus faecalis, Pediococcus acidilactici	Tamil Nadu, India	[113]

(Continued)

Table 4-3 (Continued)

Substrate	Products	Probiotic microorganisms	Region	Reference
	Adaidosa	Leuconostoc sp., Pediococcus sp., Streptococcus spp.	South India, India	[112]
	Kallappam	Lactobacillus plantarum, Lactobacillus fermentum	South India	[114]
	Selroti	Zygosaccharomyces rouxii, Leuconostoc mesenteroides, Lactobacillus curvatus, Enterococcus faecium, Saccharomyces cerevisiae, Pichia burtonii	Himachal Pradesh and Sikkim, India	[115]
	Yakju and Takju	Yeasts, Bacillus spp., Lactobacillus sp., Leuconostoc spp.	Korea	[116]
	Sake	Flavobacterium or Micrococcus spp. Achromobacter, Pseudomonas	Japan	[116]
	Marcha	Mucor praini, Absidia lichtheimi, Rhizopus oryzae	India, Bhutan, Nepal	[116]
	Lugri, Chhang	Saccharomyces cerevisiae, Bacillus amyloliquifaciens, Bacillus licheniformis Lactobacillus plantarum, Lactobacillus reuteri, Pediococcus acidilactici, Lactobacillus fermentum, Lactobacillus pentosus, Lactobacillus paraplantarum, Lactobacillus argentoratensis, Lactobacillus paracasei, Lactobacillus brevis	Himachal Pradesh, Ladakh	[13,117]
Millet	Ambali	Enterococcus faecalis, Lactobacillus mesenteroides, Lactobacillus fermentum	India	[118]
	Koozhu	Lactobacillus fermentum, Weissella paramesenteroides	Tamil Nadu	[116]
Fruits and vegetables				
Cabbage	Dhamuoi	Lactobacillus plantarum, Leuconostoc mesenteroides	Vietnam	[119]
	Gundruk	Lactobacillus spp., Pediococcus	Nepal, India	[120]
	Kimchi	Lactobacillus plantarum, Leuconostoc mesenteroides, Lactobacillus sakei, Lactobacillus brevis	Korea	[121]
	Paocai	Leuconostoc mesenteroides, Lactobacillus brevis, Lactobacillus pentosus, Lactobacillus fermentum, Lactobacillus plantarum, Lactobacillus lactis	China	[122]
	Sauerkraut	Lactobacillus mesenteroides Lactobacillus brevis, Lactobacillus plantarum, Lactobacillus rhamnosus	Worldwide	[123]
Cucumber	Jiang-gua	Enterococcus casseliflavus, Weissella cibaria, Leuconostoc lactis, Lactobacillus plantarum, Weissella hellenica	Taiwan	[124]
	Khalpi	Pediococcus pentosaceus, Lactobacillus plantarum	Nepal	[125]
Olives	Table olives	Lactococcus sp., Lactobacillus sp., Leuconostoc sp., Pediococcus sp., Enterococcus sp.	Spain	[126]
Raddish	Sinki	Leuconostoc fallax, Lactobacillus plantarum, Lactobacillus casei, Lactobacillus brevis	Nepal, India	[127]
Bamboo shoots	Mesu	Pediococcus pentosaceus, Lactobacillus plantarum, Lactobacillus curvatus, Pediococcus pentosaceus, Leuconostoc citreum, Lactobacillus brevis	India, Nepal, and Bhutan	[128]
	Soibum	Enterococcus durans, Leuconostoc lactis, Lactobacillus plantarum, Leuconostoc fallax, Leuconostoc mesenteroides, Lactobacillus brevis	Manipur, India	[128]
	Soidon	Leuconostoc lactis, Leuconostoc fallax, Lactobacillus brevis	Manipur, India	[128]
	Ekung	Tetragenococcus halophilus, Lactobacillus brevis, Lactobacillus casei, Lactobacillus plantarum	Arunachal Pradesh, India	[129]
	Eup	Lactobacillus fermentum, Lactobacillus plantarum	India	[129]
	Hirring	Lactococcus lactis, Lactobacillus plantarum	Arunachal Pradesh, India	[129]
	Naw-mai-dong	Lactobacillus fermentm, Pediococcus pentosaceus, Lactobacillus brevis, Leuconostoc mesenteroides, Lactobacillus buchneri	Thailand	[130]
Meat	Jerky	Lactobacillus bulgaricus, Pediococcus acidilactici	Japan	[130]
	Salami	Macrococcus, Lactobacillus spp.	Europe	[131]

	Product	Microorganisms	Location	Ref
	Sausages	Lactobacillus plantarum, Lactobacillus curvatus, Lactobacillus sakei	Worldwide	[132]
	Arjia	Weissella cibaria, Leuconostoc citreum, Enterococcus durans, Enterococcus faecium, Pediococcus pentosaceus, Enterococcus hirae, Leuconostoc mesenteroides, Enterococcus faecalis	India	[133]
	Chartayshya	Weissella cibaria, Pediococcus pentosaceus, Enterococcus faecalis, Leuconostoc citreum, Enterococcus hirae, Leuconostoc mesenteroides, Enterococcus faecium	Uttrakhand, India West Nepal	[133]
	Honoheingrain	Saccharomyces cerevisiae, Bacillus stearothermophilus, Lactobacillus brevis, Debaryomyces hansenii Lactobacillus plantarum, Micrococcus, Leuconostoc mesenteroides, Staphylococcus spp., Enterococcus faecium	Assam, India	[134]
	Jama or Geemaljuma	Enterococcus faecalis, Pediococcus pentosaceus, Enterococcus faecium, Leuconostoc citreum, Leuconostoc mesenteroides, Weissella cibaria, Enterococcus durans, Enterococcus hirae	West India	[133]
Fish				
	Fish sauce	Lactobacillus brevis, Bacillus spp., Pediococcus cerevisiae	Southeast Asia	[135]
	Bagoong	Lactobacillus coryneformis, Lactobacillus plantarum, Aspergillus oryzae	Philippines	[136]
	Gnuchi	Saccharomycopsis spp., Candida chiropterorumm, Lactococcus lactis, Pediococcus pentosaceus, Lactococcus plantarum, Enterococcus faecalis, Candida bombicola, Leuconostoc mesenteroides	Sikkim, India	[97]
	Hentak	Lactococcus plantarum, Saccharomycopsis Enterococcus faecium, Lactobacillus fructosus, Lactococcus. amylophilus, Micrococcus, Bacillus subtilis, Lactococcus lactis subsp. cremoris	Manipur, India	[137]
	Karati, Bordia, and Lashim	Lactobacillus plantarum, Bacillus subtilis, Leuconostoc mesenteroides, Lactococcus lactis subsp. cremoris, yeast Candida	Assam, India	[138]
	Ngari	Micrococcus, Lactobacillus coryniformis, Lactobacillus fructosus, Lactococcus lactis, Lactobacillus amylophilus, Lactococcus plantarum, Enterococcus faecium, Lactobacillus plantarum, Saccharomycopsis spp.	Manipur, India	[137]
	Shidal	Escherichia coli, Bacillus spp., Micrococcus spp.	Manipur, India	[139]
	Suka ko macha	Saccharomycopsis spp., Candida bombicola, Candida chiropterorum Lactococcus lactis subsp. cremoris, Pediococcus pentosaceus Lactococcus lactis subsp. lactis, Lactococcus plantarum, Leuconostoc mesenteroides, Enterococcus faecalis, Enterococcus faecium	Sikkim, India	[140]
	Sidra	Candida chiropterorum, Weissella confuse, Lactococcus lactis subsp. cremoris, Lactococcus plantarum, Enterococcus faecalis, Pediococcus pentosaceus, Leuconostoc mesenteroides, Lactococcus lactis subsp. lactis, Enterococcus faecium Lactococcus lactis subsp. lactis, Lactococcus lactis subsp. lactis, Candida bombicola, Candida chiropterorum	Sikkim, India	[141]
	Sukuti	Saccharomycopsis spp., Pediococcus pentosaceus, Lactococcus lactis subsp. cremoris, Leuconostoc mesenteroides, Lactococcus lactis subsp. lactis, Lactococcus lactis subsp. lactis, Candida bombicola, Candida chiropterorum	Sikkim, India	[214]
	Tungtap	Micrococcus, Lactobacillus fructosus, Lactobacillus puhozzihi, Lactococcus plantarum, Lactococcus lactis subsp. cremoris, Candida and Saccharomycopsis Enterococcus faecium, Lactobacillus amylophilus, Lactococcus plantarum, Lactobacillus coryniformis subsp. Torquens, Bacillus subtilis	Meghalaya, India	[215]

L. brevis, *L. casei*, *Streptococcus suis*, and *Pediococcus* sp. are some microorganisms involved in nondairy fermentation [142].

4.4 Interactions between probiotics and components of fermented foods

Nowadays, the intake of fast food, aging factors, and antibiotics causes an imbalance in humans' gut microbiota. Therefore it is necessary to improve the immune system and maintain the intestine homeostasis with a balanced diet. Fermented foods are enriched with metabolites, namely, vitamins, amino acids, minerals, phytochemicals, etc. [143] that help in digestion, nutrient absorption, balance the microbiota, and enhance the immune system [144]. Although probiotics can stimulate fermented foods functional and nutritional quality by contributing to the nutrients, bioactive compounds, namely, EPSs, organic acids, and conjugated linoleic acid (CLA) [143,145], come from microbial metabolism. For this reason, probiotics, prebiotics, and dietary fibers should be used, which could help in maintaining the intestine homeostasis [146].

With the increasing demand for food like fermented food that could promote some health benefits to the host, the effect of food formulations on probiotics properties is worthwhile to be considered because of the variety of delivery vehicles available to the consumers. A clinical study dates back to 1983, showing that *Lactobacillus* therapy's different outcomes on intestinal disorders depend upon a lot of variation in *Lactobacillus* preparation [147]. For instance, the first comparative study reported by Isolauri [148] demonstrated that Lactobacillus GG could reduce the course of acute diarrhea from fermented milk or as a freeze-dried powder. In another study, it has been shown that the same strain belonging to different sources may have other probiotics properties [149]. Leclerc [150] has demonstrated similar immune-stimulation effects of a *Bifidobacterium* in yogurt or capsules on a human adult's upper respiratory tract. Therefore a different form of fermented foods or type of matrix used to the vehicle the probiotics can alter the beneficial microorganisms' viability and efficacy.

The beneficial effects of fermented foods or microorganisms on the host were reported earlier, without knowing the specific components. Scientists are now trying to avoid this "black box" approach and attempt to identify these bioactive ingredients in fermented foods. The term "Probioactive" refers to a bioactive compound present in fermented foods or supplements, and it gives information about the presence and activity of the probiotics strain. It can be of two types: first, direct synthesis by the probiotic bacteria, and second, by modification of the food matrix by the probiotics culture. Former includes EPSs, bacteriocins, or enzymes, and the later includes, a peptide released from the hydrolysis of milk [151]. The pro-bioactive depends on the food matrix, but it can vary between different strains of bacteria.

Consequently, various comparative studies have advanced our knowledge in this area, but the literature lacks comparative clinical studies that directly relate the probiotic efficacy in a different matrix on human health. International Scientific Association for Probiotics and Prebiotics, a vibrant and nonprofit organization, is actively investigating the role of food matrix and its components, after fermentation, on the probiotic properties in terms of health benefits to the host. It suggested that a meta-analysis of the data by comparing clinical studies conducted by different research groups on the same strain in other fermented foods or supplements could provide a robust relation between probiotics and prebiotic.

A variety of food vehicles, which support their number of the various probiotics by delivering it to the human intestine, are available for the consumer. Dairy products are among the highest known food source for probiotics. Dairy products, due to high-fat content, are responsible for protecting the probiotics strain during digestion. For instance, Yakult, a dairy-based product, has been used for years in Japan [152]. Like-wise cereal-based products are also reported to be enriched in bioactive components along with the probiotic microorganisms. They can only be utilized by adding the fermentable substrate for the starter culture (such as LAB) during food formulation (probiotics), which could help the growth of potential probiotics. Therefore it is obvious to recognize the beneficial effects of both probiotics and prebiotics.

Some other grain-based food products have also been found to promote the quality of the probiotics by their characteristic resistance to gastric acid in the gut. Carbohydrates such as fructooligosaccharides, galactooligosaccharides, lactulose, soybean oligosaccharides, and inulin have the potential to resist digestive enzymes due to their chemical structures, hence exerting beneficial effects by probiotics. Then, in the large intestine, they are fermented by the saccharolytic bacteria; this process supports *Bifidobacterium*'s growth and consequently changes the gut microflora by increasing the growth of probiotics bacteria [146].

Ultimately, the whole grain-based food product is best for probiotics formulation due to its many beneficial effects from probiotic microorganisms and grains products such as bioactive components, nondigestible carbohydrates, and soluble fibers and phytochemicals [153]. In Asia and Africa, fermentation has been done in whole grain components to produce porridges, beverages, and gruels. In Western countries, bread is made by adding a prefermented starter to whole wheat dough. In the European market, various cereal and cereal components-based probiotics foods have been debuted. Some examples include Muesli probiotics flakes in Portugal, CornyActiv cereal bars in Germany, Weetaflakes whole-wheat breakfast cereals in France, and whole grain porridge in the United Kingdom [152]. Hagiwara has utilized cereal components to grow LAB [154], and Molin [155] used probiotics nutrition for making healthy drinks to feed the patients. Kontula [156] studied *L. rhamnosus*, *L. plantarum*, and *L. lactis* strains to utilize oat bran carbohydrates and oligosaccharides for fermentation, *Bifidobacterium*, and *Lactobacillus* were also reviewed by the same group using Lichenase hydrolysates of oat β-glucan and xylan as substrates [157]. Because of this beneficial feature, these grain-based probiotic microorganisms would reduce many diseases like the risk of cardiovascular disease, type 2 diabetes, obesity, and cancer prevention [158].

4.5 Impact of probiotics on the human gut microbiome

Just after the birth, microbes from the mother's intestine, vagina, or nearby environment enter and colonize an individual's gut. These microbes create a favorable niche and dominate the gut in a few weeks, ultimately leading to native microflora formation there [159,160]. The gut microbiome is dynamic, generally getting affected by several factors such as pH, medication, age, health, and diet of the host. Under normal conditions, the beneficiary microorganisms are prevalent and help in the normal gut homeostasis. These microbes play a vital role in energy regulation, immunity, neutralization of toxic compounds, and resistance against pathogens [161]. Among the different factors, the food diet is considered a significant contributor to gut microflora because of their regular exposure to the dietary originated bacteria. The gut microbes residing in the gut offer resistance to the food ingested bacteria, but some organisms can still get integrated with these communities transiently. Fermented foods and beverages contribute approximately one-third of human nutrition. They contain several bacteria that enter the GI tract upon consumption. LAB are generally used in the preparation of plant and animal-based fermented foods [162]. The LAB often converts carbohydrates to lactic acid and a mixture of different acids in rare cases. The species which are prevalent in dairy fermented products are well adapted to utilize lactose and dairy proteins. In contrast, the LAB, which is found in plant-based foods, can use complex plant-derived carbohydrates such as cellulose and xylans [163]. The orally ingested bacteria range from 10^8 to 10^{12} colony forming units (CFUs) per day in fermented foods and probiotics. These bacteria pass through the stomach, duodenum, and small intestine, eventually reaching the colon. During their journey, they face low pH and pepsin stressors in the stomach, causing cell death. Some of the bacterial cells manage to reach the small intestine, exposed to alkaline pH, pancreatin, lipase, etc. Once the cells arrive colon, they get favorable conditions to grow and increase in numbers.

The ingested bacteria modulate gut microflora mainly at three levels: food chain, direct fitness, and indirect fitness. These bacteria convert carbohydrates to many SCFAs like butyric acid, propionic acid, acetic acid, lactic acid, etc. [164]. The SCFAs integrate with metabolic pathways utilized by gut microbes, thus altering these pathways' final products [165]. They also produce EPSs that act as a growth substrate for other gut bacteria [166]. At the second level, the probiotic microorganisms either promote or inhibit microbial cells growth in the gut. They modify the physicochemical conditions of the gut and thus directly influencing resident communities there. They compete for growth substrate and inhibit the growth of pathogenic bacteria such as *Salmonella enterica* serovar *typhimurium* or *Clostridium difficile* [167]. They have also been observed to produce precursors of vitamins; for example, menaquinone (vitamin K) precursors by *Propionibacterium freudenreichii* promote Bifidobacteria's growth in vitro [168]. Some of the ingested bacteria also secrete bacteriocins that remove harmful pathogens from the GI tract [169]. In the third mechanism, the probiotic bacteria regulate the host mechanisms to influence the diversity and composition of gut microbiota. The human body produces mucus, antimicrobial peptides, and sIgA for the protection of intestinal epithelium. It has been observed that probiotic bacteria increase the amount of fecal sIgA [170]. Also, probiotic bacteria increase mucin secretion in mice [171].

4.5.1 Probiotics in the prevention and treatment of clinical diseases

The use of antibiotics has increased the chances of antibiotic resistance in microorganism and the emergence of the so-called superbugs. Probiotics are known to treat several

GIT-associated diseases, which offers a potential alternative to antibiotics. Potential probiotics to prevent/treat GI conditions like acute diarrhea, antibiotic-associated diarrhea, IBS, inflammatory bowel disease, and gastroenteritis have been reported [172,173].

The mechanism of action underlying the probiotics properties includes inhibition of pathogenic bacterial adhesion, secretion of bioactive metabolites, modulation of innate and adaptive systems. The role of probiotics for other diseases like diabetes, obesity, anxiety, arthritis, cancer, etc., has been reported but needs further investigation to identify the optimal probiotics, prebiotic, and dose for specific diseases.

4.5.1.1 Probiotics and COVID-19

A new emerging pandemic caused by the coronavirus, COVID-19, caused approximately 406,353 deaths worldwide until May 2020. COVID-19 can affect the lungs and airways. China's National Health Commission and National Administration of Traditional Chinese Medicine suggested using probiotics in patients with severe COVID-19 [174]. The meta-analysis shows that probiotics (*L. rhamnosus* GG) [175,176] play a role in reducing the duration of respiratory tract infection [177,178]. Some COVID-19 patients in China are also reported to have microbial dysbiosis, cured by *Lactobacillus* and *Bifidobacterium* [179]. A novel and more rational approach, using probiotics to combat COVID-19, depends upon a further understanding of the SARS-Cov-2 and its effect on the gut microbiota.

4.5.1.2 Obesity

Obesity is abnormal fat deposition on the body, generally above 20% of the body weight. Obesity has also been linked with other metabolic diseases such as type 2 diabetes, hypertension, and cardiac disorders. It is considered a pandemic disease affecting almost every age group, that is, children, adults, and old age persons. The global diabetes prevalence in 2019 is estimated to be 9.3% (463 million people), rising to 10.2% (578 million) by 2030 and 10.9% (700 million) by 2045 [8]. In India, approximately 135 million individuals alone are suffering from this disease. In recent studies, it has been observed that gut microbiota is mainly composed of genera *Porphyromonas*, *Prevotella*, *Bacteroides*, *Ruminococcus*, *Clostridium*, *Lactobacillus*, *Eubacteria*, and *Bifidobacteria*, etc. [180]. Studies have shown alteration of healthy gut microbiota in humans and animals in diseased conditions, and probiotic feed with high fat diet leads to a decrease in Gram-positive bacteria phyla Firmicutes and Actinobacteria in mice [181,182]. Also, reports suggest that microbiota plays an important role in energy harvest from diet and fat storage [183]. In a study, 16S rRNA amplicon sequencing of gut microbiota in obese mice revealed an increase in the abundance of species Firmicutes and a reduction in Bacteroides [184]. In another study, a reduction in Gram-negative bacteria (Bacteroidetes) and increment in Gram-positive bacteria (Firmicutes) with less diverse microflora were observed [184]. We also know that probiotic microorganisms present in the gut can digest the complex carbohydrates and produce SCFAs. These SCFAs serve as an energy source and signaling molecules for the epithelial microflora. They bind to G-protein coupled receptors, GPR41 and GPR43, present in adipose and intestinal tissues and suppress fat accumulation in the body [185]. Probiotic bacteria like *L. rhamnosus*,

L. acidophilus, and *L. plantarum*, produce CLA, which has been shown to reduce the process of adipogenesis and lipogenesis, thereby reducing lipid deposition in the body [186,187].

4.5.1.3 Diabetes mellitus

Diabetes mellitus (DM) is defined as a metabolic disease characterized by hyperglycemic blood condition. It is mainly caused due to the lack of insulin production or insulin sensitivity in the body. DM is classified as type 1 and type 2 diabetes, type 1 diabetes occurs due to the destruction of pancreatic β-cells by T-cells leading to no insulin production, whereas type 2 diabetes is associated with reduced insulin production or insulin sensitivity by receptors [188]. In India, 77 million people were suffering from DM in 2019, which is anticipated to increase to 101 million by 2030 [189]. Also, an increase in *Lactobacillus* and *Escherichia* species was noted in the patients treated with metformin [190]. Since an individual's diet intake also affects the gut microbiome, so one should take care of the quality of the ingested food. The consumption of grain and fiber-rich food increases microbes' diversity in the human gut, such as a *Prevotella* species [191]. It has been observed that low consumption of carbohydrates also decreases the number of bacteria utilizing it as a substrate in the gut, eventually reducing SCFA [192]. Probiotics also reduce the occurrence of gestational diabetes [193]. Further, the fermented papaya possesses prophylactic potential and thus can be used in the treatment of DM.

4.6 Ethnic fermented probiotics products translated to industry and their advantages over nonfermented products

The history and legacy of fermented food are so rich and diverse, not only in India but also across the world. Almost in all the ancient civilizations of the world, dairy and fermented food consumption is evident. Although these food products were high shelf-life and preservation was likely because of using fermentation techniques. The implementation of these methods readily enhances the desirable attributes such as unique flavor, textures, appearances, and functionalities compared to the raw ingredients from which they are made. The preparation and consumption of fermented food have been deliberately produced as a balanced source for vitamins, minerals, calories, and other nutrients for ages, even way before advancing food science.

Across several parts of the world, fermented dairy food products are quite famous and almost untouched by industrialization, such as in Mongolia, several aspects of Africa, and the Indian subcontinent. In India, several probiotics products have been identified and being produced on an industrial scale in a controlled way, such as *dahi* (yogurt), *idlis, dosa, kanji, achaar*, and *dhokla*, etc. [6,194] (Table 4–3). However, the importance of fermented food used in the Sikkim region has been described by Rai [195]. The following are their brief details about their origin and usages.

4.6.1 *Dahi*

It is also being traded as yogurt. Several by-products can be prepared from *dahi*, such as *chaas* and *lassi*, which is considered a delicious appetizer. *Dahi* is primarily organized by mixing starter cultures (a cocktail of LAB) containing *Lactobacillus* strains procured from milk. Live active cultures of LAB impart several beneficial effects to one's digestive system [194]. But according to the health experts, homemade *Dahi* is best for consumption once it is fermented for 24 hours. Fermentation brings about the utilization of most lactose (milk sugar) and imparts beneficial properties to it. It is also considered a functional food for its antidiarrheal, anticarcinogenic, and cholesterol-lowering properties [196].

4.6.2 *Idlis* and *dosa*

Basic ingredients of *idlis* and *dosa* are rice and urad dal, which are prepared by overnight fermentation. The fermented grains are then ground to prepare the batter. Fermentation provides a safe ground for beneficial bacteria to grow, which subsequently makes the *idlis* and *dosa* relatively healthy for the human gut. It is a staple food in South India but also quite popular in other parts of the country as well. Especially, since these foods are low in calories and are quite beneficial for good health, some LAB strains such as *L. plantarum* and *L. lactis* found in *idli* could produce vitamin B_{12}, and β-galactosidase enzymes increases the probiotic activity and improve health disorders [194]. Whereas in dosa *L. plantarum* has been isolated that was found to retard the growth of foodborne organisms.

4.6.3 *Kaanji*

This is a traditional Punjabi drink (Northwestern India), which is made up of black carrots, mustard seeds, and sea salt in water. For its preparation, the ingredients are left to ferment for around a week, which leads to the growth of beneficial bacteria. This drink is good for health, but some people should avoid this drink for excess salt intake.

4.6.4 *Achaar*

It is also traded as a pickle in India. Based on the raw material (fruits and vegetables) used, its preparation method changes slightly. Commonly people avoid pickles, considering it a big lousy lump of oil, salt, and spices. The preparation of *achaar* involves the inclusion of a tasty concoction of vegetables and fruits mixed with sugar, salt, herbs, and oil, which is allowed to ferment under the sun. This unique procession ultimately favors good bacteria's growth and gives a delicious and mouth water nature to it. It has been reported that fermentation could occur by strains like *L. plantarum, L. casei, L. rhamnosus, L. pentosus, L. casei,* and *Leuconostoc mesenteroides* [196].

4.6.5 *Dhokla*

This is one of the most delicious foods used as snacks, a popular food originating from Gujarat. Due to its awesome taste and flavor, this food is quite popular across the country's length and breadth. The dish is made using a batter of fermented gram flour, dal or rice, and curd, which imparts additional probiotics attributes. This food is the best diet and training choice for one's fitness or weight loss plan. Some LAB strains like *L. fermentum*, *Leuconostoc mesenteroides*, *Pichia silvicola*, and *S. faecalis* have also been reported in this fermented food. Due to low glycemic content, it is beneficial to cure diabetes and cardiovascular diseases [194].

4.6.6 Koozh

It is a kind of porridge, which is quite popular in the southern part of India, especially in Tamil Nadu. It is made in a clay pot from *Kezhvaragu* or *Cumbu* flour and broken rice (also known as *noiyee* in Tamil) in a clay pot. Usually, *Koozh* is made in large batches, which gives a sour tang flavor if fermented. It is considered very nutritious and imparts health-promoting properties such as preventing several intestinal disorders such as diarrhea, constipation, etc. [197,198]. In previous studies, some probiotics strains have been reported in this fermented food, such as *Weissella paramesentroides*, *L. plantarum*, and *L. fermentum* [194].

4.6.7 Toddy

It is also called palm wine. It is an alcoholic beverage obtained as the sap of various palm tree species such as the palmyra, date palms, coconut palms, etc. It is not only famous in India but also quite popular in other parts of the world. Probiotic strains reported in palm wine are *Lactobacillus fermentum* and *Bacillus subtilis* and were characterized for their adhesion property with epithelial cells to remove pathogens [194].

4.7 Challenges in the probiotics industry and future prospects

With advancements in molecular and microbial techniques, our knowledge of commensal microbiota inside the human gut has enhanced considerably. Probiotics for therapeutic uses have been driven by a medical condition called "dysbiosis," whereby imbalance in normal gut microflora is observed following antibiotic treatment that can induce chronic diseases in a patient [199]. With the rising trend of probiotics as major dietary supplements, probiotics have been the fastest-growing functional food industry. The worldwide market value of probiotics dietary supplements is expected to increase from US$3.3 billion in 2015 to US$7 billion by 2025 (https://www.statista.com). The market value of probiotics is dependent on the efficacy of the constituent microorganisms [200]. Increasing awareness regarding wellness has motivated people to emphasize healthy eating for disease prevention and maintaining healthy lifestyles. This shift in interest to healthy eating has led to a sudden boom in the functional food market. Ongoing researches focus on developing novel probiotics with boosted health benefits. Academicians and clinicians have been working and reporting on beneficial microbial probiotics species on a large scale. The concept of "synbiotics" has developed over time for enhancing the efficacy of probiotics, which is a combination of probiotics and prebiotic fibers [201]. Synbiotics proved to be useful in treating IBS symptoms, rheumatoid arthritis, and other related disorders, and propose a potential solution for overuse of antibiotics. However, for probiotics, prebiotics, and synbiotics to be widely accepted for health benefits, there are numerous challenges to be surpassed.

4.7.1 Viability and stability of probiotics

The probiotic bacteria must remain viable during passage through the GIT and during storage. Foremost existing challenges in the administration of the probiotic that restrict their beneficial effects are due to harsh processing and storage parameters such as pH, temperature, osmotic pressure, etc., harsh GIT conditions like low pH, the effect of digestive enzymes, effects of the immune system, antagonism from pathogenic bacteria, and finally colonization in the colon [202]. Microencapsulation has been suggested as a recent, effective approach for probiotic delivery and survival from degradation during manufacturing, storage, and passage through the GIT [203]. The coating materials used for encapsulating the core include polysaccharides, proteins, and lipids. Encapsulated probiotics are finding popularity in maintaining probiotic bacteria's viability and stability in a harsh physical environment. Nonetheless, enhancing the efficacy of these encapsulated probiotics in exerting their positive effects is still in progress [204]. Moreover, maintaining the production cost budget will be challenging in future probiotics formulation techniques. Commercial probiotics are blends of bacterial strains in definite CFUs, so consumers should be well-versed on the product shelf-life, certifying the bacterial viability in the consumed dose [205].

4.7.2 Safety and associated risks

Commercial probiotics available in the market are derivatives of fermented foods and healthy human microflora [206]. Probiotics are thus regarded as safe for consumption. They have been known to confer many clinical benefits associated with GIT infections and immune modulation [199]. However, although probiotics are known to promote health in adults, some adverse effects have also been documented. Reports showed that some people are susceptible to its adverse effects, and those at high risk include immunosuppressed, critically ill patients, and young infants. There have been infections after the consumption of certain probiotics, such as bacterial sepsis, infective endocarditis, bacteremia, fungemia, and other localized infections. Moreover, the mechanisms underlying clinical effects conferred by probiotics are mostly unknown [207]. Diverse theories on mode of action are available, stating that probiotics function by stimulating immunomodulation, defense from physiological stress, subduing pathogens, and enhancing gut epithelial barrier, among others [208]. These theories are poorly supported by the *in vivo* studies, as they do not consider other physiological factors governing the interaction between microbe–microbe and microbe–host inside the GIT. As the potential strains being used in a probiotic formulation come from varied niches like feces and fermented foods, they do not necessarily comply with the endogenous microbes of the human gut. The ingested probiotic strains might be subjected to pressure from a distinct microenvironment of the gut and immune dynamics, consequently giving rise to virulence factors [199]. Different probiotic formulations, dosage, and means of administration can have varied effects on individuals depending on the variable host factors. This inadequate knowledge and evidence supporting the safety of clinical applications pose a limitation on the probiotics industry. Considering all these, careful safety assessment before clinical use of probiotics is necessary, particularly in immune-compromised, critically ill persons and children [208].

4.7.3 Unregulated market

A significant limitation lies in the lack of regulation of the probiotics market due to the unavailability of standard international accord for screening probiotic microbial species [205]. Furthermore, there is no approval for any probiotic formulations for therapeutic practices by the chief health authorities like European Food Safety Authority and the USFDA [208]. The enactment of the "Dietary Supplement Health and Education Act, 1994" by the USFDA endorsed probiotics' marketing as health supplements under flexible regulations [199]. Unlike probiotics as proper therapeutic drugs that undergo strict guided rules, there are no stringent guidelines for proving the safety, quality, and value of probiotics as a dietary supplement [206]. Consequently, probiotics in dietary supplements are marketed based on specific attributes like safety, the viability in the GIT, and food taste maintenance [208]. This lack of standard regulatory practice leads to many false and ineffective probiotics products in the market. Consumers in India and many other Asia-Pacific countries are often misled and thrown into a dilemma regarding the acceptance of probiotic products [209]. Recent years have seen the advertising and promotion of fortified cereals, yogurt, and ice creams with no discrete information on their composition, safety, and efficacy. So, the dearth of product standardization practices is one of the key obstacles in the progress of the probiotics market. A high-throughput screening method for safeguarding the selected probiotics needs to be formulated. Previous probiotic researches were concerned with the bacteria only. With a better understanding of the host immune system mechanisms, there has been a development in the importance of considering host–bacteria interaction rather than the bacteria alone for probiotics' effectiveness [199].

Safeguarding the supply of wholesome, nutritious food contributes to global security. The massive growth prospects in probiotics give a sense of a promising future [210]. With an upsurge in people's interest in wellness and healthy lives, the benefits of traditional fermented foods are being translated into therapeutic innovations. It has been established that intestinal microbes influence most health aspects through their antagonistic reactions against pathogens and inducing immunomodulation [211]. Considering a series of challenges faced by the probiotics industry, emphasis should be given to overcoming the limitations and improving probiotics' functionality. Recent findings on probiotics have led to the introduction of "Next-Generation Probiotics (NGP)," which are health-promoting bacteria apart from the traditional *Lactobacilli* and *Bifidobacteria* [212]. Potential candidates for NGP are being screened from the gut microflora, which includes genera like *Bacteroides* and *Clostridium* [213].

Provided the technological challenges in NGP formulations, NGP may be marketed solely in the form of probiotics. Recombinant probiotics with desired and improved characteristics showcase the scope for emergent therapeutics. Genetic engineering will enable fortifying the qualities of existing probiotic strains and construct entirely novel and enriched probiotics [214]. However, new findings on novel probiotic bacterial species or strains do not ensure their safety based on their history as traditional probiotic strains. Safety concerns need to be addressed as a priority for releasing genetically modified organisms. It is essential to evaluate the emerging probiotic strains before creating a formulation carefully to guarantee their complete safety and efficacy. The testing should embrace the provision for strain resistance to common antibiotics and nontransmission of resistance genes to pathogenic bacteria [214]. Modern molecular-based techniques may enable adequate documentation, safety, and quality assessment of probiotics strains. There is a scope for exploring human microbiota for potential strains that can be engineered and employed as alternative therapeutics [211]. There are recent theories on enhancing probiotics' functionality by identifying and optimizing the level of "probioactives," which are bioactive components in foods and supplements formed by the direct activity of probiotics bacteria [215]. More understanding of these probioactives can serve as a key to perform meta-analyses for clinical benefits.

Further, proactive content can be enhanced by fermentation tools. Considering the inconsistency in results and reports on current case studies, it is recommended to conduct thorough, extensive research on potential probiotics concerning dosage, administration, and efficacy. Also, it is suggested to study the antagonistic effects in high-risk individuals through extensive clinical trials and for the development of novel therapeutics [207].

4.8 Conclusions and perspectives

Fermented food products provide several health benefits to the host, such as the availability of nutrients, inhibiting pathogenic bacteria's growth, and enhancing immunity. The probiotics diversity of some traditional fermented food products is still untapped. Although the current available studies in the literature are limited, the metagenomic approach, whole-genome, and marker-based metagenomic sequencing could provide a robust tool for assessing the species diversity and the functionality in various fermented foods, and beverages. Recent research on fermented foods is focused on probiotics and how the food components affect probiotics properties. A meta-analysis of the data and *in vivo* studies could provide better insight into the probiotics and food component interaction.

Acknowledgments

NB is thankful to ICMR, Govt. of India for "ICMR-Senior Research Fellowship" Proposal ID:No.3/1/2(8)/Obs./2021-NCD-II. KD is thankful to CSIR, Govt. of India for "Senior Research Fellowship" Grant (CSIR-NET JRF award no: 31/054(0139)/2019-EMR-I/CSIR-NET JRF JUNE 2017). SM acknowledges ICMR, GoI for SRF award (No. 45/17/2020-/BIO/BMS). RK is thankful to DST, Govt. of India for financial assistance under the DST

INSPIRE Faculty Scheme (DST/INSPIRE/04/2014/001280), Science and Engineering Research Board Start-up research grant no. SRG/2019/001071, NMHS project of MoEF&CC grant no. GBPNI/NMHS-2018-19/SG/178, and DST-TDT project no. DST/TDT/WM/2019/43. KD is thankful to CSIR, Govt. of India for "Research Fellowship" Grant (CSIR-NET JRF award no: 31/054(0139)/2019-EMR-I/CSIR-NET JRF JUNE 2017). Authors also acknowledge financial support from CSIR MLP 0137, MLP 0143, and MLP 0201. This manuscript represents CSIR-IHBT communication no. 4673.

References

[1] E. Dimidi, S.R. Cox, M. Rossi, K. Whelan, Fermented foods: definitions and characteristics, impact on the gut microbiota and effects on gastrointestinal health and disease, Nutrients 11 (2019) 1806.

[2] J.P. Tamang, P.D. Cotter, A. Endo, N.S. Han, R. Kort, S.Q. Liu, et al., Fermented foods in a global age: east meets west, Comprehensive Reviews in Food Science and Food Safety 19 (2020) 184–217.

[3] J.B. Prajapati, B.M. Nair, The history of fermented foods, Fermented Functional Foods, Handbook of Fermented Functional Foods, CRC Press, Boca Raton, New York, London, Washington DC, (2003), pp. 1–25.

[4] N. Mota de Carvalho, E.M. Costa, S. Silva, L. Pimentel, T.H. Fernandes, M.E. Pintado, Fermented foods and beverages in human diet and their influence on gut microbiota and health, Fermentation 4 (2018) 90.

[5] M.L. Marco, D. Heeney, S. Binda, C.J. Cifelli, P.D. Cotter, B. Foligne, et al., Health benefits of fermented foods: microbiota and beyond, Current Opinion in Biotechnology 44 (2017) 94–102.

[6] C. Hill, F. Guarner, G. Reid, G.R. Gibson, D.J. Merenstein, B. Pot, et al., Expert consensus document: The international scientific association for probiotics and prebiotics consensus statement on the scope and appropriate use of the term probiotic, Nature Reviews Gastroenterology & Hepatology 11 (2014) 506–514.

[7] M.M. Natanzi, S.M.H. Ghaderian, Z. Khodaii, Iron absorption improvement: an additional health benefit for certain probiotics, in vitro and in vivo study, Acta Medica Mediterranea 33 (2017) 295–300.

[8] M. Azad, A. Kalam, M. Sarker, T. Li, J. Yin, Probiotic species in the modulation of gut microbiota: an overview, Biomed Research International (2018).

[9] T.R. Callaway, T.S. Edrington, R.C. Anderson, R.B. Harvey, K.J. Genovese, C.N. Kennedy, et al., Probiotics, prebiotics and competitive exclusion for prophylaxis against bacterial disease, Animal Health Research Reviews 9 (2008) 217.

[10] R. Nagpal, A. Kumar, M. Kumar, P.V. Behare, S. Jain, H. Yadav, Probiotics, their health benefits and applications for developing healthier foods: a review, FEMS Microbiology Letters 334 (2012) 1–15.

[11] Y. Shi, Y. Dong, W. Huang, D. Zhu, H. Mao, P. Su, Fecal microbiota transplantation for ulcerative colitis: a systematic review and meta-analysis, PLoS One 11 (2016) e0157259.

[12] J. Behnsen, E. Deriu, M. Sassone-Corsi, M. Raffatellu, Probiotics: properties, examples, and specific applications, Cold Spring Harbor Perspective in Medicine 3 (2013) a010074.

[13] N. Baliyan, K. Dindhoria, A. Kumar, A. Thakur, R. Kumar, Comprehensive substrate-based exploration of probiotics from undistilled traditional fermented alcoholic beverage 'Lugri', Frontiers in Microbiology 12 (2021).

[14] D.C. Donohue, Safety of probiotics, Asia Pacific Journal of Clinical Nutrition 15 (2006).

[15] E. Salvetti, S. Torriani, G.E. Felis, The genus Lactobacillus: a taxonomic update, Probiotics and Antimicroial Proteins 4 (2012) 217–226.

[16] K. Khalid, An overview of lactic acid bacteria, International Journal of Biosciences 1 (2011) 1–13.

[17] K. Papadimitriou, A. Alegria, P.A. Born, M. De Angelis, M. Gobbetti, M. Kleerebezem, et al., Stress physiology of lactic acid bacteria, Microbiology and Molecular Biology Reviews 80 (2016) 837–890.

[18] K. Todar, Bacterial Resistance to Antibiotics, Todar's Online Textbook of Bacteriology, 2011, p. 4.

[19] M. Briges, The classification of Lactobacilli by means of physiological tests, Microbiology 9 (1953) 234–248.

[20] M.I. Masood, M.I. Qadir, J.H. Shirazi, I.U. Khan, Beneficial effects of lactic acid bacteria on human beings, Critical Reviews in Microbiology 37 (2011) 91–98.

[21] M.E. Stiles, W.H. Holzapfel, Lactic acid bacteria of foods and their current taxonomy, International Journal of Food Microbiology 36 (1997) 1–29.

[22] G. Klein, A. Pack, C. Bonaparte, G. Reuter, Taxonomy and physiology of probiotic lactic acid bacteria, International Journal of Food Microbiology, 41, 1998, pp. 103–125.

[23] P. Vos, G. Garrity, D. Jones, N.R. Krieg, W. Ludwig, F.A. Rainey, et al., Bergey's Manual of Systematic Bacteriology: Volume 3: The Firmicutes, Springer Science & Business Media, 2011, p. 3.

[24] J. Zheng, S. Wittouck, E. Salvetti, C.M. Franz, H.M. Harris, P. Mattarelli, et al., A taxonomic note on the genus *Lactobacillus*: description of 23 novel genera, emended description of the genus *Lactobacillus beijerinck* 1901, and union of *Lactobacillaceae* and *Leuconostocaceae*, International Journal of Systematic and Evolutionary Microbiology 70 (2020) 2782–2858.

[25] A. Homayouni, P. Bastani, S. Ziyadi, S. Mohammad-Alizadeh-Charandabi, M. Ghalibaf, A.M. Mortazavian, et al., Effects of probiotics on the recurrence of bacterial vaginosis: a review, Journal of Lower Genital Tract Disease 18 (2014) 79–86.

[26] M. Ortiz-Lucas, A. Tobias, J.J. Sebastián, P. Saz, Effect of probiotic species on irritable bowel syndrome symptoms: a bring up to date meta-analysis, Revista Española de Enfermedades Digestiva 105 (2013) 19–36.

[27] Y.K. Lee, S. Salminen, Handbook of Probiotics and Prebiotics, 2009, pp. 123–139.

[28] A. García-Ruiz, D.G. de Llano, A. Esteban-Fernández, T. Requena, B. Bartolomé, M.V. Moreno-Arribas, Assessment of probiotic properties in lactic acid bacteria isolated from wine, Food Microbiology 44 (2014) 220–225.

[29] K.W. Lee, J.Y. Park, H.D. Sa, J.H. Jeong, D.E. Jin, H.J. Heo, et al., Probiotic properties of Pediococcus strains isolated from jeotgals, salted and fermented Korean sea-food, Anaerobe 28 (2014) 199–206.

[30] K. Damodharan, Y.S. Lee, S.A. Palaniyandi, S.H. Yang, J.W. Suh, Preliminary probiotic and technological characterization of Pediococcus pentosaceus strain KID7 and in vivo assessment of its cholesterol-lowering activity, Frontiers in Microbiology 6 (2015) 768.

[31] P.P. Balgir, B. Kaur, T. Kaur, A preliminary clinical evaluation of probiotics Pediococcus acidilactici MTCC5101 and Bacillus coagulans MTCC492 on young anemic women, International Journal of Fermented Foods 3 (2014) 45–59.

[32] B. Kaur, N. Garg, A. Sachdev, B. Kumar, Effect of the oral intake of probiotic Pediococcus acidilactici BA28 on Helicobacter pylori causing peptic ulcer in C57BL/6 mice models, Applied Biochemistry and Biotechnology 172 (2014) 973–983.

[33] K.H. Schleifer, R. Kilpper-Bälz, Transfer of Streptococcus faecalis and Streptococcus faecium to the genus Enterococcus nom. rev. as Enterococcus faecalis comb. nov. and Enterococcus faecium comb. nov, International Journal of Systematic and Evolutionary Microbiology 34 (1984) 31–34.

[34] K.H. Schleifer, R. Kilpper-Bälz, Molecular and chemotaxonomic approaches to the classification of Streptococci, Enterococci and Lactococci: a review, Systematic and Applied Microbiology 10 (1987) 1–19.

[35] P. Švec, C.M. Franz, The Family Enterococcaceae, Lactic Acid Bacteria: Biodiversity and Taxonomy, 2014, pp. 171–173.

[36] C.M. Franz, M. Huch, H. Abriouel, W. Holzapfel, A. Gálvez, Enterococci as probiotics and their implications in food safety, International Journal of Food Microbiology 151 (2011) 125–140.

[37] O. Uriot, S. Denis, M. Junjua, Y. Roussel, A. Dary-Mourot, S. Blanquet-Diot, Streptococcus thermophilus: From yogurt starter to a new promising probiotic candidate? Journal of Functional Foods 37 (2017) 74–89.

[38] S. Drouault, J. Anba, G. Corthier, Streptococcus thermophilus is able to produce a β-galactosidase active during its transit in the digestive tract of germ-free mice, Applied and Environmental Microbiology 68 (2002) 938–941.

[39] C. Rodríguez, M. Medici, A.V. Rodríguez, F. Mozzi, G.F. de Valdez, Prevention of chronic gastritis by fermented milks made with exopolysaccharide-producing Streptococcus thermophilus strains, Journal of Dairy Science 92 (2009) 2423–2434.

[40] J.M. Saavedra, N.A. Bauman, J.A. Perman, R.H. Yolken, I. Oung, Feeding of Bifidobacterium bifidum and Streptococcus thermophilus to infants in hospital for prevention of diarrhoea and shedding of rotavirus, The Lancet 344 (1994) 1046–1049.

[41] E.L. Garvie, Genus Leuconostoc van Tieghen 1878, 198AL emendmut. Ed. Char. Hucker and Pederson 1930, 66', Bergey's Manual of Systematic Bacteriology 2 (1986) 1071–1075.

[42] A. Endo, S. Okada, Reclassification of the genus Leuconostoc and proposals of Fructobacillus fructosus gen. nov., comb. nov., Fructobacillus durionis comb. nov., Fructobacillus ficulneus comb. nov. and Fructobacillus pseudoficulneus comb. nov, International Journal of Systematic and Evolutionary Microbiology 58 (2008) 2195–2205.

[43] J.P. Euzéby, List of bacterial names with standing in nomenclature: a folder available on the Internet, International Journal of Systematic and Evolutionary Microbiology 47 (1997) 590–592.

[44] H.J. Buckenhüskes, Selection criteria for lactic acid bacteria to be used as starter cultures for various food commodities, FEMS Microbiology Reviews 12 (1993) 253–271.

[45] K.H. Steinkraus, Fermentations in world food processing, Comprehensive Reviews in Food Science and Food Safety 1 (2002) 23–32.

[46] J.C. Ogier, E. Casalta, C. Farrokh, A. Saïhi, Safety assessment of dairy microorganisms: the Leuconostoc genus, International Journal of Food Microbiology 126 (2008) 286–290.

[47] N. Agarwal, D.N. Kamra, L.C. Chaudhary, A. Sahoo, N.N. Pathak, Selection of Saccharomyces cerevisiae strains for use as a microbial feed additive, Letters in Applied Microbiology 31 (2000) 270–273.

[48] J.L. Balcázar, I. De Blas, I. Ruiz Zarzuela, D. Vendrell, A.C. Calvo, I. Marquez, et al., Changes in intestinal microbiota and humoral immune response following probiotic administration in brown trout (Salmo trutta), British Journal of Nutrition 97 (2007) 522–527.

[49] Y.K. Nakamura, S.T. Omaye, Metabolic diseases and pro-and prebiotics: mechanistic insights, Nutrition & Metabolism 9 (1) (2012) 1–9.

[50] V. Fusco, G.M. Quero, G.S. Cho, J. Kabisch, D. Meske, H. Neve, et al., The genus Weissella: taxonomy, ecology and biotechnological potential, Frontiers in Microbiology 6 (2015) 155.

[51] M.D. Collins, J. Samelis, J. Metaxopoulos, S. Wallbanks, Taxonomic studies on some Leuconostoc-like organisms from fermented sausages: description of a new genus Weissella for the Leuconostoc paramesenteroides group of species, Journal of Applied Bacteriology 75 (1993) 595–603.

[52] J. Björkroth, L.M.T. Dicks, A. Endo, The genus Weissella in lactic acid bacteria: biodiversity and taxonomy, Wiley Online Library (2014) 417–428.

[53] J. Björkroth, W. Holzapfel, Genera Leuconostoc, Oenococcus and Weissella, The Prokaryotes 4 (2006) 267–319.

[54] J.M.R. Tingirikari, D. Kothari, R. Shukla, A. Goyal, Structural and biocompatibility properties of dextran from Weissella cibaria JAG8 as food additive, International Journal of Food Sciences and Nutrition 65 (2014) 686–691.

[55] M.S. Kang, J. Chung, S.M. Kim, K.H. Yang, J.S. Oh, Effect of Weissella cibaria isolates on the formation of Streptococcus mutans biofilm, Caries Research 40 (2006) 418–425.

[56] Y. Cai, Y. Benno, T. Nakase, T.K. Oh, Specific probiotic characterization of Weissella hellenica DS-12 isolated from flounder intestine, The Journal of General and Applied Microbiology 44 (1998) 311–316.

[57] Y.J. Moon, J.R. Soh, J.J. Yu, H.S. Sohn, Y.S. Cha, S.H. Oh, Intracellular lipid accumulation inhibitory effect of Weissella koreensis OK 1–6 isolated from kimchi on differentiating adipocyte, Journal of Applied Microbiology 113 (2012) 652–658.

[58] J.A. Park, P.B. Tirupathi Pichiah, J.J. Yu, S.H. Oh III, Y.S. Daily, Cha, Anti-obesity effect of kimchi fermented with Weissella koreensis OK 1–6 as starter in high-fat diet-induced obese C57 BL/6J mice, Journal of Applied Microbiology 113 (2012) 1507–1516.

[59] H. Abriouel, L.L. Lerma, M.D.C. Casado Munoz, B.P. Montoro, J. Kabisch, R. Pichner, et al., The controversial nature of the Weissella genus: technological and functional aspects vs whole genome analysis-based pathogenic potential for their application in food and health, Frontiers in Microbiology 6 (2015) 1197.

[60] K. Yadav, A. Bhardwaj, G. Kaur, R. Iyer, S. De, R.K. Malik, Potential of Lactococcus lactis as a probiotic and functional lactic acid bacteria in dairy industry, International Journal of Probiotics and Prebiotics 4 (2009) 219–228.

[61] W. Kim, The genus Lactococcus Lactic acid Bacteria, Biodiversity and Taxonomy, John Wiley Sons, Chichester, UK, 2014, pp. 429–443.

[62] M. Gobbetti, P. Ferranti, E. Smacchi, F. Goffredi, F. Addeo, Production of angiotensin-I-converting-enzyme-inhibitory peptides in fermented milks started by Lactobacillus delbrueckii subsp. bulgaricus SS1 and Lactococcus lactis subsp. cremoris FT4, Applied and Environmental Microbiology 66 (2000) 3898–3904.

[63] F.M. Elshaghabee, N. Rokana, R.D. Gulhane, C. Sharma, H. Panwar, Bacillus as potential probiotics: status, concerns, and future perspectives, Frontiers in Microbiology 8 (2017) 1490.

[64] T. Hosoi, K. Kiuchi, Production and probiotic effects of natto, Bacterial Spore Formers: Probiotics and Emerging Applications Eds. Ricca, E.; Henriques, A.O.; Cutting, S.M., Horizon Bioscience, (2004), pp. 143–153.

[65] T.V. Horosheva, V. Vodyanoy, I. Sorokulova, Efficacy of Bacillus probiotics in prevention of antibiotic-associated diarrhoea: a randomized, double-blind, placebo-controlled clinical trial, JMM Case Reports 1 (2014) e004036.

[66] L. Pacifico, J.F. Osborn, E. Bonci, S. Romaggioli, R. Baldini, C. Chiesa, Probiotics for the treatment of Helicobacter pylori infection in children, World Journal of Gastroenterology 20 (2014) 673.

[67] S.M. Cutting, Bacillus probiotics, Food Microbiology 28 (2011) 214–220.

[68] S. Hempel, S.J. Newberry, A.R. Maher, Z. Wang, J.N. Miles, R. Shanman, et al., Probiotics for the prevention and treatment of antibiotic-associated diarrhea: a systematic review and meta-analysis, JAMA 307 (2012) 1959–1969.

[69] E. Stackebrandt, B.J. Tindall, Appreciating microbial diversity: rediscovering the importance of isolation and characterization of microorganisms, Environmental Microbiology 2 (2000) 9–10.

[70] F. Turroni, C. Peano, D.A. Pass, E. Foroni, M. Severgnini, M.J. Claesson, et al., Diversity of bifidobacteria within the infant gut microbiota, PLoS One 7 (5) (2012) 36957.

[71] F. Turroni, S. Duranti, F. Bottacini, S. Guglielmetti, Van Sinderen, M. Ventura, Bifidobacterium bifidum as an example of a specialized human gut commensal, Frontiers in Microbiology 5 (2014) 437.

[72] V. Delcenserie, B. Taminiau, F. Gavini, M.A. De Schaetzen, I. Cleenwerck, M. Theves, et al., Detection and characterization of Bifidobacterium crudilactis and B. mongoliense able to grow during the manufacturing process of French raw milk cheeses, BMC Microbiology 13 (2013) 239.

[73] W.H. Holzapfel, P. Haberer, R. Geisen, J. Björkroth, U. Schillinger, Taxonomy and important features of probiotic microorganisms in food and nutrition, The American Journal of Clinical Nutrition 73 (2001) 365s–373s.

[74] I. Masneuf, J. Hansen, C. Groth, J. Piskur, D. Dubourdieu, New hybrids between Saccharomyces sensu stricto yeast species found among wine and cider production strains, Applied and Environmental Microbiology 64 (1998) 3887–3892.

[75] A. van der Aa Kühle, L. Jespersen, The taxonomic position of Saccharomyces boulardii as evaluated by sequence analysis of the D1/D2 domain of 26S rDNA, the ITS1–5.8 S rDNA-ITS2 region and the mitochondrial cytochrome-c oxidase II gene, Systematic and Applied Microbiology 26 (2003) 564–571.

[76] Z. Kurugöl, G. Koturoğlu, Effects of Saccharomyces boulardii in children with acute diarrhoea, Acta Paediatrica 94 (2005) 44–47.

[77] T. Kelesidis, C. Pothoulakis, Efficacy and safety of the probiotic Saccharomyces boulardii for the prevention and therapy of gastrointestinal disorders, Therapeutic Advances in Gastroenterology 5 (2012) 111–125.

[78] C.H. Choi, S.Y. Jo, H.J. Park, S.K. Chang, J.S. Byeon, S.J. Myung, A randomized, double-blind, placebo-controlled multicenter trial of Saccharomyces boulardii in irritable bowel syndrome: effect on quality of life, Journal of Clinical Gastroenterology 45 (2011) 679–683.

[79] H. Szajewska, A. Skórka, Saccharomyces boulardii for treating acute gastroenteritis in children: updated meta-analysis of randomized controlled trials, Alimentary Pharmacology & Therapeutics 30 (2009) 960–961.

[80] J.B. Thygesen, H. Glerup, B. Tarp, Saccharomyces boulardii fungemia caused by treatment with a probioticum, Case Reports, bcr0620114412 (2012).

[81] A.J. Marsh, C. Hill, R.P. Ross, P.D. Cotter, Fermented beverages with health-promoting potential: past and future perspectives, Trends in Food Science & Technology 38 (2014) 113–124.

[82] R. Levin, Probiotics-the road map, International Journal of Probiotics & Prebiotics 6 (2011) 133.

[83] L.F. Rettger, H.A. Cheplin, A Treatise on the Transformation of the Intestinal Flora: With Special Reference to the Implantation of Bacillus Acidophilus, 13, Yale University Press, 1921.

[84] R.W. Hutkins, Micro-organism and metabolism, in: Microbiology and Technology of Fermented Foods., Ed. Robert W. Hutkins (Ed.), 2228, John Wiley & Sons, (2008), Feb.

[85] M.J. Slashinski, S.A. McCurdy, L.S. Achenbaum, S.N. Whitney, A.L. McGuire, "Snake-oil,""quack medicine," and "industrially cultured organisms": biovalue and the commercialization of human microbiome research, BMC Medical Ethics 13 (2012) 28.

[86] S. Panahi, M.A. Fernandez, A. Marette, A. Tremblay, Yogurt, diet quality and lifestyle factors, European Journal of Clinical Nutrition 71 (2017) 573–579.

[87] S. Bansal, M. Mangal, S.K. Sharma, R.K. Gupta, Non-dairy based probiotics: a healthy treat for intestine, Critical Reviews in Food Science and Nutrition 56 (2016) 1856–1867.

[88] G. Macori, P.D. Cotter, Novel insights into the microbiology of fermented dairy foods, Current Opinion in Biotechnology 49 (2018) 172–178.

[89] M. Chavan, Y. Gat, M. Harmalkar, R. Waghmare, Development of non-dairy fermented probiotic drink based on germinated and ungerminated cereals and legume, LWT 91 (2018) 339–344.

[90] P. Kandylis, K. Pissaridi, A. Bekatorou, M. Kanellaki, A. Koutinas, Dairy and non-dairy probiotic beverages, Current Opinion in Food Science 7 (2016) 58–63.

[91] F. Akabanda, J. Owusu-Kwarteng, K. Tano-Debrah, R.L. Glover, D.S. Nielsen, L. Jespersen, Taxonomic and molecular characterization of lactic acid bacteria and yeasts in nunu, a Ghanaian fermented milk product, Food Microbiology 34 (2013) 277–283.

[92] M. Kahala, M. Mäki, A. Lehtovaara, J.M. Tapanainen, R. Katiska, M. Juuruskorpi, et al., Characterization of starter lactic acid bacteria from the Finnish fermented milk product viili, Journal of Applied Microbiology 105 (2008) 1929–1938.

[93] R. Wu, L. Wang, J. Wang, H. Li, B. Menghe, J. Wu, et al., Isolation and preliminary probiotic selection of lactobacilli from koumiss in Inner Mongolia, Journal of Basic Microbiology 49 (2009) 318–326.

[94] S. Dewan, J.P. Tamang, Microbial and analytical characterization of Chhu-A traditional fermented milk product of the Sikkim Himalayas, Journal of Scientific and Industrial Research 65 (2006) 747–752.

[95] J.P. Tamang, N. Tamang, S. Thapa, S. Dewan, B. Tamang, H. Yonzan, et al., Microorganisms and nutritional value of ethnic fermented foods and alcoholic beverages of North East India, Indian Journal of Traditional Knowledge 11 (2012) 7–25.

[96] A. Blandino, M.E. Al-Aseeri, S.S. Pandiella, D. Cantero, C. Webb, Cereal-based fermented foods and beverages, Food Research International 36 (2003) 527–543.

[97] K.A. Hachmeister, D.Y.C. Fung, Tempeh: a mold-modified indigenous fermented food made from soybeans and/or cereal grains, Critical Reviews in Microbiology 19 (1993) 137–188.

[98] R. Di Cagno, P. Filannino, M. Gobbetti, Novel fermented fruit and vegetable-based products, Novel Food Fermentation Technologies. Eds. R . Di Cagno, P. Filannino, M. Gobbetti. Springer International Publishing, ISBN: 9783319424552., Springer, (2016), pp. 279–291.

[99] N.R. Reddy, M.D. Pierson, S.K. Sathe, D.K. Salunkhe, L.R. Beuchat, Legume-based fermented foods: their preparation and nutritional quality, Critical Reviews in Food Science and Nutrition 17 (1983) 335–370.

[100] P.K. Sarkar, J.P. Tamang, P.E. Cook, J.D. Owens, Kinema—a traditional soybean fermented food: proximate composition and microflora, Food Microbiology 11 (1994) 47–55.

[101] K. Jeyaram, W.M. Singh, T. Premarani, A.R. Devi, K.S. Chanu, N.C. Talukdar, et al., Molecular identification of dominant microflora associated with 'Hawaijar'—a traditional fermented soybean (Glycine max (L.)) food of Manipur, India, International Journal of Food Microbiology 122 (2008) 259–268.

[102] R. Chettri, J.P. Tamang, Bacillus species isolated from tungrymbai and bekang, naturally fermented soybean foods of India, International Journal of Food Microbiology 197 (2015) 72–76.

[103] T.A. Singh, K.R. Devi, G. Ahmed, K. Jeyaram, Microbial and endogenous origin of fibrinolytic activity in traditional fermented foods of northeast India, Food Research International 55 (2014) 356–362.

[104] P.S. Panesar, S.S. Marwaha, Biotechnology in Agriculture and Food Processing: Opportunities and Challenges, CRC Press, 2013.

[105] R. Ekıncı, The effect of fermentation and drying on the water-soluble vitamin content of tarhana, a traditional Turkish cereal food, Food Chemistry 90 (2005) 127–132.

[106] M. Prakash, R. Ravi, A. Dattatreya, K.K. Bhat, Sensory profiling and positioning of jilebi samples by multivariate analysis, Journal of Food Quality 27 (2004) 418–427.

[107] N. Thakur, T.C. Bhalla, Characterization of some traditional fermented foods and beverages of Himachal Pradesh, Indian Journal of Traditional Knowledge 3 (2004) 325–335.

[108] K. Sharma, S. Attri, G. Goel, Selection and evaluation of probiotic and functional characteristics of autochthonous lactic acid bacteria isolated from fermented wheat flour dough babroo, Probiotics and Antimicrobial Proteins 11 (2019) 774–784.

[109] N. Ben Omar, F. Ampe, Microbial community dynamics during production of the Mexican fermented maize dough pozol, Applied and Environmental Microbiology 66 (2000) 3664–3673.

[110] J.K. Chavan, S.S. Kadam, L.R. Beuchat, Nutritional improvement of cereals by fermentation, Critical Reviews in Food Science and Nutrition 28 (1989) 349–400.

[111] C.V. Ramakrishnan, The use of fermented foods in India Symposium on Indigenous Fermented Foods, Bangkok, Thailand 1977.

[112] Savitri, T.C. Bhalla, Traditional foods and beverages of Himachal Pradesh, Indian Journal of Traditional Knowledge 6 (2007) 17–24.

[113] H. Yonzan, J.P. Tamang, Traditional processing of selroti—a cereal based ethnic fermented food of the Nepalis, Indian Journal of Traditional Knowledge 8 (2009) 110–114.

[114] J.P. Tamang, N. Thapa, B. Tamang, A. Rai, R. Chettri, Microorganisms in fermented foods and beverages, Health Benefits of Fermented Foods and New York, NY: CRC Press, Taylor and Francis Group., Ed. J.P. Tamang, (2015), pp. 1–110.

[115] N. Sharma, S. Handa, A. Gupta, Comprehensive study of different traditional fermented foods/beverages of Himachal Pradesh to evaluate their nutrition impact on health and rich biodiversity of fermenting microorganisms, International Journal of Research in Applied Natural and Social Science 1 (2013) 19–28.

[116] R. Satish Kumar, P. Kanmani, N. Yuvaraj, K.A. Paari, V. Pattukumar, V. Arul, Traditional Indian fermented foods: a rich source of lactic acid bacteria, International Journal of Food Sciences and Nutrition 64 (2013) 415–428.

[117] K.H. Steinkraus, Classification of fermented foods: worldwide review of household fermentation techniques, Food Control 8 (1997) 311–317.

[118] J.P. Tamang, Food culture of Sikkim. Sikkim Study Series, vol. 4, Information and Public Relations Department, Government of Sikkim, Gangtok, 2005.

[119] J.S. Lee, G.Y. Heo, J.W. Lee, Y.J. Oh, J.A. Park, Y.H. Park, et al., Analysis of kimchi microflora using denaturing gradient gel electrophoresis, International Journal of Food Microbiology 102 (2005) 143–150.

[120] P.M. Yan, W.T. Xue, S.S. Tan, H. Zhang, X.H. Chang, Effect of inoculating lactic acid bacteria starter cultures on the nitrite concentration of fermenting Chinese paocai, Food Control 19 (2008) 50–55.

[121] B. Viander, M. Mäki, A. Palva, Impact of low salt concentration, salt quality on natural large-scale sauerkraut fermentation, Food Microbiology 20 (2003) 391–395.

[122] Y.S. Chen, H.C. Wu, H.Y. Lo, W.C. Lin, W.H. Hsu, C.W. Lin, et al., Isolation and characterisation of lactic acid bacteria from jiang-gua (fermented cucumbers), a traditional fermented food in Taiwan, Journal of the Science of Food and Agriculture 92 (2012) 2069–2075.

[123] N.R. Dahal, T.B. Karki, B. Swamylingappa, Q. Li, G. Gu, Traditional foods and beverages of Nepal—a review, Food Reviews International 21 (2005) 1–25.

[124] A. Hurtado, C. Reguant, A. Bordons, N. Rozès, Lactic acid bacteria from fermented table olives, Food Microbiology 31 (2012) 1–8.

[125] J.P. Tamang, Himalayan Fermented Foods: Microbiology, Nutrition, and Ethnic Value, CRC Press, 2009.

[126] B. Tamang, J.P. Tamang, U. Schillinger, C.M. Franz, M. Gores, W.H. Holzapfel, Phenotypic and genotypic identification of lactic acid bacteria isolated from ethnic fermented bamboo tender shoots of North East India, International Journal of Food Microbiology 121 (2008) 35–40.

[127] B. Tamang, J.P. Tamang, Lactic acid bacteria isolated from indigenous fermented bamboo products of Arunachal Pradesh in India and their functionality, Food Biotechnology 23 (2009) 133–147.

[128] G. Dhavises, Microbial studies during the pickling of the shoot of bamboo, Bambusa arundinacea, Wild., and of pak sian, Gynandropsis pentaphylla, DC. DCMS Thesis Kasetsart University, Bangkok, 1972.

[129] C. Zhao, X. Zhao, Z. Lu, J. Huang, S. He, H. Tan, et al., Production of fermented pork jerky using Lactobacillus bulgaricus, LWT-Food Science and Technology 72 (2016) 377–382.

[130] M. Sidira, P. Kandylis, M. Kanellaki, Y. Kourkoutas, Effect of immobilized Lactobacillus casei on volatile compounds of heat treated probiotic dry-fermented sausages, Food Chemistry 178 (2015) 201–207.

[131] K. Oki, A.K. Rai, S. Sato, K. Watanabe, J.P. Tamang, Lactic acid bacteria isolated from ethnic preserved meat products of the western Himalayas, Food Microbiology 28 (2011) 1308–1315.

[132] J. Chakrabarty, G.D. Sharma, J.P. Tamang, Traditional technology and product characterization of some lesser-known ethnic fermented foods and beverages of North Cachar hills district of Assam, Indian Journal of Traditional Knowledge 13 (2014) 706–715.

[133] P. Saisithi, Traditional fermented fish: Fish sauce production, Fisheries Processing, Springer, 1994, pp. 111–131.

[134] M.S.D. Olympia, Fermented fish products in the Philippines, Application of Biotechnology to Traditional Fermented Foods 131 (1992).

[135] N. Thapa, J. Pal, J.P. Tamang, Microbial diversity in Ngari, Hentak and Tungtap, fermented fish products of North-East India, World Journal of Microbiology and Biotechnology 20 (2004) 599.

[136] N. Thapa, J. Pal, J.P. Tamang, Microbiological profile of dried fish products of Assam, Indian Journal of Fisheries 54 (2007) 121–125.

[137] A.U. Muzaddadi, Minimisation of fermentation period of shidal from barbs (Puntius spp.), Fishery Technology 52 (2015) 34–41.

[138] N. Thapa, J. Pal, J.P. Tamang, Phenotypic identification and technological properties of lactic acid bacteria isolated from traditionally processed fish products of the Eastern Himalayas, International Journal of Food Microbiology 107 (2006) 33–38.

[139] A.U. Muzaddadi, Minimisation of fermentation period of shidal from barbs (Puntius spp.), Fishery Technology 52 (2015) 34–41.

[140] N. Thapa, J. Pal, J.P. Tamang, Phenotypic identification and technological properties of lactic acid bacteria isolated from traditionally processed fish products of the Eastern Himalayas, International Journal of Food Microbiology 107 (2006) 33–38.

[141] N. Thapa, J. Pal, J.P. Tamang, Microbial diversity in ngari, hentak and tungtap, fermented fish products of North-East India, World Journal of Microbiology and Biotechnology 20 (2004) 599.

[142] W. Zhao, Y. Liu, M. Latta, W. Ma, Z. Wu, P. Chen, Probiotics database: a potential source of fermented foods, International Journal of Food Properties 22 (2019) 198–217.

[143] F. Leroy, L. De Vuyst, Fermented food in the context of a healthy diet: how to produce novel functional foods? Current Opinion in Clinical Nutrition & Metabolic Care 17 (2014) 574–581.

[144] D.T. Gordon, Nutraceuticals and functional foods-intestinal health through dietary fiber, prebiotics, and probiotics, Food Technology 56 (2002) 23.

[145] M.L. Clements, M.M. Levine, P.A. Ristaino, V.E. Daya, T.P. Hughes, Exogenous lactobacilli fed to man-their fate and ability to prevent diarrheal disease, Progress in Food & Nutrition Science 7 (1983) 29–37.

[146] E. Isolauri, T. Rautanen, M. Juntunen, P. Sillanaukee, T. Koivula, A human Lactobacillus strain (Lactobacillus casei sp strain GG) promotes recovery from acute diarrhea in children, Pediatrics 88 (1991) 90–97.

[147] L. Grześkowiak, E. Isolauri, S. Salminen, M. Gueimonde, Manufacturing process influences properties of probiotic bacteria, British Journal of Nutrition 105 (2011) 887–894.

[148] H. Meng, Y. Lee, Z. Ba, J. Peng, J. Lin, A.S. Boyer, et al., Consumption of Bifidobacterium animalis subsp. lactis BB-12 impacts upper respiratory tract infection and the function of NK and T cells in healthy adults, Molecular Nutrition & Food Research 60 (2016) 1161–1171.

[149] P.L. Leclerc, S.F. Gauthier, H. Bachelard, M. Santure, D. Roy, Antihypertensive activity of casein-enriched milk fermented by Lactobacillus helveticus, International Dairy Journal (2002) 995–1004.

[150] L. Dornblaser, Probiotics and prebiotics: what in the world is going on? Cereal Foods World 52 (2007) 20.

[151] L. Marquart, E.A. Cohen, Increasing whole grain consumption, Food Science and Nutrition 59 (2005) 24–32.

[152] Y. Hagiwara, M. Tetsuji, Y. Morimasa, Process for preparing food products containing a lactic acid bacteria-fermented product of a cereal germ, United States Patent 4,056,637, 1977.

[153] N. Molin, C.E. Albertsson, S. Bengmark, K. Larsson, A nutritional composition and method for the preparation thereof, 1993.

[154] P. Kontula, A. von Wright, T. Mattila-Sandholm, Oat bran β-gluco-and xylo-oligosaccharides as fermentative substrates for lactic acid bacteria, International Journal of Food Microbiology 45 (1998) 163–169.

[155] J. Jaskari, P. Kontula, A. Siitonen, H. Jousimies-Somer, T. Mattila-Sandholm, K. Poutanen, Oat β-glucan and xylan hydrolysates as selective substrates for Bifidobacterium and Lactobacillus strains, Applied Microbiology and Biotechnology 49 (1998) 175–181.

[156] R. Clemens, P. Pressman, Heyday in grain land, Food Technology 60 (2006) 18.

[157] C. Palmer, E.M. Bik, D.B. DiGiulio, D.A. Relman, P.O. Brown, Development of the human infant intestinal microbiota, PLoS Biology 5 (2007).

[158] R.I. Mackie, A. Sghir, H.R. Gaskins, Developmental microbial ecology of the neonatal gastrointestinal tract, The, American Journal of Clinical Nutrition 69 (1999) 1035s–1045s.

[159] F. Sommer, F. Bäckhed, The gut microbiota—masters of host development and physiology, Nature Reviews Microbiology 11 (2013) 227–238.

[160] H. Bachmann, M.J.C. Starrenburg, D. Molenaar, M. Kleerebezem, J.E.T. van Hylckama, Vlieg, Microbial domestication signatures of Lactococcus lactis can be reproduced by experimental evolution, Genome Research 22 (2012) 115–124.

[161] R.J. Siezen, J.E. Hylckama, Vlieg, Genomic diversity and versatility of Lactobacillus plantarum, a natural metabolic engineer, Microbial Cell Factories 10 (2011) 1–13.

[162] F. Bäckhed, J.K. Manchester, C.F. Semenkovich, J.I. Gordon, Mechanisms underlying the resistance to diet-induced obesity in germ-free mice, Proceedings of the National Academy of Sciences United States of America 104 (2007) 979–984.

[163] M. Derrien, J.E. van Hylckama, Vlieg, Fate, activity, and impact of ingested bacteria within the human gut microbiota, Trends in Microbiology 23 (2015) 354–366.

[164] P.M. Ryan, R.P. Ross, G.F. Fitzgerald, N.M. Caplice, C. Stanton, Sugar-coated: exopolysaccharide producing lactic acid bacteria for food and human health applications, Food and Function 6 (2015) 679–693.

[165] K.M. Ng, J.A. Ferreyra, S.K. Higginbottom, J.B. Lynch, P.C. Kashyap, S. Gopinath, et al., Microbiota-liberated host sugars facilitate post-antibiotic expansion of enteric pathogens, Nature 502 (2013) 96–99.

[166] K. Isawa, K. HoJo, N. YoDA, T. KAMiYAMA, S. Makino, M. Saito, et al., Isolation and identification of a new bifidogenic growth stimulator produced by Propionibacterium freudenreichii ET-3, Bioscience, Biotechnology, and Biochemistry 66 (2002) 679–681.

[167] E. Riboulet-Bisson, M.H. Sturme, I.B. Jeffery, M.M. O'Donnell, B.A. Neville, B.M. Forde, et al., Effect of Lactobacillus salivarius bacteriocin Abp118 on the mouse and pig intestinal microbiota, PLoS One 7 (2012).

[168] L.Y. Kwok, Z. Guo, J. Zhang, L. Wang, J. Qiao, Q. Hou, et al., The impact of oral consumption of Lactobacillus plantarum P-8 on faecal bacteria revealed by pyrosequencing, Beneficial Microbes 6 (2015) 405–413.

[169] C. Caballero-Franco, K. Keller, C. De Simone, K. Chadee, The VSL#3 probiotic formula induces mucin gene expression and secretion in colonic epithelial cells, American Journal of Physiology-Gastrointestinal and Liver Physiology 292 (2007) G315–G322.

[170] E.P. Culligan, C. Hill, R.D. Sleator, Probiotics and gastrointestinal disease: successes, problems and future prospects, Gut Pathogens 1 (2009) 19.

[171] T.L. Whyand, M.E. Caplin, Review of the evidence for the use of probiotics in gastrointestinal disorders, Journal of Gastroenterology, Pancreatology & Liver Disorders 1 (2014) 1–9.

[172] National health committee of the people's republic of china, national administration of traditional Chinese medicine. Diagnostic and therapeutic guidance for 2019 novel coronavirus disease (version 5). http://www.nhc.gov.cn/yzygj/s7653p/202002/d4b895337e19445f8d728fcaf1e3e13a/files/ab6bec7-f93e64e7f998d802991203cd6.pdf (accessed 15.04.20).

[173] R.P. Laursen, I. Hojsak, Probiotics for respiratory tract infections in children attending day care centers —a systematic review, European Journal of Pediatrics 177 (2018) 979–994.

[174] M. Quick, Cochrane commentary: probiotics for prevention of acute upper respiratory infection, Explore 11 (2015) 418–420.

[175] S. King, J. Glanville, M.E. Sanders, A. Fitzgerald, D. Varley, Effectiveness of probiotics on the duration of illness in healthy children and adults who develop common acute respiratory infectious conditions: a systematic review and meta-analysis, British Journal of Nutrition 112 (2014) 41–54.

[176] Q. Hao, B.R. Dong, T. Wu, Probiotics for preventing acute upper respiratory tract infections, Cochrane Database of Systematic Reviews (2015).

[177] K. Xu, H. Cai, Y. Shen, Q. Ni, Y. Chen, S. Hu, et al., Management of corona virus disease-19 (COVID-19): the Zhejiang experience, Journal of Zhejiang University (Medical Science) 49 (2020) 1.

[178] Global and regional diabetes prevalence estimates for 2019 and projections for 2030 and 2045: results from the International Diabetes Federation Diabetes Atlas, ninth ed., Diabetes Research and Clinical Practice. Available from: https://doi.org/10.1016/j.diabres.2019.107843.

[179] Global status report on non-communicable diseases. Available from: http://www.who.int/nmh/publications/ncd-status-report-2014/en/ 2014.

[180] N. Kobyliak, O. Virchenko, T. Falalyeyeva, Pathophysiological role of host microbiota in the development of obesity, Nutrition Journal 15 (2015) 43.

[181] F. Bäckhed, H. Ding, T. Wang, L.V. Hooper, G.Y. Koh, A. Nagy, et al., The gut microbiota as an environmental factor that regulates fat storage, Proceedings of the National Academy of Sciences United States of America 101 (2004) 15718–15723.

[182] C. Huttenhower, Structure, function and diversity of the healthy human microbiome, Nature 486 (2012) 207.

[183] P.J. Turnbaugh, R.E. Ley, M.A. Mahowald, V. Magrini, E.R. Mardis, J.I. Gordon, An obesity-associated gut microbiome with increased capacity for energy harvest, Nature 444 (2006) 1027.

[184] D. Raoult, Probiotics and obesity: a link? Nature Reviews Microbiology 7 (2009) 616.

[185] H.Y. Lee, J.H. Park, S.H. Seok, M.W. Baek, D.J. Kim, K.E. Lee, et al., Human originated bacteria, Lactobacillus rhamnosus PL60, produce conjugated linoleic acid and show anti-obesity effects in diet-induced obese mice, Biochimica et Biophysica Acta (BBA)-Molecular and Cell Biology of Lipids 1761 (2006) 736–744.

[186] S. Kishino, J. Ogawa, Y. Omura, K. Matsumura, S. Shimizu, Conjugated linoleic acid production from linoleic acid by lactic acid bacteria, Journal of the American Oil Chemists Society 79 (2002) 159–163.

[187] P. Saeedi, I. Petersohn, P. Salpea, B. Malanda, S. Karuranga, N. Unwin, et al., Global and regional diabetes prevalence estimates for 2019 and projections for 2030 and 2045: Results from the International Diabetes Federation Diabetes Atlas, Diabetes Research and Clinical Practice 157 (2019) 107843.

[188] K. Forslund, F. Hildebrand, T. Nielsen, G. Falony, E. Le Chatelier, S. Sunagawa, et al., Disentangling type 2 diabetes and metformin treatment signatures in the human gut microbiota, Nature 528 (2015) 262–266.

[189] L.A. David, C.F. Maurice, R.N. Carmody, D.B. Gootenberg, J.E. Button, B.E. Wolfe, et al., Diet rapidly and reproducibly alters the human gut microbiome, Nature 505 (2014) 559–563.

[190] E.D. Sonnenburg, J.L. Sonnenburg, Starving our microbial self: the deleterious consequences of a diet deficient in microbiota-accessible carbohydrates, Cell Metabolism 20 (2014) 779–786.

[191] L.F.G. Arango, H.L. Barrett, L.K. Callaway, H.D. McIntyre, M.D. Nitert, Probiotics in the prevention of gestational diabetes mellitus (GDM), Nutrition and Diet in Maternal Diabetes, Nutrition and Health, Humana Press, Cham, 2018, pp. 275–288.

[192] A. Chaudhary, D.K. Sharma, A. Arora, Prospects of Indian traditional fermented food as functional foods, Indian Journal of Agriculture Science 88 (2018) 1496–1501.

[193] A.K. Rai, R. Kumar, Potential of microbial bio-resources of Sikkim Himalayan region, ENVIS Bulletin 23 (2015) 99–105.

[194] S.S. Behera, A.F. El Sheikha, R. Hammami, A. Kumar, Traditionally fermented pickles: How the microbial diversity associated with their nutritional and health benefits? Journal of Functional Foods 70 (2020) 103971.

[195] S. Sarkar, Probiotics as functional foods: documented health benefits, Nutrition and Food Science (2013).

[196] S. Ilango, U. Antony, Assessment of the microbiological quality of koozh, a fermented millet beverage, African Journal of Microbiology Research 8 (2014) 308–312.

[197] D. Kothari, S. Patel, S.K. Kim, Probiotic supplements might not be universally-effective and safe: a review, Biomedicine Pharmacotherapy 111 (2019) 537–547.

[198] M. Yadav, P. Shukla, Efficient engineered probiotics using synthetic biology approaches: a review, Biotechnology Applied Biochemistry 67 (2020) 22–29.

[199] R. Kumar, U. Sood, V. Gupta, M. Singh, J. Scaria, R. Lal, Recent advancements in the development of modern probiotics for restoring human gut microbiome dysbiosis, Indian Journal of Microbiology 60 (2019) 12–25.

[200] K. Feng, R.M. Huang, R.Q. Wu, Y.S. Wei, M.H. Zong, R.J. Linhardt, et al., A novel route for double-layered encapsulation of probiotics with improved viability under adverse conditions, Food Chemistry 310 (2020) 125977.

[201] M. Yao, J. Xie, H. Du, D.J. McClements, H. Xiao, L. Li, Progress in microencapsulation of probiotics: a review, Comprehensive Reviews in Food Science and Food Safety 19 (2020) 857–874.

[202] S.B. Chauhan, V.S.R. Chauhan, Microencapsulation to enhance the storage stability of Lactobacillus rhamnosus GG, The Pharma Innovation Journal 8 (2019) 757–762.

[203] G. Sharma, S.H. Im, Probiotics as a potential immunomodulating pharmabiotics in allergic diseases: current status and future prospects, Allergy Asthma & Immunology Research 10 (2018) 575–590.

[204] C. De Simone, The unregulated probiotic market, Clinical Gastroenterology and Hepatology 17 (2019) 809–817.

[205] F. Sotoudegan, M. Daniali, S. Hassani, S. Nikfar, M. Abdollahi, Reappraisal of probiotics' safety in human, Food Chemistry Toxicology (2019).

[206] J. Suez, N. Zmora, E. Segal, E. Elinav, The pros, cons, and many unknowns of probiotics, Nature Medicine 25 (2019) 716–729.

[207] S. Raghuwanshi, S. Misra, R. Sharma, P. Bisen, Probiotics: nutritional therapeutic tool, Journal of Probiotics Health 6 (2018) 2.

[208] M. Zommiti, M.L. Chikindas, M. Ferchichi, Probiotics—live biotherapeutics: a story of success, limitations, and future prospects—not only for humans, Probiotics Antimicrobial Proteins (2019) 1–24.

[209] S.F. Mazhar, M. Afzal, A. Almatroudi, S. Munir, U.A. Ashfaq, M. Rasool, et al., The prospects for the therapeutic implications of genetically engineered probiotics, Journal of Food Quality (2020).

[210] P. Langella, F. Guarner, R. Martín, Next-generation probiotics: from commensal bacteria to novel drugs and food supplements, Frontiers in Microbiology 10 (2019) 1973.

[211] M.H. Saarela, Safety aspects of next generation probiotics, Current Opinion in Food Science 30 (2019) 8–13.

[212] M.L.Y. Wan, S.J. Forsythe, H. El-Nezami, Probiotics interaction with foodborne pathogens: a potential alternative to antibiotics and future challenges, Critical Review in Food Science and Nutrition 59 (2019) 3320–3333.

[213] C.P. Champagne, A.G. da Cruz, M. Daga, Strategies to improve the functionality of probiotics in supplements and foods, Current Opinion in Food Science 22 (2018) 160–166.

[214] A.Y. Tamime, R.K. Robinson, Tamime and Robinson's Yoghurt, Science and Technology, Elsevier, 2007.

[215] L. Quigley, O. O'Sullivan, T.P. Beresford, R.P. Ross, G.F. Fitzgerald, P.D. Cotter, Molecular approaches to analysing the microbial composition of raw milk and raw milk cheese, International Journal of Food Microbiology 150 (2011) 81–94.

5

Fructooligosaccharides production and the health benefits of prebiotics

Clarisse Nobre, Lívia S. Simões, Daniela A. Gonçalves, Paulo Berni, José A. Teixeira

CENTRE OF BIOLOGICAL ENGINEERING (CEB), UNIVERSITY OF MINHO, BRAGA, PORTUGAL

5.1 Introduction

The concept of prebiotics has been changing since 1995, when it was firstly introduced by Gibson and Roberfroid [1]. The most recent definition, published by the International Scientific Association for Probiotics and Prebiotics (ISAPP) in 2017, defines prebiotic as "a substrate that is selectively utilized by host microorganisms conferring a health benefit" [2]. The consensus definition does not specify the microorganisms, rather it requires a health-promoting bacterium with a selective metabolism. The concept of prebiotics was expanded and, although it remains mostly confined to nondigestible oligosaccharides, it also includes noncarbohydrate substances as prebiotic candidates, such as phenolics and phytochemicals, conjugated linoleic acid, and polyunsaturated fatty acid. The prebiotic application is now also expanded to other body sites than the gastrointestinal tract, such as the urogenital tract and upper gastrointestinal tract including the mouth and skin. The new definition applies to prebiotics not only for humans, but also for use by animals, since animal gut microbiota regulation, to prevent and treat disease, is also of high significance. Thus, other categories rather than food prebiotics were also included, such as feed and cosmetics. Finally, the new definition aims to bring out an adequate use of the term, not only for the scientific community, but also for regulatory agencies, the food industry, consumers, and health professionals, so that consistency and clarity may be achieved in reports, product labels, and regulatory oversight of the category [2,3].

The acceptance of a prebiotic largely varies with its level of evidence. Fructans, galactooligosaccharides (GOS), and lactulose are the most well studied and tested prebiotics. Other prebiotic candidates lack sufficient data confirming the health-promoting benefit in humans. Nevertheless, since compounds that do not act as prebiotics for humans may be effective prebiotics for animals, the *in vivo* tests have to be performed within the target host. Among the emergent prebiotics are the polyphenols, isomaltooligosaccharides (IMO), resistant starch, xylooligosaccharides (XOS), soybean oligosaccharides (SOS), maltooligosaccharides (MOS), glucooligosaccharides, lactosucrose, raffinose, and stachyose.

Although the prebiotic definition has been undergoing modifications over the years, considering more or fewer ingredients as prebiotics, fructans were always considered as prebiotics, meeting all the criteria and proven effects demonstrated both in animals and in human studies [4]. In this vein, the industrial production of fructooligosaccharides (FOS), both from inulin hydrolysis or enzymatic synthesis, has been of enormous interest for the scientific and industrial communities, as observed from the numerous research studies and patents on the subject [5]. On the other hand, most emergent prebiotics are not yet produced on a large industrial scale [6].

To establish a successful oligosaccharides production process, it is crucial to develop high-yield and high-content upstream processes which will minimize the costs associated with the downstream treatment. Therefore, this chapter critically explores emergent biotechnological processes being developed for production of FOS at high yield and content (Fig. 5–1). Fructans occurrence, chemical structure, technological properties, and market are also discussed. Finally, the role of prebiotics in control of human disease is discussed, emphasizing the most recent studies carried both with the well-established prebiotics FOS, GOS, and lactulose, and with emerging prebiotics.

5.2 Fructans occurrence, chemical structure, technological properties, and market

Fructans occur in many plants and vegetables. Among plants with higher fructans content are chicory, Jerusalem artichoke, garlic, and onion. Other examples include asparagus, wheat, bananas, rye, leeks, dahlia, carambola, june plum, and blue agave [7]. Structurally, fructans are carbohydrates constituted mostly by linked units of fructosyl–fructose monomers, in a β-configuration, which do not allow their metabolization by the human digestive enzymes. Instead, fructans reach the colon intact where they are metabolized specifically by bifidobacteria [4].

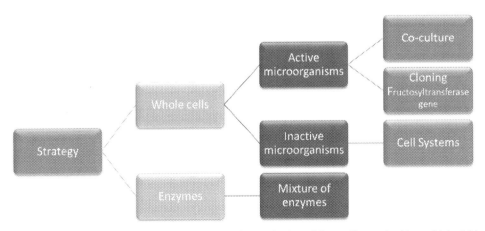

FIGURE 5–1 Overview of the biotechnological strategies for production of fructooligosaccharides at high yield and content.

According to the type of these glycosidic linkages, fructans may be classified into three main types: inulin-type, levan-type, and branched group ("inulin neo-series"). The inulin-type are mainly, if not exclusive, constituted by (2→1) fructosyl–fructose linkages. The levan-type are constituted mainly, or exclusively, by (2→6) fructosyl–fructose linkages, and the branched group neo-series contains both (2→1) and (2→6) fructosyl–fructose linkages (e.g., graminan). In the neo-series, one linear chain is linked to the carbon 1 [C1] of the fructose residue (as in the case of inulin-type), and the other branch is linked to the carbon 6 [C6] of the glucose residue [8].

Industrially, inulin is mainly extracted from chicory roots. The degree of polymerization (DP) of chicory inulin varies from 2 to ~60 units. The partial hydrolysis of inulin by endo-inulinase enzymes produces oligofructose, which are short-chain molecules (DP from 2 to 7 units, $DP_{av} = 4$), including a mixture of $G_{py}F_n$ [glucopyranosyl-(fructofuransoyl)$_n$-fructose] and $F_{py}F_n$ [fructopyranosyl-(fructofuranosyl)$_n$-fructose], such as inulobiose (F_2), inulotriose (F_3), and inulotetraose (F_4) [9].

Since most plants contain low amounts of fructans and their harvest is season limited, they have been alternatively produced by microbial sources in higher yield and cost-effective processes. Fructans are enzymatically synthesized by fructosyltransferases (β-fructofuranosidase, EC 3.2.1.26 or β-D-fructosyltransferase, EC 2.4.1.9) through transfructosilation of sucrose. Fructans produced by fructosyltransferases are known as FOS. FOS synthesized with commercial fructosyltransferase have been mainly assigned as GF_n inulin-type (2 < DP < 4, $DP_{av} = 3.6$), namely 1-kestose (GF_2), nystose (GF_3), and 1^F-fructofuranosylnystose (GF_4). Since FOS derive from sucrose, a starting α-D-glucose moiety is present in most of the synthesized FOS. Although FF_n inulin-type, levan-type (e.g., 6-kestose and 6-nystose), neo-series (e.g., neo-kestose and neo-nystose), and the sucrose isomer blastose [Fru-β(2→6)-Glc] have also been identified in mixtures obtained by enzymatic synthesis, they are available in much smaller or trace amounts [10,11].

Fructans are found in more than 500 food products and beverages commercially available. They are mainly used as bulking ingredients in dairy and bakery products, but also in fruit preparations and juices, toppings, fiber-milk, ice-creams, jelly sweets, fruit gums, toffees, chocolates, cereal-based products, meat, dietary supplements, and infant foods.

FOS have been used as low-caloric sugar substitutes in dietetic food and dietary supplements, food aimed at diabetics, targeting infants, pregnant, lactating women, and the elderly, among other vulnerable consumers [7,12]. Additionally, FOS have been used as ingredients in food formulations to modify the texture or creaminess, for instance, as an emulsifier in margarine and as an additive in ice-cream to avoid the formation of ice crystals [5]. Fructans are also currently marketed as feed additives for poultry in Japan and the United States [8]. Fructans incorporation in food greatly improves the technological and nutritional properties of the final products, namely the organoleptic characteristics, product shelf-life due to the high moisture-retaining capacity, without increasing water activity, reduction of browning due to Maillard reactions, and enable fiber incorporation within liquid foods. FOS is 0.3–0.6 times as sweet as sucrose [13].

Other applications of FOS include incorporation into: (1) edible starch films—resulting in films with higher solubility and elongation, and decreased water vapor permeability (e.g., FOS exerted a plasticizing effect on films, decreasing their glass transition temperature) [14]; (2) carriers for delivery of active compounds, or texture modifiers—FOS addition in the biopolymeric system resulted in the formation of stronger gels, with smaller pores, smaller bead sizes, more cohesive structures, and higher viscosity [15]; and (3) microcapsules with probiotics—promoting the survival of probiotic cells along the gastrointestinal tract and reducing the loss in cell viability [16].

The global market for prebiotics was valued at US$3.4 billion in 2018, with a growth expectancy of US$8.34 billion in 2026, at a 10.1% compound annual growth rate [17]. The European prebiotic market size was valued at over US$250 million in 2019, reaching over US$1.9 billion, with an expectancy to increase above 9.5% by 2026 [18]. The largest consumers are Japan, the United States, Europe, China, Canada, Brazil, and India [5].

The main FOS manufacturers, obtained from sucrose conversion, include the Beghin-Meiji, subsidiary of the Tereos Group, France (Actilight and Profeed); Cheil Foods and Chemicals Inc., Korea (Oligo-sugar); GTC Nutrition, the United States (NutraFlora); Meiji Seika Kaisha Ltd., Japan (Meioligo); and Victory Biology Engineering. Ltd., China (Prebiovis ScFOS) [7]. Commercially available FOS are being produced through fructosyltransferases extracted from *Aureobasidium pullulans* and *Aspergillus niger*. Fructans derived from chicory are mainly produced in the Netherlands by Sensus (Frutalose) and Belgium by Beneo-Orafti (Raftilose) and Cosucra (Fibrulose) [7].

5.3 Fructooligosaccharides production

Numerous microorganisms have been reported as fructosyltransferase producers. The most studied ones are fungi, predominantly belonging to the *Aspergillus*, *Aureobasidium*, and *Penicillium* genera [12,19,20]. Systems using purified enzymes have been mainly used to synthesize FOS. The process is run in a two-step process, the first one being a fermentation step for microorganism growth and enzyme production, followed by purification of the enzyme, and the second one for FOS synthesis with the purified enzymes. FOS have been produced either by submerged fermentation (SmF) or solid-state fermentation (SSF) when other than pure sucrose solutions are used as substrate source.

Many efforts have been made to use agro-industrial wastes as sucrose substitutes, toward the development of more sustainable processes, aligned with the principles of a circular bioeconomy. Among the agro-industrial wastes being exploited are sucrose-rich solutions (sugar cane molasses, beet molasses, agave sirups, spent osmotic solutions) fruit peels, leaves and pomaces (e.g., from mango, banana, pineapple, orange, and date), bagasse (e.g., from sugar cane, agave, corn, coconut, cassava), and coffee processing by-products [21].

The main drawback while producing FOS using the enzymatic method is the low yields and purity degree achieved. Processes using pure enzymes generally do not attain more than 50%–60% conversion of sucrose into FOS, with FOS purities around 50% (w/w) [7]. During the synthesis, a large amount of glucose is released, which not only induces the enzyme activity inhibition, leading to low sucrose conversion yields, but also results in more than 10% unreacted sucrose [22,23]. Also, the hydrolytic activity of the same enzymes ultimately hydrolyzes back the produced FOS to the single monomer forms fructose and glucose. Since fructose, glucose, and sucrose are nonprebiotic sugars, the inclusion of the produced mixtures into diabetic and dietetic food needs a prior purification step, to achieve values between 90% and 100% of pure FOS.

Downstream treatments have been suggested to purify the FOS mixtures produced, namely, membrane techniques using advanced ultra- and nanofiltration systems [24,25], activated charcoal [26], ion exchange and simulated moving bed (SMB) chromatography [27–30], and microbial treatment using successive fermentations. Microbial treatment is applied in the FOS mixture produced, in a second fermentation. Glucose removal microorganisms such as *Zymomonas mobilis*, *Bacillus coagulans*, *Saccharomyces cerevisiae*, *Pichia heimii*, and *Pichia pastoris* have been used [31–35].

Industrially, FOS are mainly purified by SMB and only a few commercial products exhibit FOS contents above 90% [6,31,36,37]. Downstream processes such as ultrafiltration, adsorption onto activated charcoal, and evaporation are further applied to remove the enzymes, remove color, and concentrate the sugars, respectively [6]. There have been several efforts for solutions envisaging the increase in purity of the FOS achieved during the production process itself, avoiding a few costly downstream steps. In addition, if glucose is removed during the FOS reaction process, an increase in the reaction rate and a high yield of FOS can be obtained (once glucose inhibits the reaction).

Since the large-scale production of FOS remains economically unfeasible due to the high cost of purified enzymes and low FOS yields, in the next section, we present and discuss the current state of the art on biotechnological processes being developed for achievement of FOS at high yield and high content during the production process itself. The following strategies will be discussed: use of (1) whole cells of active microorganisms, including microorganism's immobilization, (2) whole cells of inactive microorganisms, (3) mixture of enzymes in the same fermentation, (4) coculture of microorganisms, and (5) enzyme engineering. Table 5–1 summarizes the most relevant works using the different approaches, as well as the FOS content, yield, productivity, and purity achieved.

5.3.1 Whole cells of active microorganisms

The use of whole cells of active microorganisms instead of the purified enzymes avoids both the first fermentation for enzyme production, and also the enzyme purification step. Fermentations run with microorganisms, under optimized conditions, are able to produce at the same time the enzymes necessary for the biotransformation and the FOS itself. This strategy has the advantage of FOS being produced in a single bioprocess. The use of solid catalysts, such as the whole cells, lends itself ideally to continuous processing [6].

The main variables affecting enzymes and FOS production are related to the composition of the fermentative broth, namely the substrate and nitrogen sources and amount, the addition of different mineral salts, and supplementation with some amino acids; and the operational parameters such as temperature, pH, aeration rates, agitation, and fermentation time.

Fermentations with microorganisms are characterized by lower initial sucrose concentrations, generally ranging from 100 to 250 g/L and by the lower temperatures ranging between

Table 5-1 Fructooligosaccharides content, yield, productivity, and purity achieved using different production strategies.

Strategy		Microorganisms/enzymes	Content (g_{FOS}/L)	Yield (w/w)	Productivity (g_{FOS}/L/h)	Purity % (w/w)	References
Whole cells	Active microorganisms	Aureobasidium pullulans		0.64 ± 0.0			[38]
		A. pullulans	118 ± 1.6	0.63 ± 0.03	4.8 ± 1.4	54.0 ± 1.6	[31]
		Aspergillus ibericus MUM 03.49	118 ± 4	0.64 ± 0.02	3.1 ± 0.1	56 ± 3	[39]
		Penicillium citreonigrum URM 4459	126.3 ± 0.1	0.65 ± 0.06	2.28 ± 0.08	61 ± 0	[40]
		Penicillium expansum MUM 02.14	117.7	0.58	3.25		[41]
		Aspergillus japonicas ATCC 20236	116.3	0.69	4.84		[42]
		A. pullulans	126.5 ± 9.6	0.61 ± 0.08	3.51 ± 0.93	49.5 ± 0.5	[43]
		A. pullulans	108.17 ± 8.83	0.52 ± 0.05	4.33 ± 1.53	43.2 ± 0.4	[43]
		A. japonicas ATCC 20236	138.73 ± 2.03	0.66 ± 0.01	6.61 ± 0.09		[44]
		P. expansum	208.8	0.87	5.41		[45]
		A. japonicas ATCC 20236	371.89	0.87	10.44	53.1	[46]
		Aspergillus sp. N74	314.60			51.11	[47]
	Inactive microorganisms	Aspergillus niger CGMCC 6640	155	0.38			[48]
		Microbacterium paraoxydans	339	0.56			[49]
		Cladosporium cladosporioides (CF_2 15)	184	0.310			[11]
		Penicillium sizovae (CK1)					[11]
Enzymes		Glucose oxidase (from A. niger) and fructosyltransferase (from A. pullulans)				90.5	[50]
		Glucose oxidase and β-fructofuranosidase (from A. pullulans KFCC)				98.32	[51]
		Glucose oxidase and β-fructofuranosidase (from A. japonicus CCRC 93007 or A. niger ATCC 20611)				> 90.0	[52]
		Glucose isomerase and β-fructofuranosidase (from A. pullulans DSM 2404)	276			69.0	[53]
		Fructosyltransferase and glycosidase (from Penicillium rugulosum)	650			83.8	[54]
		A. japonicus CCRC 93007 or A. pullulans ATCC 9348 and Gluconobacter oxydans ATCC 23771			> 160	83–87	[55]
Coculture		A. pullulans and Saccharomyces cerevisiae	32.9 ± 8.8			60.9 ± 0.4	[31]
		A. pullulans CCY 27-1-94 and S. cerevisiae 11982	119	0.59	5.9	67.0	[37]
		A. ibericus MUM 03.49 and S. cerevisiae YIL162W	137.7 ± 0.1	0.70		93.8 ± 0.7	[56]
		A. japonicus TU 21 and Pichia heimi BCRC 20410	186	0.62		98.2	[34]
Cell systems		β-Fructofuranosidase [FopA (A178P)] overexpressed in A. niger ATCC 20611				> 59.0	[57]
		β-Fructofuranosidase1 gene overexpressed in Aureobasidium melanogenum		0.66			[58]
		Fructosyltransferase from Aspergillus oryzae displayed on the cell surface of Yarrowia lipolytica	155			70.5	[59]
Enzyme engineering		A. melanogenum with the CREA gene encoding glucose repressor in the β-fructofuranosidase disrupted		0.58			[60]
		Y. lipolytica overexpressing an endo-inulinase gene from A. niger	546.6	0.91	15.18	90.3	[61]
		Use of an inulin-binding module (IBM) from a cycloinulinooligosaccharide fructanotransferase of Bacillus macerans CFC1 fused into either N- or C-terminal of an endo-inulinase from A. niger	717.3	0.912	358.6	91.4	[62]
		S. cerevisiae strain with the heterologous endo-inulinase gene expressed and the inherent invertase gene SUC2 disrupted	180.2 ± 0.8	0.9	7.51 ± 0.03		[63]

25°C and 40°C. Reactions conducted with the pure enzymes or inactivated microorganisms may use up to 770 g/L sucrose and temperatures between 55°C and 60°C [5]. The high sugar concentration may cause osmotic stress and suppress microorganism growth. On the other hand, in fermentations run at concentrations below 100 g/L a large amount of sucrose is used for growth of microbial cells instead of FOS synthesis [20]. The use of low temperatures may contribute to an improved energetic efficiency of the process.

Both organic and inorganic sources of nitrogen have been applied in FOS fermentation, such as yeast extract, corn steep liquor, $NaNO_3$, and NH_4NO_3 [64]. Fermentations are generally run under acidic conditions. Agitation influences oxygen transfer and induces shear stress and abrasive forces in the cells, which can influence fungal morphology. The agitation of 200 rpm in a stirred-tank reactor is suitable for the formation of rounded pellets, while higher agitations may damage the pellets, contributing to free filamentous mycelia and reseed cells growth, which ultimately can result in an increase in FOS production by the enzymes being produced by the highest biomass level [39].

The use of *A. pullulans* for FOS production via sucrose fermentation has been studied using a one-stage method. The fermentation process with the whole cells of *A. pullulans* was firstly developed and optimized using experimental design tools. The optimum fermentation conditions achieved were 32°C and 385 rpm. Starting with 200 g/L of sucrose, a yield of 64.1 $g_{FOS}/g_{initial\ Sucrose}$ was obtained [38]. The same strain was then tested at reduced amounts of salts to improve the efficiency of the subsequent downstream purification steps, involving ion-exchange resins, such as SMB chromatography, and to reduce the associated costs. Similar fermentation yields were obtained with the optimized culture media (0.63 ± 0.03 $g_{FOS}/g_{initial\ sucrose}$) [31].

An *Aspergillus ibericus* MUM 03.49 isolated from Portuguese wine grapes was also used to produce FOS in one-step fermentation. A yield of 0.64 ± 0.02 $g_{FOS}/g_{initial\ Sucrose}$ was obtained in a 2-L bioreactor, at 38 hours, with a content of 118 ± 4 g/L in FOS and a purity of $56\% \pm 3\%$. The optimum fermentation conditions were 37.0°C and a pH of 6.2, starting from 200 g/L of sucrose. Interestingly, after analysis by gas chromatography-mass spectrometry, besides GF_2, GF_3, and GF_4, it was also identified [Fru(2→6)Glc] (possibly blastose) and a reducing trisaccharide (possibly [Fru(β2→6)Glc(α1↔β2)Fru],neo-kestose) in the obtained mixture [10].

Penicillium strains have also been used to produce FOS using the whole cells of the active fungi. A strain of *P. citreonigrum* URM 4459 produced high yield, content, productivity, and purity of FOS, namely 0.65 ± 0.06 $g_{FOS}/g_{initial\ Sucrose}$, 126.3 ± 0.1 g/L, 2.28 ± 0.08 $g_{FOS}/L/h$, and $61\% \pm 0\%$, respectively. For this strain, a high amount of a reducing trisaccharide, possibly neo-kestose, was identified in the FOS produced mixture [40]. *P. expansum* MUM 02.14 achieved a FOS yield of 0.58 $g_{FOS}/g_{initial\ sucrose}$ [41].

In general, the use of microorganisms instead of pure enzymes seems to attain high FOS yields, resulting in faster and more economical bioprocesses. Targeting to improve the above-mentioned FOS production processes, the whole cells of active microorganisms have been immobilized. The immobilization of microorganisms or enzymes either by entrapment or binding into carriers may introduce many technological advantages to the FOS production process. By immobilization, the thermal, chemical, and mechanical resistance of the microorganism is improved. It also allows the reuse of the biomass or enzymes in repeated production cycles, by enabling downstream separation and recovery from liquid mixtures, and avoids biomass washout in the continuous mode. Process inhibition may also be reduced, such as inhibition by the high amount of glucose released in the mixture. All these features may have a high economic impact in the overall costs of the process.

Among the different methods for cell immobilization, entrapment in gels, cross-linking, and the use of supporting carriers (e.g., synthetic materials, minerals, and agro-industrial wastes) are the most reported. Some carriers may even be used as substrate sources under SSF processes [65]. The main factors affecting the immobilization mechanism are the microorganism itself and the porosity, the water adsorption index (WAI), and the critical humidity point (CHP) of the carrier used. Several authors have reported that high porosity, high WAI, and low CHP increases the amount of immobilized cells [42,43].

A. pullulans was immobilized in several types of carriers, such as: (1) synthetic materials—reticulated polyurethane foam, vegetal scourer, polyester fiber scourer, light expanded clay aggregate, polyester staple fiber, and fiber glass wool; (2) agro-industrial by-products—mandarin peel, banana peel, walnut shell, almond shell, pistachio shell, and chestnut shell; and (3) and mineral carriers—zeolite and pumice stone. The immobilization of cells was shown to improve the FOS bioprocess. The synthetic reticulated polyurethane foam and the agro-industrial walnut shell achieved the best results. As compared to fermentations run with free cells, reticulated polyurethane foam increased FOS concentration, purity degree, and yield by 15%, 8%, and 12% (w/w), respectively, and immobilization in walnut shell resulted in values increasing by 27%, 10%, and 25% (w/w). Interestingly, although one of the lowest immobilization efficiencies were obtained with walnut shell [immobilizing less than 20% (w/w) of the total cells], this material achieved the highest FOS yields, which may be associated with the high cell growth obtained using this carrier [43].

A. japonicus ATCC 20236 was also immobilized in vegetal fiber [42] and in several lignocellulosic materials, including brewer's spent grain, wheat straw, corn cobs, coffee husks, cork oak, and loofa sponge [44]. For this microorganism, immobilization with the synthetic material induced a higher yield of produced FOS (0.69 $g_{FOS}/g_{initial.sucrose}$) than with agro-industrial by-products (0.66 $g_{FOS}/g_{initial.sucrose}$ obtained with corn cobs). Outstanding results were obtained from repeated batch fermentation with *P. expansum* immobilized into synthetic fiber. In the three initial fermentation cycles, yields of 87%, 72%, and 44% of FOS were obtained, while fructofuranosidase activity was kept constant up to six cycles [45].

SmF has been claimed to be more effective compared to SSF for fructosyltransferase and FOS production [66]. However, a SSF work conducted with coffee silver skin as solid matrix and *A. japonicus* ATCC 20236 obtained a remarkable FOS yield of 87 $g_{FOS}/g_{initial.sucrose}$ from 240 g/L of initial sucrose [46]. Nevertheless, SSF scale-up remains challenging due to the existing gradients inside the reactor in temperature, pH, moisture, oxygen, and inoculum [67].

5.3.2 Whole cells of inactive microorganisms

The immobilization of the inactive cells has similar advantages to those previously described for active microorganisms, for example, higher resistance to shear forces, straightforward enzyme/product separation, and easier enzyme recovery and reuse. As compared to the use of pure enzymes it provides several technological advantages to the bioprocess, since: (1) it does not require enzyme purification from microbe cells; (2) higher productivity efficiency can be obtained, because the enzymes are protected in native cell bags without the need for carriers for immobilization, and sucrose can easily pass through the cell membrane to react with the enzyme with the FOS being quickly released; and (3) cells can be reused easily without the use of expensive enzyme immobilization process [48]. To facilitate sucrose access to the enzymes, the cells may be permeabilized. Dry *A. niger* CGMCC 6640 mycelia containing fructofuranosidases as a whole-cell catalyst was used to produce FOS. The fermentation run at pH 7.0°C and 33°C, using a content of 60 g/L whole cell and 600 g/L of initial sucrose, attained an amount of 314.60 g/L of FOS with 51.11% purity, at 40 hours fermentation [48]. *Microbacterium paraoxydans* cells were sucrose-induced before reacting with 40% (w/v) sucrose. A maximum FOS amount of 155 g/L with 0.38 $g_{FOS}/g_{initial\ sucrose}$ yield was attained by this microorganism synthesizing mainly 1-kestose and nystose [49].

Higher enzyme activity was determined for *Cladosporium cladosporioides* (CF_2 15) mycelium-bound enzymes, as compared to extracellular enzymes from the crude extract. Starting from 600 g/L of sucrose, a maximum FOS amount of 339 g/L [yield of 56% (w/w), after 72 hours] was obtained with the lyophilized mycelium-bound cells. The main FOS in the reaction mixtures were identified as 1-kestose (158 g/L), nystose (97 g/L), 1^F-fructosyl-nystose (19 g/L), 6-kestose (12 g/L), neo-kestose (10 g/L), and a disaccharide (34 g/L) that after its purification and nuclear magnetic resonance spectroscopy analysis was identified as blastose [Fru-$\beta(2\rightarrow6)$-Glc] [11]. Entrapment of mycelial *A. oryzae* IPT-301 glutaraldehyde-cross-linked was shown to increase the transfructosylation activity by about 70% when compared to the free cells not glutaraldehyde-cross-linked [68].

Finally, mathematical modeling of the reaction mechanisms involved in the FOS production process has been proposed. Several models have been developed envisaging the maximization of productivity, yields, and purity of the FOS achieved, assessing different operational modes, such as batch and fed-batch [69–74].

5.3.3 Mixed enzymes and cell systems

FOS have been simultaneously produced and purified by a combination of the action of β-fructofuranosidase from a variety of microorganisms (FOS producers) with other enzymes able to convert glucose or fructose into other metabolites (small saccharides removers). To remove the large amount of glucose released by fructosyltransferase during fermentation glucose oxidase has been applied, which is able to convert glucose into gluconic acid. This mechanism requires sufficient oxygen, as shown in the following equation:

$$\text{Glucose} + O_2 \rightarrow \text{Gluconic acid} + H_2O_2$$

The application of glucose oxidase with fructosyltransferase from *A. pullulans* [50] or mycelia of *A. japonicus* CCRC 93007 [52] concentrated FOS up to 90% (w/w) in a single fermentation. The system was shown to accumulate a higher content of GF_3 and produce GF_4

in trace amounts as compared to that produced by fructosyltransferases/furanosidases [50]. Envisaging the transference of the process to an industrial scale, mycelia from *A. japonicus* TIT-KJ1 with *A. niger* ATCC 16888 (with glucose oxidase activity) were separately entrapped in calcium alginate. The reaction was run in an airlift bioreactor to enhance oxygen transfer, without damaging the beads. The FOS mass fraction reached up to 90% (w/w) and the initial rate of transfructosylation was increased almost twofold without supplying glucose oxidase. To remove the produced gluconic acid, either cationic ion exchange resins or precipitation with a slurry of $CaCO_3$ into calcium gluconate has been applied [50,52].

Other enzymes such as glucose isomerase, involved in the isomerization of glucose to fructose, were also evaluated to remove glucose. Although the results had shown that the system was not useful, due to its equilibrium characteristics [50], the addition of commercial glucose isomerase to a crude enzyme β-fructofuranosidase from *A. pullulans* DSM 2404 was able to increase the FOS yield from 62% to 69% (w/w) [53]. Interestingly, crude extract of *Penicillium rugulosum* demonstrated that the crude enzyme acted as a mixed enzyme system of fructosyltransferase and glycosidase. The mixed enzyme converted 775 g/L of sucrose into 650 g/L of FOS in 10 hours, at 55°C and pH 5.5, with a purity of 83.8% (w/w) [54].

Another cell system, consisting of the mycelia of *A. japonicus* CCRC 93007 or *A. pullulans* ATCC 9348 and *Gluconobacter oxydans* ATCC 23771, with glucose dehydrogenase activity, was reported to produce FOS. A bioreactor, equipped with a microfiltration module to discharge the FOS produced, was run in continuous mode. The system produced more than 80% FOS on a dry weight basis [55].

5.3.4 Coculture of microorganisms

The use of coculture of microorganisms, that is, one microorganism able to synthesize FOS and another able to consume nonprebiotic sugars, allows the integration of the production and purification processes in a single step, avoiding downstream FOS purification steps. The integration process design allows effective conversion of sucrose into FOS, sustainable process management, reduces energy-resource consumption and waste formation, and increases economic efficiency [6].

Besides all the advantages of using whole cells of active microorganisms, this strategy allows the removal of glucose during the FOS production process itself, which results in a decrease in enzyme inhibition by this sugar. Thus, FOS may be obtained not only with higher purity but also in higher content as compared to fermentations using only the FOS producer strain. Still, mixtures of FOS obtained from active microorganisms contain a high salts content that can influence the composition of the final products. Thus, a demineralization step might be necessary to prevent undesired organoleptic characteristics.

The integrated production and purification process using cocultures was explored using a coculture of *A. pullulans* (FOS producer strain) with *S. cerevisiae*. The *S. cerevisiae* can be conveniently used to remove mono- and disaccharides from the mixtures obtained from the enzymatic production of FOS without hydrolyzing β-linked saccharides, such as FOS. Nevertheless, in coculture strategy, *S. cerevisiae* was shown to compete with *A. pullulans* by the substrate leading to lower final fermentation yields as compared to monoculture. The purity of FOS in the mixture increased only 5% (up to 58%, w/w) [31]. The influence of cell immobilization and the time of the inoculation of the *S. cerevisiae* was further studied. The fermentation run with *A. pullulans* (immobilized in reticulated polyurethane foam) while inoculated at 10-hour fermentation with *S. cerevisiae* (encapsulated in calcium alginate beads) enhanced FOS production from 4.9 to 5.9 g_{FOS}/L/h and purity from 53.5% to 67.0% (w/w), when compared to a control fermentation of immobilized *A. pullulans* in monoculture [37].

To overcome the competition by the substrate, *S. cerevisiae* YIL162W, with the gene responsible for sucrose hydrolysis disrupted, was applied in a fermentation run with *A. ibericus*. Coculture fermentation with simultaneous inoculation of the strains, run with free cells, led to FOS mixtures with 97.4% ± 0.2% (w/w) purity in a shake-flask. *S. cerevisiae* YIL162W reduced the amount of glucose to residual values (2 g/L) which had an effective impact on the activity of the β-fructofuranosidase enzyme from the *A. ibericus*, thus raising the FOS fermentation yield from 0.64, attained with a single culture, to values up to 0.70 ± 0.00 g_{FOS}/$g_{initial\ sucrose}$ after 45 hours of fermentation [56].

Examples of other integrated production–purification processes are membrane reactors, incorporating ultra- and nanofiltration membranes such as ceramic membranes [75,76]. Membrane reactors have been developed to achieve continuous reuse of the enzyme, removal of glucose, and simultaneous replacement of sucrose. However, they are limited by membrane fouling, which decreases productivity and increases operating costs.

5.3.5 Cloning fructosyltransferase gene for the production of fructooligosaccharides

The information on a microorganism's genome sequence can provide an important contribution for understanding the production of enzymes and metabolites by those organisms. Genomic studies allow to find new genes, as the ones encoding fructosyltransferase, in microorganisms not yet reported as FOS producers [20,77,78]. Strain development and genetic manipulation may improve enzyme production, enzyme activity, and maximize the capacity to produce a particular metabolite. Engineered enzymes have been homologously and heterologously expressed [57,79]. Classical mutagenesis by physical and chemical agents has also been applied to FOS producer enzymes, for example, irradiation with gamma rays [77,80].

Fructosyltransferase enzymes have been also engineered to alter their function, targeting the production of FOS with defined composition profile, such as synthesis of 6-kestose [81]. Another study shifted the product profile from levan-type FOS to a range of new oligo- and polysaccharides with higher DP [79]. The β-fructofuranosidase (fopA) from *Aspergillus fijiensis* ATCC 20611, originally deposited as *A. niger* ATCC 20611 [57] and later reclassified as *A. japonicus*, has been extensively used for commercial FOS production [57]. Molecular techniques have been used to improve the production of this enzyme (58% increase) with homologous expression [57]. The enzyme was also heterologously expressed in *S. cerevisiae* [82] and *P. pastoris* [83]. In this last study, the enzyme was engineered targeting the production of FOS in a profile similar to that of the commercial prebiotic Actilight product [83]. Overexpression of the β-fructofuranosidase gene in the *Aureobasidium melanogenum* has been shown to increase the enzyme activity from 281.7 to 557.7 U/mL. Within 7 hours of the transfructosylation reaction, a yield of 0.66 $g_{FOS}/g_{sucrose}$ was achieved [58].

Genes encoding fructosyltransferases, fructosylfuranosidades, and endo-inulinases from fungi and bacteria have been characterized and cloned both in wild-type strains and recombinant ones [20,84,85]. *Yarrowia lipolytica* has been used to reduce glucose released during FOS synthesis, by conversion of glucose into high-value-added erythritol. Fructosyltransferase from *A. oryzae* was displayed on the cell surface of *Y. lipolytica*. Dry cells were used to convert 800 g/L of sucrose into FOS, achieving a yield of 60% and productivity of 160 g_{FOS}/L/h. Glucose was converted to erythritol at a yield of 50%–55%. The use of whole cells improved fructosyltransferase stability and increased at least 10-fold the recycling cycles, with 10% of fructosyltransferase activity lost. Although the glucose and fructose monosaccharides were efficiently eliminated during fermentation, the mixture obtained still contained 9% nonprebiotic sucrose, together with 70.5% FOS and 20.5% erythritol [59].

Another possible strategy to avoid β-fructofuranosidase repression by glucose is disruption of the gene encoding glucose repression. This gene was disrupted in an *A. melanogenum* resulting in increased β-fructofuranosidase activity. Whole cells of the disrupted fungi were further used to convert 350 g/L of sugars from cane molasses into FOS, attaining a yield of 0.58 $g_{FOS}/g_{molasses\ sugar}$ within 4 hours [60].

New engineered endo-inulinases have been created as an alternative to the natural enzymes from microorganisms. An optimized endo-inulinase gene from *A. niger* was overexpressed in an engineered *Y. lipolytica*. Inulin was further hydrolyzed in a second stage, at a higher temperature, in a 10-L bioreactor, reaching a FOS yield, purity, and productivity of 0.91 g_{FOS}/g_{inulin}, 90.3%, and 15.18 g_{FOS}/L/h, respectively [61]. Further, an inulin-binding module from a cycloinulinooligosaccharide fructanotransferase of *Bacillus macerans* CFC1 was fused into the C-terminal of the endo-inulinase from *A. niger* and expressed in the *Y. lipolytica*. Within this strategy, FOS was produced in only 2 hours, increasing the process productivity to 358.6 g_{FOS}/L/h [62].

Inulooligosaccharides have been produced in a one-step bioprocess integrating endo-inulinase production, FOS fermentation, and non-FOS sugars removal into the same bioreactor. One study used a recombinant *S. cerevisiae* strain, in which a heterologous endo-inulinase gene was expressed, and the inherent invertase gene SUC2 was disrupted. The bioprocess, ran at 40°C with initial 200 g/L of chicory inulin, presented a maximum titer, yield, and productivity of 180.2 ± 0.8 g/L, 0.9 g_{FOS}/g_{inulin}, and 7.51 ± 0.03 g_{FOS}/L/h, respectively [63]. Another study optimized the gene encoding endo-inulinase from *Pseudomonas mucidolens* and cloned it into a shuttle vector that was further transfected into *Bacillus subtilis* WB800-R, which had the gene sacC encoding levanase deleted. A crude extract of inulin was converted in a maximum FOS concentration and yield of 64.84 ± 0.72 g/L and 75.38%, respectively [86].

5.4 Role of prebiotics in control of human diseases

Over the past decades, the potential of prebiotic molecules to benefit human health and well-being has been demonstrated in numerous researches [87–90]. As the complex interaction between the host and intestinal microbiota impacts a wide variety of biological processes, for instance, the host metabolism, the gut–brain axis, immune system functions, and hormone levels, the extended interest in prebiotics by the scientific community goes far beyond its fermentation [91].

The beneficial effects on human health through prebiotics intake have been mainly related to the stimulation of healthful bacteria growth, that promotes increased production of short-chain fatty acids (SCFAs), which are quickly absorbed in the large intestine [92]. The SCFAs are further metabolized in several human body tissues, for example, butyrate is used in the colonic epithelium, propionate is targeted to the liver, whereas acetate is allocated to the liver, muscle, and other peripheral tissues [93]. Studies demonstrated that SCFAs may influence dendritic cell biology, epithelial integrity, and IgA antibody response, and promote the formation of mucin, antimicrobial peptides, and tight junction proteins [94,95]. In addition, they were also associated with the stimulation of water absorption and sodium molecules, with enhanced mineral absorption, colonic blood flow, and oxygen uptake, and with a decrease in the colon's pH and bile acid solubility [96]. In this regard, studies have shown that gut microbiota plays an essential role in the prevention of disease progression, which suggests that the microbiome modulation is a suitable therapeutic strategy [89].

Information on the performance mechanisms of prebiotics has been collected through: human trials and animal feeding studies (i.e., in vivo studies); test samples of blood or tissue obtained from humans or animals (i.e., *ex vivo* studies); and in experiments carried out with isolated cells, grown under controlled conditions (i.e., *in vitro* studies) [97]. These methods, when associated with advanced molecular biology techniques, continue to allow progress in research involving prebiotics, establishing a relationship between microbiome alteration and its positive effects in the treatment of diseases [98].

5.4.1 Fructooligosaccharides as prebiotics

Commonly, FOS promote the growth of some *Bifidobacterium* and *Lactobacillus* species and simultaneously reduce the number of harmful Gram-positive and Gram-negative bacteria present in the colon such as *Bacteroides* species, *Clostridium* species, *Escherichia coli*, *Fusobacterium* species, *Salmonella* species, *Listeria* species, *Shigella* species, *Campylobacter*, and *Vibrio cholera* [1,99]. The beneficial effects of bifidobacteria on human health have been related to the SCFAs produced upon fermentation of prebiotics, mainly acetic and lactic acids, which by acidification of the colon induces an antimicrobial effect. Additionally, bifidobacteria growth may also result in a decrease in blood ammonia and cholesterol levels. On the other hand, lactobacilli have been related to the enhancement of nutrient bioavailability, improved immunity, and a reduction in the symptoms associated with lactose intolerance [100,101].

Intake of food supplemented with FOS has been linked with the treatment of several human diseases and conditions [91,102,103]. Most FOS health benefits evidence is well-established, such as increased mineral absorption and reduction of osteoporosis risk

[104–107], reduced risk of cancer [108–110], prevented cardiovascular disease [111], enhanced immune system [112–114], improved responses to allergic activity [115–117], and attenuated irritable bowel syndrome (IBS) [118,119] and Crohn disease [120].

Nonetheless, recent research has been addressing FOS intervention with different pathological scenarios. In this regard, some of these studies are following discussed. Depression is a mental disorder with a high global incidence that influences societal and personal conditions. FOS extracted from *Morinda officinalis* were administered to depressed rats for 3 weeks (50 mg/kg). FOS treatment was found to alleviate depression-like behaviors and repaired intestinal epithelia damage by the appearance of beneficial bacteria (e.g., *Acinetobacter, Barnesiella, Coprococcus, Dialister, Lactobacillus,* and *Paenibacillus*) and disappearance of depression-associated bacteria (e.g., *Anaerostipes, Oscillibacter, Proteobacteria,* and *Streptococcus*). Interestingly, FOS consumption stimulated the growth of *Cyanobacteria*, a bacteria phylum known to secrete metabolites with antidepressive properties [121].

Alzheimer disease is the most common cause of neurodegenerative dementia in the world [122]. The prebiotic effect of *Morinda officinalis* FOS in model rodents (administered in a dosage of 100 mg/kg per day for 8 weeks) was also studied in Alzheimer disease. The results demonstrated that FOS administration can ameliorate learning and memory abilities. It improved oxidative stress and inflammation disorder, and regulated the synthesis and secretion of neurotransmitter [123]. Although Alzheimer disease is a disease of multifactorial causes, it was characterized by extensive oxidative stress and inflammation due to the β-amyloid plaques generated during the disease [124]. Administration of FOS in mice [2.5%–5% (w/w) diet] was found to improve brain β-amyloid, β-secretase, cognitive function, and plasma antioxidant levels, which may attenuate the development of Alzheimer disease [124]. Altered mouse microbiota gut composition was reversed by administration of 2% (w/w) FOS in drinking water, for 6 weeks. The growth of *Lactobacillus* was stimulated and the expression levels of synaptic plasticity markers were upregulated, resulting in an ameliorated cognitive deficit [125].

Currently, chemotherapy and radiation therapy are the major treatments for cancer patients, which often lead to undesirable side effects [126]. Thus, the administration of FOS as a cancer prevention or treatment is an attractive alternative. Evidence has demonstrated that the effect of prebiotics in cancer patients was mainly mediated by gut microbial modification [127,128]. The underlying mechanisms have been mainly explained by: (1) the production of SCFA end-products, which stimulated apoptosis in colonic cancer cell lines and promoted healthy bacteria growth in the colon; (2) the changing of the activities of the enzymes from the lactic acid bacteria that have been linked to carcinogenesis; and (3) the shift of protein and lipid bacteria metabolism to saccharolytic [100,127,129]. FOS supplementation presented protective effects on intestinal barrier function. Pre- and treatment with FOS attenuated intestinal mucositis in mice, which is a side effect in patients undergoing chemotherapy, by reducing inflammatory infiltrate, intestinal permeability, and oxidative stress [130].

Recently, gut microbiota modification through FOS intervention has been also shown to contribute to the suppression of the progression of hearing loss. In the inner ear of mice fed with 10% (w/w) of FOS for 8 weeks, gene expression of neurotrophin, brain-derived neurotrophic factor, its receptor, tyrosine kinase receptor b, and the SCFAs receptor, free fatty acid receptor 3, were increased by FOS [131].

5.4.2 Galactooligosaccharides as prebiotics

The earliest source of GOS for humans is the mother's milk. Women naturally produce several specific forms of GOS during the *puerperium*, the period following childbirth, lasting approximately 6 weeks, composing around 25% of mother's milk carbohydrates [132]. This implies that the first, and maybe the most important health benefit of GOS, is the formation of the earliest human microbiome itself, acting as an essential prebiotic [133]. In this way, human milk's GOS has been commonly known as the "*bifidus* factor" since the 1950s due to its specific bifidogenic activity [134–136]. Breast milk GOS is efficiently used as a carbon source by *Bifidobacterium* and *Bacteroides* strains that are the precursors to babies' microbiomes. In fact, *Bifidobacteria longum* subspecies *infantis* is predominantly present in the intestine of infants, and is adapted for producing specific enzymes for GOS metabolism. Moreover, GOS can bind to pathogenic bacteria and viruses and inhibit infections inside the gut due to avoidance of its anchorage on cellular surface receptors [133]. Some GOS also can be absorbed into the bloodstream and link to monocytes, lymphocytes, and endothelial cells, and have a systemic effect in the maturation of the infant immune system [137]. For this reason, the addition of GOS in infant formula is a very promising approach that is currently being evaluated for new products that are intended to mimic breast milk.

Prebiotic effects of GOS and FOS, in contemporary diseases, have been studied in several clinical conditions linked to the gut–brain axis, that is, the crosstalk that allows the microbiome to exchange information with the brain and affect mood [138]. Trials have consistently proved the benefits of GOS consumption to counter depression, anxiety, and stress [138]. Despite not being completely understood, mood benefits can be observed, independently of sex, age, or health/disease condition. For example, healthy volunteers, men and women, between 18 and 45 years old, who daily consumed 5.5 g of GOS for 3 weeks, significantly reduced the salivary cortisol awakening response compared with placebo or FOS. Participants also showed decreased attentional vigilance to negative versus positive information after GOS intake [139]. Moreover, in patients with IBS *trans*-GOS stimulated gut *Bifidobacterium* spp., while reduced *Clostridium perfringens* changed the *Bacteroides/Prevotella* ratio, and effectively relieved the IBS symptoms, such as flatulence, bloating, and anxiety [140]. Therefore, the gut–brain axis crosstalk is a very important subject for better understanding GOS effects on ameliorating disease symptoms.

The microbiome is essential for innate immune system maturation, while the education of the immune response, that is, the adaptive immune system, is a lifelong process that is also microbiome-mediated. Alterations to the innate and adaptive immune systems throughout our lives are called immunosenescence. Meantime, some pre- and probiotics, for example, GOS, can modulate this microbiome–immunesenescene *ballet* in many aging-related diseases [141]. GOS have also been shown to suppress photoaging in hairless albino mice, after exposure to ultraviolet radiation for 8 weeks [142]. GOS were shown to prevent skin aging by influencing important gene, cytokine, and protein levels associated with healthy skin characteristics, namely water-holding capacity, transepidermal water loss, skin elasticity, and moisture. Notably, the mitogen-activated protein kinase (MAPK) pathway was involved in this protection, since GOS intake inhibited MAPK phosphorylation. Moreover, the collagen breakdown was regulated and the collagen's production gene, COL-1, was highly expressed [142].

Detailed mechanisms of how GOS affect health and reverse many different diseases are currently being clarified. These mechanistic evidences are all related to the microbiome–host functions. Some key elements are demonstrated mainly in the regulation of the microbiome's antiinflammatory and immunomodulatory functions. Recently, after GOS treatment, an ulcerative colitis murin's model mice showed an increment of *Bacteroides fragilis* and a decrease of *Verruco microbia* and *Proteobacteria*. The ingestion of GOS decreased IL-6, IL-18, IL-13, and IL-33 secretion and also mRNA expression in the colon due to inhibition of NF-κB activation [135]. Furthermore, the internal mechanism underlining the anticolitis effect of GOS, caused by bacterial infection, has been recently explored by studying its regulatory effect on miRNAs. GOS were shown to protect colon cells against lipopolysaccharide-induced injury, attenuating *Helicobacter hepaticus*-induced colitis via miR-19b immunomodulation [143]. Although many researchers are elucidating how GOS can promote health and help to control certain diseases, much more remains to be done regarding regulation, dose recommendations, and disease treatment protocols.

5.4.3 Lactulose as a prebiotic

The effects of lactulose on human health are dependent on its administered dose. For instance, at low dose (i.e., 10 g per day) it is used as a prebiotic due to the bifidogenic effect; at moderate dose (i.e., range between 20–40 g per day) it is used in chronic constipation treatment, due to its laxative effect; and at high dose (i.e., above 60 g per day) it presents a detoxifying effect and is used in hepatic encephalopathy treatment [144].

Regarding lactulose as a prebiotic, it has been shown to promote *Lactobacillus* and *Bifidobacterium* growth and to suppress the levels of pathogenic microorganisms [100]. Some authors have reported positive effects of lactulose consumption in disease treatments. For instance, lactulose intragastrically administered in rats (0.25 g/kg diluted to 0.75 mL/kg) has been shown to ameliorate cerebral ischemia–reperfusion injury by inducing endogenous hydrogen, which activated the expression of NF-E2-related factor 2 (Nrf2), inhibited neuron apoptosis, and improved brain oxidative stress [145].

Lactulose has been also tested to revert vaginal dysbiosis, a common condition among women, since a healthy vaginal microbiota is characterized by low diversity and colonization by *Lactobacillus* spp. [146]. Experiments run with freshly collected vaginal bacteria shown that lactulose is able to promote commensal *Lactobacillus* growth and dominance and induces a healthy acidity, partially through lactic acid production. In assays conducted with isolated strains, the growth of *Candida albicans*, associated with bacterial vaginosis, was also suppressed [147]. The effects of lactulose on renal function and chronic kidney disease have also been recently studied with a model of adenine-induced rats [148]. Doses of 3.0% and 7.5% lactulose-containing diets, fed for 4 weeks, were shown to adjust the gut microbiota and to ameliorate the chronic kidney disease progression by suppressing uremic toxin production [148].

5.4.4 Emergent prebiotics

With the new definition of prebiotics, a large range of compounds, including noncarbohydrate ones, are now signaled as potential prebiotics. This opened the opportunity of studying innovative ingredients with potential functional properties. Nevertheless, the host−microbiome interactions, the underlying mechanisms, and the actual benefits of the emergent prebiotics remain poorly understood [87]. In addition, studies conducted with humans are still scarce. Table 5−2 summarizes a few studies conducted with emergent prebiotics on the treatment of specific diseases, including information on the dosage administered, model used, and evidence found. Some of the prebiotic candidates are herein discussed in more detail.

Noncarbohydrate plant bioactive compounds are emerging as great prebiotics since their ability to modulate the microbiome is being uncovered. For example, dietary polyphenols can reach the colon and interact with bacteria enzymes, promoting or regulating its growth and consequently producing health benefits [4]. Despite being nonfermentable carbohydrates, these plants' secondary metabolism compounds, including catechins, flavonoids, anthocyanins, tannins, and resveratrol, are only partially absorbed in the small intestine, and the remaining compounds are subject to several interactions and interconversions that, although not completely understood, seem to play a role in countering metabolic syndrome-related diseases [159]. Studies conducted with polyphenols in animals and *in vitro* have described many health benefits associated with gut microbiota regulation, such as protection against cancer, obesity, insulin resistance, hepatic inflammation, sleep deprivation, and atherosclerosis [160−162]. Seventeen reviewed human clinical trials have demonstrated that polyphenols from different foods, for example, green tea, black tea, red wine, cocoa, pomegranate, and blackcurrant, improved mainly the markers associated with cardiovascular disease and colon cancer [160]. Nevertheless, trials with larger samples, longer duration, and high-throughput molecular techniques are needed to provide more conclusive results on the impact of polyphenols in human health.

Isomaltooligosaccharides (IMO) are carbohydrates consisting of glucose units linked mainly by α-(1→6) and α-(1→4) glucosidic bonds, and may also include α-(1→2) and α-(1→3) bounds. IMO include isomaltose, panose, isopanose, isomaltotriose, nigerose, kojibiose, long-chain IMO, and cyclo-IMO [163]. They occur in natural sources (e.g., honey, beer, and fermented food including rice miso, soy sauce, and sake) and are commercially synthesized with enzymes from starch. IMO have been suggested as promising prebiotics, but their *in vivo* digestibility is poorly characterized [164−166]. *In vitro* studies have demonstrated that IMO stimulates bifidobacteria and lactobacilli growth and inhibits pathogens, such as *Clostridium difficile*, allowing restoration of gut microbiota dysbiosis [163,166]. Nevertheless, amylolytic enzymes from the small intestine are capable of

Table 5-2 Examples of emerging prebiotics involved in disease prevention, including dosage, model used, and main evidence found.

Prebiotic	Disease	Dosage	Model	Evidence	References
Stachyose	Type II diabetes mellitus (T2DM)	5000 mg/kg body weight once a day	Adult rats	A total of 21 differential metabolites were identified in urinary samples, mainly associated with citrate cycle, carbohydrate and amino acid metabolism, which help to explain the mechanisms of stachyose against T2DM. The insulin level was adjusted through regulation of the energy metabolism, gut microbiota, and inflammatory response	[149]
IMO	Irritable bowel syndrome	2 mL of 5% IMO solution twice daily for 14 days	Male rats	Reduction of colorectal distension and increased pain threshold. Damage on the ileal epithelial ultrastructure was attenuated	[150]
XOS	Prediabetic	2 g per day for 8 weeks	Adult human	Reduction of bacteria associated with prediabetics (genera *Howardella*, *Enterorhabdus*, and *Slackia*). Tendency to reduce insulin levels	[151]
	Colon cancer	100 g/kg for 45 days	Adult rats	Increase in the bifidobacterial population and reduction of the lipid peroxidation level. Enhanced glutathione-S-transferase and catalase activities in the colonic mucosa	[152]
	Obesity	7% (w/w) for 6 weeks	Male mice	*Firmicutes* and *Bacteroidetes* associated with obesity were reduced, and *Actinobacteria* decreased. Decreased gene expression of markers of adipogenesis and fat synthesis	[153]
SOS	Cardiovascular disorders	200 mg/kg body weight daily for 30 days	Male rats	Higher antioxidant enzyme activities and decreased lipid peroxidation. Cardiac contractile function was recovered and reduced the myocardial specific enzymes (i.e., creatine kinase, aspartate transaminase, and lactate dehydrogenase)	[154]
	Gestational diabetes mellitus	10 g per day	97 pregnant women	Reduced oxidative stress and alleviated insulin resistance	[155]
Lactosucrose	Inflammatory bowel disease	250 mg/kg per day for 21 days	Female rats	Promoted the production of Th2-type cytokines, rebalanced the ratio of Th1/Th2, enhanced IL-4 production due to the increase in CD86 expression in the gut	[156]
Raffinose	Metabolic diseases	1 μg/mL	Cell culture differentiated (3T3-L1 preadipocytes, L6 skeletal myocytes, and HepG2 hepatocytes)	Inhibited lipid accumulation in differentiated HepG2 and 3T3-L1 cells, and enhanced the secretion of adiponectin in 3T3-L1 adipocytes. Attenuated the expression of biomarkers involved in fatty acid synthesis. Demonstrated antidiabetic potential through improved glucose uptake	[157]
Resistant starch	Type II diabetic	42 g of resistant starch per 100 g of oat, during 9 weeks treatment	High-fat diet-induced type II diabetic rats	Oat-resistant starch ameliorated insulin resistance and inflammation throughout microbiome modulation (promoted *Clostridium* and *Butyricicoccus*; and reduced the *Bacteroides*, *Lactobacillus*, *Oscillospira*, and *Ruminococcus*)	[158]

IMO, Isomaltooligosaccharides; *SOS*, soybean oligosaccharides; *XOS*, xylooligosaccharides.

hydrolyzing short-chain IMO, although hydrolysis and uptake occur more distal than maltose or glucose, which may alter the glucose-mediated endocrine response, but avoiding them to reach the large intestine microbiota and meet the criteria to be considered as prebiotics [167,168]. For the large-chain IMO, fermentation to SCFA occurs in a more proximal location than the resistant starch [164]. The health benefits associated with IMO have been reviewed. Evidence from human trials includes reduced frequency of stomach discomfort, constipation prevention, reduced glycemic response, reduction in serum triglyceride and increased HDL-cholesterol levels, improved mineral absorption, and increased levels of lactate and acetate [163]. An *in vivo* investigation also suggests that IMO consumption may reduce the effects of IBS as it repaired damage to the intestinal epithelial ultrastructure of a rat model [150]. IMO exhibit novel properties, stimulating insulin and incretin hormones secretion to a similar extent as dextrose when administered as an oral solution in healthy adults [164].

Stachyose is a water-soluble tetrasaccharide composed of one glucose, one fructose, and two galactose units, which naturally occurs in many vegetables and plants [149]. Stachyose can significantly decrease the blood glucose level in hyperglycemic rats [169]. Recently, stachyose treatment in type II diabetes mellitus rats showed antidiabetes activity, through the regulation of energy metabolism, gut microbiota changes, and inflammation [149]. It also enabled partial microbiota recovery after enterotoxigenic *E. coli* infection, by promoting *Akkermansia muciniphila* and *Bifidobacterium* proliferation, improved intestinal morphology, by increasing the expression of tight junction proteins, and reduced the level of intestinal inflammation [170].

5.5 Conclusions and perspectives

The growing consumer demand for functional foods has been reflected in the prosperity of the prebiotic market. Still, production at a large scale of prebiotics, such as FOS, remains challenging. The main drawback to the prebiotics production process is the low yields obtained, which requires time-consuming multistep downstream processes. Presently, new cost-effective strategies to produce FOS have been explored. The use of whole cells of active microorganisms, instead of pure enzymes, reduces the process to a single fermentation and generally leads to higher FOS yields. Additionally, microorganism immobilization enables work in continuous mode with a positive impact on the FOS yield. For enhancement of FOS purity, the use of a mixture of enzymes integrated in the production process has been successfully applied. Therefore, the integration of the production−purification process in one-step fermentation using a coculture of microorganisms seems to be a powerful strategy to increase FOS fermentation yields, productivity, and purity. Finally, genetic engineering allows targeting of the specific type and/or size of the FOS to be produced. However, research is still needed to improve the microbial FOS production process.

Regarding human health, several findings of prebiotic consumption-derived health benefits have been demonstrated by reliable research, including *in vitro* tests, animal studies, and human trials. Prebiotics can hinder progression of some diseases and, eventually, can be used as treatments in some specific diseases. Many positive effects of prebiotics on the 21st century's important diseases and health conditions are already established, for example, in the cases of Alzheimer disease, diabetes, overweight, obesity, cancer, IBS, mood alterations, and cardiovascular and immune functions. The mechanisms of prebiotics can produce benefits are related to: (1) promotion of the microbiomes' beneficial bacteria growth; (2) inhibition of unfavorable intestinal microbiota; (3) improvement of intestinal barrier function; and (4) enhancement of essential immune functions. Despite the prebiotic effects of most studied compounds on specific diseases having been revealed, there remains a lack of information regarding several key roles in microbiologic, physiologic, and biochemical pathways. Some issues such as ethnic-, gender-, and age-related variations in the gut microbiome should be considered in order to use more efficient strategies for screening prebiotic candidates. Nonetheless, there is a scientific consensus that prebiotics should have widespread consumption, representing the utmost support for human health and well-being.

Acknowledgments

The authors acknowledge the Portuguese Foundation for Science and Technology under the scope of the strategic funding of UIDB/04469/2020 unit and the Project ColOsH 02/SAICT/2017 (POCI-01-0145-FEDER-030071), and also by the project cLabel+ (POCI-01-0247-FEDER-046080) co-financed by Compete 2020, Lisbon 2020, Portugal 2020, and the European Union, through the European Regional Development Fund (ERDF).

References

[1] G.R. Gibson, M.B. Roberfroid, Dietary modulation of the human colonic microbiota: introducing the concept of prebiotics, The Journal of Nutrition 125 (6) (1995) 1401–1412.

[2] G.R. Gibson, R. Hutkins, M.E. Sanders, S.L. Prescott, R.A. Reimer, S.J. Salminen, et al., Expert consensus document: the international scientific association for probiotics and prebiotics (ISAPP) consensus statement on the definition and scope of prebiotics, Nature Reviews Gastroenterology & Hepatology 14 (8) (2017) 491–502.

[3] R.W. Hutkins, J.A. Krumbeck, L.B. Bindels, P.D. Cani, G. Fahey, Y.J. Goh, et al., Prebiotics: why definitions matter, Current Opinion in Biotechnology 37 (2016) 1–7.

[4] K.P. Scott, R. Grimaldi, M. Cunningham, S.R. Sarbini, A. Wijeyesekera, M.L.K. Tang, et al., Developments in understanding and applying prebiotics in research and practice—an ISAPP conference paper, Journal of Applied Microbiology 128 (4) (2020) 934–949.

[5] M.J. Sánchez-Martínez, S. Soto-Jover, V. Antolinos, G.B. Martínez-Hernández, A. López-Gómez, Manufacturing of short-chain fructooligosaccharides: from laboratory to industrial scale, Food Engineering Reviews 12 (2) (2020) 149–172.

[6] A. Kruschitz, B. Nidetzky, Downstream processing technologies in the biocatalytic production of oligosaccharides, Biotechnology Advances 43 (2020) 107568.

[7] C. Nobre, M.Â. Cerqueira, L.R. Rodrigues, A.A. Vicente, J.A. Teixeira, Chapter 19 – Production and extraction of polysaccharides and oligosaccharides and their use as new food additives, Industrial Biorefineries & White Biotechnology, Elsevier B.V., 2015, pp. 653–679.

[8] N. Benkeblia, Fructooligosaccharides and fructans analysis in plants and food crops, Journal of Chromatography A 1313 (2013) 54–61.

[9] M.B. Roberfroid, Inulin-type fructans: functional food ingredients, The Journal of Nutrition 137 (11 Suppl) (2007) 2493S–2502S.

[10] C. Nobre, S.C. Sousa, S.P. Silva, A.C. Pinheiro, E. Coelho, A.A. Vicente, et al., In vitro digestibility and fermentability of fructo-oligosaccharides produced by *Aspergillus ibericus*,", Journal of Functional Foods 46 (2018) 278–287.

[11] P. Zambelli, L. Fernandez-Arrojo, D. Romano, P. Santos-Moriano, M. Gimeno-Perez, A. Poveda, et al., Production of fructooligosaccharides by mycelium-bound transfructosylation activity present in *Cladosporium cladosporioides* and *Penicillium sizovae*, Process Biochemistry 49 (12) (2014) 2174–2180.

[12] V. Bali, P.S. Panesar, M.B. Bera, R. Panesar, Fructo-oligosaccharides: production, purification and potential applications, Critical Reviews in Food Science and Nutrition 55 (11) (2015) 1475–1490.

[13] A. Franck, Technological functionality of inulin and oligofructose, The British Journal of Nutrition 87 (Suppl 2) (2002) S287–S291.

[14] G.T. Bersaneti, J. Mantovan, A. Magri, S. Mali, M.A.P.C. Celligoi, Edible films based on cassava starch and fructooligosaccharides produced by *Bacillus subtilis* natto CCT 7712, Carbohydrate Polymers 151 (2016) 1132–1138.

[15] K.C.G. Silva, A.C.K. Sato, Biopolymer gels containing fructooligosaccharides, Food Research International 101 (2017) 88–95.

[16] N. Liao, B. Luo, J. Gao, X. Li, Z. Zhao, Y. Zhang, et al., Oligosaccharides as co-encapsulating agents: effect on oral *Lactobacillus fermentum* survival in a simulated gastrointestinal tract, Biotechnology Letters 41 (2) (2019) 263–272.

[17] Reports and Data, Prebiotic Ingredients Market Analysis, By Type, By Source (Roots, Vegetables, Grains, Others), By Bacterial Activity, By Functionality, By Application (Fortified Food and Beverages, Dietary Supplements, Animal Feed, Instant Formula and Others), Forecasts to 2026 [Online]. Available from: <https://www.reportsanddata.com/report-detail/prebiotic-ingredients-market#-utm_source = globenewswire&utm_medium = referral&utm_campaign = john16oct2019>, 2019 (accessed 25.05.20).

[18] Graphical Research, Europe Prebiotics Market Size By Ingredient (Inulin, GOS, FOS, MOS), By Application (Animal Feed, Food & Beverages [Dairy, Cereals, Baked Goods, Fermented Meat, Dry Foods], Dietary Supplements [Food, Nutrition, Infant Formulations]), Application Potential, Competitive Market Share & Forecast, 2020– 2026. <https://www.graphicalresearch.com/industry-insights/1309/europe-prebiotics-marke>, 2020 (accessed 27.05.20).

[19] A.L. Dominguez, L.R. Rodrigues, N.M. Lima, J.A. Teixeira, An overview of the recent developments on fructooligosaccharide production and applications, Food and Bioprocess Technology 7 (2) (2014) 324–337.

[20] M.A. Ganaie, A. Lateef, U.S. Gupta, Enzymatic trends of fructooligosaccharides production by microorganisms, Applied Biochemistry and Biotechnology 172 (4) (2014) 2143–2159.

[21] O. de la Rosa, A.C. Flores-Gallegos, D. Muñíz-Marquez, C. Nobre, J.C. Contreras-Esquivel, C.N. Aguilar, Fructooligosaccharides production from agro-wastes as alternative low-cost source, Trends in Food Science and Technology 91 (2019) 139–146.

[22] K. Nishizawa, M. Nakajima, H. Nabetani, Kinetic study on transfructosylation by β-fructofuranosidase from *Aspergillus niger* ATCC 20611 and availability of a membrane reactor for fructooligosaccharide production, Food Science and Technology Research 7 (1) (2001) 39–44.

[23] P.T. Sangeetha, M.N. Ramesh, S.G. Prapulla, Fructooligosaccharide production using fructosyl transferase obtained from recycling culture of *Aspergillus oryzae* CFR 202, Process Biochemistry 40 (3–4) (2005) 1085–1088.

[24] R.C. Kuhn, L. Palacio, P. Prádanos, A. Hernández, F.M. Filho, Selection of membranes for purification of fructooligosaccharides, Desalination and Water Treatment 27 (1–3) (2011) 18–24.

[25] M.J.L. Alles, I.C. Tessaro, C.P.Z. Noreña, Concentration and purification of yacon (*Smallanthus sonchifolius*) root fructooligosaccharides using membrane technology, Food Technology and Biotechnology 53 (2) (2015) 190–200.

[26] C. Nobre, J.A. Teixeira, L.R. Rodrigues, Fructo-oligosaccharides purification from a fermentative broth using an activated charcoal column, New Biotechnology 29 (3) (2012) 395–401.

[27] C. Nobre, P. Suvarov, G. De Weireld, Evaluation of commercial resins for fructo-oligosaccharide separation, New Biotechnology 31 (1) (2014) 55–63.

[28] D. Campos, L. Mescua, A. Aguilar-Galvez, R. Chirinos, R. Pedreschi, Effect of Yacon (*Smallanthus sonchifolius*) fructooligosaccharide purification technique using activated charcoal or ion exchange fixed bed column on recovery, purity and sugar content, International Journal of Food Science and Technology 52 (12) (2017) 2637–2646.

[29] P. Suvarov, A. Kienle, C. Nobre, G. De Weireld, A. Vande Wouwer, Cycle to cycle adaptive control of simulated moving bed chromatographic separation processes, Journal of Process Control 24 (2) (2014) 357–367.

[30] C. Nobre, J.A. Teixeira, L. Rodrigues, A. Severino, C. Retamal, G. De Weireld, et al., Operating conditions of a simulated moving bed chromatography unit for the purification of fructo-oligosaccharides, 6th International Conference on Simulation and Modelling in the Food and Bio-Industry 2010, FOODSIM 2010, 2010, pp. 216–218.

[31] C. Nobre, C.C. Castro, A.-L. Hantson, et al., Strategies for the production of high-content fructo-oligosaccharides through the removal of small saccharides by co-culture or successive fermentation with yeast, Carbohydrate Polymers 136 (2016) 274–281.

[32] Y. Yang, J. Wang, D. Teng, F. Zhang, Preparation of high-purity fructo-oligosaccharides by *Aspergillus japonicus* β-fructofuranosidase and successive cultivation with yeast, Journal of Agricultural and Food Chemistry 56 (8) (2008) 2805–2809.

[33] R. Fan, J.P. Burghardt, T. Xiong, P. Czermak, Removal of small-molecular byproducts from crude fructo-oligosaccharide preparations by fermentation using the endospore-forming probiotic *Bacillus coagulans*, Fermentation 6 (1) (2020) 6.

[34] D.C. Sheu, J.Y. Chang, C.Y. Wang, C.T. Wu, C.J. Huang, Continuous production of high-purity fructooligosaccharides and ethanol by immobilized *Aspergillus japonicus* and *Pichia heimii*, Bioprocess and Biosystems Engineering 36 (11) (2013) 1745–1751.

[35] R. Crittenden, M. Playne, Purification of food-grade oligosaccharides using immobilised cells of *Zymomonas mobilis*, Applied Microbiology and Biotechnology 58 (3) (2002) 297–302.

[36] C. Nobre, J.A. Teixeira, L.R. Rodrigues, New trends and technological challenges in the industrial production and purification of fructo-oligosaccharides, Critical Reviews in Food Science and Nutrition 55 (10) (2015) 1444–1455.

[37] C.C. Castro, C. Nobre, G. De, et al., Microbial co-culturing strategies for fructo-oligosaccharide production, New Biotechnology 51 (2019) 1–7.

[38] A. Dominguez, C. Nobre, L.R. Rodrigues, A.M. Peres, D. Torres, I. Rocha, et al., New improved method for fructooligosaccharides production by *Aureobasidium pullulans*, Carbohydrate Polymers 89 (4) (2012) 1174–1179.

[39] C. Nobre, E.G. Alves Filho, F.A.N. Fernandes, E.S. Brito, S. Rodrigues, J.A. Teixeira, et al., Production of fructo-oligosaccharides by *Aspergillus ibericus* and their chemical characterization, LWT 89 (2018) 58–64.

[40] C. Nobre, A.K.C. do Nascimento, S.P. Silva, E. Coelho, M.A. Coimbra, M.T.H. Cavalcanti, et al., Process development for the production of prebiotic fructo-oligosaccharides by *Penicillium citreonigrum*, Bioresource Technology 282 (2019) 464–474.

[41] M.B. Prata, S.I. Mussatto, L.R. Rodrigues, J.A. Teixeira, Fructooligosaccharide production by *Penicillium expansum*, Biotechnology Letters 32 (6) (2010) 837–840.

[42] S.I. Mussatto, C.N. Aguilar, L.R. Rodrigues, J.A. Teixeira, Colonization of *Aspergillus japonicus* on synthetic materials and application to the production of fructooligosaccharides, Carbohydrate Research 344 (6) (2009) 795–800.

[43] C.C. Castro, C. Nobre, M.-E. Duprez, et al., Screening and selection of potential carriers to immobilize *Aureobasidium pullulans* cells for fructo-oligosaccharides production, Biochemical Engineering Journal 118 (2017) 82–90.

[44] S.I. Mussatto, C.N. Aguilar, L.R. Rodrigues, J.A. Teixeira, Fructooligosaccharides and β-fructofuranosidase production by *Aspergillus japonicus* immobilized on lignocellulosic materials, Journal of Molecular Catalysis B: Enzymatic 59 (1–3) (2009) 76–81.

[45] S.I. Mussatto, M.B. Prata, L.R. Rodrigues, J.A. Teixeira, Production of fructooligosaccharides and β-fructofuranosidase by batch and repeated batch fermentation with immobilized cells of *Penicillium expansum*, European Food Research and Technology 235 (1) (2012) 13–22.

[46] S.I. Mussatto, L.F. Ballesteros, S. Martins, D.A.F. Maltos, C.N. Aguilar, J.A. Teixeira, Maximization of fructooligosaccharides and β-fructofuranosidase production by *Aspergillus japonicus* under solid-state fermentation conditions, Food and Bioprocess Technology 6 (8) (2013) 2128–2134.

[47] O.F. Sánchez, A.M. Rodriguez, E. Silva, L.A. Caicedo, Sucrose biotransformation to fructooligosaccharides by *Aspergillus* sp. N74 free cells, Food and Bioprocess Technology 3 (5) (2010) 662–673.

[48] X.-A. Zeng, K. Zhou, D. Liu, C.S. Brennan, M. Brennan, J. Zhou, et al., Preparation of fructooligosaccharides using *Aspergillus niger* 6640 whole-cell as catalyst for bio-transformation, LWT – Food Science and Technology 65 (2016) 1072–1079.

[49] S. Ojha, N. Rana, S. Mishra, Fructo-oligosaccharide synthesis by whole cells of *Microbacterium paraoxydans*, Tetrahedron: Asymmetry 27 (24) (2016) 1245–1252.

[50] J.W. Yun, S.K. Song, The production of high-content fructo-oligosaccharides from sucrose by the mixed-enzyme system of fructosyltransferase and glucose oxidase, Biotechnology Letters 15 (6) (1993) 573–576.

[51] J.W. Yun, M.G. Lee, S.K. Song, Batch production of high-content fructo-oligosaccharides from sucrose by the mixed-enzyme system of β-fructofuranosidase and glucose oxidase, Journal of Fermentation and Bioengineering 77 (2) (1994) 159–163.

[52] D.C. Sheu, P.J. Lio, S.T. Chen, C.T. Lin, K.J. Duan, Production of fructooligosaccharides in high yield using a mixed enzyme system of β-fructofuranosidase and glucose oxidase, Biotechnology Letters 23 (18) (2001) 1499–1503.

[53] J. Yoshikawa, S. Amachi, H. Shinoyama, T. Fujii, Production of fructooligosaccharides by crude enzyme preparations of β-fructofuranosidase from *Aureobasidium pullulans*, Biotechnology Letters 30 (3) (2008) 535–539.

[54] C. Barthomeuf, H. Pourrat, Production of high-content fructo-oligosaccharides by an enzymatic system from *Penicillium rugulosum*, Biotechnology Letters 17 (9) (1995) 911–916.

[55] D.-C. Sheu, K.-J. Duan, C.-Y. Cheng, J.-L. Bi, J.-Y. Chen, Continuous production of high-content fructooligosaccharides by a complex cell system, Biotechnology Progress 18 (6) (2002) 1282–1286.

[56] C. Nobre, D.A. Gonçalves, J.A. Teixeira, L.R. Rodrigues, One-step co-culture fermentation strategy to produce high-content fructo-oligosaccharides, Carbohydrate Polymers 201 (2018) 31–38.

[57] J. Zhang, C. Liu, Y. Xie, N. Li, Z. Ning, N. Du, et al., Enhancing fructooligosaccharides production by genetic improvement of the industrial fungus *Aspergillus niger* ATCC 20611, Journal of Biotechnology 249 (2017) 25–33.

[58] T. Aung, H. Jiang, G.L. Liu, Z. Chi, Z. Hu, Z.M. Chi, Overproduction of a β-fructofuranosidase1 with a high FOS synthesis activity for efficient biosynthesis of fructooligosaccharides, International Journal of Biological Macromolecules 130 (2019) 988–996.

[59] L. Zhang, J. An, L. Li, H. Wang, D. Liu, N. Li, et al., Highly efficient fructooligosaccharides production by an erythritol-producing yeast *Yarrowia lipolytica* displaying fructosyltransferase, Journal of Agricultural and Food Chemistry 64 (2016) 3828–3837.

[60] S. Zhang, H. Jiang, S. Xue, N. Ge, Y. Sun, Z. Chi, et al., Efficient conversion of cane molasses into fructooligosaccharides by a glucose derepression mutant of *Aureobasidium melanogenum* with high β-fructofuranosidase activity, Journal of Agricultural and Food Chemistry 67 (49) (2019) 13665–13672.

[61] Y.-Z. Han, C.-C. Zhou, Y.-Y. Xu, J.-X. Yao, Z. Chi, Z.-M. Chi, et al., High-efficient production of fructo-oligosaccharides from inulin by a two-stage bioprocess using an engineered *Yarrowia lipolytica* strain, Carbohydrate Polymers 173 (2017) 592–599.

[62] W. Mao, Y. Han, X. Wang, X. Zhao, Z. Chi, Z. Chi, et al., A new engineered endo-inulinase with improved activity and thermostability: application in the production of prebiotic fructo-oligosaccharides from inulin, Food Chemistry 294 (2019) 293–301.

[63] D. Wang, F.-L. Li, S.-A. Wang, A one-step bioprocess for production of high-content fructo-oligosaccharides from inulin by yeast, Carbohydrate Polymers 151 (2016) 1220–1226.

[64] M.A. Ganaie, H. Soni, G.A. Naikoo, L. Taynara, S. Oliveira, H.K. Rawat, et al., Screening of low cost agricultural wastes to maximize the fructosyltransferase production and its applicability in generation of fructooligosaccharides by solid state fermentation, International Biodeterioration & Biodegradation 118 (2017) 19–26.

[65] P.K. Sadh, S. Duhan, J.S. Duhan, Agro-industrial wastes and their utilization using solid state fermentation: a review, Bioresources and Bioprocessing 5 (1) (2018) 1–15.

[66] M.C.R. Mano, I.A. Neri-Numa, J.B. da Silva, B.N. Paulino, M.G. Pessoa, G.M. Pastore, Oligosaccharide biotechnology: an approach of prebiotic revolution on the industry, Applied Microbiology and Biotechnology 102 (1) (2018) 17–37.

[67] M. El-Bakry, J. Abraham, A. Cerda, R. Barrena, S. Ponsá, T. Gea, et al., From wastes to high value added products: novel aspects of SSF in the production of enzymes, Critical Reviews in Environmental Science and Technology 45 (18) (2015) 1999–2042.

[68] M.C.P. Gonçalves, S.A.V. Morales, E.S. Silva, A.E. Maiorano, R.F. Perna, T.G. Kieckbusch, Entrapment of glutaraldehyde-crosslinked cells from *Aspergillus oryzae* IPT-301 in calcium alginate for high transfructosylation activity, Journal of Chemical Technology & Biotechnology (2020).

[69] K.J. Duan, J.S. Chen, D.C. Sheu, Kinetic studies and mathematical model for enzymatic production of fructooligosaccharides from sucrose, Enzyme and Microbial Technology 16 (4) (1994) 334–339.

[70] R. Fekih-Salem, L. Dewasme, C.C. Castro, C. Nobre, A.-L. Hantson, A. Vande Wouwer, Sensitivity analysis and reduction of a dynamic model of a bioproduction of fructo-oligosaccharides, Bioprocess and Biosystems Engineering 42 (11) (2019) 1793–1808.

[71] R. Fekih-Salem, A. Vande Wouwer, C. De Castro, C. Nobre, A.-L. Hantson, Parameter identification of the fermentative production of Fructo-oligosaccharides by *Aureobasidium pullulans*, in: 2015 19th International Conference on System Theory, Control and Computing, ICSTCC 2015 - Joint Conference SINTES 19, SACCS 15, SIMSIS 19, 2015, pp. 43–48.

[72] O. Rocha, C. Nobre, A. Dominguez, D. Torres, N. Faria, L. Rodrigues, et al., A dynamical model for the fermentative production of fructooligosaccharides, Computer Aided Chemical Engineering 27 (C) (2009) 1827–1832.

[73] J. Schorsch, C.C. Castro, L.D. Couto, C. Nobre, M. Kinnaert, Optimal control for fermentative production of fructo-oligosaccharides in fed-batch bioreactor, Journal of Process Control 78 (2019) 124–138.

[74] K.H. Jung, J.W. Yun, K.R. Kang, J.Y. Lim, J.H. Lee, Mathematical model for enzymatic production of fructo-oligosaccharides from sucrose, Enzyme and Microbial Technology 11 (8) (1989) 491–494.

[75] J.P. Burghardt, L.A. Coletta, R. van der Bolt, M. Ebrahimi, D. Gerlach, P. Czermak, Development and characterization of an enzyme membrane reactor for fructo-oligosaccharide production, Membranes 9 (11) (2019) 20.

[76] K. Nishizawa, M. Nakajima, H. Nabetani, A forced-flow membrane reactor for transfructosylation using ceramic membrane, Biotechnology and Bioengineering 68 (1) (2000) 92−97.

[77] A.E. Maiorano, R.M. Piccoli, E.S. Da Silva, et al., Microbial production of fructosyltransferases for synthesis of pre-biotics, Biotechnology Letters 30 (11) (2008) 1867−1877.

[78] S. Han, T. Ye, S. Leng, L. Pan, W. Zeng, G. Chen, et al., Purification and biochemical characteristics of a novel fructosyltransferase with a high FOS transfructosylation activity from *Aspergillus oryzae* S719, Protein Expression and Purification 167 (2020) 105549.

[79] M. De Abreu, M. Alvaro-Benito, J. Sanz-Aparicio, F.J. Plou, M. Fernandez-Lobato, M. Alcalde, Synthesis of 6-kestose using an efficient β-fructofuranosidase engineered by directed evolution, Advanced Synthesis and Catalysis 355 (9) (2013) 1698−1702.

[80] M.I. Rajoka, A. Yasmeen, Improved productivity of β-fructofuranosidase by a derepressed mutant of *Aspergillus niger* from conventional and non-conventional substrates, World Journal of Microbiology and Biotechnology 21 (4) (2005) 471−478.

[81] R. Beine, R. Moraru, M. Nimtz, S. Na'amnieh, A. Pawlowski, K. Buchholz, et al., Synthesis of novel fructooligosaccharides by substrate and enzyme engineering, Journal of Biotechnology 138 (1−2) (2008) 33−41.

[82] K.M. Trollope, J.F. Görgens, H. Volschenk, Semirational directed evolution of loop regions in *Aspergillus japonicus* β-fructofuranosidase for improved fructooligosaccharide production, Applied and Environmental Microbiology 81 (20) (2015) 7319−7329.

[83] G. Coetzee, E. van Rensburg, J.F. Görgens, Evaluation of the performance of an engineered β-fructofuranosidase from *Aspergillus fijiensis* to produce short-chain fructooligosaccharides from industrial sugar streams, Biocatalysis and Agricultural Biotechnology 23 (2020) 101484.

[84] G.L. Liu, Z. Chi, Z.M. Chi, Molecular characterization and expression of microbial inulinase genes, Critical Reviews in Microbiology 39 (2) (May-2013) 152−165.

[85] M.A. Rodríguez, O.F. Sánchez, C.J. Alméciga-Díaz, Gene cloning and enzyme structure modeling of the *Aspergillus oryzae* N74 fructosyltransferase, Molecular Biology Reports 38 (2) (2011) 1151−1161.

[86] R. Jiang, Y. Qiu, W. Huang, L. Zhang, F. Xue, H. Ni, et al., One-step bioprocess of inulin to product inulo-oligosaccharides using *Bacillus subtilis* secreting an extracellular endo-inulinase, Applied Biochemistry and Biotechnology 187 (1) (2019) 116−128.

[87] I.A. Neri-Numa, G.M. Pastore, Novel insights into prebiotic properties on human health: a review, Food Research International 131 (2020) 108973.

[88] T. Klancic, R.A. Reimer, Gut microbiota and obesity: impact of antibiotics and prebiotics and potential for musculoskeletal health, Journal of Sport and Health Science 9 (2) (2020) 110−118.

[89] A.G. Colantonio, S.L. Werner, M. Brown, The effects of prebiotics and substances with prebiotic properties on metabolic and inflammatory biomarkers in individuals with type 2 diabetes mellitus: a systematic review, Journal of the Academy of Nutrition and Dietetics 120 (4) (2020) 587−607.e2.

[90] I.H.R. Paiva, E. Duarte-Silva, C.A. Peixoto, The role of prebiotics in cognition, anxiety, and depression, European Neuropsychopharmacology 34 (2020) 1−18.

[91] J. Behera, J. Ison, S.C. Tyagi, N. Tyagi, The role of gut microbiota in bone homeostasis, Bone 135 (2020) 115317.

[92] E.M.M. Quigley, Prebiotics and probiotics in digestive health, Clinical Gastroenterology and Hepatology 17 (2) (2019) 333−344.

[93] M. Sabater-Molina, E. Larque, F. Torrella, S. Zamora, Dietary fructooligosaccharides and potential benefits on health, Journal of Physiology and Biochemistry 65 (3) (2009) 315−328.

[94] P. Markowiak, K. Śliżewska, "Effects of probiotics, prebiotics, and synbiotics on human health, Nutrients 9 (2017) 9.

[95] B.K. Rodiño-Janeiro, M. Vicario, C. Alonso-Cotoner, R. Pascua-García, J. Santos, A review of microbiota and irritable bowel syndrome: future in therapies, Advances in Therapy 35 (3) (2018) 289–310.

[96] A. Rivière, M. Selak, D. Lantin, F. Leroy, L. De Vuyst, Bifidobacteria and butyrate-producing colon bacteria: importance and strategies for their stimulation in the human gut, Frontiers in Microbiology 7 (2016) 1–21.

[97] M. Rossi, Seyed S. Mirbagheri, A. Keshavarzian, F. Bishehsari, "Nutraceuticals in colorectal cancer: a mechanistic approach, European Journal of Pharmacology 833 (2018) 396–402.

[98] C.M. Whisner, C.M. Weaver, Prebiotics and bone, Understanding the Gut-Bone Signaling Axis, Advances in Experimental Medicine and Biology, Springer International Publishing, 2017, pp. 201–224.

[99] F. Respondek, P. Gerard, M. Bossis, L. Boschat, A. Bruneau, S. Rabot, et al., Short-chain fructo-oligosaccharides modulate intestinal microbiota and metabolic parameters of humanized gnotobiotic diet induced obesity mice, PLoS One 8 (2013) 8.

[100] A. Ashwini, H.N. Ramya, C. Ramkumar, K.R. Reddy, R.V. Kulkarni, V. Abinaya, et al., Reactive mechanism and the applications of bioactive prebiotics for human health: review, Journal of Microbiological Methods 159 (2019) 128–137.

[101] M.M.A. Scheid, Y.M.F. Moreno, M.R. Maróstica Jr., G.M. Pastore, Effect of prebiotics on the health of the elderly, Food Research International 53 (1) (2013) 426–432.

[102] G.T. Choque Delgado, W.M. da, S.C. Tamashiro, M.R.M. Junior, Y.M.F. Moreno, G.M. Pastore, The putative effects of prebiotics as immunomodulatory agents, Food Research International 44 (10) (2011) 3167–3173.

[103] L.A. David, C.F. Maurice, R.N. Carmody, D.B. Gootenberg, J.E. Button, B.E. Wolfe, et al., Diet rapidly and reproducibly alters the human gut microbiome, Nature 505 (7484) (2014) 559–563.

[104] A.R. Lobo, C. Colli, T.M.C.C. Filisetti, Fructooligosaccharides improve bone mass and biomechanical properties in rats, Nutrition Research 26 (8) (2006) 413–420.

[105] B.R. Martin, M.M. Braun, K. Wigertz, R. Bryant, Y. Zhao, W.H. Lee, et al., Fructo-oligosaccharides and calcium absorption and retention in adolescent girls, Journal of the American College of Nutrition 29 (4) (2010) 382–386.

[106] Y. Wang, T. Zeng, S. e Wang, W. Wang, Q. Wang, H.X. Yu, Fructo-oligosaccharides enhance the mineral absorption and counteract the adverse effects of phytic acid in mice, Nutrition 26 (3) (2010) 305–311.

[107] C. Yan, S. Zhang, C. Wang, Q. Zhang, A fructooligosaccharide from *Achyranthes bidentata* inhibits osteoporosis by stimulating bone formation, Carbohydrate Polymers 210 (2019) 110–118.

[108] P. Garcia-Peris, C. Velasco, M. Hernandez, M.A. Lozano, L. Paron, C. De La Cuerda, et al., Effect of inulin and fructo-oligosaccharide on the prevention of acute radiation enteritis in patients with gynecological cancer and impact on quality-of-life: a randomized, double-blind, placebo-controlled trial, European Journal of Clinical Nutrition 70 (2) (2016) 170–174.

[109] S. Zheng, P. Steenhout, D. Kuiran, W. Qihong, W. Weiping, C. Hager, et al., Nutritional support of pediatric patients with cancer consuming an enteral formula with fructooligosaccharides, Nutrition Research 26 (4) (2006) 154–162.

[110] M.C. Boutron-Ruault, P. Marteau, A. Lavergne-Slove, A. Myara, M.F. Gerhardt, C. Franchisseur, et al., Effects of a 3-mo consumption of short-chain fructo-oligosaccharides on parameters of colorectal carcinogenesis in patients with or without small or large colorectal adenomas, Nutrition and Cancer 53 (2) (2005) 160–168.

[111] L.B. Richards, M. Li, B.C.A.M. van Esch, J. Garssen, G. Folkerts, The effects of short-chain fatty acids on the cardiovascular system, PharmaNutrition 4 (2) (2016) 68–111.

[112] G.T.C. Delgado, R. Thomé, D.L. Gabriel, W.M.S.C. Tamashiro, G.M. Pastore, Yacon (*Smallanthus sonchifolius*)-derived fructooligosaccharides improves the immune parameters in the mouse, Nutrition Research 32 (11) (2012) 884–892.

[113] J.O. Lindsay, K. Whelan, A.J. Stagg, P. Gobin, H.O. Al-Hassi, N. Rayment, et al., Clinical, microbiological, and immunological effects of fructo-oligosaccharide in patients with Crohn's disease, Gut 55 (3) (2006) 348–355.

[114] Y. Nakamura, S. Nosaka, M. Suzuki, S. Nagafuchi, T. Takahashi, T. Yajima, et al., Dietary fructooligosaccharides up-regulate immunoglobulin A response and polymeric immunoglobulin receptor expression in intestines of infant mice, Clinical and Experimental Immunology 137 (1) (2004) 52–58.

[115] S. Fujitani, K. Ueno, T. Kamiya, T. Tsukahara, K. Ishihara, T. Kitabayashi, et al., Increased number of CCR4-positive cells in the duodenum of ovalbumin-induced food allergy model NC/jic mice and anti-allergic activity of fructooligosaccharides, Allergology International 56 (2) (2007) 131–138.

[116] A. Yasuda, K.I. Inoue, C. Sanbongi, R. Yanagisawa, T. Ichinose, T. Yoshikawa, et al., Dietary supplementation with fructooligosaccharides attenuates airway inflammation related to house dust mite allergen in mice, International Journal of Immunopathology and Pharmacology 23 (3) (2010) 727–735.

[117] A. Yasuda, K. Ichiro Inoue, C. Sanbongi, R. Yanagisawa, T. Ichinose, M. Tanaka, et al., "Dietary supplementation with fructooligosaccharides attenuates allergic peritonitis in mice, Biochemical and Biophysical Research Communications 422 (4) (2012) 546–550.

[118] M. Olesen, E. Gudmand-Høyer, Efficacy, safety, and tolerability of fructooligosaccharides in the treatment of irritable bowel syndrome, American Journal of Clinical Nutrition 72 (6) (2000) 1570–1575.

[119] D. Paineau, F. Payen, S. Panserieu, G. Coulombier, A. Sobaszek, I. Lartigau, et al., The effects of regular consumption of short-chain fructo-oligosaccharides on digestive comfort of subjects with minor functional bowel disorders, British Journal of Nutrition 99 (2) (2008) 311–318.

[120] J.L. Benjamin, C.R.H. Hedin, A. Koutsoumpas, S.C. Ng, N.E. McCarthy, A.L. Hart, et al., Randomised, double-blind, placebo-controlled trial of fructo-oligosaccharides in active Crohn's disease, Gut 60 (7) (2011) 923–929.

[121] L. Chi, I. Khan, Z. Lin, J. Zhang, M.Y.S. Lee, W. Leong, et al., Fructo-oligosaccharides from *Morinda officinalis* remodeled gut microbiota and alleviated depression features in a stress rat model, Phytomedicine 67 (2020) 153157.

[122] M.L.B. Pulido, J.B.A. Hernández, M.Á.F. Ballester, C.M.T. González, J. Mekyska, Z. Smékal, Alzheimer's disease and automatic speech analysis: a review, Expert Systems with Applications 150 (2020) 113213.

[123] D. Chen, X. Yang, J. Yang, G. Lai, T. Yong, X. Tang, et al., Prebiotic effect of fructooligosaccharides from *Morinda officinalis* on Alzheimer's disease in rodent models by targeting the microbiota-gut-brain axis, Frontiers in Aging Neuroscience 9 (2017) 1–28.

[124] C.H. Yen, C.H. Wang, W.T. Wu, H.L. Chen, Fructo-oligosaccharide improved brain β-amyloid, β-secretase, cognitive function, and plasma antioxidant levels in D-galactose-treated Balb/cJ mice, Nutritional Neuroscience 20 (4) (2017) 228–237.

[125] J. Sun, S. Liu, Z. Ling, F. Wang, Y. Ling, T. Gong, et al., Fructooligosaccharides ameliorating cognitive deficits and neurodegeneration in APP/ps1 transgenic mice through modulating gut microbiota, Journal of Agricultural and Food Chemistry 67 (10) (2019) 3006–3017.

[126] G. Chen, C. Li, K. Chen, Chapter 6—Fructooligosaccharides: a review on their mechanisms of action and effects, Studies in Natural Products Chemistry 48 (2016) 209–229.

[127] M. Song, A.T. Chan, J. Sun, Influence of the gut microbiome, diet, and environment on risk of colorectal cancer, Gastroenterology 158 (2) (2020) 322–340.

[128] R.K. Purama, M. Raman, P. Ambalam, S. Pithva, C. Kothari, M. Doble, Chapter 30—Prebiotics and probiotics in altering microbiota: implications in colorectal cancer, Immunity and Inflammation in Health and Disease: Emerging Roles of Nutraceuticals and Functional Foods in Immune Support, Academic Press, 2017, pp. 403–413.

[129] D. Davani-Davari, M. Negahdaripour, I. Karimzadeh, M. Seifan, M. Mohkam, S.J. Masoumi, et al., Prebiotics: definition, types, sources, mechanisms, and clinical applications, Foods 8 (3) (2019) 1–27.

[130] F.M.P. Galdino, M.E.R. Andrade, P.A.V. de Barros, S. de, V. Generoso, J.I. Alvarez-Leite, et al., Pretreatment and treatment with fructo-oligosaccharides attenuate intestinal mucositis induced by 5-FU in mice, Journal of Functional Foods 49 (2018) 485–492.

[131] T. Kondo, S. Saigo, S. Ugawa, M. Kato, Y. Yoshikawa, N. Miyoshi, et al., Prebiotic effect of fructo-oligosaccharides on the inner ear of DBA/2 J mice with early-onset progressive hearing loss, Journal of Nutritional Biochemistry 75 (19) (2020) 108247.

[132] G.T. Macfarlane, H. Steed, S. Macfarlane, Bacterial metabolism and health-related effects of galacto-oligosaccharides and other prebiotics, Journal of Applied Microbiology 104 (2) (2008) 305–344.

[133] G. Boehm, B. Stahl, Oligosaccharides from milk, The Journal of Nutrition 137 (3) (2007) 847S–849S.

[134] C. Kunz, H. Egge, Chapter 1—From bifidus factor to human milk oligosaccharides: a historical perspective on complex sugars in milk, Prebiotics and Probiotics in Human Milk: Origins and Functions of Milk-Borne Oligosaccharides and Bacteria, Elsevier Inc., 2017, pp. 3–16.

[135] H. Chu, X. Tao, Z. Sun, W. Hao, X. Wei, Galactooligosaccharides protects against DSS-induced murine colitis through regulating intestinal flora and inhibiting NF-κB pathway, Life Sciences 242 (2020) 117220.

[136] S. Musilova, V. Rada, E. Vlkova, V. Bunesova, Beneficial effects of human milk oligosaccharides on gut microbiota, Beneficial Microbes 5 (3) (2014) 273–283.

[137] A. Kulinich, L. Liu, Human milk oligosaccharides: the role in the fine-tuning of innate immune responses, Carbohydrate Research 432 (2016) 62–70.

[138] A.M. Taylor, H.D. Holscher, A review of dietary and microbial connections to depression, anxiety, and stress, Nutritional Neuroscience 23 (3) (2020) 237–250.

[139] K. Schmidt, P.J. Cowen, C.J. Harmer, G. Tzortzis, S. Errington, P.W.J. Burnet, Prebiotic intake reduces the waking cortisol response and alters emotional bias in healthy volunteers, Psychopharmacology 232 (10) (2015) 1793–1801.

[140] D.B.A. Silk, A. Davis, J. Vulevic, G. Tzortzis, G.R. Gibson, Clinical trial: the effects of a trans-galactooligosaccharide prebiotic on faecal microbiota and symptoms in irritable bowel syndrome, Alimentary Pharmacology and Therapeutics 29 (5) (2009) 508–518.

[141] T.N. Tibbs, L.R. Lopez, J.C. Arthur, The influence of the microbiota on immune development, chronic inflammation, and cancer in the context of aging, Microbial Cell 6 (8) (2019) 324–334.

[142] M.G. Suh, G.Y. Bae, K. Jo, J.M. Kim, K.B. Hong, H.J. Suh, Photoprotective effect of dietary galacto-oligosaccharide (GOS) in hairless mice via regulation of the MAPK signaling pathway, Molecules 25 (7) (2020) 1–9.

[143] J. Sun, W. Liang, X. Yang, Q. Li, G. Zhang, Cytoprotective effects of galacto-oligosaccharides on colon epithelial cells via up-regulating miR-19b, Life Sciences 231 (2019) 116589.

[144] D. Pranami, R. Sharma, H. Pathak, Lactulose: a prebiotic, laxative and detoxifying agent, Drugs and Therapy Perspectives 33 (5) (2017) 228–233.

[145] X. Zhai, X. Chen, J. Shi, D. Shi, Z. Ye, W. Liu, et al., Lactulose ameliorates cerebral ischemia-reperfusion injury in rats by inducing hydrogen by activating Nrf2 expression, Free Radical Biology and Medicine 65 (2013) 731–741.

[146] D.E. Soper, Bacterial vaginosis and surgical site infections, American Journal of Obstetrics and Gynecology 222 (3) (2020) 219–223.

[147] S.L. Collins, A. Mcmillan, S. Seney, C. Van Der Veer, R. Kort, M.W. Sumarah, Promising prebiotic candidate established by evaluation of lactitol, lactulose, raffinose, and oligofructose for maintenance of a Lactobacillus-dominated vaginal microbiota, Applied and Environmental Microbiology 84 (5) (2018) 1–15.

[148] M. Sueyoshi, M. Fukunaga, M. Mei, A. Nakajima, G. Tanaka, T. Murase, et al., Effects of lactulose on renal function and gut microbiota in adenine-induced chronic kidney disease rats, Clinical and Experimental Nephrology 23 (7) (2019) 908–919.

[149] L. Liang, G. Liu, G. Yu, F. Zhang, R.J. Linhardt, Q. Li, Urinary metabolomics analysis reveals the antidiabetic effect of stachyose in high-fat diet/streptozotocin-induced type 2 diabetic rats, Carbohydrate Polymers 229 (2020) 115534.

[150] W. Wang, H. Xin, X. Fang, H. Dou, F. Liu, D. Huang, et al., Isomalto-oligosaccharides ameliorate visceral hyperalgesia with repair damage of ileal epithelial ultrastructure in rats, PLoS One 12 (4) (2017) 1–14.

[151] J. Yang, P.H. Summanen, S.M. Henning, M. Hsu, H. Lam, J. Huang, et al., Xylooligosaccharide supplementation alters gut bacteria in both healthy and prediabetic adults: a pilot study, Frontiers in Physiology 6 (Aug) (2015) 1–11.

[152] A.A. Aachary, D. Gobinath, K. Srinivasan, S.G. Prapulla, Protective effect of xylooligosaccharides from corncob on 1,2-dimethylhydrazine induced colon cancer in rats, Bioactive Carbohydrates and Dietary Fibre 5 (2) (2015) 146–152.

[153] J. Long, J. Yang, S.M. Henning, S.L. Woo, M. Hsu, B. Chan, et al., Xylooligosaccharide supplementation decreases visceral fat accumulation and modulates cecum microbiome in mice, Journal of Functional Foods 52 (2018) 138–146.

[154] M. Zhang, S.L. Cai, J.W. Ma, Evaluation of cardio-protective effect of soybean oligosaccharides, Gene 555 (2) (2015) 329–334.

[155] B.B. Fei, L. Ling, C. Hua, S.Y. Ren, Effects of soybean oligosaccharides on antioxidant enzyme activities and insulin resistance in pregnant women with gestational diabetes mellitus, Food Chemistry 158 (2014) 429–432.

[156] Y. Zhou, Z. Ruan, X. Zhou, X. Huang, H. Li, L. Wang, et al., Lactosucrose attenuates intestinal inflammation by promoting Th2 cytokine production and enhancing CD86 expression in colitic rats, Bioscience, Biotechnology and Biochemistry 79 (4) (2015) 643–651.

[157] P. Muthukumaran, G. Thiyagarajan, R. Arun, et al., Raffinose from *Costus speciosus* attenuates lipid synthesis through modulation of PPARs/SREBP1c and improves insulin sensitivity through PI3K/AKT, Chemico-Biological Interactions 284 (2018) 80–89.

[158] Y. Zhu, L. Dong, L. Huang, Z. Shi, J. Dong, Y. Yao, et al., Effects of oat β-glucan, oat resistant starch, and the whole oat flour on insulin resistance, inflammation, and gut microbiota in high-fat-diet-induced type 2 diabetic rats, Journal of Functional Foods 69 (2020) 103939.

[159] E. Azzini, J. Giacometti, G.L. Russo, Antiobesity effects of anthocyanins in preclinical and clinical studies, Oxidative Medicine and Cellular Longevity 2017 (2017) 1–11.

[160] M. Moorthy, N. Chaiyakunapruk, S.A. Jacob, U.D. Palanisamy, Prebiotic potential of polyphenols, its effect on gut microbiota and anthropometric/clinical markers: a systematic review of randomised controlled trials, Trends in Food Science and Technology 99 (2020) 634–649.

[161] E.O. Igwe, K.E. Charlton, Y.C. Probst, K. Kent, M.E. Netzel, A systematic literature review of the effect of anthocyanins on gut microbiota populations, Journal of Human Nutrition and Dietetics 32 (1) (2019) 53–62.

[162] M.C. Rodríguez-Daza, L. Daoust, L. Boutkrabt, G. Pilon, T. Varin, S. Dudonné, et al., Wild blueberry proanthocyanidins shape distinct gut microbiota profile and influence glucose homeostasis and intestinal phenotypes in high-fat high-sucrose fed mice, Scientific Reports 10 (2020) 1.

[163] W. Sorndech, K.N. Nakorn, S. Tongta, A. Blennow, Isomalto-oligosaccharides: recent insights in production technology and their use for food and medical applications, LWT 95 (2018) 135–142.

[164] F.B. Subhan, Z. Hashemi, M.C. Archundia Herrera, K. Turner, S. Windeler, M.G. Gänzle, et al., Ingestion of isomalto-oligosaccharides stimulates insulin and incretin hormone secretion in healthy adults, Journal of Functional Foods 65 (2020) 103730.

[165] Y. Hu, V. Winter, M. Gänzle, In vitro digestibility of commercial and experimental isomalto-oligosaccharides, Food Research International 134 (2020) 109250.

[166] D.P. Singh, J. Singh, R.K. Boparai, J.H. Zhu, S. Mantri, P. Khare, et al., Isomalto-oligosaccharides, a prebiotic, functionally augment green tea effects against high fat diet-induced metabolic alterations via preventing gut dysbacteriosis in mice, Pharmacological Research 123 (2017) 103–113.

[167] T. Kaneko, T. Kohmoto, H. Kikuchi, M. Shiota, H. Iino, T. Mitsuoka, Effects of isomaltooligosaccharides with different degrees of polymerization on human fecal bifidobacteria, Bioscience, Biotechnology, and Biochemistry 58 (12) (1994) 2288–2290.

[168] Y. Hu, C.M.E. Heyer, W. Wang, R.T. Zijlstra, M.G. Gänzle, Digestibility of branched and linear α-gluco-oligosaccharides in vitro and in ileal-cannulated pigs, Food Research International 127 (2020) 108726.

[169] R. Zhang, J. Zhou, Z. Jia, Y. Zhang, G. Gu, Hypoglycemic effect of *Rehmannia glutinosa* oligosaccharide in hyperglycemic and alloxan-induced diabetic rats and its mechanism, Journal of Ethnopharmacology 90 (1) (2004) 39–43.

[170] M. Xi, Q. Yao, W. Ge, Y. Chen, B. Cao, Z. Wang, et al., Effects of stachyose on intestinal microbiota and immunity in mice infected with enterotoxigenic *Escherichia coli*, Journal of Functional Foods 64 (22) (2020) 103689.

6

Production of food enzymes

Qinghua Li[1,2], Guoqiang Zhang[1,2,3], Guocheng Du[2,4]

[1]NATIONAL ENGINEERING LABORATORY FOR CEREAL FERMENTATION TECHNOLOGY, JIANGNAN UNIVERSITY, WUXI, P.R. CHINA [2]SCHOOL OF BIOTECHNOLOGY AND KEY LABORATORY OF INDUSTRIAL BIOTECHNOLOGY, MINISTRY OF EDUCATION, JIANGNAN UNIVERSITY, WUXI, P.R. CHINA [3]JIANGSU PROVISIONAL RESEARCH CENTER FOR BIOACTIVE PRODUCT PROCESSING TECHNOLOGY, JIANGNAN UNIVERSITY, WUXI, P.R. CHINA [4]THE KEY LABORATORY OF CARBOHYDRATE CHEMISTRY AND BIOTECHNOLOGY, MINISTRY OF EDUCATION, JIANGNAN UNIVERSITY, WUXI, P.R. CHINA

6.1 Introduction

Enzymes are special proteins produced by organisms with a catalytic activity which plays an important role in different physiological activities. Animals, plants, and microorganisms synthesize a broad variety of enzymes. The demand for industry has made the development and application of enzymatic engineering a rapid process, and many research results have been obtained with the unremitting efforts of researchers. Ever since the French chemists, Anselme Payen and Jean-Francois Persoz successfully isolated diastase in 1833 [1], enzyme research has been initiated and rapidly growing. The natural properties, biochemical characteristics, and mechanisms of enzyme catalysis are progressively revealed and understood in the research process.

As enzyme research develops, the production and application of enzymes is also becoming ever more extensive and has been widely used in the food, feed, medication, fine chemicals, medical care, and environment fields [2], which played a crucial role in improving the technical level of relevant industries, saving energy and reducing consumption, and protecting the environment. The application of enzymes in the food industry has attracted considerable attention, like people paying increasing attention to food quality and safety. With the rapid development of the food industry, the application of enzyme preparations in the food industry is becoming ever more in-depth and extensive [3]. The application of food enzymes in the food industry, mainly reflected in the following: (1) production of food raw materials; (2) improving food quality; (3) preparation of bioactive components; (4) improving food safety; (5) food safety inspection, etc. Table 6−1 shows the commonly used enzymes and their key roles in various food industries.

Table 6-1 Overview of enzymes and their roles in food industries [4–7].

Enzyme	Production microorganisms	Food industries	Role
α-Acetolactate decarboxylase	Bacillus amyloliquefaciens, Bacillus subtilis, Saccharomyces cerevisiae	Brewing industry	Reduce the content of diacetyl in beer and improve beer flavor
Glucose oxidase	Aspergillus niger, Aspergillus oryzae, Penicillium chrysogenum	Bakery industry	Dough strengthening
Laccases	A. oryzae, Trametes hirsute, Trichoderma reesei	Fruit industry Brewing industry	Clarification of juices, flavor enhancer (beer)
Lipoxygenase	S. cerevisiae	Bakery industry	Dough strengthening, bread whitening
Fructosyltransferase	A. niger, A. oryzae, Kluyveromyces lactis	Functional foods	Synthesis of fructose oligomers
Transglutaminase	Streptomyces mobaraensis	Bakery industry Meat industry	Modification of viscoelastic properties, dough processing, meat processing
Transglucosidase	A. niger, B. amyloliquefaciens or subtilis, S. cerevisiae	Functional foods	To transfer the liberated glucose to another glucose or maltose molecule
Amylases	A. niger, A. oryzae, T. reesei, B. amyloliquefaciens, subtilis, licheniformis or stearothermophilus	Starch industry Bakery industry Fruit industry Brewing industry	Starch liquefaction and saccharification, flour adjustment, juice treatment, low-calorie beer
Proteases	Aspergillus melleus, A. niger, A. oryzae, B. amyloliquefaciens, subtilis, licheniformis or stearothermophilus, K. lactis, Rhizomucor miehei, T. reesei	Dairy industry Bakery industry Functional foods Meat industry Aquatic product Brewing industry	Milk clotting, low-allergenic infant food formulation, flavor improvement in milk and cheese, meat tenderizer
Glucanase	A. niger, A. oryzae, B. amyloliquefaciens or subtilis, Penicillium, T. reesei	Feed industry	Viscosity reduction in barley and oats, enhanced digestibility
Asparaginase	A. niger, A. oryzae	Bakery industry	Reduce the content of acrylamide
α-Galactosidase	A. niger, S. cerevisiae	Feed industry	Viscosity reduction in lupins and grain legumes, enhanced digestibility
Glucoamylase	A. niger, Rhizopus oryzae, T. reesei	Brewing industry	Saccharification
Invertase	S. cerevisiae	Sugar industry	Sucrose hydrolysis, production of invert sugar sirup
Chymosin	A. oryzae, Mucor miehei, B. subtilis	Dairy industry	Causes raw milk to coagulate and provides conditions for the discharge of whey
Papain	Pichia pastoris	Brewing industry Meat industry	Hydrolyze the protein in beer to avoid turbidity caused by cold storage, meat tenderizer
Glutaminase	B. amyloliquefaciens or subtilis	Beverage industry	Catalytic deamidation
Lactase	A. oryzae, Bacillus circulans, K. lactis	Dairy industry	Lactose hydrolysis, whey hydrolysis
Lipase	A. niger, A. oryzae, R. oryzae, Pseudomonas fluorescens, Candida lipolytica	Dairy industry Bakery industry	Cheese flavor, emulsification for dough
Lysozyme	Streptomyces violaceoruber	Aquatic product Meat industry Dairy industry	Preservation of food increase intestinal antiinfectivity, digestion of casein
Hemicellulases	A. niger, A. oryzae, P. pastoris	Feed industry Dairy industry	Promote the digestion and absorption of fat and protein in animals, improve the texture and softness of bread
Bromelain		Fruit industry Brewing industry	Selective hydrolysis of fibrin, clarification of beer
Pectinase	A. niger, T. reesei	Fruit industry	Juice clarification

(Continued)

Table 6-1 (Continued)

Enzyme	Production microorganisms	Food industries	Role
Peptidase	A. niger, A. oryzae, B. amyloliquefaciens, or subtilis	Bakery industry Dairy industry	Hydrolysis of proteins, cheese ripening
Arabino-furanosidase	A. niger, P. pastoris	Sugar industry	Hydrolyze glycoside bonds on side chains of arabinoxylan
Phospholipase	A. niger, A. oryzae, S. vialoceoruber, T. reesei, P. pastoris	Bakery industry	Emulsification for dough
Xylanases	A. niger, A. oryzae, B. amyloliquefaciens, licheniformis or subtilis, S. violaceoruber, T. reesei	Bakery industry	Viscosity reduction, enhanced digestibility
Tannase	A. niger	Beverage industry Brewing industry	Clarification
Urease	Lactobacillus fermentum	Food analysis and testing	Detection of heavy metal content in food

6.2 Applications of food enzymes in food industry

6.2.1 Food processing of grains and oils

During the baking process, the batter or dough is subjected to a series of enzymatic reactions. The alpha-amylase was most widely used in baked foods with relatively high inactivation temperature (65°C~75°C). Present uses of alpha-amylase include malt-amylase, bacterial-amylase, and fungal-amylase in baked bread. Different sources of alpha-amylase have special purposes. Malt-amylase could enhance the flavor, structure, and color of baked products. Fungal-amylase brings a positive effect on the quality of the dough and prolongs the baked food's shelf-life. The bacterial-amylase is resistant to high temperatures but it produces a high amount of soluble dextrin during baking, making the final product sticky. Dough properties and composition are key factor in the baking industry. The texture of the dough is largely determined by gluten (a protein that forms the basic network in wheat). It is known that glucose oxidase (GOD) can catalyze glucose to glucose acid and hydrogen peroxide. Hydrogen peroxide can oxidize the sulfhydryl group (-SH) in gluten into a disulfide bond (-S-S-) as a strong oxidant, thus improving the strength of the dough and its processing performance. Hydrogen peroxide can also oxidize in flour carotene, lutein, and other plant pigments to enhance food color and brightness [8]. Besides, lipase, lipoxygenase, laccase, xylanase, and pentosanase may improve the color, taste, elasticity, and other baking product qualities. The synergistic effect of multiple enzymes is often the result of the final quality of food products, and most of the enzyme preparations themselves are susceptible to multiple factors. Therefore the influence of many factors on the activity of preparations for enzymes must be considered simultaneously, and it is best to combine several preparations for enzymes [9].

Transglutaminase (TG) can catalyze the cross-linking of protein molecules and the connection between protein and amino acids to form a protein network structure, alter the properties of noodle products such as protein plasticity, water retention, and stretch resistance, thereby improving the texture and structure of flour products [10]. Protease can be used to break down the gluten and make the dough soft to improve the dough's viscoelasticity and fluidity [11].

Plant oils are mainly present in the cytoplasm of oilseeds and are surrounded by stable/strong cell walls. During oil pressing, hemicellulase, cellulase, pectinase, and protease are used to hydrolyze the cell wall, lipids, and their complexes (lipopolysaccharides and lipoprotein). The oil could be released from the complex and the yield of oil pressing could be significantly improved [12]. Moreover, the unrefined oil contains a certain amount of free fatty acids which can easily produce a pungent smell and affect the oil flavor. Lipase can catalyze the fatty acid and glyceride synthesis into deacidifying triglycerides. Presently, lipase has been developed for oil modification to produce 1,2- or 1,3-diglyceride by hydrolysis of triglyceride. For example, 1,2-diglyceride has a greater emulsification effect than triglyceride, and in the food industry, it has a higher application value. Additionally, 1,3-diglyceride is a kind of functional grease because it can be absorbed into the small intestine and does not form fat again in the human body [13].

6.2.2 Dairy industry

The dairy industry is an important part of the food industry. The enzymes used in the dairy industry mainly include chymosin, galactosidase, lipase, etc., which are used in cheese, baby milk powder, lactose dairy products, butter flavoring, etc. The key to cheese production is the chymosin with good performance. In the past, chymosin was extracted from the gastric juice of slaughtered cows. With the growth of the cheese market, the yield of chymosin is far from enough to meet the demand for cheese production. The heterologous expression of chymosin is now used for industrial cheese making. The rennet produced by fermentation can be prepared by introducing the calf chymosin gene into microorganisms. Camel chymosin has been used to produce a cheese with a high-fat content, which describes an ideal property of limited hydrolysis to produce a soft, chewy product with a low bitterness [14]. At present, more than 85% of chymosin in cheese production comes from microorganisms.

Lactose affects milk sweetness, solubility, and causes lactose intolerance symptoms. To solve the problem of lactose intolerance, lactase can catalyze the hydrolysis of lactose into galactose and glucose [15]. Butter has a unique aroma, mainly derived from volatile substances such as fatty acids, esters, and amines. Lipase can partially degrade butter to produce more fatty substances such as fatty acids, which enhances its flavor [16].

6.2.3 Meat products processing

In the meat processing industry, enzymes can be applied to improve tissue structure, transform low-value proteins, tenderize meat, and so on. Collagen is a fibrous protein with strong mechanical strength, cross-linked by secondary bonds. Its content has an important impact on the quality of meat. Proteases can partially hydrolyze collagen and play a role in tenderizing meat [17]. The tenderization process that relies on proteolytic enzyme-containing plant extracts has the problem of overtenderization triggered by excessive hydrolysis. Therefore microbial proteases are exploited for meat tenderization. It had been reported that aspartic protease from *Rhizomucor miehei* showed good effects on tenderness and yield [18].

TG is a useful enzyme that catalyzes the formation of a covalent bond between the amino group of lysine and the γ-hydroxyl amide group of glutamic acid, leading to protein polymerization. Through the catalytic action of TG, ε-(γ-glutamyl)-lysine isopeptide bonds are formed to improve the protein's water absorption, water retention, viscosity, and gelation. At the same time, the functional properties such as adhesion and emulsification help to improve the appearance, flavor, and texture of protein foods [19]. The role of enzymes in meat processing is to: (1) improve the texture of meat products; (2) improve the utilization of meat raw materials; and (3) develop new healthy meat foods or protein foods [10]. As the degree of hydrolysis increases, bitter taste peptides are easily produced when producing meat protein peptides, which affect the quality of the product. The use of proteases that do not generate bitter peptides through hydrolysis is one way to solve this problem. Carboxypeptidase may also be used to continue the hydrolysis of bitter peptides to remove the product bitter taste [20].

6.2.4 Food analysis and testing

With the continuous improvement of people's safety awareness, food safety testing has been highly regarded. The development of rapid detection and analysis methods for the types and contents of key ingredients in diet products has become one of the research hotspots. Due to the change of food components, conventional chemical detection is difficult. To solve this problem, the enzyme detection method was developed for the determination of ingredients in the food items, which is simple, fast, and highly sensitive. For example, urease, catalase, GOD, and phosphatase are commonly used enzymes for detection. Among them, the urease enzyme is highly specific, very efficient, and it is highly sensitive to heavy metal ions during the detection process. The enzyme sensor combined with electrochemical analysis can accurately, quickly, and conveniently determine specific substances [21]. Besides, the application of PCR [22] and enzyme-linked immunosorbent assay [23] technology may further shorten the test time, improve efficiency, and accuracy in food detection.

6.2.5 Functional foods

Functional food is defined as a natural food or processed food containing known or unknown biologically active substances, in which the content of biologically active substances should be clear to ensure that the functional food is effective, nontoxic, and clinical trials prove that it helps prevent or treat chronic diseases [24]. The protein hydrolysate degraded and prepared by various proteases from plants and microorganisms has a significant role in the production of healthy foods, baby foods, and soft drinks. Proteins used in the preparation of hydrolysates are relatively extensive, such as casein, gelatin, soy protein, whey, fish protein, and meat protein. Although alkaline proteases had been used in the preparation of protein hydrolysates, acid proteases, and neutral proteases also have great potential for development and application.

The production of functional oligosaccharides is an emerging field of application and development of food enzymes [2]. Currently, the oligosaccharides that have been widely used as food raw materials mainly include isomalto-oligosaccharide, xylooligosaccharides, fructooligosaccharides, and galactooligosaccharides. The production of different types of oligosaccharides requires corresponding enzymes. For example, oligosaccharides are mainly produced by the hydrolysis of xylan in cellulosic materials by xylanase [25]. Fructosidase hydrolyzes sucrose to fructose and glucose and then transfers fructosyl to sucrose to synthesize fructooligosaccharide. The production of galactooligosaccharides is based on the synthesis of galactosidase [26].

6.3 Current progress and challenges in food enzymes production

The application of enzymes in food is becoming more extensive with the increasing variety of the food processing industry [27]. The overview of enzyme production and related technologies is shown in Fig. 6–1. Due to the involvement of extreme environments in food processing, some enzymes cannot meet the requirements. So there is a need to develop and produce new high-efficiency enzymes adapted to various food processing conditions. In recent years, great progress has been made in the discovery of new enzymes based on traditional techniques. Still, the high time-consumption and low screening efficiency seriously hinder the acquisition of new enzymes. The screened enzymes often need to be modified at the molecular level to satisfy the application conditions of the food processing industry [28]. At present, molecular modifications are focused on combination methods of rational design and irrational design [29]. It is necessary to develop an efficient modification of the enzyme

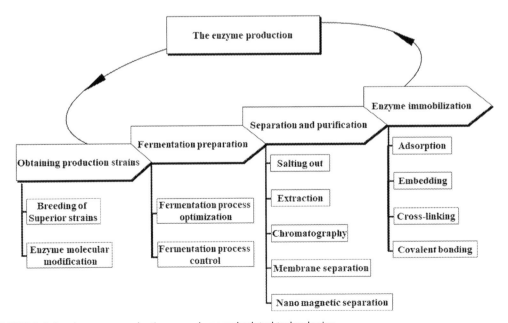

FIGURE 6–1 Food enzymes production procedures and related technologies.

molecular structure to find the relationship between protein structure and function. It would help to establish a theoretical system by cross-combining computational biology, statistics, and molecular biology.

To improve the catalytic performance and environmental adaptability of enzyme proteins, various genetic engineering techniques for enzyme protein molecule modification have been widely used. Directed evolution, which mimics the natural evolution mechanism, plays a key role in the modification of enzyme protein stability. It can obtain mutants that meet the needs without prior understanding of the spatial structure and mechanism of protein molecules. To improve the thermal stability of lipases expressed in *Pichia pastoris*, Yu et al. used two rounds of error-prone PCR and DNA shuffling technology, respectively [30]. The dissolution temperature is increased by 22°C, and the thermal stability is increased by 46 times at 60°C. Directed evolution has been widely used in the modification of enzymes. However, there are still some issues to resolve, such as heavy screening, and most of the mutant characteristics have not improved. As more and more crystal structures of enzymes have been resolved, scholars have gradually developed semirational and rational transformation methods that combine the crystal structure and catalytic properties of enzymes. Semirational design is based on a certain understanding of the molecular structure and function of the enzyme. Rational design is based on a well understanding of the structure and function of the enzyme protein, and mainly uses the site-directed mutation technique and site-directed saturation mutation technique to transform the enzyme molecule.

The efficient enzyme expression system is one of the main limitations of industrial enzyme production. It is affected by multiple factors, such as cell growth characteristics, intracellular/extracellular expression patterns, and posttranslational modification. It is crucial to choose an efficient expression system from diverse hosts. For example, the *Bacillus* expression system has the advantage of high-efficiency secretion ability, low protease activity, and inexpensive and broad fermentation medium [31]. Yeast expression system has the advantages of protein posttranslational modification (such as glycosylation), high heat tolerance, and high-density fermentation [32]. The filamentous fungal expression system can efficiently express macromolecular eukaryotic proteins and has a strong ability to secrete proteins with complicated posttranslational processing [33]. The lactic acid bacteria expression system is characterized as a safe expression host in the field of food or medicine [34]; however, its expression level is low [35]. Therefore, to select the appropriate expression system according to the target enzyme is the prerequisite for efficient enzyme expression.

During the heterologous expression of microbial enzymes, modification problems such as protein misfolding, insolubility, or lack of glycosylation after folding always result in the formation of inactive proteins. Routine methods are to adjust the synthesis rate and folding efficiency of the enzyme protein, such as optimizing the promoter and signal peptide, optimizing the induction system, and coexpression with molecular chaperones [36]. Given the lack of glycosylation, it mainly occurs in the expression process of proteins derived from eukaryotic cells. This problem can be solved by choosing eukaryotic expression hosts such as yeast and filamentous fungi [32,33]. *Saccharomyces cerevisiae* is considered to be one of the safest hosts, and the enzyme preparations expressed in this host are widely employed in food and medicine fields [37]. Molecular modification methods (such as fusion chaperone, gene knockout, etc.) can increase the high-level expression of recombinant protein in *S. cerevisiae*. It had been reported that the endoglucanase of *Paenibacillus barcinonensis* was successfully expressed in *S. cerevisiae* using different regions of cell wall proteins as translation fusion partners [38]. Meanwhile, the regulation of the pathway regulator Hac1p can activate the secretory pathway and promote the secretion of recombinant protein in *S. cerevisiae*. So far, a few filamentous fungal hosts have been developed for the production of recombinant protein. *Aspergillus niger* and *Aspergillus oryzae* are food-safe hosts and are suitable for the production of food and pharmaceutical-grade enzyme preparations [39,40]. A variety of recombinant proteins were successfully expressed in *A. niger* or *A. oryzae* [41,42].

Microorganisms require different fermentation environments. A suitable growth environment is conducive to efficient enzyme expression. The culture media required for different stages of microbial growth are also different. The main factors affecting the culture medium include carbon and nitrogen sources, trace element contents, and inorganic salts. The external environment of fermentation, such as temperature, pH, and dissolved oxygen, will affect the growth and metabolism of microorganisms, thereby affecting the production of enzyme proteins. Dissolved oxygen is one of the key factors for aerobic microorganisms. Sufficient oxygen can maintain the normal growth and metabolism of microorganisms. At present, the strategies used to optimize the fermentation process of microorganisms are response surface design strategy, Plackett–Burman design strategy, and orthogonal experiment design strategy [43].

Separation and purification is another important step in enzyme production. Enzyme purification methods are usually based on the molecular weight, spatial structure, charge properties, solubility, and specific binding sites of the enzyme. Under normal circumstances, it often requires a combination of multiple purification methods to purify enzymes with higher purity. When designing the purification scheme of the target enzyme, the influence of different purification methods and the order of purification on the purification effect should also be considered. Traditional protein separation technology has a long history, and the relevant theoretical basis is also relatively clear. Salting-out is the most commonly used separation method in the separation of active proteins. It specifically refers to the process of adding inorganic salts to the solution to reduce protein solubility. The most commonly used salt for protein salting-out is ammonium sulfate. It has good precipitation and gentlest effect on the protein and can keep its biological activity. Besides, the separation of ion-exchange chromatography and hydrophobic chromatography is closer to physiological conditions, which is conducive to the maintenance of protein biological activity. The combination of these two chromatographic techniques is widely used in the separation of proteins. With the development of technology and the emergence of new materials, some new membrane separation technologies, such as ion-exchange membrane separation [44] and affinity membrane separation [45], have been developed. However, due to the relatively complicated separation principle, it is still in the stage of exploration and research.

Immobilized enzyme technology also plays an important role in food enzyme applications. The immobilized enzyme technology is helpful to recycle the enzyme, improve the properties of the enzyme, and reduce the catalytic cost. The immobilized enzyme technology includes adsorption, embedding, cross-linking, and covalent bonding. For example, the adsorption method can avoid the chemical modification of the enzyme and can maintain the vitality of the enzyme to a large extent. The preparation method of the embedding method is relatively simple, and the recovery rate of enzyme activity after immobilization is also high. However, traditional immobilization methods also have certain limitations. For example, substrate and product diffusion are not efficient [46], the microenvironment affects the kinetic properties of enzymes [47,48], which limits the further application of immobilized enzymes. Therefore it is necessary to develop new immobilized enzyme technology to overcome the above deficiencies. For example, intelligent carrier immobilized enzymes that are sensitive to environmental factors (temperature, pH, ions, and solvents) can undergo solid−liquid phase transitions to maintain the activity of enzymes and maintain the diffusivity of substrates and products [49,50]. The substrate and the product can be efficiently diffused, which helps the diffusion of the product and reduces the product inhibition. A suitable immobilization method can improve the binding capacity and catalytic effect of the enzyme to a certain extent. Therefore, to obtain an excellent intelligent carrier, the molecular design is combined with the properties of various polymer materials. For example, the introduction of appropriate reactive groups makes the hyperbranched polymer have a unique supramolecular structure effect to achieve ideal environmental responsiveness.

6.4 Advanced technologies for food enzymes production

With the wide application of enzymes in the food field, food processing has put forward higher requirements on the properties of enzymes. It is of great significance to obtain new and efficient food enzymes suitable for various food processing. The main related advanced technologies are shown in Fig. 6−2.

6.4.1 High-throughput screening technology

The traditional screening approaches are time-consuming, labor-intensive, and have low efficiency. High-throughput screening technology plays a key role in the screening of various food enzymes due to its advantages of high automation, high sensitivity, high specificity, and high throughput [51]. The huge screening library helps to improve the possibility of screening the target phenotype strain. In a study, Liu et al. used droplet-microfluidic technology to successfully screen 12 positive mutants with significantly improved antibiotic resistance from a library containing 10^9 mutants in 13 hours [51]. At present, many effective high-throughput screening technologies have been successfully developed.

High-throughput screening technology based on the microfluidic droplet platform is developing rapidly [52]. The microfluidic droplet platform has a compartment with good separation of single cells, which can be used as a microreactor for cell culture, protein expression analysis, metabolite detection, and omics analysis. It provides favorable technical support for high-throughput screening of food enzymes. Compared with traditional screening technology, droplet-microfluidic technology shows higher throughput and accuracy, which greatly reduces the screening time and cost. It can analyze up to 1.5 million samples

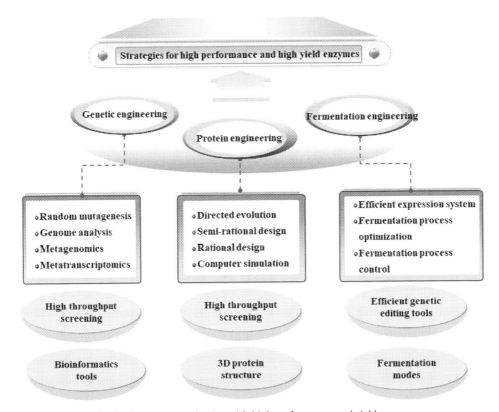

FIGURE 6–2 Strategies for food enzymes production with high performance and yield.

simultaneously and can introduce environmental disturbances quickly [53]. The dynamic microenvironment control system provides a powerful tool for obtaining ideal microbial strains. Fluorescence signal detection has high sensitivity and is particularly suitable as a detection signal for a microreactor of the microfluidic droplet platform. Besides, the development of fluorescence-activated cell sorting (FACS) has greatly improved the screening efficiency. At the single-cell level, cells are sorted according to fluorescence intensity, and each screening flux can reach 10^9 mutants [54]. For enzymes that have no fluorescence by themselves, target enzymes can be selected by adding a substrate to produce fluorescence. Based on microfluidic technology and FACS technology, a round of high-throughput screening can enrich positive clones by 196 times.

6.4.2 Strategies for efficient modification of food enzymes

Many in silico tools such as SWISS-MODEL [55], HotSpot Wizard [56], ProSAR [57], and SCHEMA have been established and improved. They can be used to analyze the three-dimensional structure of the enzyme or perform a homology model to effectively guide the enzyme modification [29]. ASRA (adaptive substituent reordering algorithm) can be used to identify enzyme variants with the desired characteristics in the focused mutation library. It is an alternative to the traditional quantitative structure−activity relationship method. Combined with some experimental methods, such as iterative saturation mutagenesis [58], the prediction of enzyme characteristics can be realized [59]. SCHEMA is a structure-oriented computational algorithm that can guide the reorganization of multiple proteins to obtain chimeric enzymes with better thermal stability or substrate specificity [60]. For instance, based on the application of this technique, human arginase I and II were recombined to form a catalytically active chimera [61].

Besides, Discovery Studio (DS) is a powerful software for enzyme engineering, including simulation of three-dimensional structures, computational chemistry, and drug design. It is widely used in the characterization of interactions between small-molecule ligands and biological macromolecules, homology modeling, molecular mechanics calculation, molecular dynamics simulation, and drug design [62]. The development of computer simulation technology has accelerated the pace of the rational transformation of enzyme proteins. Homologous modeling technology provides a new method for protein structure prediction and achieves the perfect combination of biotechnology and computer technology. At the same time, DS has also been widely developed and applied in the process of structural and functional relationship analysis, molecular design, and is constantly showing great vitality [63]. When studying a target protein, even if there is no information about its three-dimensional structure, the homologous structure can be used as a template to predict its structure utilizing molecular simulation technology, which has been gradually employed in the rational transformation of enzyme molecules [64]. In addition to changing the amino acid or loop composition of the enzyme molecule through genetic manipulation, it is also possible to introduce exogenous molecules or forces to stabilize the conformation of the enzyme molecule to improve its stability. For example, self-aggregating amphoteric peptides contain a certain number of hydrophilic and hydrophobic amino acids, which can spontaneously form regular nanostructures.

6.4.3 Development of safe and efficient expression hosts

As the enzymes have been applied in the food industry widely, it has greatly promoted the development of the food industry. However, the safety issues of enzymes also affect food safety. Therefore, the safety evaluation of enzymes for food is an unavoidable problem in the field of food processing. The biological enzymes used in the food industry should be derived from food-grade microorganisms. The biological enzymes produced by host bacteria such as *Escherichia coli* and *P. pastoris* have potential safety issues. Besides, the use of nonfood-grade selectable markers in a food-grade host, bacterial expression systems, such as antibiotic resistance genes [65]. Therefore the development of food-grade expression systems has become necessary for the production of food-grade enzymes.

At present, food-grade microorganisms mainly include *Bacillus subtilis*, lactic acid bacteria, yeast, and some *Aspergillus*. The following are some of the advantages of *B. subtilis* expression system: (1) there is no obvious codon preference, and the expression products are not easy to form inclusion bodies; (2) it has a strong protein secretion function, which is conducive to the recovery and purification of the target protein and other subsequent operations; (3) the genetic background research is relatively detailed and has many years of industrial production technology basis. It has been proved that B. subtilis can express a variety of soluble and bioactive proteins. However, this expression system also has deficiencies: (1) the recombinant plasmid is unstable; (2) the molecular cloning efficiency is low; (3) production of a large number of extracellular proteases. Therefore the B. subtilis expression system, as an effective tool for exogenous protein expression, still needs to be improved. It is one of the urgent problems to develop with integrated carriers, nutrition-deficient carriers, and new Bacillus and filamentous fungi expressing host strains with excellent characters and establish a safe, efficient, and stable food enzyme platform system employing molecular biology and genetics.

In recent years, the development and modification of filamentous fungi such as A. niger and A. oryzae have been favored by researchers because of their protein secretion capacity and complicated posttranslational processing. Filamentous fungi are often used in the production of recombinant enzymes, organic acids, and food additives [66]. Therefore the development of filamentous fungi such as A. niger and A. oryzae into food-grade expression systems has great potential and significance. However, problems such as low yield of heterologous protein expression [67], unclear metabolic regulation mechanism [68], and difficult genetic manipulation have seriously hampered the development and application of filamentous fungal expression systems. Recently, the successful application of CRISPR/Cas9 editing technology in filamentous fungi has accelerated the pace of research [69], but it is still at the initial stage of genetic editing. Therefore developing a variety of highly efficient genetic manipulation tools and improving editing efficiency is an important part of development of expression systems.

6.4.4 Optimization of food enzyme fermentation process

The optimization of the fermentation process plays a significant role in the enzyme fermentation industry. It is a key technical issue in the field of fermentation engineering to study the relatively unified fermentation process optimization theory of high yield, high substrate conversion rate, high production intensity, and successfully applied it to industrial practice [43]. Different microorganisms need different fermentation conditions to boost the production of beneficial enzyme fermentation products due to their metabolism. At present, the optimization of the fermentation process is mainly based on the understanding of the function of microorganisms to model and optimize the fermentation process. When the macroenvironmental conditions change, microorganisms can automatically adjust the intracellular metabolic network to adapt to changes in the environment. Therefore the optimization of growth environment, cultivation mode, and induction strategy can affect the metabolism flow direction and flux of microbial metabolism by regulating the intracellular microenvironment and the extracellular macroenvironment. However, due to the complexity of the bacterial metabolic network, the diversity of model parameters, and the difficulty of measuring intracellular substances, the establishment of fermentation models has become difficult in the modeling fermentation process. Also, the inaccuracy of the model parameters or deviations in the fermentation process will also lead to the deterioration of the optimal control results.

At present, with the development of modeling and detection technologies, there are many models for biological process control and optimization, such as a metabolic network model and fermentation kinetic model [70]. Process state variables are generally expressed in the form of concentration or product activity, such as cell concentration, substrate concentration, and various product concentrations. This type of model uses several equations to describe the characteristics of biological processes, and the operating variables are also written into the equations of state in the form of variables. Several of the most important kinetic model parameters in the equation of states can be described by choosing different empirical formulas or models, such as the specific proliferation rate of bacteria, the specific consumption rate of substrates, and the specific production rate of metabolites. This model reflects the apparent dynamic characteristics of the process and cannot consider all the reaction networks involved in biological processes. Therefore there is a long way for the exploration of fermentation process modeling.

6.5 Conclusions and perspectives

Food enzymes have been widely used in various food industries. The characteristics of their mild conditions, low energy consumption, and less pollution are in line with the current world theme of "green, safety, and environmental protection." With the continuous upgrading of the food processing industry, the types and performance requirements of enzymes are increasing and improving. Therefore it is of great significance to use ultra-high-throughput screening technology to obtain new enzymes with excellent properties that can meet the needs of food processing. At the same time, a semirational or rational design and modification strategy, combining the crystal structure of the enzyme molecule and computer-aided simulation, will be used more widely to improve the catalytic properties of the enzyme. Besides, because enzyme production is still relatively expensive relative to some chemical additives, it is important to develop safe, economical, and efficient expression systems and optimize existing fermentation processes.

The development of efficient food enzymes, the establishment of food-level large-scale expression systems, optimization of enzyme fermentation and cleaning processes, and the establishment of food enzyme safety evaluation systems will play a key role in improving food quality and safety, which will also produce huge economic and social benefits.

References

[1] A. Payen, J.F. Persoz, Mémoire sur la diastase, les principaux produits de ses réactions, et leurs applications aux arts industriels, Annales de chimie et de physique 53 (1833) 73–92.

[2] J.M. Choi, S.S. Han, H.S. Kim, Industrial applications of enzyme biocatalysis: current status and future aspects, Biotechnology Advances 33 (2015) 1443–1454.

[3] S. Li, X. Yang, S. Yang, Technology prospecting on enzymes: application, marketing and engineering, Computational and Structural Biotechnology Journal 2 (2012) e201209017.

[4] P. Singh, S. Kumar, Microbial enzyme in food biotechnology, Enzymes in Food Biotechnology (2019) 19–28.

[5] P. Fernandes, F. Carvalho, Microbial enzymes for the food industry, Biotechnology of Microbial Enzymes (2017) 513–544.

[6] R.C. Ray, C.M. Rosell, Microbial Enzyme Technology in Food Applications, CRC Press, 2017.

[7] H. Rastogi, S. Bhatia, Future prospectives for enzyme technologies in the food industry, Enzymes in Food Biotechnology (2019) 845–860.

[8] T. Yang, Y. Bai, F. Wu, Combined effects of glucose oxidase, papain and xylanase on browning inhibition and characteristics of fresh whole wheat dough, Journal of Cereal Science 60 (2014) 249–254.

[9] I. Alaunyte, V. Stojceska, A. Plunkett, Improving the quality of nutrient-rich Teff (Eragrostis tef) breads by combination of enzymes in straight dough and sourdough breadmaking, Journal of Cereal Science 55 (2012) 22–30.

[10] A.L.C. Gaspar, S.P. de Góes-Favoni, Action of microbial transglutaminase (MTGase) in the modification of food proteins: a review, Food Chemistry 171 (2015) 315–322.

[11] J. Mamo, F. Assefa, The role of microbial aspartic protease enzyme in food and beverage industries, Journal of Food Quality 2018 (2018).

[12] M.M. Yusoff, M.H. Gordon, K. Niranjan, Aqueous enzyme assisted oil extraction from oilseeds and emulsion de-emulsifying methods: a review, Trends in Food Science & Technology 41 (2015) 60–82.

[13] M. Mukherjee, Human digestive and metabolic lipases—a brief review, Journal of Molecular Catalysis B: Enzymatic 22 (2003) 369–376.

[14] N. Bansal, M. Drake, P. Piraino, Suitability of recombinant camel (*Camelus dromedarius*) chymosin as a coagulant for Cheddar cheese, International Dairy Journal 19 (2009) 510–517.

[15] A. Lamri, A. Poli, N. Emery, The lactase persistence genotype is associated with body mass index and dairy consumption in the DESIR study, Metabolism 62 (2013) 1323–1329.

[16] A. Tomasini, G. Bustillo, J.M. Lebeault, Fat lipolyzed with a commercial lipase for the production of Blue cheese flavour, International Dairy Journal 3 (1993) 117–127.

[17] K. Ryder, M. Ha, A.E.D. Bekhit, Characterisation of novel fungal and bacterial protease preparations and evaluation of their ability to hydrolyse meat myofibrillar and connective tissue proteins, Food Chemistry 172 (2015) 197–206.

[18] Q. Sun, F. Chen, F. Geng, A novel aspartic protease from *Rhizomucor miehei* expressed in *Pichia pastoris* and its application on meat tenderization and preparation of turtle peptides, Food Chemistry 245 (2018) 570–577.

[19] J. Stangierski, R. Rezler, G. Lesnierowski, Analysis of the effect of heating on rheological attributes of washed mechanically recovered chicken meat modified with transglutaminase, Journal of Food Engineering 141 (2014) 13–19.

[20] L. Kupski, M.I. Queiroz, E. Badiale-Furlong, Application of carboxypeptidase A to a baking process to mitigate contamination of wheat flour by ochratoxin A, Process Biochemistry 64 (2018) 248–254.

[21] E. Mahmoudi, H. Fakhri, A. Hajian, High-performance electrochemical enzyme sensor for organophosphate pesticide detection using modified metal-organic framework sensing platforms, Bioelectrochemistry 130 (2019) 107348.

[22] H.A. Al-Kahtani, E.A. Ismail, M.A. Ahmed, Pork detection in binary meat mixtures and some commercial food products using conventional and real-time PCR techniques, Food Chemistry 219 (2017) 54–60.

[23] L. Xu, X. Suo, Q. Zhang, ELISA and chemiluminescent enzyme immunoassay for sensitive and specific determination of lead (II) in water, food and feed samples, Foods 9 (2020) 305.

[24] M.C. Hueda, Functional Food: Improve Health through Adequate Food, BoD – Books on Demand, 2017.

[25] F. Kallel, D. Driss, F. Bouaziz, Production of xylooligosaccharides from garlic straw xylan by purified xylanase from *Bacillus mojavensis* UEB-FK and their in vitro evaluation as prebiotics, Food and Bioproducts Processing 94 (2015) 536–546.

[26] M. Frenzel, K. Zerge, I. Clawin-Rädecker, Comparison of the galacto-oligosaccharide forming activity of different β-galactosidases, LWT—Food Science and Technology 60 (2015) 1068–1071.

[27] P.S. Panesar, Enzymes in Food Processing: Fundamentals and Potential Applications, IK International Pvt Ltd, 2010.

[28] P. Dheeran, S. Kumar, Y.K. Jaiswal, Characterization of hyperthermostable α-amylase from *Geobacillus* sp. IIPTN, Applied Microbiology and Biotechnology 86 (2010) 1857–1866.

[29] J. Damborsky, J. Brezovsky, Computational tools for designing and engineering biocatalysts, Current Opinion in Chemical Biology 13 (2009) 26–34.

[30] X.W. Yu, R. Wang, M. Zhang, Enhanced thermostability of a *Rhizopus chinensis* lipase by in vivo recombination in *Pichia pastoris*, Microbial Cell Factories 11 (2012) 102.

[31] K. Terpe, Overview of bacterial expression systems for heterologous protein production: from molecular and biochemical fundamentals to commercial systems, Applied Microbiology and Biotechnology 72 (2006) 211.

[32] E. Çelik, P. Çalık, Production of recombinant proteins by yeast cells, Biotechnology advances 30 (2012) 1108–1118.

[33] O.P. Ward, Production of recombinant proteins by filamentous fungi, Biotechnology Advances 30 (2012) 1119–1139.

[34] E. García-Fruitós, Lactic acid bacteria: a promising alternative for recombinant protein production, Microbial Cell Factories 11 (2012).

[35] W. Zhang, W. Chuan, C.Y. Huang, Construction and secretory expression of β-galactosidase gene from *Lactobacillus bulgaricus* in *Lactococcus lactis*, Biomedical and Environmental Sciences 25 (2012) 203–209.

[36] M. Vitikainen, H.L. Hyyryläinen, A. Kivimäki, Secretion of heterologous proteins in *Bacillus subtilis* can be improved by engineering cell components affecting post-translocational protein folding and degradation, Journal of Applied Microbiology 99 (2005) 363–375.

[37] A.L. Demain, P. Vaishnav, Production of recombinant proteins by microbes and higher organisms, Biotechnology Advances 27 (2009) 297–306.

[38] M. Mormeneo, F.J. Pastor, J. Zueco, Efficient expression of a *Paenibacillus barcinonensis* endoglucanase in *Saccharomyces cerevisiae*, Journal of Industrial Microbiology & Biotechnology 39 (2012) 115–123.

[39] T. Gauthier, X. Wang, J.S. Dos Santos, Trypacidin, a spore-borne toxin from *Aspergillus fumigatus*, is cytotoxic to lung cells, PLoS One 7 (2012) e29906.

[40] V. Ramamoorthy, S. Shantappa, S. Dhingra, veA-dependent RNA-pol II transcription elongation factor-like protein, RtfA, is associated with secondary metabolism and morphological development in *Aspergillus nidulans*, Molecular Microbiology 85 (2012) 795–814.

[41] M.J. Kwon, T.R. Jørgensen, B.M. Nitsche, The transcriptomic fingerprint of glucoamylase over-expression in *Aspergillus niger*, BMC Genomics 13 (2012) 701.

[42] B.A. van den Berg, M.J. Reinders, M. Hulsman, Exploring sequence characteristics related to high-level production of secreted proteins in *Aspergillus niger*, PLoS One 7 (2012) e45869.

[43] V. Singh, S. Haque, R. Niwas, Strategies for fermentation medium optimization: an in-depth review, Frontiers in Microbiology 7 (2017) 2087.

[44] H.C. Chiu, C.W. Lin, S.Y. Suen, Isolation of lysozyme from hen egg albumen using glass fiber-based cation-exchange membranes, Journal of Membrane Science 290 (2007) 259–266.

[45] F. Wolman, D.G. Maglio, M. Grasselli, One-step lactoferrin purification from bovine whey and colostrum by affinity membrane chromatography, Journal of Membrane Science 288 (2007) 132–138.

[46] X.J. Huang, A.G. Yu, J. Jiang, Surface modification of nanofibrous poly(acrylonitrile-co-acrylic acid) membrane with biomacromolecules for lipase immobilization, Journal of Molecular Catalysis B: Enzymatic 57 (2009) 250–256.

[47] J. Zhu, G. Sun, Lipase immobilization on glutaraldehyde-activated nanofibrous membranes for improved enzyme stabilities and activities, Reactive and Functional Polymers 72 (2012) 839–845.

[48] S. Kumar, I. Haq, J. Prakash, Improved enzyme properties upon glutaraldehyde cross-linking of alginate entrapped xylanase from *Bacillus licheniformis*, International Journal of Biological Macromolecules 98 (2017) 24–33.

[49] J. Huang, Y. Liu, P. Zhang, A temperature-triggered fiber optic biosensor based on hydrogel-magnetic immobilized enzyme complex for sequential determination of cholesterol and glucose, Biochemical Engineering Journal 125 (2017) 123–128.

[50] Q. Zhang, F. Zheng, W.J. Qin, Preparation and application of novel thermo-sensitive matrix-based immobilized enzyme for fast and highly efficient proteome research, Chinese Journal of Analytical Chemistry 44 (2016) 1692–1697.

[51] G. Du, Q. Fang, J.M. den Toonder, Microfluidics for cell-based high throughput screening platforms—a review, Analytica Chimica Acta 903 (2016) 36–50.

[52] E.M. Payne, D.A. Holland-Moritz, S. Sun, High-throughput screening by droplet microfluidics: perspective into key challenges and future prospects, Lab on a Chip 20 (2020) 2247–2262.

[53] N. Szita, K. Polizzi, N. Jaccard, Microfluidic approaches for systems and synthetic biology, Current Opinion in Biotechnology 21 (2010) 517–523.

[54] M. Schallmey, J. Frunzke, L. Eggeling, Looking for the pick of the bunch: high-throughput screening of producing microorganisms with biosensors, Current Opinion in Biotechnology 26 (2014) 148–154.

[55] A. Waterhouse, M. Bertoni, S. Bienert, SWISS-MODEL: homology modelling of protein structures and complexes, Nucleic Acids Research 46 (2018) W296–W303.

[56] A. Pavelka, E. Chovancova, J. Damborsky, HotSpot wizard: a web server for identification of hot spots in protein engineering, Nucleic Acids Research 37 (2009) W376–W383.

[57] R.J. Fox, S.C. Davis, E.C. Mundorff, Improving catalytic function by ProSAR-driven enzyme evolution, Nature Biotechnology 25 (2007) 338–344.

[58] M.T. Reetz, J.D. Carballeira, A. Vogel, Iterative saturation mutagenesis on the basis of B factors as a strategy for increasing protein thermostability, Angewandte Chemie International Edition 45 (2006) 7745–7751.

[59] X. Feng, J. Sanchis, M.T. Reetz, Enhancing the efficiency of directed evolution in focused enzyme libraries by the adaptive substituent reordering algorithm, Chemistry – A European Journal 18 (2012) 5646–5654.

[60] P. Heinzelman, R. Komor, A. Kanaan, Efficient screening of fungal cellobiohydrolase class I enzymes for thermostabilizing sequence blocks by SCHEMA structure-guided recombination, Protein Engineering, Design & Selection 23 (2010) 871–880.

[61] P.A. Romero, E. Stone, C. Lamb, SCHEMA-designed variants of human arginase I and II reveal sequence elements important to stability and catalysis, ACS Synthetic Biology 1 (2012) 221–228.

[62] V. Temml, T. Kaserer, Z. Kutil, Pharmacophore modeling for COX-1 and-2 inhibitors with LigandScout in comparison to Discovery Studio, Future Medicinal Chemistry 6 (2014) 1869–1881.

[63] D. Studio, Discovery Studio, Accelrys [2.1], 2008.

[64] J. Damborsky, J. Brezovsky, Computational tools for designing and engineering enzymes, Current Opinion in Chemical Biology 19 (2014) 8–16.

[65] T. Maischberger, I. Mierau, C.K. Peterbauer, High-level expression of Lactobacillus β-galactosidases in *Lactococcus lactis* using the food-grade, nisin-controlled expression system NICE, Journal of Agricultural and Food Chemistry 58 (2010) 2279–2287.

[66] P.J. Punt, N. van Biezen, A. Conesa, Filamentous fungi as cell factories for heterologous protein production, Trends in Biotechnology 20 (2002) 200–206.

[67] F. Alberti, G.D. Foster, A.M. Bailey, Natural products from filamentous fungi and production by heterologous expression, Applied Microbiology and Biotechnology 101 (2017) 493–500.

[68] Y. Zhao, J. Lim, J. Xu, Nitric oxide as a developmental and metabolic signal in filamentous fungi, Molecular Microbiology 113 (2020) 872–882.

[69] T.Q. Shi, G.N. Liu, R.Y. Ji, CRISPR/Cas9-based genome editing of the filamentous fungi: the state of the art, Applied Microbiology and Biotechnology 101 (2017) 7435–7443.

[70] C. González-Figueredo, R.A. Flores-Estrella, O.A. Rojas-Rejón, Fermentation: metabolism, kinetic models, and bioprocessing, Current Topics in Biochemical Engineering, Intechopen, 2018, pp. 1–17.

7

Production of fibrinolytic enzymes during food production

Ali Muhammed Moula Ali[1], Sri Charan Bindu Bavisetty[2], Maria Gullo[3], Sittiwat Lertsiri[4], John Morris[5], Salvatore Massa[6]

[1]DEPARTMENT OF FOOD SCIENCE AND TECHNOLOGY, FACULTY OF FOOD-INDUSTRY, KING MONGKUT'S INSTITUTE OF TECHNOLOGY LADKRABANG, BANGKOK, THAILAND
[2]DEPARTMENT OF FERMENTATION TECHNOLOGY, FACULTY OF FOOD-INDUSTRY, KING MONGKUT'S INSTITUTE OF TECHNOLOGY LADKRABANG, BANGKOK, THAILAND
[3]DEPARTMENT OF LIFE SCIENCES, UNIVERSITY OF MODENA AND REGGIO EMILIA, REGGIO EMILIA, ITALY [4]DEPARTMENT OF BIOTECHNOLOGY, FACULTY OF SCIENCE, MAHIDOL UNIVERSITY, BANGKOK, THAILAND [5]KMITL RESEARCH AND INNOVATION SERVICES, KING MONGKUT'S INSTITUTE OF TECHNOLOGY LADKRABANG, BANGKOK, THAILAND
[6]DEPARTMENT OF AGRICULTURAL, FOOD AND ENVIRONMENTAL SCIENCES, UNIVERSITY OF FOGGIA, FOGGIA, ITALY

7.1 Introduction

Fibrinolytic enzymes (FEs) are protease which can degrade fibrin mesh of thrombus clots. These proteases are classified as hydrolases (EC 3) and subclass of peptidases (EC 3.4) [1]. FEs aid in the disintegration of blood clots (thrombus), thus maintaining uniform blood flow in blood vessels. Thrombi are majorly responsible for a group of disorders of the heart, brain, and blood vessels, causing pulmonary embolism and deep vein thrombosis [2,3]. As per the data of World Health Organization, CVDs caused 17.9 million deaths worldwide in 2016, representing 31% of global deaths and predicted to cause more than 20 million deaths per year by 2020 [4]. The fibrinolytic system maintains homeostasis by dissolving the fibrin mesh formed from fibrinogen as a result of thrombin activation and maintains an uninterrupted blood flow at the site of vascular injury dissolving these clots [5]. Unhydrolyzed clots would otherwise result in several heart ailments such as myocardial infarction, ischemic stroke, etc. [6].

Anticoagulants such as warfarin, dabigatran, apixaban, or rivaroxaban have been used traditionally to dissolve these clots [7,8]. Alternatively, FE activators such as urokinase, streptokinase, and tissue plasminogen activators (tPA) have been employed for treating thrombosis [9,10]. However, these treatment options pose the risk of fatal complications like bleeding due to hemorrhage, especially for patients suffering from intracerebral hemorrhage [11–13]. Due to the undesired side effects exhibited during clinical therapy, and the high cost of production, a demand for alternative economic, sustainable, and reliable sources is on the rise. Several studies have been conducted on alternative sources for FE production, purification, and characterization from microorganisms, algae, insects, plants, snakes, and earthworms [14–16]. Microbial FEs have particularly gained an edge over the others due to their substrate specificity. Due to their availability from widespread sources, FEs, especially of microbial origin, have attracted a great deal of attention in the last decade [17]. Proteases are assured for their marginal side effects to patients with inborn health complications. Hence, drug screening programs are focusing on finding microbial origin FEs owing to their fibrin specificity and low production costs [13,18].

Currently, thrombolytic therapy by administering FEs has manifested successful outcome due to their significant role in the disintegration of blood clots, hence retaining a uniform blood flow [19]. This property of FEs makes them desirable in thrombolytic therapy. Economic feasibility is one of the primary factors required for the commercial success of FE [20]. This can be achieved either by biotechnological interventions such as cloning and overexpression of a strong FE-producing gene, site-directed mutagenesis, inoculum strain improvement through genetic modifications, or fermentative production. Bioprocessing through fermentation is one of the most crucial means of boosting yield. Numerous reports have been published on the production, optimization, purification, and characterization of FEs. Statistical optimization has been conducted on many variables, including sustainable substrates from agroindustrial wastes and residues for improved yield and process economics [21–24].

7.2 Mechanism of blood clotting and fibrinogenesis

Before understanding the fibrinolytic mechanism, an understanding of the cascade of blood coagulation mechanism is a must. The mechanism of action of blood clotting is a complex process consisting of intrinsic, extrinsic, and some common mechanisms [25]. The intrinsic pathway consists of factors I (fibrinogen), II (prothrombin), IX (Christmas factor), X (Stuart–Prower factor), XI (plasma thromboplastin), and XII (Hageman factor). The extrinsic pathway comprises factor I, II, VII (stable factor), and X. The common pathway consists of I, II, V, VIII, and X. All these factors circulate in the blood in the form of zymogens or proenzyme waiting to be activated into serine proteases. These serine proteases later lead to a cascade of catalysis, ultimately leading to the activation of fibrinogen.

The intrinsic pathway is activated as a result of exposure to endothelial collagen. The pathway is triggered with the activation of factor XII (inactive serine protease) to XIIa (active serine protease). XIIa, in turn, catalyzes the activation of XI to XIa followed by activation of factor X (Prothrombinase) to Xa (Thrombinase). From this point onwards, the intrinsic pathway works in synchronization with the extrinsic pathway where Xa activates prothrombin (zymogen) to thrombin, leading to the activation of fibrin from its inactive form (fibrinogen).

The extrinsic pathway is comparatively shorter. It initiates with the activation of factor VIIa in turn activating factor Xa. From this point onwards, the pathway merges with that of intrinsic, leading to the development of a fibrin clot. The common pathway begins with factor X

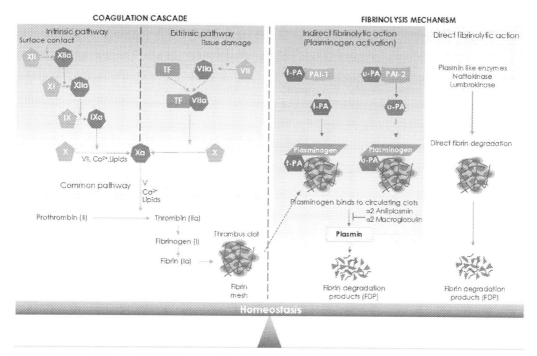

FIGURE 7–1 Coagulation and fibrinolytic cascades. The figure represents a delicate homeostatic balance between blood coagulation and fibrinolysis maintained in physiological conditions. The coagulation cascade consists of intrinsic and extrinsic pathways which leads to the formation of thrombus clots as a physiological response to injury or tissue damage. However, development of thrombi as a result of diseased conditions like CVDs is counteracted by the fibrinolysis mechanism of the body. tPA (tissue plasminogen activator) and uPA (urokinase plasminogen activator) are two of the naturally occurring indirect type fibrinolytic enzymes which bind to any potential circulating clots and break them down to FDP (fibrin degradation products). However, direct type of fibrinolytic action is observed exclusively in enzyme isolated form nonphysiological sources like microbial (nattokinase, streptokinase, etc.) which have the ability to directly degrade thrombus to FDP (fibrin degradation products).

(prothrombinase). The enzyme tenase coverts inactive X (prothrombinase)to active Xa. Tenase in the extrinsic pathway consists of factor VII, factor III (tissue factor), and Ca^{2+}. Whereas, tenase in the intrinsic pathway consists of VIII, factor IXa, a phospholipid, and Ca^{2+}. After activating Xa, tenase activates factor II (prothrombin) to IIa (thrombin) which in turn, activates fibrinogen to fibrin. At the same time, platelets adhere to the damaged blood vessel releasing factor III (Fig. 7–1). The fibrin mesh envelops up over the adhered platelets forming a thrombus clot [26].

7.3 Mechanism of action of FEs: fibrinolysis and thrombolysis

Thrombolysis implies to the mechanism of thrombus (fibrin mesh wrapped around adhered platelets) dissolution, while, fibrinolysis refers to the disintegration of fibrin mesh around a blood clot. As discussed in the above section, fibrin is the final product of the coagulation pathway. Fibrin undergoes breakdown by an enzyme called plasmin. Plasmin is activated from plasminogen. When circulating plasminogen bind to a potential blood clot, they become activated by adopting an open conformation. This cleavage is initiated by tPA or urokinase-type plasminogen activator (uPA), Kallikrein, and factor XII [10]. Plasmin ultimately breaks down the fibrin into soluble fibrin degradation products (FDP) [27] (Fig. 7–1).

Fibrinolysis is distinguished into two major categories. The indirect category consists of the plasminogen activators (PAs) such as tPA (naturally present in the blood), uPA, and streptokinase (microbial source). These activators work in indirect fibrin hydrolysis by activating plasminogen to plasmin, eventually leading to fibrin hydrolysis. The direct type includes plasmin-like enzymes such as nattokinase (NK) and lumbrokinase,. These enzymes are known to directly degrade fibrin, thereby leading to a complete dissolution of thrombi [28,29].

7.4 Classification of FEs

Based on the mechanism of action, FEs have been categorized into three main groups: (1) serine protease, (2) metalloprotease, and (3) serine metalloprotease. Endopeptidase-like serine proteases cleave peptide bonds in which serine serves as a nucleophilic amino acid at the enzyme's active site [30]. Serine proteases exhibit both direct and indirect fibrinolytic activity. Serine proteases like plasmin, trypsin, and brinase dissolve thrombin by direct hydrolysis [31]. An inhibitor of serine protease, namely, phenyl methyl sulfonyl fluoride (PMSF) is commonly used to identify FEs belonging to the class of serine protease [32]. FEs with a direct mode of action pose negative effects such as toxicity and disintegration of essential fibrin clots, responsible for homeostasis [33–35]. On the other hand, serine proteases with indirect action such as tPA and uPA provide safer modes of treatment but, however, they are still known to pose a risk hemorrhage and reocclusion [15,36].

Metalloproteases are the second class of FEs, which need divalent metal ions such as Zn^{2+}, Mg^{2+}, Ca^{2+}, Hg^{2+}, or Co^{2+} for their activity [37–39]. A small number of literature have reported the activation of fibrinolytic metalloproteases through metal ions such as K^+, Na^+, and Ca^{2+} [40,41]. The third category of FE belongs to the class of serine metalloproteases, which have properties of both serine proteases and metalloproteases [16,42,43]. Based on the specificity of FEs, these enzymes are further classified into four groups: (1) first-generation thrombolytic agents: streptokinase, urokinase (nonspecific) [44,45]; (2) second-generation: recombinant tissue tPA (specific), saruplase [46,47] or prourokinase (nonspecific) [48,49], anistreplase [50], Alteplase [51]; (3) third-generation: tenecteplase (tnk-tPA), reteplase, monteplase, lanoteplase, pamiteplase, staphylokinase (specific) [52]; (4) fourth-generation: plasminogen activator inhibitors (nonspecific) like desmoteplase [53,54].

7.5 Purification of FEs

Purification of enzymes is conducted to make them free of contaminants, increase their stability and shelf life. Moreover, conformational studies related to its structure and biochemical properties such as kinetic and thermodynamic parameters can solely be performed in its pure form

[55]. FE, in its pure form, is required in devising formulations for industrial and clinical. Purification of FEs has been performed either form (1) cell supernatants from fermentative media; (2) cell supernatants of transformed bacterial cells after they have been cloned with a desirable and overexpressing FE-producing gene and have been engineered to overexpress the enzyme [40,56]. Purification of the crude enzyme extracts was conducted through concentration by varying percentages of ammonium sulfate, ethanol or acetone precipitation, ultrafiltration, ultracentrifugation, and dialysis, either in combinations or separately. Ammonium sulfate precipitation is the most commonly used purification technique to obtain crude extracts, followed chromatographic separation at several stages. A wide variety of chromatographic techniques such as anion exchange, cation exchange, gel filtration, affinity column chromatography, fast protein liquid chromatography (FPLC), hydrophobic interaction chromatography, high-performance liquid chromatography (HPLC), and chromatofocusing were used. Other techniques such as three-phase partitioning (combination of ammonium sulfate and t-butanol for protein precipitation from crude extracts) and AOT [sodium di (2-ethylhexyl) sulfosuccinate]/isooctane reverse micelles system and aqueous two-phase systems (ATPS) [polyethylene glycol (PEG)/sulfate] PEG/sodium sulfate ATPS have also been reported. Specific activity was studied in terms of unit of measurement for the purified enzyme.

7.6 Biochemical characterization of FEs

7.6.1 Physiochemical properties of FEs

7.6.1.1 Molecular weight, optimum pH, temperature, and inhibitors

The physicochemical properties of enzymes commonly studies include molecular mass (kDa), pH, and temperature at which the enzyme exhibited its highest activity. Usually, the molecular weights of purified FEs spanned from as low as 14 kDa to as high as 97 kDa [57–59]. Nevertheless, the average range of molecular weight documented was from 27 to 29 kDa [37,60]. The optimal pH for maximal enzymatic activity was either neutral or near to alkaline, that is, from 6 to 7 and 8 to 9 [38,39,61,62]. However, very few enzymes preferred acidic or basic conditions. Crude enzyme extracted from *Bacillus tequilensis*, CFR15-protease extracts from *Bacillus amyloliquefaciens* MCC2606 and CK protease from *Bacillus* sp. of strain CK 11-4 showed optimal activity at pH 10.5 and protease KSK from *Lactobacillus plantarum* KSK-II were documented to possess optimal pH of 10 [15,43,63,64]. On the other hand, crude enzyme extracts from *Staphylococcus* sp. strain AJ exhibited optimal at pH 2.5 to 3 [65], and crude enzyme extracts from fermented shrimp paste exhibited optimal pH at 3 to 7 [66]. The average range of optimal temperature reported was from 30°C to 50°C [63,67,68]. The maximum and the minimum optimal temperature documented were 70°C (CK protease from *Bacillus* sp. strain CK 11-4) [64] and 20°C (*Streptomyces venezuelae* and FVP-I protease from *Flammulina velutipes*) [69,70].

7.6.1.2 Fibrinogenolytic, fibrinolytic, and plasminogen activation activity

Fibrinogen is mainly comprised three major polypeptide chains, namely, Aα, Bβ, and γ chains. After cleavage of fibrinopeptides An and B, fibrin protofibrils are generated with α, β, and γ chains. These protofibrils are consequently polymerized to form fibrin [71]. Fibrinogen has the ability to selectively hydrolyzed either Aα, Bβ, and γ chains individually or all together [60]. A similar hydrolysis mechanism of fibrin has been reported to occur at α, β, and γ chains [72]. The fibrinolytic activity is generally assessed using the fibrinolytic assay [73]. FE after purification is mixed human fibrin and human thrombin which were incubated previously. After incubation, the aliquots are evaluated for their pattern of cleavage with the aid of SDS PAGE [73]. The fibrinogenolytic activity was generally tested using the fibrin plate assay in which the cleavage patterns were observed on SDS PAGE after fibrinogen was incubated with purified FE [65].

The activity of plasminogen is determined based on the zone of lysis measured on plates with and without plasminogen [74]. The presence of a zone of lysis indicates an indirect mode of action by activating plasminogen to plasmin. However, if the enzyme was able to demonstrate strong fibrinolytic activity in both plates with and without plasminogen, the enzyme is known to possess dual characteristic, wherein the enzyme can directly hydrolyze fibrin clots or show indirect mechanism by activating plasminogen, for example, streptokinase, urokinase, and tPA [75]. The activity of several FEs documented for their fibrinogenolytic, fibrinolytic, and plasminogen activation has been elaborated in Table 7–1. Majority of the FEs demonstrated strong Aα fibrinolytic activity followed by Bβ and γ chain fibrinolysis. Nevertheless, FEs isolated from *Bacillus subtilis* JS2, *B. amyloliquefaciens* CB1, *Staphylococcus* sp. strain AJ exhibited no γ chain lysis [39,76]. *Bacillus* sp. nov. SK006 was the sole bacterial species that exhibited strong Bβ fibrinolytic activity [77].

7.6.2 Amidolytic properties of FEs

Amidolytic activity was determined to understand the enzyme's mechanisms of cleavage. The amidolytic activity of isolated FEs was assessed with the help of various synthetic chromogenic substrates. The specificity of the FEs toward the substrates was expressed in the form of color development, which was measured in a spectrophotometer at respective wavelengths. Some of the chromogenic substrates reported are *N*-succinyl-ala-ala-pro-phe-*p*-nitroanilide, MeO-Suc-Arg-Pro-Tyr-pNA (substrate for serine proteases like subtilin and chymotrypsin), *N*-benzoyl-Phe-Val-Arg-pNA, H-D-Phe-Pip-Arg-pNA (substrate for trypsin and thrombin), *N*-benzoyl-Pro-Phe-Arg p-NA, *N*-(*p*-Tosyl)-Gly-Pro-Lys-pNA D-Val-Leu-Lys-pNA (substrate for plasmin) [37,60], D-Val-leu-Arg-pNA (substrate for kallikrein), and pyro-Glu-Gly-Arg-pNA (substrate for urokinase) [38,76,78–80]. Most of the characterized enzyme manifested a substrate specificity toward *N*-succinyl-ala-ala-pro-phe-*p*-nitroanilide, hence categorizing them into the class of serine proteases as they were bestowed with substrate specificity toward serine proteases like subtilin and chymotrypsin [37,39,60] Nevertheless, protease DFE27 isolated from *B. subtilis* DC27 demonstrated a distinctive substrate specificity toward D-Val-Leu-Lys-pNA showing that DFE27 has functions of a tPA due to its substrate specificity to plasmin [38]. In addition, the specificity of FEs was also tested against natural substrates like fibrin, casein, gelatin, hemoglobin, keratin, bovine serum albumin, globulin,

Table 7.1 Fibrinolytic, fibrinogenolytic, and plasminogen activation activity of various proteases.

Source	Enzyme	Mechanism	Mode of action	Reference
Bacillus velezensis BS2	AprEBS2	Strong α-fibrinogenase, moderate β-fibrinogenase, and mild some γ-fibrinogenase activity	Direct	[37]
Bacillus amyloliquefaciens Jxnuwx-1	—	Both fibrinogen and fibrinolytic activity with the highest degrading activity towards the Aα chains, followed by Bβ chains and γ chains	Direct and indirect	[60]
Bacillus subtilis JS2	AprEJS2	Strong α-fibrinogenase degradation followed by moderate β-fibrinogenase but no γ-fibrinogenase activity	Direct	[39]
B. amyloliquefaciens MCC2606	CFR15	Strong β-fibrinogenase activity	Direct	[43]
B. amyloliquefaciens RSB34	AprE34	Strong α-fibrinogenase degradation followed by moderate β-fibrinogenase but no γ-fibrinogenase activity	Direct	[91]
B. subtilis ZA400	BsfA	Higher activity to break down the γ-bond and γ—γ bond in the fibrin	Direct	[61]
B. amyloliquefaciens CB1	AprECB1	Strong Aα and Bβ fibrinolytic ability but no effect on γ chain	Direct	[76]
B. amyloliquefaciens MJ5-41	AprE5-41	Degraded Aα and Bβ chains but not the γ chain of fibrinogen	Direct	[95]
Staphylococcus sp. strain AJ	AJ	Strong Aα fibrinogenolytic ability but no effect on Bβ and γ chains	Direct	[65]
Bacillus sp. nov. SK006	—	Strong Bβ fibrinolytic ability followed by weak γ chain cleavage but no Aα fibrinolytic ability	Direct and indirect	[77]
Bacillus licheniformis KJ-31	bpKJ-31	Strong Aα and fibrin(ogen) lytic activity	Direct	[90]
B. subtilis DC33	Subtilisin FS33	Strong Aα and Bβ fibrinogen lytic activity	Direct and indirect	[105]
B. subtilis TP-6	TPase	Strong Aα and Bβ fibrinogenolytic activity	Direct and indirect	[100]
Cordyceps militaris	—	Rapidly hydrolyzed the fibrin α chain, followed by the γ—γ chains. It also hydrolyzed the β chain, but more slowly Aα, bβ, and γ chains of fibrinogen were also cleaved very rapidly	Direct	[123]

fibrinogen, fibrin, collagen, azoalbumin, and morpholinopropane sulfonic acid [15,42,81–84]. Fibrin was considered to be reference standard demonstrating 100% activity.

7.7 Sources of FEs

The cost-ineffectiveness and undesirable side effects due to the clinical use of traditional Fes have prompted researchers to come up with alternative cheap and effective sources for these enzymes. FEs have been isolated from diverse sources such as microbes, fungi, algae, plants, insects, snake venoms, and earthworms [14,16,85,86]. Microbial originated FEs have gained considerable interest due to minimal cost of production, comparatively high quantitative production of enzymes due to mass culture, biochemical diversity, and feasibility for genetic manipulation through biotechnological approaches [16,18]. Enhanced production along with economic viability is a prerequisite for the commercial value of FEs [20]. The majority of the studies conducted on microbial production of FEs are the ones isolated from fermented foods as they are believed to be rich in protease-producing bacteria [87]. FEs with a broad variety of biochemical characteristics have been recorded from a variety of microbes. Among them all, *Bacillus* sp. was the widely reported isolate among several fermented foods worldwide. FEs with diverse biochemical characteristics, for example, molecular weight, Optimal pH, and temperature for enzyme activity and specific activity have been recorded from a variety of fermented food around the world such as *Cheonggukjang, Doenjang, Douche, Dosa, Gembus, jeotgals, kishk, Natto*, etc. [37,43,60,78,88,89]. Table 7–2 illustrates information on FEs, from fermented food sources, their respective microbe producing the enzyme, biochemical, and specific activity. Some of the well-known traditional fermented foods have been discussed with respect to the production of FEs.

7.7.1 *Jeotgal*—traditional Korean fermented food

Jeotgal is a traditional Korean fermented food prepared from different types and parts of fish. Usually, the raw materials are mixed with salt and preserved for several months before consumption [37]. Several studies have reported the preparation of *jeotgals* from sea squirt [37], *Baekhajot* (pickled shrimp), *Gonjaengijot* (pickled opossum shrimp), *Myolchijot* (pickled anchovy), and *Jabojot* (pickled offal fish) [84], *saeu* (small shrimp) [39], shrimp, anchovy, and yellow corniva [90]. Due to its high salt concentration, several halophilic and halotolerant bacteria have been isolated. Halophiles such as *Tetragenococcus* sp. or *Halanaerobium* sp. and halotolerant organisms such as *Staphylococcus* sp. are the commonly isolated organisms. However, some *Bacillus* spp. with strong fibrinolytic activity and salt tolerance were studied for their potential as starter cultures [39]. Several enzymes have been reported to be isolated from *Jeotgals*, for example, JP-I, JP-II [84], AprEBS2 [37], AprEJS2 [39], and bpKJ-31 [90]. FEs purified from *jeotgals* showed their maximal activity at alkaline pH. The molecular weights ranged from 27 to 37 kDa. FEs form *jeotgals* have been purified using several techniques, for example, JP-I and JP-II were purified using ethanol precipitation (75%), AprEBS2 and AprEJS2 were purified using affinity chromatography with a HiTrap IMAC FF column, whereas bpKJ-31 was purified using ammonium sulfate precipitation (75%), anion-exchange chromatography (DEAE-Sepharose FF column), and gel filtration chromatography (HiPrep 16/60 Sephacryl S-200 HR column) [90]. The specific activities of the purified enzymes ranged from as high as 7245 U/mg to 243 U/mg. JP-I and JP-II belong to the class of metalloproteases, whereas AprEJS2, bpKJ-31 belong to the class of serine proteases as they showed substrate specificity toward N-Succ-Ala-Ala-Pro-Phe-pNA. AprEBS2, AprEJS2, bpKJ-31, all showed strong fibrinogenase activity. However, AprEBS2 showed moderate β-fibrinogenase mild γ-fibrinogenase, whereas, AprEJS2 showed moderate β-fibrinogenase and no γ-fibrinogenase action.

7.7.2 *Douche*—Chinese traditional fermented black soybean food

Douchi is one of the popular traditional fermented soybean foods from China. Several FEs producing bacterial isolates have been identified from *Douchieg. B. subtilis* DC33,

Table 7-2 Sources and physiochemical properties of fibrinolytic enzymes.

Source	Bacterial/fungal strain	Enzyme	Optimal pH and temperature	Molecular mass	Specific activity	Reference
Korean fermented foods						
Jeotgal, fermented *Gonjaengijot* (pickled opossum shrimp)	—	JP-I and JP-II	8.1, 50	36	7245, 2394	[84]
Jeotgal fermented sea squirt (munggae)	Bacillus velezensis BS2	AprEBS2	8, 37	27	131[a]	[37]
Jeotgals from salted *saeu* (small shrimp)	Bacillus subtilis JS2	AprEJS2	8, 40	27	—	[39]
Jeotgal, fermented seafood	Bacillus licheniformis KJ-31	bpKJ-31	9, 40	37	243	[90]
Doenjang, fermented soy food	Bacillus amyloliquefaciens RSB34	AprE34	8, 40	27	83[a]	[91]
Doenjang, traditional Korean fermented food	Bacillus sp. DJ-4	Subtilin DJ-4	—	29	—	[92]
Cheonggukjang, fermented soy food	B. subtilis HK176	AprE176	8, 40	27	217	[78]
Cheonggukjang, fermented soy food	B. subtilis HK176	M179	—	—	480	[78]
Cheonggukjang, fermented soy food	B. amyloliquefaciens CB1	AprECB1	6, 40	28	1584	[76]
Cheonggukjang, fermented soy food	B. amyloliquefancies CH51	AprE51	6, 45	27	342	[93]
Cheonggukjang, fermented soy food	B. subtilis CH-3	AprE2	—	29	—	[94]
Chungkook-Jang, fermented soybean sauce	Bacillus sp. strain CK 11-4	CK	10.5, 70	28	143	[64]
Meju fermented soy food	B. amyloliquefaciens MJ5-41	AprE5-41	7, 45	27	104	[95]
Korean salt-fermented Anchovy-jeot	Staphylococcus sp. strain AJ	AJ	2.5–3.0, 85	26	72	[65]
Kimchi	B. subtilis ZA400	BsfA	6, 30–37	28.4	4.5	[61]
Korean traditional soybean paste	Bacillus sp. KDO-13	—	8, 50	44	931	[96]
Indonesian traditional fermented foods						
Indonesian traditional fermented foods	—	NatTK, NatOC, NatWT, and DFEG169A	—	—	—	[97]
Soybean-based fermented food	Stenotrophomonas sp.	—	—	—	—	[98]
Gembus, fermented soybean	Bacillus pumilus 2. g	—	7, >60	20	1442	[88]
Tempeh, fermented soybean	Fusarium sp. BLB	FP	—	—	—	[99]
Tempeh, fermented soybean	B. subtilis TP-6	TPase	7, 50	27.5	1197	[100]
Chinese fermented foods						
Douche, fermented black soya bean	B. amyloliquefaciens Jxnuwx-1	—	7.6, 41	29	1240	[60]
Douche, fermented black soya bean	B. subtilis DC27	DFE27	7, 45	29	11274	[38]
Douche, fermented black soya bean	Strain XY-1	DFE	—	—	21.33	[101]
Douche, fermented black soya bean	B. subtilis LD-8547	DFE	—	—	21750[b]	[102]
Douche, fermented black soya bean	B. subtilis DC-2	—	—	—	1165.58	[103]
Douche, fermented black soya bean	B. amyloliquefaciens DC-4	Subtilisin DFE	—	28	—	[104]
"Ba-bao Douchi," a traditional soybean-fermented food in China	B. subtilis DC33	Subtilisin FS33	8, 55	30	15,495	[105]
Distillery isolate from "daqu," a fermentative agent used in Chinese liquor and vinegar	Rhizopus microsporus var. tuberosus	—	7, 37	24.5	1645.2	[106]
Chinese fermented soybean paste	B. amyloliquefaciens LSSE-62	—	—	—	—	[107]
Indian fermented foods						
Dosa batter, a fermented Indian food	B. amyloliquefaciens MCC2606	CFR15	10.5, 45	32	584	[43]
Fermented rice	Bacillus cereus IND5	—	8, 50	47	364[c]	[108]

(Continued)

Table 7−2 (Continued)

Source	Bacterial/fungal strain	Enzyme	Optimal pH and temperature	Molecular mass	Specific activity	Reference
Japanese fermented foods						
Natto, fermented soya food	B. subtilis RJAS19	Nattokinase	—	—	—	[109]
Fermented red bean	B. subtilis	—	9, 60	29.93	6284	[110]
Natto, fermented soya food	B. subtilis natto B-12	B-12 nattokinase	8, 40	29	5316	[111]
Natto, fermented soya food	Bacillus firmus NA-1 p	—	—	—	—	[112]
skipjack "Shiokara," salt-fermented food	—	Katsuwokinase (KK)	1–10, 37	35	—	[113]
Vegetable cheese natto	—	Nattokinase	6–12, 30–40	35	—	[114]
Nonfermented foods						
Vietnamese traditional fermented soybean paste products	B. amyloliquefaciens	—	—	—	—	[115]
Pigeon pea	B. subtilis 14714, B.subtilis 14715, B. subtilis 14716, and B. subtilis 14718	Nattokinase	—	—	—	[116]
Kishk, a traditional Egyptian food	Lactobacillus plantarum KSK-II	KSK-II	10, 50	28	—	[15]
Asian fermented shrimp paste	Bacillus sp. nov. SK006	—	7.2, 30	43–46	—	[77]
Fermented shrimp paste	—	—	3–7, 30–40	18	—	[66]
Indian rice (Xanthomonas oryzae IND3)	B. subtilis HQS-3	—	—	—	2296a	[23]
Rice	B. cereus IND1	—	8, 60	29.5	960	[117]
Cooked Indian rice	Paenibacillus sp. IND8.	—	—	—	4418	[118]
Bovine milk	Streptococcus agalactiae EBL-31	Streptokinase	—	—	147.08	[119]
Edible mushroom	Cordyceps militaris	—	7.2, 37	28	1467	[120]
Edible mushroom	Pleurotus ferulae	—	(4, 5, 8), 50	20	1253	[68]
Pigeon pea	B. subtilis 14714, B. subtilis 14715, B. subtilis 14716 and B. subtilis 14718	Nattokinase	—	—	—	[116]
Fish scales	Pseudoalteromonas sp. IND11	—	—	—	1573	[121]
Chickpeas	—	Nattokinase	—	—	—	[122]

^{a}mU μl^{-1}.
^{b}U mL^{-1}.
^{c}U g^{-1}.

B. amyloliquefaciens Jxnuwx-1, *B. subtilis* DC27, Strain XY-1, *B. subtilis* LD-8547, *B. subtilis* DC-2, and *B. amyloliquefaciens* DC-4 [38,60,101–105]. The FE isolated from strain XY-1 and *B. subtilis* LD-8547 was named DFE, the one isolated from *B. subtilis* DC27 was designated as DFE27. Whereas enzymes named subtilisin DFE and subtilisin FS33 were isolated from *B. amyloliquefaciens* DC-4 and *B. subtilis* DC33, respectively. Most of the enzymes exhibited their highest activity at neutral to alkaline pH ranging from 7 to 8. The optimal temperature for maximal enzyme activity ranged from 29°C to 30°C. DFE, DFE27, subtilisin FS33, and subtilisin DFE belonged to the class of serine proteases. Whereas FE, DFE isolated for *B. subtilis* LD-8547 and the enzyme isolated from *B. amyloliquefaciens* Jxnuwx-1 exhibited serine metalloprotease properties. All the enzymes studies from *Douchi* isolates showed their substrate specificity toward *N*-Succ-Ala-Ala-Pro-Phe-pNA, which serves as a substrate for serine proteases like subtilin and chymotrypsin, whereas, DFE27 showed specificity toward D-Val-Leu-Lys-pNA which serves as a substrate for kallikrein [38]. Finally, the purification of this enzyme was commonly performed using ammonium sulfate precipitation (30%−80%), ion-exchange chromatography using ion-exchange chromatography (DEAE-Sepharose FF anion exchange column, ultrafiltration, UNOsphere Q ion-exchange column, CM-Sepharose FF ion exchange column, DEAE-Sepharose FF ion exchange column) and gel filtration chromatography (Superdex 75 column, Sephadex G-50 column), hydrophobic interaction chromatography (Phenyl-Sepharose 6 Fast Flow column), HPLC and FPLC systems.

7.7.3 *Tempeh*—traditional Indonesian soybean-fermented food

Tempeh is a popular Indonesian food fermented soybean food. A protease named FP produced by Fusarium sp. BLB and TPase from *B. subtilis* TP-6 were two of the enzymes isolated from *Tempeh*. TPase had an optimal pH and temperature of 7°C and 50°C and a molecular weight of 27.5 kDa belonging to the class of serine proteases. TPase was found to be quite stable at acidic pH but showed a significant decrease in its enzyme activity at alkaline pH. Purification of the crude enzyme was carried out through ammonium sulfate precipitation (40%−70%), hydrophobic interaction chromatography (octyl Sepharose), and ion exchange (SP Sepharose). TPase showed an *N*-terminal sequence of AESVPYGVSQIKAPALHSQGFTGS, which was similar to subtilisin K54 from *B. subtilis* K-54 (97%), subtilisin BPN from *B. amyloliquefaciens* (95%), and NK from *B. natto* (82%). TPase showed a high substrate specificity toward *N*-Succ-Ala-Ala-Pro-Phe-pNA. The *N*-terminal sequence of FDP showed that TPase could cleave at the glutamine and aspartic acid residues at the α- and β chains of fibrinogen generating D and E fragments. This cleavage pattern is similar to the action of human plasmin. TPase was found to have both direct and indirect mechanisms of action. Another enzyme named FP produced by *Fusarium* sp. BLB was isolated studies from *Tempeh*. Molecular characterization of the enzyme was performed by isolating and sequencing the gene encoding the FP enzyme. The cDNA of FP showed a sequence of 250 amino acids. This enzyme was found to possess approximately twofold stronger fibrinolytic activity than NK.

7.7.4 *Doenjang*—traditional Korean fermented food

Doenjang is a traditional fermented soy food from Korea. Kim and Choi [92] studied the characterization of enzyme subtilisin DJ produced from *Bacillus* sp. strain DJ-4, an isolate from *doenjang* subtilisin DJ-4. The enzyme's optimal pH, temperature, and molecular weight were found to be 10, 40°C, and 29 kDa, respectively. However, the enzyme was found to be stable at a wide range of temperatures from acidic (pH 4) to alkaline (pH 11). The enzyme was purified using ultrafiltration, ion-exchange chromatography (DEAE-Sepharose CL-B6), and gel filtration (Toyopearl HW-55F). DJ-4 was classified as a serine protease based on its specificity toward chromogenic substrate D-Val-Leu-Lys-pNA. The *N*-terminal sequence of DJ-4 was found to be AQSVPYGVSQIKAP, similar to that of subtilisin BPN. Subtilisin DJ-4 was found also to possess indirect fibrinolytic activity with plasmin-like action. Yao, Liu, Shim, Lee, Kim, and Kim [91] characterized another FE named AprE34 produced from a *doenjang* isolate, *B. amyloliquefaciens* RSB34. The optimal pH, temperature, and molecular weight of AprE34 were found to be 8, 40°C, and 27 kDa, respectively. This enzyme was cloned and overexpressed in *Escherichia coli* expression systems and purified using column chromatography HiTrap IMAC FF column. AprE34 showed substrate specificity toward *N*-succinyl-ala-ala-pro-phe-*p*-nitroanilide proving that this enzyme belongs to the class of serine proteases.

7.7.5 *Cheonggukjang*—Korean fermented soy food

Cheonggukjang is a popular traditional fermented Korean soybean food. *B. subtilis* is a common starter culture used for the fermentation of *cheonggukjang*. Usually, whole soybeans are washed and soaked in water for about 12 hours, followed by cooking at 100°C for 2 hours. After cooling, the cooked soybeans are incubated with starter cultures at 40°C for about 2 days. Several enzymes have been isolated from starter cultures of *cheonggukjang*, for example, AprE176 and M179 from *B. subtilis* HK176, AprECB1 from *B. amyloliquefaciens* CB1, and AprE51 from *B. amyloliquefaciens*, CH51 and CK from *Bacillus* sp. strain CK 11-4 [64,76,78,93,94]. One of the major disadvantages of *cheonggukjang* is its sensitivity toward the gastrointestinal tract and thermolabile property's acidic conditions. These disadvantages have prompted several researchers to search for suitable microencapsulation techniques [124]. AprE176 and M179 were studied for their molecular characterization by cloning and overexpression. Ni-NTA column chromatography was used as a purification technique for these enzymes [78]. However, ammonium sulfate precipitation was commonly involved, followed by hydrophobic interaction chromatography and ion-exchange chromatography [64,76]. AprECB1 exhibited strong α-fibrinogenase and β-fibrinogenase but no γ-fibrinogenase activity.

7.7.6 Natto

Natto is a traditional fermented Japanese food prepared from boiled soybeans. The oriental population has consumed natto for 1000 years. The starter microorganism for the production of *natto* is *B. subtilis natto* previously called *B. natto*. FE named NK was traditionally discovered in *natto*. *Natto* has been used historically as a remedy for many ailments like beriberi, fatigue, dysentery, thrombotic diseases, hypertension, osteoporosis, and osteoporosis [125]. NK belongs to the class of serine proteases. This enzyme was reported to have stronger thrombolytic activity compared to that of plasmin and increases the production of plasmin from plasminogen [126]. Some of the microorganisms producing NK include *B. subtilis* RJAS19, *B. subtilis natto* B-12, *B. subtilis* 14714, *B. subtilis* 14715, *B. subtilis* 14716, and *B. subtilis* 14718 isolated from different food sources. Kumar et al. [109] studied the characterization of this enzyme. The purification techniques include ammonium sulfate precipitation (40%–60%), dialysis and ion-exchange chromatography. Wang et al. [127] purified B-12 NK using ammonium sulfate precipitation (30%–80%), dialysis, gel filtration chromatography ultrafiltration, and hydrophobic interaction chromatography.

7.7.7 Asian fermented shrimp paste

Several different types of shrimp pastes are available and are widely popular in the Southeast Asian region which exhibits thrombolytic potential. It is generally prepared by salting shrimp at 10%–15% concentration. After squashing, it is incubated for 2 days at room temperature. This partially squashed matter is homogenized into a paste and dried under the sun for about 30 days. The final product is packed and sold. Hua et al. [77] identified *Bacillus* sp. nov. SK006 responsible for producing FE from fermented shrimp paste which showed an optimal pH and temperature of 7.2, 30°C and a molecular weight between 43 and 46 with both direct and indirect fibrinolytic activity. Wong and Mine [66] characterized a FE that showed stability at acidic pH widely desired in industrial set up and an optimal temperature between 30°C and 40°C. Hua et al. [77] used a combination of dialysis, ion exchange gel filtration, and gel filtration column to obtain a specific activity of 11 U/mg. Wong and Mine [66] used a combination of filtration, dialysis, gel filtration, ultrafiltration, and reversed-phase chromatography.

7.7.8 Kimchi- fermented Korean food

Kimchi is a traditional fermented Korean food. It is prepared from different types of vegetables, for example, cabbage, garlic, ginseng, bell peppers, cucumber, etc. It is usually served as a side dish in Korean cuisine. Several microbes producing FEs have been isolated from *kimchi* such as *B. amyloliquefaciens*, *B. brevis*, and *Micrococcus luteus* producing 2.6, 1.5, and 2.0 plasmin unit/mL, respectively [128]. An enzyme named BsfA was identified in kimchi produced by *B. subtilis* ZA400 which showed an optimal pH and temperature of 6 and 30°C–37°C, respectively [61]. The purification methods used for the recovery of this enzyme were PEG concentration, Ni-NTA chromatography purification which showed a specific activity of 4.5 U/mg. BsfA exhibited strong $\gamma-\gamma$ bond fibrinolytic activity.

7.7.9 Skipjack

A FE named katsuwokinase (KK) was isolated from skipjack (shiokara or *Katsuwonus pelamis*), a traditional fermented food with salt. This fermented food is usually prepared with various types of marine organisms, especially shrimp. The meat is usually fermented in salt (10%) and malted rice (30%), homogenized into a paste, packed, and fermented for about a month. The very first report of FE from skipjack was reported by Sumi et al. [113]. The crude extract of KK showed a specific activity of 45/µg. This enzyme showed stability at a wide range from pH from 1 to 10 and temperature of 37°C.

7.7.10 Kisk

Kisk is a traditional Egyptian traditional fermented food. It is a homemade product made of parboiled wheat and milk. Wheat is initially boiled, dried, and powdered. Milk is separated fermented in clay pots and mixed with the powdered wheat to make a paste called *hamma*. This *hamma* is fermented for about 24 hours and later diluted with sour salted milk and water. Alternatively, the *hamma* is added with fresh milk and allowed to ferment for 24 hours. The dough mass thus formed is later sundried. KSK-07 from *B. megaterium* KSK-07 was isolated from this fermented product with a molecular mass of 28.5 kDa, optimal pH, and temperature of 8–10 and 50°C, respectively [85]. KSK-07 belonged to the family of serine proteases and exhibited direct protease action. Another study by the same group isolated a FE called KSK-II produced by [15] *L. plantarum* KSK-II with a molecular weight of 43.6 kDa and an optimal PH and temperature of 10 and 50°C, respectively. However, KSKI-II belonged to the family of metalloprotease, unlike the previous enzyme isolated from *kisk*.

7.7.11 *Meju*- traditional Korean fermented soy product

Meju is a traditional Korean fermented soybean food [95]. It is one of the important components used for the preparation of soy sauce and *doenjang*. Jo et al. [95] isolated a FE (AprE5-41) producing strain *B. amyloliquefaciens* MJ5-41 which produced a 27 kDa enzyme with an optimal pH and temperature of 7 and 45°C. AprE5-41 showed strong Aα and Bβ fibrinogenase activity. It was classified into the class of serine protease due to its substrate specificity towards *N*-succinyl-Ala-Ala-Pro-Phe *p*-nitroanilide. Purification techniques like ammonium sulfate (80%), dialysis, and column chromatographic techniques resulted in a specific activity of 104 U/mg. AprE5-41 exhibited strong α-fibrinogenase and β-fibrinogenase activity but γ-fibrinogenase activity.

7.7.12 Fermented Indian products

Fermented rice and fermented batter are studied for the production of FEs [43,108]. *Dosa* is prepared by soaking black gram lentils (*Vigna mungo*) overnight. The soaked lentils are then ground into a fine paste and allowed for spontaneous fermentation, which constitutes the batter for *dosa*. This batter is cooked similar to pancakes and consumed mostly as breakfast in India.CFR15 produced from *B. amyloliquefaciens* MCC2606 and crude extract from *B. cereus* IND5 showed their maximal activity at alkaline pH and higher temperatures of around 45°C and 50°C, respectively. CFR15 purified through ammonium sulfate precipitation, ion-exchange, and gel filtration chromatography exhibited a specific activity of 584 U/mg. This enzyme belonged to the class of serine metalloproteases as it was inhibited by both PMSF and EDTA. CFR15 also showed an unusual strong β-fibrinogenase activity, unlike any other FE.

7.8 Production of FEs

7.8.1 Biotechnological approaches

Microbial enzymes have always been of great importance in the development of industrial bioprocess. In order to develop novel, sustainable, and economically viable production processes, there is a need for improved and versatile enzymes. Modern molecular techniques are used to discover new microbial enzymes whose catalytic properties can be improved by various techniques. Most industrial enzymes produced are recombinant forms produced from bacteria and fungi [129].

For many decades, FEs like subtilisins have been used as a model enzyme for exploring studies related to protein engineering and protein folding machinery facilitated by chaperones. Subtilisins are enzymes with great commercial value, which makes them the perfect prototypes to study the physicochemical, pharmacological, and clinical properties of these proteins. Such studies demand the protein requirements in large quantities, which can be attained through cloning, expression, and purification to depend on an ample recombinant source for further analysis [37,130]. Adequate quantities of enzymes can support further studies such as site-directed mutagenesis experiments, protein engineering, substrate specificity, stability, safety, plasma half-life, etc. [131–135]. FEs possess a broad range of applications; however, the one intended to be used for medical applications such as treatment for thrombosis and CVD require enhanced activity and stability. As mentioned above, site-directed mutagenesis is one technique employed to produce enzymes with improved physicochemical properties and catalytic functions [136]. An error-prone PCR was performed on the aprE176 to build mutants with enhanced fibrinolytic activities [78]. Meng et al. [137] studied the mechanism of action of protein expressed through cloning of bacillopeptidase F gene. They especially deciphered the protein's catalytic mechanisms and the role of C-terminal end in retaining the enzyme activity. NK is another widely studies FE in terms of biotechnological approach. NK comes with advantages like cost-effectiveness and oral administration without side effects for clinical practices. Sustainable methods like cloning, overexpression systems, and/or mutations in E. coli have been exploited for NK-specific strains [61,138,139].

Traditional fermentation comes with a number of challenges such as time consumption, slow rate of fermentation, high salinity, and quality issues. Hence, biotechnological manipulations have been used to construct engineered strains with enhanced fibrinolytic activity. These strains can further be used as starter cultures in fermentation [37,76,95]. Strain improvement of a suitable strain has been conducted through mutagenesis and screening of random mutants. Random mutagenesis is performed to induce mutations in the chosen strains by UV, ethidium bromide, and ethyl methyl sulfonate. Several studies have been conducted on strain improvement through random mutagenesis [70] conducted strain improvement studies on *S. venezuelae* by inducing mutations through UV, ethidium bromide, and ethyl methyl sulfonate. The mutants were picked based on the highest thrombinase activity exhibited improved halo and thermotolerant properties compared to their wild strains. The mutant strains are known to show improved specific growth rates and tolerance to lactose concentrations during initial stages of fermentation.

A mixture of biotechnological approaches, along with optimization of culture media, has also been effectively employed for augmented production of FE [24,91]. A detailed description of significant parameters used in cloning and expression systems has been summarized in Table 7–3.

7.8.2 Fermentative approach

One of the main limitations for the commercial use of microbial enzymes is its high production. As reviewed earlier, biotechnological approaches are one of the successful ways to cope up with this issue. Apart from this, fermentative approaches like (1) selection of ideal strain, (2) alternative sustainable and economical sources for fermentative media, (3) optimization of fermentation conditions like nutrition (carbon, nitrogen, metal ion source), temperature, pH, time, etc. have also been researched for their role in cutting down the cost of production

Table 7.3 Cloning and expression parameters used for fibrinolytic enzyme production.

Bacterial strain	Gene	Primer	Cloning host	Cloning vector	Expression host	Expression vector	Reference
Bacillus velezensis BS2	aprEBS2	CH51-F (5′-AGGATCCCAAGAGAGCGATTGCGCTGTGTAC-3′, BamHI site underlined)	Bacillus subtilis WB600	pHY300PLK	Escherichia coli BL21 (DE3)	pETBS2	[37]
B. subtilis JS2	aprEJS2	CH51-R (5′-AGAATTCTTCAGAGGGAGCCACCCGTCGATCA-3′, EcoRI site underlined) CH51-F (5′-AGGATCCCAAGAGAGCGATTGCGCTGTGTAC-3′, BamHI site underlined) and CH51-R (5′-AGAATTCTTCAGAGGGAGCCACCCGTCGATCA-3′, EcoRI site underlined)	B. subtilis WB600	pHY300PLK	E. coli BL21 (DE3)	pHYJS2	[39]
Bacillus amyloliquefaciens RSB34	aprE34	CH51-F (5′-AGGATCCCAAGAGAGCGATTGCGCTGTGTAC-3′, BamHI site underlined) CH51-R (5′-AGAATTCTTCAGAGGGAGCCACCCGTCGATCA-3′, EcoRI site underlined)	E. coli DH5α	pGEM-T	E. coli BL21(DE3)	pET26b (+)	[91]
M179 mutants of B. subtilis HK176	aprE176	51F (5′-AGGATCCCAAGAGAGCGATTGCGGCTGTGTAC-3′, BamHI site underlined) 51 R (5′-AGAATTCTTCAGAGGGAGCCACCCGTCGATCA-3′, EcoRI site underlined)	E. coli DH5α	pHY300PLK	E. coli BL21(DE3)	pET26b (+)	[78]
B. subtilis ZA400	bsfA	ZA400-F (5′-GGATCCGATGAGAAGCAAAAATTGTGGAT-3′, BamHI site underlined) ZA400-R (5′-CTCGAGTTGTGCAGCTGCTTGTACG-3′, XhoI site underlined)	E. coli DH5α	pGEM-T Easy	E. coli BL21(DE3) pLysS	pET26b (+)	[61]
B. amyloliquefaciens CB1	aprECB1	CH51-F (5′-AGGATCCCAAGAGAGCGATTGCGGCTGTGTAC-3′, BamHI site underlined) CH51-R (5′-AGAATTCTTCAGAGGGAGCCACCCGTCGATCA-3′, EcoRI site underlined)	E. coli DH5 α	pHY300PLK	B. subtilis WB600	pHY300PLK	[76]
B. subtilis PTCC 1023	Subtilisin gene	F (5′-CCGCTCGAGATGAGAAGCAAAAATTGTGG-3′, XhoI site underlined) R (5′-CGCGGATCCTTATTGTGCAGCTGCTTGTAC-3′)BamHI site underlined	E. coli BL21 (DE3)	pET-15b	E. coli BL21 (DE3)	pET-15b	[130]
B. amyloliquefaciens LSSE-62	—	aprF (5′-CCGTGAGAGGCAAAAAGGTATGGATCA-3′) aprR (5′-ATTTACTGAGCTGCCGCCTGTACGTTG-3′)	E. coli DH5α	pUC19	—	—	[107]
B. amyloliquefaciens MJ5-41	aprE5-41	51 F, (5′-AGGATCCCAAGAGAGCGATTGCGGCTGTGTAC-3′, BamHI site underlined) 51 R, (5′-AGAATTCTTCAGAGGGAGCCACCCGTCGATCA-3′, EcoRI site underlined)	E. coli DH5α	pHY300PLK	B. subtilis WB600	pHY300PLK	[95]
B. amyloliquefaciens CH51	aprE51	CH51-F (5′-AGGATCCCAAGAGAGCGATTGCGGCTGTGTAC-3′, BamHI site underlined) CH51-R (5′-AGAATTCTTCAGAGGGAGCCACCCGTCGATCA-3′, EcoRI site underlined)	E. coli DH5α	pGEM-T Easy	B. subtilis WB600	pHY300PLK	[93]
Staphylococcus sp. strain AJ	AJ	F-(5′-GGAATTCCATATGGTAATATTACCTAATAATAATAGAC-3′, NdeI site underlined) R-(5′-CCGCTCGAGTTACTGAAATATTTATATCAGGTATA-3′, XhoI site underlined)	—	pGEM-T Easy	—	—	[65]
Fusarium sp. BLB	FP	FP-(5′-GGCGACTTTCCCTTCCTCATCGTGAGCAT-3′) RP-(5′-TCACCCTGGCAAGAGTCCTTGCCACC-3′)	—	pT7 Blue T-vector	—	—	[99]
B. subtilis CH-3	aprE2	FP-(5′-GCGAATTCGCCGCATCTGTGTCTTTG-3′, EcoRI site underlined) RP-(5′-GCGAATTCGAGAACAGAAAGCCGCT-3′ EcoRI site underlined)	E. coli DH5α	pGEM-T Easy	B. subtilis ISW1214 and WB600	pHY300PLK	[94]
B. subtilis TP-6	TPase	F-[5′-GTGAGA(A/G)GCAAAAA(A/G)(G/T)T(A/G)TGGATCAG-3′] 5′-[A(A/T)TGTGC(A/T)GCTGCTTGTACGTTGA T(C/T)]	E. coli DH5a	pGEM-T Easy	E. coli M15	pQE30	[100]
B. amyloliquefaciens DC-4	DFE	P1, 5′-TCACAGCTTTTCTCGGTC-3′ P2, 5′-TGATCCGATTACGAATGC-3′	E. coli JM109	pGEM-T	B. subtilis WB600	pSUGV4	[104]
B. amyloliquefaciens DC-4	DFE	Primer 1: 5′-GCGAGTCCGTCCTTAC-3′, Primer 2: 5′-TTACTGAGCTGCCGCCTGT-AC-3′	E. coli JM109	pGEM-T	—	—	[140]

[141]. Below discussed are some of the substantial variables for optimal production of FEs through fermentation:

7.8.2.1 Optimization of temperature
The physiological and nutritional needs of microorganisms vary significantly. Therefore optimization of fermentation conditions is mandatory. Temperature is one of the crucial variables for the fermentative production of FEs. Most of the starter cultures applied for fermentation showed an optimum temperature between 30°C and 40°C for enzyme activity [20,24,142]. However, thermophilic bacteria like *B. cereus* IND1 at 60°C [117], and *Shewanella* sp. IND20 at 50°C [59].

7.8.2.2 pH
An optimum pH of the medium aids in retaining homeostasis of electric charges inside membranes and proteins regulation of protons pumps and transportation of nutrients within the cell membrane [142]. An unoptimized pH can lead to anomalies in formation of structure, disulfide bonds cleavage, protein aggregation, deamination, etc., as a result of imbalanced homeostasis leading to a decrease in enzyme production [143]. The optimal pH for most of the fermentative enzymes ranged from neutral to slightly alkaline. pH was studied as one of the important parameters influencing the functions of several isolated FEs, which have been discussed as follows: Khursade et al. [142], and Taneja et al. [40] documented a maximal FE activity at 901 U/mL and 250.41 U/mL at pH 8, respectively, and a sharp decline on either side of the pH range. However, Khan et al. [144] documented maximal enzyme activity (506 U/mL) of FE ZMS-2 at pH 8.

7.8.2.3 Substrate selection
Agroindustrial wastes like corn steep powder, cow dung, cuttlefish waste, soybean meal, wheat bran, banana peel, tapioca peel, rice bran, green gram husk, soybean residue, monosodium glutamate waste liquor, soybean flour, soybean hydrolysate, soybean grits, and chickpeas have been used as substrates for fermentative production of FEs [22,23,41,108,112,117,121,122,145–147]. These agroindustrial residues claim to be economic and cut environmental pollution through utilization of waste making them eco-friendly and sustainable.

7.8.2.4 Nutrients
Nitrogen, carbon substrates, and minerals are the major nutrient sources required for fermentative production [22–24]. Complex nitrogen sources are preferentially broken down during fermentation compared to simple sources of inorganic nitrogen. This phenomenon is feasible as bacteria are equipped with proteases responsible for breaking down complex nitrogen sources [22]. On the contrary, microorganisms lack the ability to breakdown complex carbon sources due to their weaker carbon metabolism. Nevertheless, several studies have reported repressive effect as a result of the presence of nutrient (primarily carbon) or substrate concentrations above optimal conditions [148]. This phenomenon is explained due to carbon catabolite repression due to high carbon concentrations.

As mentioned above, FEs belonging to the class of metalloproteases need metal ions as cofactors for their metabolism. However, the presence of metal ions is also known to cease the growth of certain starter cultures. This inhibitory action is speculated to be occurring due to the attachment of metal ions at the enzyme's sulfhydryl groups, cleavage of disulfide bonds, or substitute of the originally present metal ion and rendering them inactive [37,63].

7.8.2.5 Shaking condition and other factors

An increase in enzyme production yield is due to the homogenous distribution of nutrients and oxygen during fermentation conditions [149]. However, extremely high rpm levels were shown to have a negative effect on the enzyme production of FE from *Schizopsullium commune* BL23 [150]. Alternatively, the use of immobilized starter culture increased the enzyme production yields as microbial immobilization is known to shield the cultures from harsh environmental conditions [151]. Extremely high and low volumes of media are also known to lead to dilution or insufficiency of nutrients. Wang et al. [111] reported maximal FE activity with 100 mL of fermentative medium for the production of NK from *Pseudomonas* sp. TKU015 than 50, 150, or 200 mL of media.

7.8.2.6 Statistical optimization of media

Optimization of media compositions and fermentative parameters is crucial for obtaining improved production yields and cost-effectiveness. Conventionally, media optimization has been performed by adjusting one variable (time, temperature, media concentration, pH, inoculum level, etc.) at a time while maintaining all the others constant. Traditional methods pose disadvantages of time consumption other than leading to false results. These shortcomings can be eliminated by the application of a statistical approach such as response surface methodology, Taguchi orthogonal array design, and fractional factorial design. Statistical approaches provide reliability, help in standardizing the nutrient composition, and understand the interaction between fermentative parameters, hence saving a lot of time and energy. Table 7–4 presents lists the bacterial strains, optimized media for fermentation, statistical approaches used for media optimization, and improved FE activity.

7.9 Applications

7.9.1 Clinical

Thrombus is the main causative agent for CVDs and fibrin being the major component of thrombus. The application of PAs can degrade fibrin. However, due to the high production cost, the demand for economical, sustainable, and safer sources of FEs is on the rise. In this regard, the FEs produced from food-grade microorganisms are considered a better alternative to PAs. Some of the FEs reported for their prospective in thrombolytic therapy, and other health-promoting properties have been stated below:

Table 7.4 Statistical methods used for optimal production of fibrinolytic enzymes through fermentation.

Bacterial strains used	Media for fermentation	Statistical methods used	FE activity (U/mL)	Reference
Serratia rubidaea KUAS001	Dextrose (1%), peptone (0.5%), yeast extract (0.5%), KH_2PO_4 (0.1%), K_2HPO_4 (0.1%) and NaCl (0.02%) (% w/v) was inoculated with 1% (v/v) inoculum	OFAT	394.9	[152]
Bacillus cereus RSA1	Glucose (10.0 g/L), soybean (10.0 g/L), $CaCl_2$ (0.5 g/L), $MgSO_4 \cdot 7H_2O$ (0.5 g/L), $NaCl_2$ (0.5 g/L), and K_2HPO_4 (5.0 g/L)	Plackett–Burman design, RSM, CCD	30.75	[83]
Bacillus subtilis WR350	Sucrose (35 g/L), corn steep powder (20 g/L) and 2 g/L $MgSO_4 \cdot 7H_2O$	L9 (34) orthogonal design, an orthogonal array of four factors with three levels	5865	[22]
Xanthomonas oryzae IND3	Glucose, sucrose, maltose, starch and xylose (1%), peptone, casein, yeast extract, gelatin and inorganic ions (1%), banana peel, green gram husk, rice bran, wheat bran, cow dung, $CaCl_2$, $MgCl_2$, $ZnSO_4$, $HgCl_2$	RSM, CCD	2296^a	[23]
B. subtilis D21-8	Glucose (18.5 g/L), yeast extract (6.3 g/L), tryptone (7.9 g/L), and NaCl (5 g/L)	RSM	3129	[24]
Stenotrophomonas maltophilia Gd2	Dextrose (10 g/L), yeast extract (10 g/L), $MgSO_4$ (0.05 g/L), KH_2PO_4 (0.05 g/L)	OFAT	1795	[142]
S. maltophilia Gd2	Dextrose (10 g/L), yeast extract (10 g/L), $MgSO_4$ (0.05 g/L), KH_2PO_4 (0.05 g/L)	Plackett–Burman media designing	3411	[142]
Streptococcus agalactiae EBL-31	Glucose (0.5 g/100 mL), yeast extract (0.5 g/100 mL), K_2HPO_4 (0.25 g/100 mL), $NaHCO_3$ (0.1 g/100 mL), $MgSO_4 \cdot 7H_2O$ (0.04 g/100 mL), $CH_3COONa \cdot 3H_2O$ (0.1 g/100 mL), $MnCl_2 \cdot 4H_2O$ (0.002 g/100 mL), and $FeSO_4 \cdot 7H_2O$ (0.002 g/100 mL)	RSM, CCRD	147.08	[119]
Bacillus sp. IND12	Cow dung as substrate, moisture (109.73%), sucrose (0.57%), and $MgSO_4$ (0.093%)	RSM	4143^a	[145]
Serratia sp. KG-2-1	Maltose (1.5%), yeast extract (4%), peptone and other trace elements	RSM, CCD	250.41	[40]
Xylaria curta	Rice chaff (10 g) and 5 mL of salt solution: KH_2PO_4 (0.5 g), $MgSO_4 \cdot 7H_2O$ (0.5 g), $MnSO_4 \cdot 7H_2O$ (0.001 g), $ZnSO_4 \cdot 7H_2O$ (0.002 g), $FeSO_4$ (0.0005 g), distilled water 10 mL	OVAT	9.22	[75]
B. cereus IND5	Cuttlefish waste and cow dung substrate, casein (1.1%) magnesium sulfate (0.1%)	OVAT and RSM, CCD	364.5^a	[108]
Serratia marcescens RSPB11	Casein (2% w/v), dextrose (1.0% w/v), $MgSO_4$ (0.02% w/v), KH_2PO_4 (0.02% w/v), NaCl (0.02% w/v), $CaCl_2$ (0.002% w/v)	Plackett–Burman design and RSM, CCD	23,910	[21]
Shewanella sp. IND20	2.0 g of Agroresidues (banana peel, cow dung, rice bran, wheat bran and green gram husk), 2 mL 0.1 M Tris–HCl buffer (pH 8.0), 10% inoculum	OFAT, e 2^5 factorial design, RSM, CCD	2751	[59]
B. cereus IND1	Beef extract (0.3%w/w), sodium dihydrogen phosphate (0.05%w/w), moisture (100%v/w)	Two-level full-factorial design, RSM, CCD	3699	[117]
Paenibacillus sp. IND8	Wheat bran substrate (2.0 g w/w), sucrose (0.5% w/w), NaH_2PO_4 (0.075% w/w), moisture (113.64%)	25 Full factorial design (first-order model), CCD	4418	[118]
strain XY-1	Peptone (11.4 g/L), sucrose (5 g/L) and $MgSO_4$ (0.5 g/L)	Plackett–Burman and RSM, BBD	21.33	[101]

(Continued)

Table 7.4 (Continued)

Bacterial strains used	Media for fermentation	Statistical methods used	FE activity (U/mL)	Reference
Bacillus sp. UFPEDA 485	soybean filtrate (2% w/v), K_2HPO_4(0.435% w/v), and 1 mL of mineral solution containing $FeSO_4 \cdot 7H_2O$ (100 mg), $MnCl_2 \cdot 4H_2O$ (100 mg), and $ZnSO_4 \cdot H_2O$ (100 mg) of distilled water 100 mL, NH_4Cl (0.1% w/v), $MgSO_4 \cdot 7H_2O$ (0.06% w/v), and glucose (1% w/v)	23 Experimental design	835	[41]
B. subtilis	Glucose (1%), Peptone (5.5%), MgSO4 (0.2%) $CaCl_2$ (0.5%)	RSM, CCRD	3194.25	[153]
B. subtilis DC-2	Soy peptone (0.25%), maltose (1%), yeast extract (0.075%), K_2HPO_4 (0.2%), NaH_2PO_4 (0.02%), $CaCO_3$ (0.15%)	RSM, CCRD	1165.58	[103]
B. subtilis	Soybean hydrolysate (6%), K_2HPO_4 (0.45%), and $CaCl_2$ (0.015%)	Plackett–Burman and RSM, BBD	7,7400	[147]

Notes: *BBD*, Box–Behnken design; *CCD*, central composite design; *CCRD*, central composite rotatable design; *OFAT*, one factor at a time; *OVAT*, One variable at a time approach; *RSM*, response surface methodology.
^aU/g.

7.9.1.1 Nattokinase

NK is considered to be a safe, effective, economic, and sustainable source for the treatment of CVD [154–156]. Several studies have shown that the blood-thinning function NK could effectively dissolve blood clots [125,157–161]. NK has been reported to exhibit a protective effect on both oxidative injury mediated arterial thrombosis and inflammation-induced venal thrombosis [162]. Enzymes like NK and lumbrokinase have been reported for their strong gastrointestinal stability [163]. NK is commercially available as a supplement in countries like Japan, Korea, Canada, Europe, and the United States [164]. NK has also been shown to exhibit antihypertensive, neuroprotective, and antiatherosclerosis effects apart from its thrombolytic effects [158,165–169]. A thorough safety data, obtained from Good Laboratory Practice compliant studies reported in 2016, stated that FEs were devoid of any kind of clastogenic nor mutagenic activity [161]. The suggested daily dosage for NK intake is two capsules (100 mg of NK/capsule) [164].

7.9.1.2 Serrapeptase

Serrapeptase exhibited strong antiinflammatory effects other than fibrinolytic effects [15,85,165]. Serrapeptase was shown to reduce swelling of the prostate glands of patients suffering from diseases like a microbial prostatovesiculitis when used in combination with along with other antiinflammatory drugs [170].

7.9.1.3 Staphylokinase

Staphylokinase, a FE produced from *Staphylococci*, was initially expected to possess a promising potential. However, it could not pass through clinical trials due to its short half-life in plasma, high levels of antistaphylokinase antibodies like IgG [171].

7.9.1.4 Streptokinase

Numerous studies have been performed on the fibrinolytic ability of streptokinase [45,70,172]. Nevertheless, streptokinase was found to exhibit in vivo immunogenic activity, which was thought to be solved by creating recombinant forms with diminished immunogenic potential [173]. Subsequently, in another independent, mutants with deleted *C*-terminal end showed diminished immunogenic potential proving it to be a promising thrombolytic agent [172].

7.9.1.5 Surfactants and antimicrobial properties

FEs like KSK-II from *L. plantarum* were shown to exhibited nonspecific hydrolysis, which was the desired property to be used as detergents such as Persil, X-Tra, and Ariel. KSK-II was shown to be a potent bloodstain remover from cotton fabrics [15]. Another exceptional attribute of KSK-II is its potential to inhibit pathogens like *S. aureus, B. cereus, Pseudomonas aeruginosa, Proteus vulgaris, E. coli,* and soilborne fungus like *Rhizoctonia solani* [174,175]. Hence, these properties of KSK-II can be utilized in the clinical and food sectors.

7.10 Conclusions and perspectives

FEs have been identified from various traditional fermented such as *DoenJang, Douche, Gonjaengijot, Jeotgal, Natto, Tempeh,* etc. Microorganisms from fermented food have been a potential source of FEs, especially the *Bacillus* sp. owing to their high specificity and fibrinolytic activity. Most of the FEs showing fibrinolytic activity have been recovered from the fermented products and have been studied for their characteristic properties. FEs such as NK, subtilisin, streptokinase, staphylokinase, and serrapeptase have shown to have potential in pharmaceutical and industrial set up. Further, to improve the efficiency and cost of commercialization production, most of the FEs have been manipulated via biotechnological approaches employing modern molecular techniques, or the fermentation process has been optimized using statistical strategies. Besides this, enhanced production FEs can facilitate exhaustive studies, including the physicochemical and biochemical characterization, in vitro and in vivo clinical trials, etc., to enhance or elaborate their applications in pharmaceutical and other conventional applications.

References

[1] A.M.M. Ali, S.C.B. Bavisetty, Purification, physicochemical properties, and statistical optimization of fibrinolytic enzymes especially from fermented foods: a comprehensive review, International Journal of Biological Macromolecules 163 (2020) 1498–1517.

[2] D. Mozaffarian, E.J. Benjamin, A.S. Go, D.K. Arnett, M.J. Blaha, M. Cushman, et al., Executive summary: heart disease and stroke statistics—2015 update: a report from the American Heart Association, Circulation 131 (4) (2015) 434–441.

[3] J. Seo, T.A. Al-Hilal, J.-G. Jee, Y.-L. Kim, H.-J. Kim, B.-H. Lee, et al., A targeted ferritin-microplasmin based thrombolytic nanocage selectively dissolves blood clots, Nanomedicine: Nanotechnology, Biology and Medicine 14 (3) (2018) 633–642.

[4] WHO, Cardiovascular diseases (CVDs), 2017.

[5] J.C. Chapin, K.A. Hajjar, Fibrinolysis and the control of blood coagulation, Blood Reviews 29 (1) (2015) 17–24.

[6] D. Collen, H.R. Lijnen, Tissue-type plasminogen activator: a historical perspective and personal account, Journal of Thrombosis and Haemostasis 2 (4) (2004) 541–546.

[7] G. Adeboyeje, G. Sylwestrzak, J.J. Barron, J. White, A. Rosenberg, J. Abarca, et al., Major bleeding risk during anticoagulation with warfarin, dabigatran, apixaban, or rivaroxaban in patients with nonvalvular atrial fibrillation, Journal of Managed Care and Specialty Pharmacy 23 (9) (2017) 968–978.

[8] M.V.S. Andrade, L.A.P. Andrade, A.F. Bispo, Ld.A. Freitas, M.Q.S. Andrade, G.S. Feitosa, et al., Evaluation of the bleeding intensity of patients anticoagulated with warfarin or dabigatran undergoing dental procedures, Arquivos Brasileiros de Cardiologia 111 (3) (2018) 394–399.

[9] J.-H. Choi, K. Sapkota, S.-E. Park, S. Kim, S.-J. Kim, Thrombolytic, anticoagulant and antiplatelet activities of codiase, a bi-functional fibrinolytic enzyme from Codium fragile, Biochimie 95 (6) (2013) 1266–1277.

[10] G.A.I., The Angiographic Investigators, The effects of tissue plasminogen activator, streptokinase, or both on coronary-artery patency, ventricular function, and survival after acute myocardial infarction, New England Journal of Medicine 329 (1993) 1615–1622.

[11] J.R. Gonzalez, M.G. Cortina, A.R. Campello, A.O. Santiago, J.H. Rocamora, E.M. Olivas, Hemorragia cerebral en pacientes en tratamiento con anticoagulantes orales, Revista de Neurologia 40 (1) (2005) 19–22.

[12] G. Zapata-Wainberg, S. Quintas, A.X.-C. Rico, L.B. Fernández, J.M. Vallejo, J.G. Culleré, et al., Prognostic factors and analysis of mortality due to brain haemorrhages associated with vitamin K antagonist oral anticoagulants, Results from the TAC Registry, Neurología 33 (7) (2018) 419–426.

[13] M.M. da Silva, T.A. Rocha, D.F. de Moura, C.A. Chagas, F.C.A. de Aguiar Júnior, N.P. da Silva Santos, et al., Effect of acute exposure in Swiss mice (Mus musculus) to a fibrinolytic protease produced by Mucor subtilissimus UCP 1262: an histomorphometric, genotoxic and cytological approach, Regulatory Toxicology and Pharmacology 103 (2019) 282–291.

[14] Y. Peng, X. Yang, Y. Zhang, Microbial fibrinolytic enzymes: an overview of source, production, properties, and thrombolytic activity in vivo, Applied Microbiology and Biotechnology 69 (2) (2005) 126–132.

[15] E. Kotb, Purification and partial characterization of a chymotrypsin-like serine fibrinolytic enzyme from Bacillus amyloliquefaciens FCF-11 using corn husk as a novel substrate, World Journal of Microbiology and Biotechnology 30 (7) (2014) 2071–2080.

[16] P.Ed.C. e Silva, R.C. de Barros, W.W.C. Albuquerque, R.M.P. Brandão, R.P. Bezerra, A.L.F. Porto, In vitro thrombolytic activity of a purified fibrinolytic enzyme from Chlorella vulgaris, Journal of Chromatography B 1092 (2018) 524–529.

[17] V.G. Nielsen, J.K. Kirklin, W.L. Holman, B.L. Steenwyk, Clot lifespan model analysis of the effects of warfarin on thrombus growth and fibrinolysis: role of contact protein and tissue factor initiation, ASAIO Journal 55 (1) (2009) 33–40.

[18] D. Cai, C. Zhu, S. Chen, Microbial production of nattokinase: current progress, challenge and prospect, World Journal of Microbiology and Biotechnology 33 (5) (2017) 84–91.

[19] A. Krishnamurthy, P.D. Belur, S.B. Subramanya, Methods available to assess therapeutic potential of fibrinolytic enzymes of microbial origin: a review, Journal of Analytical Science and Technology 9 (1) (2018) 10–21.

[20] D.N. Avhad, V.K. Rathod, Ultrasound assisted production of a fibrinolytic enzyme in a bioreactor, Ultrasonics Sonochemistry 22 (2015) 257–264.

[21] P.L. Bhargavi, R.S. Prakasham, Enhanced fibrinolytic protease production by Serratia marcescens RSPB11 through Plackett-Burman and response surface methodological approaches, Journal of Applied Biology and Biotechnology 4 (3) (2016) 006–014.

[22] R. Wu, G. Chen, S. Pan, J. Zeng, Z. Liang, Cost-effective fibrinolytic enzyme production by Bacillus subtilis WR350 using medium supplemented with corn steep powder and sucrose, Scientific Reports 9 (1) (2019) 6824–6834.

[23] P. Vijayaraghavan, M.V. Arasu, R. Anantha Rajan, N.A. Al-Dhabi, Enhanced production of fibrinolytic enzyme by a new Xanthomonas oryzae IND3 using low-cost culture medium by response surface methodology, Saudi Journal of Biological Sciences 26 (2) (2019) 217–224.

[24] S. Pan, G. Chen, R. Wu, X. Cao, Z. Liang, Non-sterile submerged fermentation of fibrinolytic enzyme by marine Bacillus subtilis harboring antibacterial activity with starvation strategy, Frontiers in Microbiology 10 (2019) 1025–1038.

[25] J.B. Lefkowitz, Coagulation pathway and physiology, in: K. Kottke-Marchant (Ed.), An Algorithmic Approach to Hemostasis Testing, Thieme Medical Publishers, New York, USA, 2008, pp. 3–12.

[26] R. Chaudhry, H.M. Babiker, Physiology, coagulation pathways, in: B.E.A. Abai (Ed.), StatPearls, StatPearls Publishing, Florida, USA, 2019.

[27] M. Nesheim, Thrombin and fibrinolysis, Chest 124 (3) (2003) 33S–39S.

[28] K. Ito, Effect of water-extractive components from funazushi, a fermented crucian carp, on the activity of fibrinolytic factors, Journal of the Science of Food and Agriculture (2020).

[29] D. Collen, H. Lijnen, Basic and clinical aspects of fibrinolysis and thrombolysis, Blood 78 (1991) 3114–3124.

[30] L. Hedstrom, Serine protease mechanism and specificity, Chemical Reviews 102 (12) (2002) 4501–4524.

[31] L. Jeffries, D. Buckley, The detection and differentiation of fibrinolytic enzymes in bacteria, Journal of Applied Bacteriology 49 (3) (1980) 479–492.

[32] V. Sekar, J.H. Hageman, Specificity of the serine protease inhibitor, phenylmethylsulfonyl fluoride, Biochemical and Biophysical Research Communications 89 (2) (1979) 474–478.

[33] N. Alkjaersig, A.P. Fletcher, S. Sherry, The mechanism of clot dissolution by plasmin, Journal of Clinical Investigation 38 (7) (1959) 1086–1095.

[34] D.B. Baruah, R.N. Dash, M. Chaudhari, S. Kadam, Plasminogen activators: a comparison, Vascular Pharmacology 44 (1) (2006) 1–9.

[35] B. Wiman, D. Collen, Molecular mechanism of physiological fibrinolysis, Nature 272 (5653) (1978) 549–550.

[36] Q. Bi, B. Han, Y. Feng, Z. Jiang, Y. Yang, W. Liu, Antithrombotic effects of a newly purified fibrinolytic protease from Urechis unicinctus, Thrombosis Research 132 (2) (2013) 135–144.

[37] Z. Yao, J.A. Kim, J.H. Kim, Characterization of a fibrinolytic enzyme secreted by Bacillus velezensis BS2 isolated from sea squirt Jeotgal, Journal of Microbiology and Biotechnology 29 (3) (2019) 347–356.

[38] Y. Hu, D. Yu, Z. Wang, J. Hou, R. Tyagi, Y. Liang, et al., Purification and characterization of a novel, highly potent fibrinolytic enzyme from Bacillus subtilis DC27 screened from Douchi, a traditional Chinese fermented soybean food, Scientific Reports 9 (1) (2019) 9235–9245.

[39] Z. Yao, J.A. Kim, J.H. Kim, Properties of a fibrinolytic enzyme secreted by Bacillus subtilis JS2 isolated from saeu (small shrimp) jeotgal, Food Science and Biotechnology 27 (3) (2018) 765–772.

[40] K. Taneja, B.K. Bajaj, S. Kumar, N. Dilbaghi, Production, purification and characterization of fibrinolytic enzyme from Serratia sp. KG-2-1 using optimized media, 3 Biotech 7 (3) (2017) 184–199.

[41] A.E. Sales, F.A. de Souza, J.A. Teixeira, T.S. Porto, A.L. Porto, Integrated process production and extraction of the fibrinolytic protease from Bacillus sp. UFPEDA 485, Applied Biochemistry and Biotechnology 170 (7) (2013) 1676–1688.

[42] A. Krishnamurthy, P.D. Belur, A novel fibrinolytic serine metalloprotease from the marine Serratia marcescens subsp. sakuensis: purification and characterization, International Journal of Biological Macromolecules 112 (2018) 110–118.

[43] Y. Devaraj, S.K. Rajender, P.M. Halami, Purification and characterization of fibrinolytic protease from Bacillus amyloliquefaciens MCC2606 and analysis of fibrin degradation product by MS/MS, Preparative Biochemistry and Biotechnology 48 (2) (2018) 172–180.

[44] K. Ouriel, A history of thrombolytic therapy, Journal of Endovascular Therapy 11 (2004) 128–133.

[45] N. Sikri, A. Bardia, A history of streptokinase use in acute myocardial infarction, Texas Heart Institute Journal 34 (3) (2007) 318–327.

[46] F. Vermeer, I. Bösl, J. Meyer, F. Bär, B. Charbonnier, J. Windeler, et al., Saruplase is a safe and effective thrombolytic agent; observations in 1698 patients: results of the PASS study, Journal of Thrombosis and Thrombolysis 8 (2) (1999) 143–150.

[47] USFDA, Biologics licence application approval of tenecplase, 2000. Available from: https://www.accessdata.fda.gov/drugsatfda_docs/appletter/2000/tenegen060200L.htm.

[48] G.J. del Zoppo, R.T. Higashida, A.J. Furlan, M.S. Pessin, H.A. Rowley, M. Gent, PROACT: a phase II randomized trial of recombinant pro-urokinase by direct arterial delivery in acute middle cerebral artery stroke, Stroke 29 (1) (1998) 4–11.

[49] A.J. Furlan, A. Abou-Chebl, The role of recombinant pro-urokinase (r-pro-UK) and intra-arterial thrombolysis in acute ischaemic stroke: the PROACT trials, Current Medical Research and Opinion 18 (2002) 44–47.

[50] K. Swedberg, I.S.S.I.S.-3 Third, International Study of Infarct Survival) Collaborative Group. ISIS-3 a randomized trial of streptokinase vs Tissue plasmino gen activator vs anistreplase and aspirin plus heparine vs aspirina lone among 41229 cases of suspected acute myocardial infarction, Lancet 339 (1992) 753–770.

[51] C.P. Semba, K. Sugimoto, M.K. Razavi, Alteplase and tenecteplase: applications in the peripheral circulation, Techniques in Vascular and Interventional Radiology 4 (2) (2001) 99–106.

[52] M. Verstraete, Third-generation thrombolytic drugs, The American Journal of Medicine 109 (1) (2000) 52–58.

[53] K. Kikuchi, N. Miura, K.-I. Kawahara, Y. Murai, M. Morioka, P.A. Lapchak, et al., Edaravone (Radicut), a free radical scavenger, is a potentially useful addition to thrombolytic therapy in patients with acute ischemic stroke, Biomedical Reports 1 (1) (2013) 7–12.

[54] A. Kumar, K. Pulicherla, K.S. Ram, K. Rao, Evolutionary trend of thrombolytics, International Journal of Bio-science and Bio-technology 2 (4) (2010) 51–68.

[55] E. Shakhnovich, Protein folding thermodynamics and dynamics: where physics, chemistry, and biology meet, Chemical Reviews 106 (5) (2006) 1559–1588.

[56] P. Katrolia, X. Liu, Y. Zhao, N.K. Kopparapu, X. Zheng, Gene cloning, expression and homology modeling of first fibrinolytic enzyme from mushroom (Cordyceps militaris), International Journal of Biological Macromolecules 146 (2020) 897–906.

[57] T.P. Nascimento, A.E. Sales, C.S. Porto, R.M. Brandão, G.M. de Campos-Takaki, J.A. Teixeira, et al., Purification of a fibrinolytic protease from Mucor subtilissimus UCP 1262 by aqueous two-phase systems (PEG/sulfate), Journal of Chromatography B: Analytical Technologies in the Biomedical and Life Sciences 1025 (2016) 16–24.

[58] H.C. Kim, B.-S. Choi, K. Sapkota, S. Kim, H.J. Lee, J.C. Yoo, et al., Purification and characterization of a novel, highly potent fibrinolytic enzyme from Paecilomyces tenuipes, Process Biochemistry 46 (8) (2011) 1545–1553.

[59] P. Vijayaraghavan, S.G. Prakash, Vincent, A low cost fermentation medium for potential fibrinolytic enzyme production by a newly isolated marine bacterium, Shewanella sp. IND20, Biotechnology Reports 7 (2015) 135–142.

[60] H. Yang, L. Yang, X. Li, H. Li, Z. Tu, X. Wang, Genome sequencing, purification, and biochemical characterization of a strongly fibrinolytic enzyme from Bacillus amyloliquefaciens Jxnuwx-1 isolated from Chinese traditional Douchi, The Journal of General and Applied Microbiology (2019).

[61] M.J. Ahn, H.J. Ku, S.H. Lee, J.H. Lee, Characterization of a novel fibrinolytic enzyme, BsfA, from Bacillus subtilis ZA400 in Kimchi reveals its pertinence to thrombosis treatment, Journal of Microbiology and Biotechnology 25 (12) (2015) 2090–2099.

[62] P.M. Mahajan, S. Nayak, S.S. Lele, Fibrinolytic enzyme from newly isolated marine bacterium Bacillus subtilis ICTF-1: media optimization, purification and characterization, Journal of Bioscience and Bioengineering 113 (3) (2012) 307–314.

[63] X. Xin, R.R. Ambati, Z. Cai, B. Lei, Purification and characterization of fibrinolytic enzyme from a bacterium isolated from soil, 3 Biotech 8 (2018) 90–98.

[64] W. Kim, K. Choi, Y. Kim, H. Park, J. Choi, Y. Lee, et al., Purification and characterization of a fibrinolytic enzyme produced from Bacillus sp. strain CK 11–4 screened from Chungkook-Jang, Applied and Environmental Microbiology 62 (7) (1996) 2482–2488.

[65] N.S. Choi, J.J. Song, D.M. Chung, Y.J. Kim, P.J. Maeng, S.H. Kim, Purification and characterization of a novel thermoacid-stable fibrinolytic enzyme from Staphylococcus sp. strain AJ isolated from Korean salt-fermented Anchovy-joet, Journal of Industrial Microbiology and Biotechnology 36 (3) (2009) 417–426.

[66] A.H. Wong, Y. Mine, Novel fibrinolytic enzyme in fermented shrimp paste, a traditional asian fermented seasoning, Journal of Agricultural and Food Chemistry 52 (4) (2004) 980–986.

[67] D. Dhamodharan, S.J.N. Jemimah, S.K. Merlyn, C.D. Subathra, Novel fibrinolytic protease producing Streptomyces radiopugnans VITSD8 from marine sponges, Marine Drugs 17 (3) (2019) 164–178.

[68] J.-H. Choi, D.-W. Kim, S. Kim, S.-J. Kim, Purification and partial characterization of a fibrinolytic enzyme from the fruiting body of the medicinal and edible mushroom Pleurotus ferulae, Preparative Biochemistry and Biotechnology 47 (6) (2017) 539–546.

[69] S.-E. Park, M.-H. Li, J.-S. Kim, K. Sapkota, J.-E. Kim, B.-S. Choi, et al., Purification and characterization of a fibrinolytic protease from a culture supernatant of Flammulina velutipes mycelia, Bioscience, Biotechnology, and Biochemistry 71 (9) (2007) 70193–70199.

[70] B. Naveena, K.P. Gopinath, P. Sakthiselvan, N. Partha, Enhanced production of thrombinase by Streptomyces venezuelae: kinetic studies on growth and enzyme production of mutant strain, Bioresource Technology 111 (2012) 417–424.

[71] A. Undas, R.A. Ariëns, Fibrin clot structure and function: a role in the pathophysiology of arterial and venous thromboembolic diseases, Arteriosclerosis, Thrombosis, and Vascular Biology 31 (12) (2011) 88–99.

[72] J.-H. Choi, K. Sapkota, S. Kim, S.-J. Kim, Starase: a bi-functional fibrinolytic protease from hepatic caeca of Asterina pectinifera displays antithrombotic potential, Biochimie 105 (2014) 45–57.

[73] X. Liu, N.K. Kopparapu, X. Shi, Y. Deng, X. Zheng, J. Wu, Purification and biochemical characterization of a novel fibrinolytic enzyme from culture supernatant of Cordyceps militaris, Journal of Agricultural and Food Chemistry 63 (8) (2015) 2215–2224.

[74] T. Astrup, S. Müllertz, The fibrin plate method for estimating fibrinolytic activity, Archives of Biochemistry and Biophysics 40 (2) (1952) 346–351.

[75] V. Meshram, S. Saxena, K. Paul, M. Gupta, N. Kapoor, Production, purification and characterisation of a potential fibrinolytic protease from Endophytic Xylaria curta by solid substrate fermentation, Applied Biochemistry and Biotechnology 181 (4) (2017) 1496–1512.

[76] Z. Yao, X. Liu, J.M. Shim, K.W. Lee, H.J. Kim, J.H. Kim, Properties of a fibrinolytic enzyme secreted by Bacillus amyloliquefaciens RSB34, isolated from Doenjang, Journal of Microbiology and Biotechnology 27 (1) (2017) 9–18.

[77] K. Heo, K.M. Cho, C.K. Lee, G.M. Kim, J.H. Shin, J.S. Kim, et al., Characterization of a fibrinolytic enzyme secreted by Bacillus amyloliquefaciens CB1 and its gene cloning, Journal of Microbiology and Biotechnology 23 (7) (2013) 974–983.

[78] H.D. Jo, H.A. Lee, S.J. Jeong, J.H. Kim, Purification and characterization of a major fibrinolytic enzyme from Bacillus amyloliquefaciens MJ5-41 isolated from Meju, Journal of Microbiology and Biotechnology 21 (11) (2011) 1166-1173.

[79] Y. Hua, B. Jiang, Y. Mine, W. Mu, Purification and characterization of a novel fibrinolytic enzyme from Bacillus sp. nov. SK006 isolated from an Asian traditional fermented shrimp paste, Journal of Agricultural and Food Chemistry 56 (4) (2008) 1451-1457.

[80] K.J. Hwang, K.H. Choi, M.J. Kim, C.S. Park, J. Cha, Purification and characterization of a new fibrinolytic enzyme of Bacillus licheniformis KJ-31, isolated from Korean traditional Jeot-gal, Journal of Microbiology and Biotechnology 17 (9) (2007) 1469-1476.

[81] C.T. Wang, B.P. Ji, B. Li, R. Nout, P.L. Li, H. Ji, et al., Purification and characterization of a fibrinolytic enzyme of Bacillus subtilis DC33, isolated from Chinese traditional Douchi, Journal of Industrial Microbiology and Biotechnology 33 (9) (2006) 750-758.

[82] S.B. Kim, D.W. Lee, C.I. Cheigh, E.A. Choe, S.J. Lee, Y.H. Hong, et al., Purification and characterization of a fibrinolytic subtilisin-like protease of Bacillus subtilis TP-6 from an Indonesian fermented soybean, Tempeh, Journal of Industrial Microbiology and Biotechnology 33 (6) (2006) 436-444.

[83] J.-S. Kim, K. Sapkota, S.-E. Park, B.-S. Choi, S. Kim, N.T. Hiep, et al., A fibrinolytic enzyme from the medicinal mushroom Cordyceps militaris, The Journal of Microbiology 44 (6) (2006) 622-631.

[84] S.J. Jeong, K. Heo, J.Y. Park, K.W. Lee, J.Y. Park, S.H. Joo, et al., Characterization of AprE176, a fibrinolytic enzyme from Bacillus subtilis HK176, Journal of Microbiology and Biotechnology 25 (1) (2015) 89-97.

[85] Y.-k Jeong, J.H. Kim, S.-w Gal, J.-e Kim, S.-s Park, K.-t Chung, et al., Molecular cloning and characterization of the gene encoding a fibrinolytic enzyme from Bacillus subtilis strain A1, World Journal of Microbiology and Biotechnology 20 (7) (2004) 711-717.

[86] K. Matsubara, K. Hori, Y. Matsuura, K. Miyazawa, Purification and characterization of a fibrinolytic enzyme and identification of fibrinogen clotting enzyme in a marine green alga, Codium divaricatum, Comparative Biochemistry and Physiology Part B: Biochemistry and Molecular Biology 125 (1) (2000) 137-143.

[87] H.P. Li, Z. Hu, J.L. Yuan, H.D. Fan, W. Chen, S.J. Wang, et al., A novel extracellular protease with fibrinolytic activity from the culture supernatant of Cordyceps sinensis: purification and characterization, Phytotherapy Research 21 (12) (2007) 1234-1241.

[88] K. Balaraman, G. Prabakaran, Production & purification of a fibrinolytic enzyme (thrombinase) from Bacillus sphaericus, Indian Journal of Medical Research 126 (5) (2007) 459-464.

[89] C. Sharma, G.E.M. Salem, N. Sharma, P. Gautam, R. Singh, Thrombolytic potential of novel thiol-dependent fibrinolytic protease from Bacillus cereus RSA1, Biomolecules 10 (3) (2020).

[90] C. Kim, K. Ri, S. Choe, A novel fibrinolytic enzymes from the Korean traditional fermented food—Jotgal: purification and characterization, Journal of Food Biochemistry (2020).

[91] E. Kotb, The biotechnological potential of subtilisin-like fibrinolytic enzyme from a newly isolated Lactobacillus plantarum KSK-II in blood destaining and antimicrobials, Biotechnology Progress 31 (2) (2015) 316-324.

[92] P.L. Bhargavi, R. Prakasham, Proteolytic enzyme production by isolated Serratia sp RSPB11: role of environmental parameters, Current Trends in Biotechnology and Pharmacy 6 (1) (2012) 55-65.

[93] S. Raveendran, B. Parameswaran, S. Beevi Ummalyma, A. Abraham, A. Kuruvilla Mathew, A. Madhavan, et al., Applications of microbial enzymes in food industry, Food Technology and Biotechnology 56 (1) (2018) 16-30.

[94] D.N. Afifah, M. Sulchan, D. Syah, et al., Purification and characterization of a fibrinolytic enzyme from Bacillus pumilus 2.g isolated from Gembus, an Indonesian fermented food, Preventive Nutrition and Food Science 19 (3) (2014) 213-219.

[95] B. Wu, L. Wu, L. Ruan, M. Ge, D. Chen, Screening of endophytic fungi with antithrombotic activity and identification of a bioactive metabolite from the endophytic fungal strain CPCC 480097, Current Microbiology 58 (5) (2009) 522–527.

[96] S.H. Kim, N.S. Choi, Purification and characterization of subtilisin DJ-4 secreted by Bacillus sp. strain DJ-4 screened from Doen-Jang, Bioscience, Biotechnology, and Biochemistry 64 (8) (2000) 1722–1735.

[97] G.M. Kim, A.R. Lee, K.W. Lee, J.Y. Park, J. Chun, J. Cha, et al., Characterization of a 27 kDa fibrinolytic enzyme from Bacillus amyloliquefaciens CH51 isolated from cheonggukjang, Journal of Microbiology and Biotechnology 19 (9) (2009) 997–1004.

[98] S.J. Jeong, G.H. Kwon, J. Chun, J.S. Kim, C.S. Park, D.Y. Kwon, et al., Cloning of fibrinolytic enzyme gene from Bacillus subtilis isolated from Cheonggukjang and its expression in protease-deficient Bacillus subtilis strains, Journal of Microbiology and Biotechnology 17 (6) (2007) 1018–1023.

[99] S.-K. Lee, D.-H. Bae, T.-J. Kwon, S.-B. Lee, H.-H. Lee, J.-H. Park, et al., Purification and characterization of a fibrinolytic enzyme form Bacillus sp. KDO-13 isolated from soybean paste, Journal of Microbiology and Biotechnology 11 (5) (2001) 845–852.

[100] E. Purwaeni, C. Riani, D.S. Retnoningrum, Molecular characterization of bacterial fibrinolytic proteins from Indonesian traditional fermented foods, The protein journal 39 (2020) 258–267.

[101] F. Nailufar, R.R. Tjandrawinata, M.T. Suhartono, Thrombus degradation by fibrinolytic enzyme of Stenotrophomonas sp. originated from Indonesian soybean-based fermented food on Wistar rats, Advances in Pharmacological Sciences 4206908 (2016) 2016.

[102] S. Sugimoto, T. Fujii, T. Morimiya, O. Johdo, T. Nakamura, The fibrinolytic activity of a novel protease derived from a tempeh producing fungus, Fusarium sp. BLB, Bioscience, Biotechnology, and Biochemistry 71 (9) (2007) 2184–2189.

[103] X. Zhang, L.J. Yun, L.B. Peng, Y. Lu, K.P. Ma, F. Tang, Optimization of Douchi fibrinolytic enzyme production by statistical experimental methods, Journal of Huazhong University of Science and Technology - Medical Science 33 (1) (2013) 153–158.

[104] J. Yuan, J. Yang, Z. Zhuang, Y. Yang, L. Lin, S. Wang, Thrombolytic effects of Douchi fibrinolytic enzyme from Bacillus subtilis LD-8547 in vitro and in vivo, BMC Biotechnology 12 (2012) 36–45.

[105] O.K. Ashipala, Q. He, Optimization of fibrinolytic enzyme production by Bacillus subtilis DC-2 in aqueous two-phase system (poly-ethylene glycol 4000 and sodium sulfate), Bioresource Technology 99 (10) (2008) 4112–4119.

[106] Y. Peng, X.J. Yang, L. Xiao, Y.Z. Zhang, Cloning and expression of a fibrinolytic enzyme (subtilisin DFE) gene from Bacillus amyloliquefaciens DC-4 in Bacillus subtilis, Research in Microbiology 155 (3) (2004) 167–173.

[107] S. Zhang, Y. Wang, N. Zhang, Z. Sun, Y. Shi, X. Cao, et al., Purification and characterisation of a fibrinolytic enzyme from Rhizopus microsporus var. tuberosus, Food Technology and Biotechnology 53 (2) (2015) 243–248.

[108] X. Wei, M. Luo, L. Xu, Y. Zhang, X. Lin, P. Kong, et al., Production of fibrinolytic enzyme from Bacillus amyloliquefaciens by fermentation of chickpeas, with the evaluation of the anticoagulant and antioxidant properties of chickpeas, Journal of Agricultural and Food Chemistry 59 (8) (2011) 3957–3963.

[109] G.D. Biji, A. Arun, E. Muthulakshmi, P. Vijayaraghavan, M.V. Arasu, N.A. Al-Dhabi, Bio-prospecting of cuttle fish waste and cow dung for the production of fibrinolytic enzyme from Bacillus cereus IND5 in solid state fermentation, 3 Biotech 6 (2) (2016) 231–236.

[110] D.J. Kumar, R. Rakshitha, M.A. Vidhya, P.S. Jennifer, S. Prasad, M.R. Kumar, et al., Production, optimization and characterization of fibrinolytic enzyme by Bacillus subtilis RJAS19, Pakistan Journal of Biological Sciences 17 (4) (2014) 529–534.

[111] C.-T. Chang, P.-M. Wang, Y.-F. Hung, Y.-C. Chung, Purification and biochemical properties of a fibrinolytic enzyme from Bacillus subtilis-fermented red bean, Food Chemistry 133 (4) (2012) 1611–1617.

[112] C. Wang, M. Du, D. Zheng, F. Kong, G. Zu, Y. Feng, Purification and characterization of nattokinase from Bacillus subtilis natto B-12, Journal of Agricultural and Food Chemistry 57 (20) (2009) 9722–9729.

[113] J.H. Seo, S.P. Lee, Production of fibrinolytic enzyme from soybean grits fermented by Bacillus firmus NA-1, Journal of Medicinal Food 7 (4) (2004) 442–449.

[114] H. Sumi, N. Nakajima, C. Yatagai, A unique strong fibrinolytic enzyme (katsuwokinase) in skipjack "Shiokara," a Japanese traditional fermented food, Comparative Biochemistry and Physiology Part B: Biochemistry and Molecular Biology 112 (3) (1995) 543–547.

[115] M. Fujita, K. Nomura, K. Hong, Y. Ito, A. Asada, S. Nishimuro, Purification and characterization of a strong fibrinolytic enzyme (nattokinase) in the vegetable cheese natto, a popular soybean fermented food in Japan, Biochemical and Biophysical Research Communications 197 (3) (1993) 1340–1347.

[116] D. Huy, P. Hao, P. Hung, Screening and identification of Bacillus sp. isolated from traditional Vietnamese soybean-fermented products for high fibrinolytic enzyme production, International Food Research Journal 23 (1) (2016) 326–331.

[117] B.H. Lee, Y.S. Lai, S.C. Wu, Antioxidation, angiotensin converting enzyme inhibition activity, nattokinase, and antihypertension of Bacillus subtilis (natto)-fermented pigeon pea, Journal of Food and Drug Analysis 23 (4) (2015) 750–757.

[118] P. Vijayaraghavan, S.G. Vincent, Statistical optimization of fibrinolytic enzyme production using agroresidues by Bacillus cereus IND1 and its thrombolytic activity in vitro, BioMed Research International 725064 (2014) 2014.

[119] P. Vijayaraghavan, S.G. Prakash, Vincent, Medium optimization for the production of fibrinolytic enzyme by Paenibacillus sp. IND8 using response surface methodology, The Scientific World Journal (2014).

[120] A. Arshad, M.A. Zia, M. Asgher, F.A. Joyia, M. Arif, Enhanced production of streptokinase from Streptococcus agalactiae EBL-31 by response surface methodology, Pakistan Journal of Pharmaceutical Sciences (2018) 1597–1602.

[121] X. Liu, N.K. Kopparapu, Y. Li, Y. Deng, X. Zheng, Biochemical characterization of a novel fibrinolytic enzyme from Cordyceps militaris, International Journal of Biological Macromolecules 94 (2017) 793–801.

[122] P. Vijayaraghavan, S.G. Vincent, Statistical optimization of fibrinolytic enzyme production by Pseudoalteromonas sp. IND11 using cow dung substrate by response surface methodology, SpringerPlus 3 (2014) 60.

[123] X. Wei, M. Luo, Y. Xie, L. Yang, H. Li, L. Xu, et al., Strain screening, fermentation, separation, and encapsulation for production of nattokinase functional food, Applied Biochemistry and Biotechnology 168 (7) (2012) 1753–1764.

[124] K.G.H. Desai, H. Jin, Park, Recent developments in microencapsulation of food ingredients, Drying Technology 23 (7) (2005) 1361–1394.

[125] H. Sumi, H. Hamada, K. Nakanishi, H. Hiratani, Enhancement of the fibrinolytic activity in plasma by oral administration of nattokinase, Acta Haematologica 84 (3) (1990) 139–143.

[126] C. Hu, C. Liu, H. Zheng, P. Zhou, Co-production of thrombolytic enzyme and gamma-polyglutamic acid by liquid-culture of Bacillus subtilis SBS, Wei Sheng Wu Xue Bao, Acta microbiologica Sinica 49 (1) (2009) 49–55.

[127] G.X. Wang, C.J. Liang, J.Y. Yao, J.L. Yuan, J.F. Wang, Effect of a novel fibrinolytic enzyme FA-I on thrombosis and thrombolysis, Sichuan Da Xue Xue Bao Yi Xue Ban 40 (2) (2009) 288–291.

[128] K.-A. Noh, D.-H. Kim, N.-S. Choi, S.-H. Kim, Isolation of fibrinolytic enzyme producing strains from kimchi, Korean Journal of Food Science and Technology 31 (1) (1999) 219–223.

[129] J.L. Adrio, A.L. Demain, Microbial enzymes: tools for biotechnological processes, Biomolecules 4 (1) (2014) 117–139.

[130] Y. Ghasemi, F. Dabbagh, A. Ghasemian, Cloning of a fibrinolytic enzyme (subtilisin) gene from Bacillus subtilis in Escherichia coli, Molecular Biotechnology 52 (1) (2012) 1–7.

[131] Y.-J. Chen, K.-P. Wu, S. Kim, L. Falzon, M. Inouye, J. Baum, Backbone NMR assignments of DFP-inhibited mature subtilisin E, Biomolecular NMR Assignments 2 (2) (2008) 131–133.

[132] M.K. Bryant, C.L. Schardl, U. Hesse, B. Scott, Evolution of a subtilisin-like protease gene family in the grass endophytic fungus Epichloë festucae, BMC Evolutionary Biology 9 (1) (2009) 168.

[133] P.O. Micheelsen, J. Vévodová, L. De Maria, P.R. Østergaard, E.P. Friis, K. Wilson, et al., Structural and mutational analyses of the interaction between the barley α-amylase/subtilisin inhibitor and the subtilisin savinase reveal a novel mode of inhibition, Journal of Molecular Biology 380 (4) (2008) 681–690.

[134] P.N. Bryan, Protein engineering of subtilisin, Biochimica et Biophysica Acta (BBA)-Protein Structure and Molecular Enzymology 1543 (2) (2000) 203–222.

[135] T. Foophow, S.-I. Tanaka, Y. Koga, K. Takano, S. Kanaya, Subtilisin-like serine protease from hyperthermophilic archaeon Thermococcus kodakaraensis with N-and C-terminal propeptides, Protein Engineering, Design and Selection 23 (5) (2010) 347–355.

[136] D.A. Estell, T.P. Graycar, J.A. Wells, Engineering an enzyme by site-directed mutagenesis to be resistant to chemical oxidation, Journal of Biological Chemistry 260 (11) (1985) 6518–6521.

[137] D. Meng, M. Dai, B.L. Xu, Z.S. Zhao, X. Liang, M. Wang, et al., Maturation of fibrinolytic bacillopeptidase F involves both hetero- and autocatalytic processes, Applied and Environmental Microbiology 82 (1) (2016) 318–327.

[138] V. Mohanasrinivasan, C.S. Devi, R. Biswas, F. Paul, M. Mitra, E. Selvarajan, et al., Enhanced production of nattokinase from UV mutated Bacillus sp, Bangladesh Journal of Pharmacology 8 (2) (2013) 110–115.

[139] H. Sumi, Y. Yanagisawa, C. Yatagai, J. Saito, Natto Bacillus as an oral fibrinolytic agent: nattokinase activity and the ingestion effect of Bacillus subtilis natto, Food Science and Technology Research 10 (1) (2004) 17–20.

[140] Y. Peng, Q. Huang, R.H. Zhang, Y.Z. Zhang, Purification and characterization of a fibrinolytic enzyme produced by Bacillus amyloliquefaciens DC-4 screened from douchi, a traditional Chinese soybean food, Comparative Biochemistry and Physiology Part B: Biochemistry and Molecular Biology 134 (1) (2003) 45–52.

[141] A.M.M. Ali, M. Gullo, A.K. Rai, S.C.B. Bavisetty, Bioconservation of iron and enhancement of antioxidant and antibacterial properties of chicken gizzard protein hydrolysate fermented by Pediococcus acidilactici ATTC 8042, Journal of the Science of Food and Agriculture 101 (7) (2021) 2718–2726.

[142] P.S. Khursade, S.H. Galande, P. Shiva Krishna, R.S. Prakasham, Stenotrophomonas maltophilia Gd2: a potential and novel isolate for fibrinolytic enzyme production, Saudi Journal of Biological Sciences 26 (7) (2019) 1567–1575.

[143] J.M. Yon, Protein aggregation, in: R.A. Meyers (Ed.), Encyclopedia of Molecular Cell Biology and Molecular Medicine, Wiley-VCH Verlag GmbH & Co., Weinheim, Germany, 2006, pp. 23–52.

[144] Z. Khan, M. Shafique, H.R. Nawaz, N. Jabeen, S.A. Naz, Bacillus tequilensis ZMS-2: a novel source of alkaline protease with antimicrobial, anti-coagulant, fibrinolytic and dehairing potentials, Pakistan Journal of Pharmaceutical Sciences 32 (4) (2019) 1913–1918.

[145] P. Vijayaraghavan, P. Rajendran, S.G. Prakash Vincent, A. Arun, N. Abdullah Al-Dhabi, M. Valan Arasu, et al., Novel sequential screening and enhanced production of fibrinolytic enzyme by Bacillus sp. IND12 using response surface methodology in solid-state fermentation, BioMed Research International 2017 (2017).

[146] W. Zeng, W. Li, L. Shu, J. Yi, G. Chen, Z. Liang, Non-sterilized fermentative co-production of poly (γ-glutamic acid) and fibrinolytic enzyme by a thermophilic Bacillus subtilis GXA-28, Bioresource Technology 142 (2013) 697–700.

[147] T.C. Chen, C.-J. Chiang, Y.-P. Chao, Medium optimization for the production of recombinant nattokinase by Bacillus subtilis using response surface methodology, Biotechnology Progress 23 (6) (2007) 1327–1332.

[148] I. Mariñas-Collado, M.J. Rivas-López, J.M. Rodríguez-Díaz, M.T. Santos-Martín, Optimal designs in enzymatic reactions with high-substrate inhibition, Chemometrics and Intelligent Laboratory Systems 189 (2019) 102–109.

[149] R.R. Chitte, S.V. Deshmukh, P.P. Kanekar, Production, purification, and biochemical characterization of a fibrinolytic enzyme from thermophilic Streptomyces sp. MCMB-379, Applied Biochemistry and Biotechnology 165 (5–6) (2011) 1406–1413.

[150] P. Pandee, Y. Dissara, Production and properties of a fibrinolytic enzyme by Schizophyllum commune BL23, Songklanakarin Journal of Science and Technology 30 (4) (2008) 447–453.

[151] E. Selvarajan, N. Bhatnagar, Nattokinase: an updated critical review on challenges and perspectives, Cardiovascular and Hematological Agents in Medicinal Chemistry 15 (2017) 128–135.

[152] M. Anusree, K. Swapna, C.N. Aguilar, A. Sabu, Optimization of process parameters for the enhanced production of fibrinolytic enzyme by a newly isolated marine bacterium, Bioresource Technology Reports 11 (2020).

[153] V. Deepak, K. Kalishwaralal, S. Ramkumarpandian, S.V. Babu, S.R. Senthilkumar, G. Sangiliyandi, Optimization of media composition for Nattokinase production by Bacillus subtilis using response surface methodology, Bioresource Technology 99 (17) (2008) 8170–8174.

[154] C. Nagata, K. Wada, T. Tamura, K. Konishi, Y. Goto, S. Koda, et al., Dietary soy and natto intake and cardiovascular disease mortality in Japanese adults: the Takayama study, The American Journal of Clinical Nutrition 105 (2) (2017) 426–431.

[155] F. Dabbagh, M. Negahdaripour, A. Berenjian, A. Behfar, F. Mohammadi, M. Zamani, et al., Nattokinase: production and application, Applied Microbiology and Biotechnology 98 (22) (2014) 9199–9206.

[156] Y. Huang, S. Ding, M. Liu, C. Gao, J. Yang, X. Zhang, et al., Ultra-small and anionic starch nanospheres: formation and vitro thrombolytic behavior study, Carbohydrate Polymers 96 (2) (2013) 426–434.

[157] M. Fujita, K. Hong, Y. Ito, S. Misawa, N. Takeuchi, K. Kariya, et al., Transport of nattokinase across the rat intestinal tract, Biological and Pharmaceutical Bulletin 18 (9) (1995) 1194–1196.

[158] M. Fujita, K. Ohnishi, S. Takaoka, K. Ogasawara, R. Fukuyama, H. Nakamuta, Antihypertensive effects of continuous oral administration of nattokinase and its fragments in spontaneously hypertensive rats, Biological and Pharmaceutical Bulletin 34 (11) (2011) 1696–1701.

[159] G.S. Jensen, M. Lenninger, M.P. Ero, K.F. Benson, Consumption of nattokinase is associated with reduced blood pressure and von Willebrand factor, a cardiovascular risk marker: results from a randomized, double-blind, placebo-controlled, multicenter North American clinical trial, Integrated Blood Pressure Control 9 (2016) 95–104.

[160] Y. Kurosawa, S. Nirengi, T. Homma, K. Esaki, M. Ohta, J.F. Clark, et al., A single-dose of oral nattokinase potentiates thrombolysis and anti-coagulation profiles, Scientific Reports 5 (2015) 11601.

[161] B.J. Lampe, J.C. English, Toxicological assessment of nattokinase derived from Bacillus subtilis var. natto, Food and Chemical Toxicology 88 (2016) 87–99.

[162] J. Xu, M. Du, X. Yang, Q. Chen, H. Chen, D.-H. Lin, Thrombolytic effects in vivo of nattokinase in a carrageenan-induced rat model of thrombosis, Acta Haematologica 132 (2) (2014) 247–253.

[163] D. Law, Z. Zhang, Stabilization and target delivery of Nattokinase using compression coating, Drug Development and Industrial Pharmacy 33 (5) (2007) 495–503.

[164] Y. Weng, J. Yao, S. Sparks, K.Y. Wang, Nattokinase: an oral antithrombotic agent for the prevention of cardiovascular disease, INternationa; Journal of Molecular Sciences 18 (523) (2017).

[165] E. Kotb, Fibrinolytic bacterial enzymes with thrombolytic activity, in: E. Kotb (Ed.), Fibrinolytic Bacterial Enzymes with Thrombolytic Activity, Springer Verlag, New York, USA, 2012, pp. 1–74.

[166] J.Y. Kim, S.N. Gum, J.K. Paik, H.H. Lim, K.-C. Kim, K. Ogasawara, et al., Effects of nattokinase on blood pressure: a randomized, controlled trial, Hypertension Research 31 (8) (2008) 1583–1588.

[167] Y.-Y. Chang, J.-S. Liu, S.-L. Lai, H.-S. Wu, M.-Y. Lan, Cerebellar hemorrhage provoked by combined use of nattokinase and aspirin in a patient with cerebral microbleeds, Internal Medicine 47 (5) (2008) 467–469.

[168] N. Fadl, H. Ahmed, H. Booles, A. Sayed, Serrapeptase and nattokinase intervention for relieving Alzheimer's disease pathophysiology in rat model, Human and Experimental Toxicology 32 (7) (2013) 721–735.

[169] J.-M. Dogné, J. Hanson, Xd Leval, D. Pratico, C.R. Pace-Asciak, P. Drion, et al., From the design to the clinical application of thromboxane modulators, Current Pharmaceutical Design 12 (8) (2006) 903–923.

[170] E. Vicari, S.V. La, C. Battiato, A. Arancio, Treatment with non-steroidal anti-inflammatory drugs in patients with amicrobial chronic prostato-vesiculitis: transrectal ultrasound and seminal findings, The Italian Journal of Urology and Nephrology 57 (1) (2005) 53–59.

[171] D. Collen, Engineered staphylokinase variants with reduced immunogenicity, Fibrinolysis and Proteolysis 12 (1998) 59–65.

[172] I. Torrèns, A.G. Ojalvo, A. Seralena, O. Hayes, J. de la Fuente, A mutant streptokinase lacking the C-terminal 42 amino acids is less immunogenic, Immunology Letters 70 (3) (2000) 213–218.

[173] V. Regnault, G. Helft, D. Wahl, D. Czitrom, A. Vuillemenot, G. Papouin, et al., Antistreptokinase platelet-activating antibodies are common and heterogeneous, Journal of Thrombosis and Haemostasis 1 (5) (2003) 1055–1061.

[174] B. Masschalck, C.W. Michiels, Antimicrobial properties of lysozyme in relation to foodborne vegetative bacteria, Critical Reviews in Microbiology 29 (3) (2003) 191–214.

[175] V. Žukaite, G. Biziulevičius, Acceleration of hyaluronidase production in the course of batch cultivation of Clostridium perfringens can be achieved with bacteriolytic enzymes, Letters in Applied Microbiology 30 (3) (2000) 203–206.

Microbial production and transformation of polyphenols

Puja Sarkar[1], Md Minhajul Abedin[1], Sudhir P. Singh[2], Ashok Pandey[3], Amit Kumar Rai[1]

[1]INSTITUTE OF BIORESOURCES AND SUSTAINABLE DEVELOPMENT, REGIONAL CENTRE, TADONG, INDIA [2]CENTER OF INNOVATIVE AND APPLIED BIOPROCESSING, MOHALI, INDIA [3]CSIR-INDIAN INSTITUTE OF TOXICOLOGY RESEARCH, LUCKNOW, INDIA

8.1 Introduction

Among numerous existing secondary metabolites from plants, polyphenols are the most widespread and ubiquitous natural compounds which are found throughout the plant kingdom [1,2]. According to their chemical structures, polyphenols are grouped into many classes as flavonoids and nonflavonoids such as stilbenes, phenolic acids, lignans, and alkylphenols [3,4]. Most of the time flavonoids are found in conjugated forms generally with sugars and organic acids [5]. In addition to alkaloid and isoprenoids, polyphenols also constitute the major class of plant secondary metabolites [6,7]. Among polyphenols, many of them are responsible for the diverse functions of plants including insects feeding deterrence, providing coloration to flowers, fruits and leaves, toxic heavy metals chelation, attraction of pollinators, protection from damage induced by UV radiation and scavenging of free radical, along with their use as pharmaceuticals [8–11]. Polyphenols are applied as pharmaceuticals, nutraceuticals, natural pigments, food colorants, preservatives, bioplastic monomers, and composites for decades [12–15].

During food fermentation, microorganism help in the production of bioactive compounds from complex organic substances present in the food [16]. In recent years due to growing demand for functional foods, which have nutritional values and promote health benefits, polyphenols are playing a major contribution in food industries. Polyphenols have beneficial effects due to their antimicrobial, anticarcinogenic, antioxidative, neuroprotective and cardio-protective properties [17]. Fermented foods prepared using plant-based raw materials are rich in polyphenols, which are derived from the food substrate [18]. Enzymes produced by starter culture during food fermentation hydrolyze bound polyphenols into free form [19,20]. The bioconversion during food fermentation improves the bioavailability, which results in enhancement of associated health benefits. The bioavailability of polyphenols is further affected by food materials and their digestion in the gastrointestinal tract, which varies according to the physiological environment of the stomach, small intestine, and colon that affect absorption of the polyphenols [21]. Naturally, phenolic compounds are present as bound form with sugar molecules, which affects their structure and functions [22,23]. Removal of these structures is important for the absorption into the gastrointestinal tract [24]. Gastrointestinal absorption depends on the structure of the nutraceutical that is transported to the target tissue affecting the physiological conditions [25]. Gut microbes composition and enzymes play an important role in the release of polyphenols resulting in enhancement of bioavailability.

Metabolic engineering is the biotechnological tool defined as the desired change in the genetic makeup of a host organism for improving the capabilities of biosynthesis of desired metabolites [26]. Synthesis of natural products through metabolic engineered microbes can be achieved very easily by manipulation of synthetic pathways for the improvement with advanced metabolic tools [27,28]. Metabolic engineering of a microorganism for polyphenols production requires knowledge of enzymes, which are involved in the biosynthetic pathway, as well as the selection of host strain and determination of targets for gene manipulation [26]. These biotechnological approaches have generated various microbial cell factories, allowing the increase in large-scale production of these metabolites in an efficient and environmentally friendly manner. Also, improvisation of enzyme characteristics is achieved by modulation of carbon flux, removing strict pathways and elimination of competing pathways by using tools like CRISPR, CRISPRi (RNAi) [29–31]. This chapter provides an overview of different types of polyphenols, its microbial production, effect of microbial fermentation on polyphenol release, and genetically engineered microbes for production of polyphenols.

8.2 Types of polyphenols and their health benefits

Polyphenols are the largest group of plant secondary metabolites generated through shikimate/phenylpropanoid pathway. The biosynthetic precursor of phenylpropanoid in plants is derivatives of cinnamic acid and p-coumaric acid (compounds with C6-C3 backbone) formed from the primary precursor aromatic amino acids, L-phenylalanine, and L-tyrosine by the action of phenylalanine ammonia lyase and tyrosine ammonia lyase, respectively, which is further transformed by the action of the biocatalyst 4 coumaroyl CoA ligase (4CL) to cinnamoyl-CoA and p-coumaroyl-CoA. Followed by this, 3 moles of malonyl CoA combine with p-coumaroyl CoA or cinnamoyl-CoA to create the backbone flavanones, stilbenes, or curcuminoid by the action of type III polyketide synthases, including curcuminoid synthase, chalcone synthase, or stilbene synthase (STS). Effect of tailoring enzymes including lipase, prenyltransferase, hydroxylases, O-methyl transferase, and wide range of polyphenolic compounds are produced including flavonoids, stilbenes, styrylpyrones, and curcuminoids [1].

Different types of polyphenols are found in plant-based foods and present in their complex bound form [32]. Polyphenols in plants protect them from oxidative stress, UV radiation, pathogens, and harsh conditions, whereas in the human body, they act as antioxidants and exert beneficial health effects, such as antidiabetic, anti-inflammatory, anticancer, antihypertensive, antibacterial, cardio-protective, and antiseptic properties [33,34]. Based on the presence number of functional groups and structural elements, polyphenols are categorized into four different classes [35]. There four different classes of polyphenols are described in this section.

8.2.1 Phenolic acids

Phenolic acids are most abundantly found nonflavonoid compounds, and a prominent class of bioactive molecules. Foods such as fruits and vegetables contain free forms of phenolic acids but grains and seeds specifically bran or hull contain bound forms that can be converted to free form through acid or alkaline hydrolysis [36,37]. They are divided into two main classes as hydroxycinnamic acids and hydroxybenzoic acids, based on distinctive carbon framework and positioning and number of hydroxyl group present on its aromatic ring [38]. Hydroxycinnamic acids are hydroxy derivative of cinnamic acids commonly found in fruits (berries, plum, grapes, and apples), vegetables, and beverages (coffee, tea, and wine), which include p-coumaric, caffeic, ferulic, and sinapic acids. Hydroxybenzoic acids, a derivative of benzoic acid, are commonly found to be present in a lower amount in edible plants with an exceptionally high concentration in few red fruits, black radish, and onions. Examples of hydroxybenzoic acids are protocatechuic acid, gallic acid, vanillic acid, and syringic acid [32,37,39–41]. Phenolic acids act as an antioxidant, which prevents free radicals formation and exerts beneficial effects toward health due to its anti-inflammatory, antimicrobial, antihyperglycemic, and antiproliferative activity [42–44].

8.2.2 Flavonoids

Flavonoids are the class of polyphenols, which are known as secondary metabolites of plants, having no less than one hydroxyl group attached to its aromatic ring [45,46]. The vast chemical diversity of flavonoids arises from its backbone structure differences, which can be modified in various ways by acetylation, aryl-migrations, glycosylation, hydroxylation, methylations, polymerizations, and prenylations. Flavanoids are divided into subclasses including flavonols, aurones, isoflavones, flavanones, flavones, flavan-3-ols (catechins and procyanidines), anthocyanidines, and tannins such as proanthocyanidins (condensed tannins). Flavonoids are dietary bioactive having various health-promoting effects, such as anti-inflammatory, anticancer, antidiabetic, antibacterial/antiviral, antiageing, antioxidant, boosting immune system, and cardio-protective benefits [45,46]. When consumed, these are metabolized extensively by the host tissue and gut microbiota, exhibiting intestinal immune function [47].

8.2.3 Stilbenes

Stilbenes are structure consisting two phenyl rings connected with a central ethylene moiety. Resveratrol (3,5,4-trihydroxystilbene) is the most studied group of plant-derived stilbene, which is a part of nonflavonoid group, applied as a medicinal ingredient as well as nutritional supplement [48]. In 1939 it was first reported from white hellebore, *Veratrum grandiflorum* initially known as phytoalexin [49,50]. After that it has been found from the root extracts of a Chinese herb *Polygonum cuspidatum* or Japanese knotweed used as a medicine, which is now the main source of commercially available resveratrol [51]. Later it was also reported to be present in grapevine *Vitis vinifera* [48] and berry fruits, such as blueberries, blackberries, and raspberries [52]. Since the discovery of resveratrol, its health benefits have been confirmed by several studies with preclinical and human trials [53]. The antioxidant capacity of resveratrol has been demonstrated through some experimental human trial studies, showing that, it can slow down the oxidation of low-density lipoprotein by copper chelators [54] and can put a halt in the lipid peroxidation [55]. Some hypothesis has shown that resveratrol might be related to the reason behind "French paradox" which is the lower occurrence of cardiovascular disease in the French population despite their high saturated fatty acid consumption [56]. Number of preclinical studies revealed the anticancer effects of resveratrol, in blocking different types of cancer [57,58]. Preclinical trials conveyed the properties of resveratrol protecting against diabetes in animal models and neurodegenerative disease such as Alzheimer's in studies with cell line and animal models [59–61]. Beneficial effect on age-related human diseases has been shown through animal models and cell cultures studies. It can help in increasing the longevity of diverse species by acting as an antiaging agent [50,62,63].

8.2.4 Lignans

Lignans are diphenolic compounds, which are present in many plants. Two cinnamic acids after dimerization form the structure of 2, 3-dibenzylbutane, which is present in lignans. There are other lignan compounds present in plants such as secoisolariciresinol, Lariciresinol Diglucoside, and Secoisolariciresinol diglucoside [64]. Linseed is a dietary source that is rich in secoisolariciresinol and amount of metairesinol [32]. Lignans exhibit several health benefits such as lowering the risk of heart disease, menopausal symptoms, osteoporosis, and prevention against breast cancer [65].

8.3 Microbial production and enhancement of polyphenols during fermentation

Most of the polyphenols extraction relies upon plant materials, which is a costly and inefficient procedure [1]. Use of organic solvent in conventional processes results in the release of bound phenolics compounds from plant materials. Extraction from the plant-derived compound cannot meet the ever-increasing demands of the growing population. Apart from that, these processes are dependent on environmental conditions, the abundance of crop species due to seasonal variations; moreover, there are complex downstream processes [66]. Production of polyphenols through microbial fermentation, such as the use of microbial cell factories, is more efficient and sustainable than conventional extraction. Microbial production of polyphenols has various advantages due to the rapid production cycle and high yields

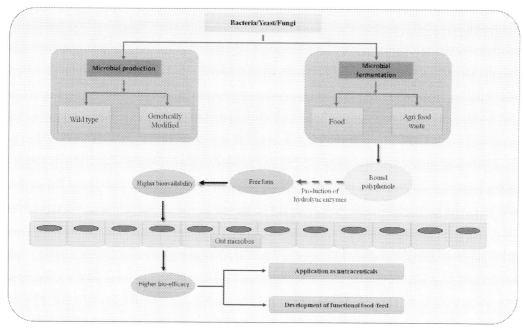

FIGURE 8–1 Effect of microbial fermentation on polyphenols.

of biomass that can be cultivated in a scale-up production process by specified media [67]. Polyphenols are also hydrolyzed and transformed from bound form to free form on microbial fermentation (Fig. 8−1). Microbial production of polyphenols and effect of microbial fermentation on polyphenol compounds in fermented foods and agri-food processing waste is covered in different subsections such as (1) polyphenol enhancement through food fermentation, (2) microbial fermentation for recovery of polyphenols from the agri-food industry waste, (3) polyphenol production by metabolic engineering, and (4) addition of polyphenols in fermented food products. The above-mentioned different aspects of microbial production of polyphenols will be discussed in detail in this chapter.

8.3.1 Polyphenol enhancement through food fermentation

Polyphenols are the most abundantly available bioactive compound with numerous health benefits which is dependent on their bioavailability after consumption. Fermentation with specific enzyme producing microorganisms hydrolyzes bound polyphenol into free form resulting in higher bioavailability and associated health benefits [19]. Carbohydrate active enzymes responsible for breakdown of bound polyphenols to their free form include amylase, β-glucosidase, xylanase pectinolytic enzymes, cellulases, hemicelluloses, and lignin degrading enzymes [19,68]. The free form of isoflavones, flavanols, and phenolic acids are groups of polyphenols that have been shown to be responsible for enhanced antioxidant properties during fermentation [69]. Enhancement of polyphenols during food fermentation is mainly reported due to on β-glucosidase enzyme activity [69–72]. Isoflavone aglycones, flavanols, acetylglycosides, and gallic acid get increased during fermentation of *cheonggukjang*, a Korean fermented soybean paste due to microbial β-glucosidase [69]. Fermentation of *cheonggukjang* (fermented soybean) by *B. subtilis* CS90 for time duration of 48 hours has decreased the levels of diadzin, genistin, and glycitin like glycosides with an increasing levels of aglycones such as daidzein, genistein, and glycitein along with flavonols such as catechin, epicatechin, and phenolic acids like vanillic acid and gallic acid [69]. Polyphenol content and antioxidant properties have also been found to increase in soybean fermented using *Bacillus subtilis* isolated from *kinema* [73]. Polyphenols such as isoflavones present in soybean have been shown to act like estrogen providing specific health benefits [74]. Similarly, the strain of *Lactobacillus rhamnosus* C6 that produces β-glucosidase leads to the formation of isoflavones like genistein and diadzein (aglycones) from genistin and diadzin (glycones) in fermented soymilk [70]. The free isoflavone content get increased in soymilk prepared by Lactic acid bacteria (LAB) including *Lactobacillus acidophilus, Lactobacillus plantarum, Lactobacillus casei, Lactobacillus fermentum,* and *Bifidobacterium animalis* [75]. Enhancement of polyphenols is reported in several plant-based food substrates by the action of enzymes produced by different microorganisms (Table 8–1). Antidiabetic, anticancer, antioxidant, and

Table 8–1 Effect of microbial fermentation for release of plant polyphenols.

Microbes involved	Plant-based food	Phenolic compound	References
Lactic acid bacteria	Soybean germ	Isoflavones and saponin	[76]
Lactobacillus plantarum	Wine	Quercetin	[77]
Lactobacillus plantarum	Cowpeas	Gallic acid, vanillic acid, quercetin, ferulic acid, p-hydroxybenzoic acid	[78]
Bacillus subtilis	Black soybean	Total phenolic, flavonoids, and aglycone contents	[79]
Bacillus natto	Soybean *natto*	Phenolic and anthocyanin increased	[80]
Bacillus subtilis	Soybean *Kinema*	Total phenolic compound content increased	[73]
Aspergillus oryzae	Soybean	Metal chelating ability of saponins and isoflavones improved	[81]
Lentinus Edodes	Cranberry pomace	Extractable phenolic contents increased	[82]
Phanerochaete chrysosporium	Apple pomace	Polyphenol concentration increased	[83]
Mycelium of *Cordyceps militaris*	Wheat	Polyphenols and citrus flavonoid rutin increased	[84]
Baker's yeast	Rye bran	Easily extractable phenolic compounds, ferulic acid	[85]
Yeast strain YM-1	Brown alga	Total phenolic compounds	[86]
Saccharomyce cerevisiae jm.20	Ashwagandharishtha	Phenolic compounds	[87]
Saccharomyces cerevisiae IFI240A	Mulberry	Anthocyanin	[88]
Saccharomyce bayanus	Blueberry	Total phenolic compound and anthocyanin	[89]
Natural Fermentation	Bush tea	Polyphenols and tannin	[90]
Skin-contact fermentation	Blueberry	Anthocyanin	[91]
Saccharomyces cerevisiae, Hanseniaspora sp.	Garcinia beverage	Polyphenols	[92,93]

immunomodulatory properties have been reported due to different plant based polyphenols [69,73,94,95].

Apart from bacterial fermentation, yeasts and filamentous fungi has also been reported for enhancement of free polyphenols [92,96]. Fermentation by filamentous fungi enhances the total phenolics and anthocyanin content along with antioxidant activity of black bean [96]. Fermentation results in breakdown of cereal's cell walls and softening of kernel structures, releasing esterified and insoluble-bound nutrients by the action of glycoside hydrolase, cellulase, xylanase, and esterase produced by fungi. Lipophilic isoflavone aglycones produced by the action of β-glucosidase enzyme liberating fungi during fermentation increases the antioxidant activity in extracts [20]. Fermented beverages produced using specific yeast starter have shown to have higher polyphenols content and enhanced antioxidant properties [92,93]. Specific strains of starter culture can be applied for production of free polyphenol and transformed bioactive metabolites enriched fermented food products.

8.3.2 Microbial fermentation for recovery of polyphenols from the agri-food industry waste

Byproducts from food and agricultural sector are the recent interest toward sustainability, which attracts researcher to look into every possible way of utilizing the residues resulting from those industries as these are an excellent source of bioactive phenolic compounds. Processes for recovering these bioactive compounds, which are primary and secondary metabolites can be utilized for several pharmacological and nutraceutical applications. These bioprocesses may significantly contribute to the reduction in problems arise from the lack of proper management of agro wastes that leads to a positive impact on the economy [97]. Wastes from industrial processing, agricultural production, and food service sector, which are the valuable sources of natural products, are often without any viable alternatives discarded into surrounding environment. Microbial decomposition of nutritional content of wastes due to high water activity produces huge quantity of methane and carbon dioxide and sanitization issues, which may have adverse effect on human health. Use of agro-waste byproducts extracted polyphenols as dietary supplements in animal feed may improve the health of animals. After processing of the wastes from agri-food sector, several value-added products rich in useful precursors of bioactive components and biomolecules can be achieved [98–100].

Microbial fermentation is a cost-effective and eco-friendly technology for the release of phenolic compounds with antioxidant property [97]. During microbial fermentation, phenolics are released as free metabolite through enzymatic action from the complex matrix of the substrate. LAB among the other microorganisms has gained great attention for showing its potential effectiveness to convert the low-cost industrial agri-food residues into valuable biofunctional products, such as phenolic compounds, bioactive protein hydrolysate, carotenoids, polyunsaturated fatty acids (PUFA), and oligosaccharides [93,97,101–103]. Application of LAB due to their generally recognized as safe (GRAS) status makes the microbial process more ecological and prevents the utilization of hazardous chemicals. LAB carries an advantage over other different bacteria used for fermentation because of its health-promoting properties [104]. The recovery of nutraceuticals from agri-food industrial waste such as dairy, bakery, fruit, brewery, fishery, and meat processing sectors is claiming as an important source of dietary supplements providing additional health benefit [97]. Fermentation of sugarbeet pulp with *Enterococcus lactis* SR1 combined with commercial halocellulose resulted in production of ferulic acid [105].

Agri-food industry to a great extent relies on solid state fermentation (SSF) for the biotransformation of wastes and extraction of bioactive components, as it is an inexpensive process developing immensity of industrially important products. Understanding the optimal operation conditions including the substrates and microorganisms used is the critical factor for efficiently developing a bioprocess under SSF [102]. Fungi and yeast are potential candidates favored as starter cultures in SSF due to their ability to adapt at low water or moisture quantity [106]. Microbial hydrolysis improvises the antioxidant activity in plant-based byproducts through increasing the phenolic compounds and flavonoids through cell wall degradation during fermentation. Polyphenols such as ellagitannins are produced from agri waste using submerged fermentation system with the help of filamentous fungi [107]. Fermentation of pineapple processing by-product with *Kluyveromyces marxianus* NRRL Y-8281 showed the highest antioxidant and anticancer properties [108]. Cranberry pomace releases phenolic aglycones on SSF using food grade fungus *Lentinus edodes* [82]. Similarly, Creosote bush leaves have been used for the production of antioxidant nutraceuticals by SSF using *Aspergillus niger* GH1 [109]. Pomegranate husk with the use of continuous bioreactor using elagitanase to hydrolyze ellagitannins has resulted in high yield of ellagic acid (175 mg/g) [110]. Residues of tea processing were used as substrate for the production of ellagic acid by *A. niger* MTCC 281 through submerged fermentation process [107,111]. Microbial process using GRAS status microbes can be an economical and eco-friendly approach for recovery of bioactive molecules from agri-food processing byproducts.

8.3.3 Polyphenol production by metabolic engineering

Plant synthesizes several types of polyphenols having the potential to help in prevention or treatment of cancer, cardiovascular, and neurodegenerative disease, but usually these polyphenols are present in mixture impeding access to individual compounds in large quantity. Further, these compounds are not produced at all the time because of the variable environmental trigger of their biosynthesis [32]. Chemical synthesis of polyphenol is not profitable because of complex reaction procedure, toxic chemical accumulation, and industrial purification process [112]. Production of several plant polyphenols can be achieved by functional integration of biosynthetic pathway into microorganisms for the synthesis of selected polyphenols as chemically distinct compounds (Table 8–2). It is a replacement and promising strategy toward eco-friendly production of individual polyphenol from renewable and economical resources [6]. Metabolite engineering of the microbial host is an alternative approach which is eco-friendly as well as can meet the global demand of natural products.

Table 8–2 Plant-derived pathways in metabolically engineered microbial host enable polyphenol production.

Polyphenols	Enzymes/ Genes	Origin of enzymes / genes	Titer value (mg/L)	Host	References
Resveratrol	4 coumarate: CoA ligases	Arabidopsis thaliana	105	Escherichia coli	[113]
	Stilbene synthases	Arachis hypogaea			
Resveratrol	Tyrosine ammonia lyases	Flavobacterium johnsoniae	59	Cornybacterium glutamicum	[63]
	4 coumarate: CoA ligases	Petroselinum			
	Stilbene synthases	Arachis hypogaea			
	aroH	E. coli			
Flavonols	4 coumarate: CoA ligases	Lithospermum erythrorhizon	33	E. coli	[114]
	Chalcone synthases, Chalcone isomerase	Glycyrrhiza echinata			
	Stilbene synthases	A. hypogaea			
	Flavone synthase	Petroselinum crispum			
	Flavanone 3-Hydroxylase, Flavonol Synthase	Citrus			
	Acetyle CoA carboxylase	C. glutamicum			
Flavanones	4 coumarate: CoA ligases	Lithospermum erythrorhizon	102	E. coli	[114]
	Chalcone synthases, Chalcone isomerase	Glycyrrhiza echinata			
	Stilbene synthases	A. hypogaea			
	Flavone synthase	Petroselinum crispum			
	Flavanone 3-Hydroxylase, Flavonol Synthase	Citrus			
	Acetyle CoA carboxylase	Cornybacterium glutamicum			
Naringenin	4 coumarate: CoA ligases	Petroselinum crispum	474	E. coli	[115]
	Chalcone synthases,	Petunia x hybrida			
	Chalcone isomerase	Medicago sativa			
	Acetyle CoA carboxylase	Photorhabdus luminescens			
	Phosphoglycerate kinase, Pyruvate dehydrogenase complex	E. coli			
Naringenin	Phenylalanine ammonia lyase, Cinnamate 4 hydroxylase, Cytochrome P450 reductase	Arabidopsis thaliana	109	Saccharomyces cerevisiae	[116]
	4 coumarate: CoA ligases, Chalcone synthases, Chalcone isomerase	Arabidopsis thaliana			
	Tyrosine ammonia lyases	Rhodobacter capsulatus			
	Tyrosine insensitive ARO4 allele (aro4 G2265)	Saccharomyces cerevisiae			
Resveratrol	AtPAL2, AtC4H, At4CL2	Arabidopsis thaliana	800	S. cerevisiae	[117]
	VvVST1	Vitis vinifera			
Resveratrol	Tyrosine ammonia lyases	Trichosporon cutaneum	304.5	E. coli	[118]
	4 coumarate: CoA ligases	Petroselinum crispum			
	Stilbene synthases	V. vinifera			
	Malonyl-CoA synthetase encoding gene (matB) and Malonate carrier protein encoding gene (matC)	R. trifolii			
	tyrA fbr and aroG fbr	E. coli K12			
p-Coumaric acid	Tyrosine ammonia lyases	Flavobacterium johnsoniaeun	1930	S. cerevisiae	[119]
	DHAP synthase protein encoding gene (aro4 K229L)	S. cerevisiae			
	Chorismate mutase (aro7 G141S)	S. cerevisiae			
	Shikimate kinase II	E. coli			
Pinosylvin	Stilbene synthases	Pinus strobus	91	E. coli	[120]
Pinosylvin	Stilbene synthases	Arachis hypogaea	121	C. glutamicum	[121]
	4 coumarate: CoA ligases	P. crispum			
	Plant-derived Chalcone isomerase and Chalcone synthases	Petunia			
Kaempferol	Flavonol synthases	Populus deltoides	66.29	S. cerevisiae	[122]
Quercetin	Stilbene synthases	Arachis hypogaea	10	C. glutamicum	[123]
	4 coumarate: CoA ligases	P. crispum			
	Plant-derived Chalcone isomerase and Chalcone synthases	Petunia			
Cyanidin-3-O-glucoside	Anthocyanidin synthase	Arabidopsis thaliana	350	E. coli	[124]
	3-O-glycosyltransferase	Pitunia hybrida			

AroH, 3-deoxy-d-arabino-heptulosonate-7-phosphate synthase; *At4CL2*, p-coumaryl-CoA ligase from Arabidopsis thaliana; *VvVST1*, resveratrol synthase from Vitis vinifera; *tyrA*, Chorismate mutase gene; *aroG*, 3-Deoxy-D-arabino-heptulosonate 7-phosphate (DAHP) synthase gene; *fbr*, feedback inhibition resistant.

Direct extraction approaches are limited due to price and availability of raw materials, its quality control, and associated purification steps [26].

Microorganisms have numerous applications for the production of value-added components for a long time [24]. During the optimization of processes, the selection of host organisms is an important factor. Bacteria are more advantageous over yeast as they are easy to manipulate and have short doubling time. Comparatively, due to higher growth rate of bacterial cell, there is high expression of enzyme and protein in comparison with yeast system [125,126]. Phenylpropanoid such as p-coumaric acid tolerance level of bacteria is higher than yeasts, which are sensitive toward their high concentration [127,128]. However, there are few drawbacks of using bacteria while expressing them as host organism. One of the disadvantages of using bacteria is its inability to conduct post translational modification, which prevents the correction of inappropriate folding of recombinant protein leading to the ill-defined functional property [129].

The first plant polyphenols produced in microorganism derived from phenylpropanoids, p-coumaric acid, and cinnamic acids were (2 S)-flavanone naringenin and pinocembrin, respectively [6]. Microorganisms can also produce aromatic amino acids and malonyl Co-A as precursor of polyphenol in a similar primary metabolism like plants; thus, microorganisms represents attractive platform for polyphenol production [6]. *Escherichia coli* is a well-defined bacterium for their application as a platform to industrial bioprocess. Microbes having GRAS status including *Lactococcus lactis*, *Corynebacterium glutamicum*, and *Streptomyces venezuelae* are the successful alternative host for microbial production of polyphenols [6,95,96]. *C. glutamicum* is an important organism for the microbial production of several plant polyphenols including flavonoids and stilbenes, either directly from glucose or supplemented phenylpropanoid precursor molecules [121,123].

Bacterial production of polyphenols such as resveratrol has been achieved by the expression of heterologous resveratrol producing biosynthetic pathway into the bacteria such as *L. lactis*, *C. glutamicum*, *E. coli*, and yeast (*Saccharomyces cerevisiae*) [67,130]. Application of *S. cerevisiae* as yeast host could be favorable depending on the desired component properties because of its easy manipulation and growth [130]. It has been reported that *Streptomyces venezuelae* is also capable of producing a broad range of phenolic secondary metabolites including flavonoids and stilbenes (resveratrol). These solely plant-based bioactive molecules are acquiring through a specially engineered form of *S. venezuelae* [131]. Application of chemical synthesis process of polyphenols such as resveratrol production is restricted despite of high yield approach due to some complexity of the synthesis process and production of unwanted byproducts [132]. Microbial production of resveratrol is reported as both in recombinant bacteria and in yeast hosts [67]. Resveratrol production has been reported from 3.1 to 812 mg/L in yeasts and 1.4 to 2.34 g/L in bacteria on engineering plant synthesis pathway in a microbial host by expressing genes for escalating its production [67,117,127,133–135]. The production of 121 mg/L pinosylvin and 158 mg/L resveratrol has been achieved in engineered *C. glutamicum* in the presence of 25 micro M cerulenin, which is a fatty acid synthesis inhibitor [118]. Application of 4 coumarate CoA ligases (4CL) from *Arabidopsis thaliana* and STS from *Arachis hypogea* in an engineered *E. coli* strain resulted in resveratrol production of 104.5 mg/L on p-coumaric acid supplementation [113]. Further, a broad range of substrate variety has been reported distinct for *A. hypogea* STS [136,137]. Ferulic acid and sinapic acid were also favorably used as substrates by *A. thaliana* 4CL [138].

Resveratrol (2.39 g/L) has been produced by *E. coli* BW27784 with STS from *V. vinifera* and 4CL from *A. thaliana* in the presence of 15 mM cerulenin and *p*-coumaric as substrates [133]. STS from *V. vinifera* allowed highest titer of resveratrol in comparison with STS from other sources [133]. Another strain of *E. coli*, *E. coli* BL21 acquired 16 mg/L (using 4CL from tobacco and STS from grapes) [67] and 80.52 mg/L (using 4CL from *A. thaliana* and STS from *A. hypogaega*) of resveratrol production in a same condition [139]. Another study reported that different strains of *E. coli* cloned with TAL from *Rhodotorula glutini* (the red yeast), 4CL from *Petroselinum crispum*, and STS from *V. vinifera* have shown low titer of resveratrol production [126]. Further, application of TAL from *R. glutini* has shown highest activity specifically in the presence of L-tyrosine [140,141].

Highest production of resveratrol has been achieved successfully from *p*-coumaric acid with STS from *V. vinifera* and 4CL from *P. crispum* [133]. A strain of *C. glutamicum* was engineered for the production of resveratrol using STS from *A. hypogaea* (peanut) and 4CL from *P. crispum* (parsley) as it had more diverse specificity of substrates in comparison with other 4CL enzymes [121]. Enhanced production of phenylpropenoic acid (*p*-coumaric acid and cinnamic acid) was achieved on the heterological expression of PAL and TAL [128,142]. Introducing a heterologous TAL gene from *Flavobacterium johnsoniae* into a strain of *C. glutamicum* with deregulated shikimate pathway has resulted in achieving 60 mg/L resveratrol from glucose [121]. Further, introduction of pathway from *R. trifolli*, which assimilates malonate and addition of down-regulation of genes (*fabD, fabB, fabH, fabI, fabF*) to inactivate the pathway consuming malonyl-CoA, increased the malonyl-CoA pool, resulting in resveratrol production from glucose in *E. coli* [118]. Metabolic engineering can be explored further for production of high value polyphenols to meet their demand by microorganisms.

8.3.4 Addition of polyphenols in fermented food products

Addition of natural products rich in phenolic compounds has been implied as an approach improving polyphenols content in dairy products like cheese, kefir, and other foods such as bread, meat, and legumes [143,144]. Addition of other products such as extracts of mint leaves, vegetable, fruits, and herbs during the development of dairy products has been applied for improvement of functional properties [143]. The hydroalcoholic extract of seed of *Vitis labrusca* var. Bordeaux rich in phenolic substance has been applied as an additive during *Petit Suisse* cheese production [144]. The polyphenol content of cheese was found to increase along with the enhancement of antioxidant property. *Aronia melanocarpa* an eastern North American berry is rich in phenolic compounds such as anthocyanins and proanthocyanins. Elevation of polyphenols has been observed with the higher antioxidant activity in *aronia kefir*, which is a fermented dairy product added with *Aronia* juice [24]. Fermentation resulted in transformation of anthocyanin with increasing antioxidant activity. Plant-based polyphenols can be explored for their application as food additive in fermented foods for enhancement of functional properties.

8.4 Effect of gut microbes on polyphenols

Polyphenols present in food are poorly absorbed but when transferred to colon during digestion they encounter the gut microbes [145]. The enzymes produced by the gut microbes degrade complex molecules into simpler form resulting in better absorption. Some preclinical studies have shown that gut microbiota-transformed polyphenol metabolites have potential as bioactive compounds and need to be studied using in vitro assay involving human cell line as well as animal model studies [146]. Differences in health-promoting effect after polyphenols intake could be due to the differences in the type of bioactive compounds production mediated by gut microorganisms, which depends on microbial diversity in the colon of an individual. Polyphenol may transform to different metabolites with higher bioactivity depending on the specific gut microbiota of each individual [147]. Bioconversion of diadzein to equol (strong estrogenic activity) takes place in certain human subjects, which depends on the microbial population of the gut [148]. Some of the microbes isolated from human and rat feces reported for equol metabolism includes *Adlercreutzia equolifaciens, Bifidobacterium longum, Asaccharobacter celatus, Enterorhabdus mucosicola*, and *Lactobacillus paracasei* [148]. Therefore, difference in biological effects of polyphenol and its metabolites is dependent on different composition and/or function of gut microbiota [149].

Non extractable bound phenolic compounds have been evaluated from diverse foods especially relevant in proanthocyanidin containing. These are described as macroantioxidants as they release antioxidants after hydrolysis or transformation by human gut microbiota. Nonextractable polyphenols include those phenolic compounds, which are not extracted using specific solvent. Gut's microbial composition is modulated when catabolizing these polyphenols leading to small-sized molecules and showing biological effects. Nonextractable polyphenols show prebiotic effect by promoting the growth of LAB such as *lactobacilli* and *bifidobacteria*. These prebiotics also positively modulate the bacteria responsible for antiobesity property including *Faecalibacterium prausnitzii* and *Akkermansia muciniphila* [150]. They also cause beneficial effect by decreasing the ratio of *Firmicutes* and *bacteriodetes* [149]. Gut microbial metabolism of most important classes of dietary polyphenols such as monomeric and oligomeric catechins (proanthocyanidins), ellagitannins, flavanones, flavonols, and isoflavones are A- and C-ring cleavage, C-ring cleavage by dioxygenases, dihydroxylation (decarboxylation or reduction reaction), and hydrogenation of alkene moieties in different polyphenols including curcumin, resveratrol, and isoflavones. Characterization of the microbial population which can help in positive modulation in the enhancement of bioactive compounds can lead to their application in food formulation with specific health benefits.

8.5 Conclusions and perspectives

Polyphenols are rich in plant-based foods, which have been studied extensively for their wide range of health benefits. Microbial fermentation using specific carbohydrate active biocatalyst has resulted in enhancement of polyphenols during food fermentation and recovery of polyphenols from agri-food processing waste. There is a significant advancement in the field of genetic engineering for polyphenol production with the application of bio-engineered microorganisms. Gut microbiota and polyphenols modulate each other by producing different metabolites and promote the growth of health-promoting microorganisms in the gut. Gut microorganisms can be a source of novel enzymes for the production of high value metabolites. Researchers are focusing on finding novel enzymes for polyphenols transformation to more bioactive form for their application in functional food industry. Studies are needed to overcome the current demand for specific polyphenols by cost-effective and eco-friendly approach to microbial fermentation resulting in higher yield.

References

[1] A. Dudnik, P. Gaspar, A.R. Neves, J. Forster, Engineering of microbial cell factories for the production of plant polyphenols with health-beneficial properties, Current Pharmaceutical Design 24 (2018) 2208–2225.

[2] A. Al-Rawahi, Phenolic constituents of pomegranate peels (Punica granatum L.) cultivated in Oman, European Journal of Medicinal Plants 4 (2014) 315–331.

[3] A. Rodriguez-Mateos, D. Vauzour, C.G. Krueger, D. Shanmuganayagam, J. Reed, L. Calani, et al., Bioavailability, bioactivity and impact on health of dietary flavonoids and related compounds: an update, Archives of Toxicology 88 (2014) 1803–1853.

[4] D. Del Rio, A. Rodriguez-Mateos, J.P.E. Spencer, M. Tognolini, G. Borges, A. Crozier, Dietary (poly)phenolics in human health: structures, bioavailability, and evidence of protective effects against chronic diseases, Antioxidants and Redox Signaling 18 (2013) 1818–1892.

[5] P.G. Anantharaju, P.C. Gowda, M.G. Vimalambike, S.V. Madhunapantula, An overview on the role of dietary phenolics for the treatment of cancers, Nutrition Journal 15 (2016) 1–16.

[6] L. Milke, J. Aschenbrenner, J. Marienhagen, N. Kallscheuer, Production of plant-derived polyphenols in microorganisms: current state and perspectives, Applied Microbiology and Biotechnology 102 (2018) 1575–1585.

[7] I.O. Minatel, C.V. Borges, M.I. Ferreira, H.A.G. Gomez, C.-Y.O. Chen, G.P.P. Lima, Phenolic compounds: functional properties, Impact of Processing and Bioavailability, in Phenolic Compounds—Biological Activity (2017). InTech.

[8] A.J. Parr, G.P. Bolwell, Phenols in the plant and in man. The potential for possible nutritional enhancement of the diet by modifying the phenols content or profile, Journal of the Science of Food and Agriculture 80 (2000) 985–1012.

[9] O.R.W. Sutherland, G.B. Russell, D.R. Biggs, G.A. Lane, Insect feeding deterrent activity of phytoalexin isoflavonoids, Biochemical Systematics and Ecology 8 (1980) 73–75.

[10] A. Michalak, Phenolic compounds and their antioxidant activity in plants growing under heavy metal stress, Polish Journal of Environmental Studies 15 (2006) 523–530.

[11] D.E. Stevenson, R.D. Hurst, Polyphenolic phytochemicals—just antioxidants or much more? Cellular and Molecular Life Sciences 64 (2007) 2900–2916.

[12] A.B. Kowska-Barczak, Acylated anthocyanins as stable, natural food colorants—a review, Polish Journal of Food and Nutrition Sciences 14 (2005) 107–116.

[13] H. Mo, Y. Zhu, Z. Chen, Microbial fermented tea—a potential source of natural food preservatives, Trends in Food Science and Technology 19 (2008) 124–130.

[14] T. Li, J. Li, W. Hu, X. Zhang, X. Li, J. Zhao, Shelf-life extension of crucian carp (Carassius auratus) using natural preservatives during chilled storage, Food Chemistry 135 (2012) 140–145.

[15] A. Harlin, Biogenic precursors for polyphenol, polyester and polyurethane resins, Handbook of Bioplastics and Biocomposites Engineering Applications, John Wiley & Sons, Inc., Hoboken, NJ, 2011, pp. 511–553.

[16] H. Xiang, D. Sun-Waterhouse, G.I.N. Waterhouse, C. Cui, Z. Ruan, Fermentation-enabled wellness foods: a fresh perspective, Food Science and Human Wellness 8 (2019) 203–243.

[17] M.L.Y. Wan, K.H. Ling, H. El-Nezami, M.F. Wang, Influence of functional food components on gut health, Critical Reviews in Food Science and Nutrition 59 (2019) 1927–1936. Taylor and Francis Inc.

[18] S.J. Hur, S.Y. Lee, Y.C. Kim, I. Choi, G.B. Kim, Effect of fermentation on the antioxidant activity in plant-based foods, Food Chemistry 160 (2014) 346–356.

[19] A.K. Rai, S. Sanjukta, K. Jeyaram, Production of angiotensin I converting enzyme inhibitory (ACE-I) peptides during milk fermentation and their role in reducing hypertension, Critical Reviews in Food Science and Nutrition 57 (2017) 2789–2800.

[20] A.K. Rai, A. Pandey, D. Sahoo, Biotechnological potential of yeasts in functional food industry, Trends in Food Science and Technology 83 (2019) 129–137.

[21] B.A. Acosta-Estrada, J.A. Gutiérrez-Uribe, S.O. Serna-Saldívar, Bound phenolics in foods, a review, Food Chemistry 152 (2014) 46–55.

[22] T.K. McGhie, M.C. Walton, The bioavailability and absorption of anthocyanins: towards a better understanding, Molecular Nutrition and Food Research 51 (2007) 702–713.

[23] T. Oksuz, E. Surek, Z. Tacer-Caba, D. Nilufer-Erdil, Phenolic contents and antioxidant activities of Persimmon and Red Beet jams produced by sucrose impregnation, Food Science and Technology 3 (2015) 1–8.

[24] X. Du, A.D. Myracle, Fermentation alters the bioaccessible phenolic compounds and increases the alpha-glucosidase inhibitory effects of aronia juice in a dairy matrix following: in vitro digestion, Food and Function 9 (2018) 2998–3007.

[25] T. Oksuz, Z. Tacer-Caba, D. Nilufer-Erdil, D. Boyacioglu, Changes in bioavailability of sour cherry (*Prunus cerasus* L.) phenolics and anthocyanins when consumed with dairy food matrices, Journal of Food Science and Technology 56 (2019) 4177–4188.

[26] S. Chouhan, K. Sharma, J. Zha, S. Guleria, M.A.G. Koffas, Recent advances in the recombinant biosynthesis of polyphenols, Frontiers in Microbiology 8 (2017) 1–16.

[27] B. Huang, J. Guo, B. Yi, X. Yu, L. Sun, W. Chen, Heterologous production of secondary metabolites as pharmaceuticals in *Saccharomyces cerevisiae*, Biotechnology Letters 30 (2008) 1121–1137.

[28] M. Jiang, G. Stephanopoulos, B.A. Pfeifer, Toward biosynthetic design and implementation of *Escherichia coli*-derived paclitaxel and other heterologous polyisoprene compounds, Applied and Environmental Microbiology 78 (2012) 2497–2504.

[29] J.A. Chemler, M.A. Koffas, Metabolic engineering for plant natural product biosynthesis in microbes, Current Opinion in Biotechnology 19 (2008) 597–605.

[30] H.M. Salis, E.A. Mirsky, C.A. Voigt, Automated design of synthetic ribosome binding sites to control protein expression, Nature Biotechnology 27 (2009) 946–950.

[31] J.A. Casas-Mollano, J. Rohr, E.-J. Kim, E. Balassa, K. van Dijk, H. Cerutti, Diversification of the Core RNA interference machinery in *Chlamydomonas reinhardtii* and the role of DCL1 in transposon silencing, Genetics 179 (2008) 69–81.

[32] K.B. Pandey, S.I. Rizvi, Plant polyphenols as dietary antioxidants in human health and disease, Oxidative Medicine and Cellular Longevity 2 (2009) 270–278.

[33] K. Ganesan, B. Xu, A critical review on polyphenols and health benefits of black soybeans, Nutrients 9 (2017) 455.

[34] S. Kumar, M.M. Abedin, A.K. Singh, S. Das, Role of phenolic compounds in plant-defensive mechanisms, in: Plant Phenolics in Sustainable Agriculture, Springer, Singapore, 2020, pp. 517–532.

[35] C. Manach, A. Scalbert, C. Morand, C. Rémésy, L. Jiménez, Polyphenols: food sources and bioavailability, American Journal of Clinical Nutrition 79 (2004) 727−747.

[36] F. Shahidi, P. Ambigaipalan, Phenolics and polyphenolics in foods, beverages and spices: antioxidant activity and health effects—a review, Journal of Functional Foods 18 (2015) 820−897.

[37] R. Tsao, Chemistry and biochemistry of dietary polyphenols, Nutrients 2 (2010) 1231−1246.

[38] H.B. Rashmi, P.S. Negi, Phenolic acids from vegetables: a review on processing stability and health benefits, Food Research International 136 (2020) 109298.

[39] F. Shahidi, M. Naczk, Phenolic compounds in fruits and vegetables, Food Phenolics: Sources, Chemistry, Effects and Applications 75107 (1995).

[40] G. Morabito, C. Migli, I. Peluso, M. Serafini, Fruit polyphenols and postprandial inflammatory stress, Polyphenols in Human Health and Disease (2014) 1107−1126.

[41] H. Abramovič, T. Košmerl, N.P. Ulrih, B. Cigić, Contribution of SO_2 to antioxidant potential of white wine, Food Chemistry 174 (2015) 147−153.

[42] F. Shahidi, Y. Zhong, Measurement of antioxidant activity, Journal of Functional Foods 18 (2015) 757−781.

[43] A. Tresserra-Rimbau, E.B. Rimm, A. Medina-Remón, M.A. Martínez-González, R. de la Torre, D. Corella, et al., Inverse association between habitual polyphenol intake and incidence of cardiovascular events in the PREDIMED study, Nutrition, Metabolism and Cardiovascular Diseases 24 (2014) 639−647.

[44] Y. Semaming, P. Pannengpetch, S.C. Chattipakorn, N. Chattipakorn, Pharmacological properties of protocatechuic acid and its potential roles as complementary medicine, Evidence-based Complementary and Alternative Medicine (2015) 2015.

[45] T. Yang Wang, Q. Li, K. Shun Bi, Bioactive flavonoids in medicinal plants: structure, activity and biological fate, Asian Journal of Pharmaceutical Sciences 13 (2018) 12−23.

[46] D. Tungmunnithum, A. Thongboonyou, A. Pholboon, A. Yangsabai, Flavonoids and other phenolic compounds from medicinal plants for pharmaceutical and medical aspects: an overview, Medicines 5 (2018) 93.

[47] R. Pei, X. Liu, B. Bolling, Flavonoids and gut health, Current Opinion in Biotechnology 61 (2020) 153−159.

[48] J.A. Baur, D.A. Sinclair, Therapeutic potential of resveratrol: the in vivo evidence, Nature Reviews Drug Discovery 5 (2006) 493−506.

[49] M. Takaoka, Resveratrol, a new phenolic compound, from *Veratrum grandiflorum*, Nippon Kagaku Kaishi 60 (1939) 1090−1100.

[50] J.H. Bauer, S. Goupil, G.B. Garber, S.L. Helfand, An accelerated assay for the identification of lifespan-extending interventions in *Drosophila melanogaster*, Proceedings of the National Academy of Sciences of the United States of America 101 (2004) 12980−12985.

[51] S. Nonomura, H. Kanagawa, A. Makimoto, Chemical constituents of polygonaceous plants. i. studies on the components of ko-j o-kon.(polygonum cuspidatum sieb. et zucc.), Yakugaku zasshi: Journal of the Pharmaceutical Society of Japan 83 (1963) 988−990.

[52] K.V. Kiselev, Perspectives for production and application of resveratrol, Applied Microbiology and Biotechnology 90 (2011) 417−425.

[53] B.P. Hubbard, D.A. Sinclair, Small molecule SIRT1 activators for the treatment of aging and age-related diseases, Trends in Pharmacological Sciences 35 (2014) 146−154.

[54] E.N. Frankel, Inhibition of human LDL oxidation by resveratrol, Lancet 341 (1993) 1103−1104.

[55] J.P. Blond, M.P. Denis, J. Bezard, Antioxidant action of resveratrol in lipid peroxidation, Sciences des Aliments (France) (1995).

[56] X. Yang, X. Li, J. Ren, From French paradox to cancer treatment: anti-cancer activities and mechanisms of resveratrol, Anti-Cancer Agents in Medicinal Chemistry (Formerly Current Medicinal Chemistry-Anti-Cancer Agents) 14 (2014) 806−825.

[57] B.B. Aggarwal, A. Bhardwaj, R.S. Aggarwal, N.P. Seeram, S. Shishodia, Y. Takada, Role of resveratrol in prevention and therapy of cancer: preclinical and clinical studies, Anticancer Research 24 (2004) 2783–2840.

[58] J.K. Kundu, Y.J. Surh, Cancer chemopreventive and therapeutic potential of resveratrol: mechanistic perspectives, Cancer Letters 269 (2008) 243–261.

[59] S. Sharma, S.K. Kulkarni, K. Chopra, Effect of resveratrol, a polyphenolic phytoalexin, on thermal hyperalgesia in a mouse model of diabetic neuropathic pain, Fundamental & Clinical Pharmacology 21 (2007) 89–94.

[60] S. Sharma, S. Sharma, R. Chourasia, A. Pandey, A.K. Rai, et al., Alzheimer's Disease : Ethanobotanical Studies. Naturally Occurring Chemicals Against Alzheimer's Disease, Academic Press, 2021, pp. 11–28.

[61] V. Vingtdeux, U. Dreses-Werringloer, H. Zhao, P. Davies, P. Marambaud, Therapeutic potential of resveratrol in Alzheimer's disease, BMC Neuroscience 9 (2008) 1–5.

[62] M. Viswanathan, S.K. Kim, A. Berdichevsky, L. Guarente, A role for SIR-2.1 regulation of ER stress response genes in determining C. elegans life span, Developmental Cell 9 (2005) 605–615.

[63] A. Braga, P. Ferreira, J. Oliveira, I. Rocha, N. Faria, Heterologous production of resveratrol in bacterial hosts: current status and perspectives, World Journal of Microbiology and Biotechnology 34 (2018) 0.

[64] S. Anjum, A. Komal, S. Drouet, H. Kausar, C. Hano, B.H. Abbasi, Feasible production of lignans and neolignans in root-derived in vitro cultures of flax (*Linum usitatissimum* L.), Plants 9 (2020) 1–19.

[65] C. Rodríguez-García, C. Sánchez-Quesada, E. Toledo, M. Delgado-Rodríguez, J.J. Gaforio, Naturally lignan-rich foods: a dietary tool for health promotion? Molecules 24 (2019) 917.

[66] I. Ignat, I. Volf, V.I. Popa, A critical review of methods for characterisation of polyphenolic compounds in fruits and vegetables, Food Chemistry 126 (2011) 1821–1835.

[67] J. Beekwilder, R. Wolswinkel, H. Jonker, R. Hall, C.H. De, Rie Vos, and A. Bovy, Production of resveratrol in recombinant microorganisms, Applied and Environmental Microbiology 72 (2006) 5670–5672.

[68] O. Gligor, A. Mocan, C. Moldovan, M. Locatelli, G. Crişan, I.C.F.R. Ferreira, Enzyme-assisted extractions of polyphenols—a comprehensive review, Trends in Food Science & Technology 88 (2019) 302–315.

[69] K.M. Cho, J.H. Lee, H.D. Yun, B.Y. Ahn, H. Kim, W.T. Seo, Changes of phytochemical constituents (isoflavones, flavanols, and phenolic acids) during cheonggukjang soybeans fermentation using potential probiotics *Bacillus subtilis* CS90, Journal of Food Composition and Analysis 24 (2011) 402–410.

[70] S. Hati, S. Vij, B.P. Singh, S. Mandal, β-Glucosidase activity and bioconversion of isoflavones during fermentation of soymilk, Journal of the Science of Food and Agriculture 95 (2015) 216–220.

[71] J.-S. Kim, S. Yoon, Isoflavone contents and β-glucosidase activities of soybeans, meju, and doenjang, Korean Journal of Food Science and Technology 31 (1999) 1405–1409.

[72] M.-Y.K.-I.S.-W.S.-H.N.-J. Shon, Biological activities of chungkugjang prepared with black bean and changes in phytoestrogen content during fermentation, Korean Journal of Food Science and Technology 32 (4) (2000) 936–941.

[73] B. Moktan, J. Saha, P.K. Sarkar, Antioxidant activities of soybean as affected by *Bacillus*-fermentation to kinema, Food Research International 41 (2008) 586–593.

[74] D.Y. Kwon, S.M. Hong, I.S. Ahn, M.J. Kim, H.J. Yang, S. Park, Isoflavonoids and peptides from meju, long-term fermented soybeans, increase insulin sensitivity and exert insulinotropic effects in vitro, Nutrition 27 (2011) 244–252.

[75] C.R. Rekha, G. Vijayalakshmi, Isoflavone phytoestrogens in soymilk fermented with β-glucosidase producing probiotic lactic acid bacteria, International Journal of Food Sciences and Nutrition 62 (2011) 111–120.

[76] J. Hubert, M. Berger, F. Nepveu, F. Paul, J. Daydé, Effects of fermentation on the phytochemical composition and antioxidant properties of soy germ, Food Chemistry 109 (2008) 709–721.

[77] J.A. Curiel, R. Muñoz, F. López de Felipe, pH and dose-dependent effects of quercetin on the fermentation capacity of Lactobacillus plantarum, LWT - Food Science and Technology 43 (2010) 926–933.

[78] M. Dueñas, D. Fernández, T. Hernández, I. Estrella, R. Muñoz, Bioactive phenolic compounds of cowpeas (Vigna sinensis L). Modifications by fermentation with natural microflora and with *Lactobacillus plantarum* ATCC 14917, Journal of the Science of Food and Agriculture 85 (2005) 297–304.

[79] M.Y. Juan, C.C. Chou, Enhancement of antioxidant activity, total phenolic and flavonoid content of black soybeans by solid state fermentation with *Bacillus subtilis* BCRC 14715, Food Microbiology 27 (2010) 586–591.

[80] Y. Hu, C. He, W. Yuan, R. Zhu, W. Zhang, L. Du, et al., Characterization of fermented black soybean natto inoculated with Bacillus natto during fermentation, Journal of the Science of Food and Agriculture 90 (2010) 1194–1202.

[81] Y.H. Huang, Y.J. Lai, C.C. Chou, Fermentation temperature affects the antioxidant activity of the enzyme-ripened sufu, an oriental traditional fermented product of soybean, Journal of Bioscience and Bioengineering 112 (2011) 49–53.

[82] D.A. Vattem, K. Shetty, Ellagic acid production and phenolic antioxidant activity in cranberry pomace (Vaccinium macrocarpon) mediated by Lentinus edodes using a solid-state system, Process Biochemistry 39 (2003) 367–379.

[83] C.M. Ajila, S.K. Brar, M. Verma, R.D. Tyagi, J.R. Valéro, Solid-state fermentation of apple pomace using *Phanerocheate chrysosporium*—Liberation and extraction of phenolic antioxidants, Food Chemistry 126 (2011) 1071–1080.

[84] Z. Zhang, G. Lv, H. Pan, et al., Production of powerful antioxidant supplements via solid-state fermentation of wheat (*Triticum aestivum* Linn.) by *Cordyceps militaris*, Food Technology and Biotechnology 50 (2012) 32–39.

[85] K. Katina, K.H. Liukkonen, A. Kaukovirta-Norja, H. Adlercreutz, S.M. Heinonen, A.M. Lampi, et al., Fermentation-induced changes in the nutritional value of native or germinated rye, Journal of Cereal Science 46 (2007) 348–355.

[86] S.H. Eom, J.H. Park, D.U. Yu, J.I. Choi, J.D. Choi, M.S. Lee, et al., Antimicrobial activity of brown alga eisenia bicyclis against methicillin-resistant *Staphylococcus aureus*, Fisheries and Aquatic Sciences 14 (2011) 251–256.

[87] J. Manwar, K. Mahadik, L. Sathiyanarayanan, A. Paradkar, S. Patil, Comparative antioxidant potential of *Withania somnifera* based herbal formulation prepared by traditional and non-traditional fermentation processes, Integrative Medicine Research 2 (2013) 56–61.

[88] M.R. Pérez-Gregorio, J. Regueiro, E. Alonso-González, L.M. Pastrana-Castro, J. Simal-Gándara, Influence of alcoholic fermentation process on antioxidant activity and phenolic levels from mulberries (Morus nigra L.), LWT - Food Science and Technology 44 (2011) 1793–1801.

[89] M.H. Johnson, A. Lucius, T. Meyer, E. Gonzalez De Mejia, Cultivar evaluation and effect of fermentation on antioxidant capacity and in vitro inhibition of α-amylase and α-glucosidase by highbush blueberry (vaccinium corombosum), Journal of Agricultural and Food Chemistry 59 (2011) 8923–8930.

[90] L.N. Hlahla, F.N. Mudau, I.K. Mariga, Effect of fermentation temperature and time on the chemical composition of bush tea (Athrixia phylicoides DC.), Journal of Medicinal Plants Research 4 (2010) 824–829.

[91] M.S. Su, P.J. Chien, Antioxidant activity, anthocyanins, and phenolics of rabbiteye blueberry (Vaccinium ashei) fluid products as affected by fermentation, Food Chemistry 104 (2007) 182–187.

[92] A.K. Rai, K.A. Anu Appaiah, Application of native yeast from Garcinia (*Garcinia xanthochumus*) for the preparation of fermented beverage: changes in biochemical and antioxidant properties, Food Bioscience 5 (2014) 101–107.

[93] A.K. Rai, M. Prakash, K.A. Anu Appaiah, production of Garcinia wine: changes in biochemical parameters, organic acids and free sugars during fermentation of Garcinia must, International Journal of Food Science and Technology 45 (2010) 1336–1339.

[94] D.Y. Kwon, J.W. Daily, H.J. Kim, S. Park, Antidiabetic effects of fermented soybean products on type 2 diabetes, Nutrition Research 30 (2010) 1–13.

[95] S. Sanjukta, A.K. Rai, D. Sahoo, Bioactive molecules in fermented soybean products and their potential health benefits, Fermented, Foods: Part II: Technological Interventions (2017) 97–121.

[96] I.H. Lee, Y.H. Hung, C.C. Chou, Solid-state fermentation with fungi to enhance the antioxidative activity, total phenolic and anthocyanin contents of black bean, International Journal of Food Microbiology 121 (2008) 150–156.

[97] L.D. Lasrado, and A.K. Rai, Recovery of Nutraceuticals From Agri-Food Industry Waste by Lactic Acid Fermentation, 185–203, 2018.

[98] X. Mao, J. Zhang, F. Kan, Y. Gao, J. Lan, X. Zhang, et al., Antioxidant production and chitin recovery from shrimp head fermentation with Streptococcus thermophilus, Food Science and Biotechnology 22 (2013) 1023–1032.

[99] A.K. Rai, H.C. Swapna, N. Bhaskar, V. Baskaran, Potential of seafood industry byproducts as sources of recoverable lipids: fatty acid composition of meat and nonmeat component of selected Indian marine fishes, Journal of Food Biochemistry 36 (2012) 441–448.

[100] A.K. Rai, K. Jeyaram, Health benefits of functional proteins in fermented foods, Health Benefits of Fermented Foods and Beverages (2015) 455–474.

[101] A.K. Rai, N. Bhaskar, V. Baskaran, Bioefficacy of EPA-DHA from lipids recovered from fish processing wastes through biotechnological approaches, Food Chemistry 136 (2013) 80–86.

[102] S. Martins, S.I. Mussatto, G. Martínez-Avila, J. Montañez-Saenz, C.N. Aguilar, J.A. Teixeira, Bioactive phenolic compounds: Production and extraction by solid-state fermentation. A review, Biotechnology Advances 29 (2011) 365–373.

[103] R. Aranday-García, A. Román Guerrero, S. Ifuku, K. Shirai, Successive inoculation of Lactobacillus brevis and Rhizopus oligosporus on shrimp wastes for recovery of chitin and added-value products, Process Biochemistry 58 (2017) 17–24.

[104] F. Leroy, L. De, Vuyst, Lactic acid bacteria as functional starter cultures for the food fermentation industry, Trends in Food Science and Technology 15 (2004) 67–78.

[105] A. Sharma, A. Sharma, J. Singh, P. Sharma, G.S. Tomar, S. Singh, et al., A biorefinery approach for the production of ferulic acid from agroresidues through ferulic acid esterase of lactic acid bacteria, 3 Biotech 10 (2020) 1–10.

[106] P.S.-N. Nigam, and A. Pandey, Biotechnology for agro-industrial residues utilisation: utilisation of agro-residues. Springer Science & Business Media, 2009.

[107] L. Sepúlveda, E. Laredo-Alcalá, J.J. Buenrostro-Figueroa, J.A. Ascacio-Valdés, Z. Genisheva, C. Aguilar, et al., Ellagic acid production using polyphenols from orange peel waste by submerged fermentation, Electronic Journal of Biotechnology 43 (2020) 1–7.

[108] M.M. Rashad, A.E. Mahmoud, M.M. Ali, M.U. Nooman, A.S. Al-Kashef, Antioxidant and anticancer agents produced from pineapple waste by solid state fermentation, International Journal of Toxicological and Pharmacological Research 7 (2015) 287–296.

[109] C.N. Aguilar, A. Aguilera-Carbo, A. Robledo, J. Ventura, R. Belmares, D. Martinez, et al., Production of antioxidant nutraceutlcals by solid-state cultures of pomegranate (Punica granatum) peel and creosote bush (*Larrea tridentata*) leaves, Food Technology and Biotechnology 46 (2008) 218–222.

[110] J. Buenrostro-Figueroa, S. Huerta-Ochoa, A. Prado-Barragán, J. Ascacio-Valdés, L. Sepúlveda, R. Rodríguez, et al., Continuous production of ellagic acid in a packed-bed reactor, Process Biochemistry 49 (2014) 1595–1600.

[111] R. Paranthaman, S. Kumaravel, K. Singaravadivel, Development of bioprocess technology for the production of bioactive compound, ellagic acid from tea waste, Curr Res Microbiol Biotechnol 1 (2013) 270–273.

[112] E. Brglez Mojzer, M. Knez Hrnčič, M. Škerget, Ž. Knez, U. Bren, Polyphenols: extraction methods, antioxidative action, bioavailability and anticarcinogenic effects, Molecules 21 (2016) 901.

[113] K.T. Watts, P.C. Lee, C. Schmidt-Dannert, Biosynthesis of plant-specific stilbene polyketides in metabolically engineered *Escherichia coli*, BMC Biotechnology 6 (2006) 22.

[114] Y. Katsuyama, N. Funa, I. Miyahisa, S. Horinouchi, Synthesis of unnatural flavonoids and stilbenes by exploiting the plant biosynthetic pathway in *Escherichia coli*, Chemistry and Biology 14 (2007) 613–621.

[115] P. Xu, S. Ranganathan, Z.L. Fowler, C.D. Maranas, M.A.G. Koffas, Genome-scale metabolic network modeling results in minimal interventions that cooperatively force carbon flux towards malonyl-CoA, Metabolic Engineering 13 (2011) 578–587.

[116] F. Koopman, J. Beekwilder, B. Crimi, A. van Houwelingen, R.D. Hall, D. Bosch, et al., De novo production of the flavonoid naringenin in engineered *Saccharomyces cerevisiae*, Microbial Cell Factories 11 (2012) 1–15.

[117] M. Li, K. Schneider, M. Kristensen, I. Borodina, J. Nielsen, Engineering yeast for high-level production of stilbenoid antioxidants, Scientific Reports 6 (2016) 1–8.

[118] J. Wu, P. Zhou, X. Zhang, M. Dong, Efficient de novo synthesis of resveratrol by metabolically engineered *Escherichia coli*, Journal of Industrial Microbiology and Biotechnology 44 (7) (2017) 1083–1095.

[119] A. Rodriguez, K.R. Kildegaard, M. Li, I. Borodina, J. Nielsen, Establishment of a yeast platform strain for production of p-coumaric acid through metabolic engineering of aromatic amino acid biosynthesis, Metabolic Engineering 31 (2015) 181–188.

[120] P.V. van Summeren-Wesenhagen, J. Marienhagen, Metabolic engineering of *Escherichia coli* for the synthesis of the plant polyphenol pinosylvin, Applied and Environmental Microbiology 81 (2015) 840–849.

[121] N. Kallscheuer, M. Vogt, A. Stenzel, J. Gätgens, M. Bott, J. Marienhagen, Construction of a *Corynebacterium glutamicum* platform strain for the production of stilbenes and (2S)-flavanones, Metabolic Engineering 38 (2016) 47–55.

[122] L. Duan, W. Ding, X. Liu, X. Cheng, J. Cai, E. Hua, et al., Biosynthesis and engineering of kaempferol in *Saccharomyces cerevisiae*, Microbial Cell Factories 16 (2017) 165.

[123] N. Kallscheuer, M. Vogt, M. Bott, J. Marienhagen, Functional expression of plant-derived O-methyltransferase, flavanone 3-hydroxylase, and flavonol synthase in *Corynebacterium glutamicum* for production of pterostilbene, kaempferol, and quercetin, Journal of Biotechnology 258 (2017) 190–196.

[124] C.G. Lim, L. Wong, N. Bhan, H. Dvora, P. Xu, S. Venkiteswaran, et al., Development of a recombinant Escherichia coli strain for overproduction of the plant pigment anthocyanin, Applied and Environmental Microbiology 81 (2015) 6276–6284.

[125] S.C. Makrides, Strategies for achieving high-level expression of genes in *Escherichia coli*, Microbiology and Molecular Biology Reviews 60 (1996) 3.

[126] J. Wu, P. Liu, Y. Fan, H. Bao, G. Du, J. Zhou, et al., Multivariate modular metabolic engineering of *Escherichia coli* to produce resveratrol from L-tyrosine, Journal of Biotechnology 167 (2013) 404–411.

[127] S.Y. Shin, N.S. Han, Y.C. Park, M.D. Kim, J.H. Seo, Production of resveratrol from p-coumaric acid in recombinant Saccharomyces cerevisiae expressing 4-coumarate:coenzyme A ligase and stilbene synthase genes, Enzyme and Microbial Technology 48 (2011) 48–53.

[128] Q. Huang, Y. Lin, Y. Yan, Caffeic acid production enhancement by engineering a phenylalanine overproducing *Escherichia coli* strain, Biotechnology and Bioengineering 110 (2013) 3188–3196.

[129] G.L. Rosano, E.A. Ceccarelli, Recombinant protein expression in *Escherichia coli*: advances and challenges, Frontiers in Microbiology 5 (2014) 172.

[130] F. Yesilirmak, Z. Sayers, Heterologous expression of plant genes, International Journal of Plant Genomics 2009 (2009) 296482.

[131] S.R. Park, J.A. Yoon, J.H. Paik, J.W. Park, W.S. Jung, Y.-H. Ban, et al., Engineering of plant-specific phenylpropanoids biosynthesis in *Streptomyces venezuelae*, Journal of Biotechnology 141 (2009) 181–188.

[132] S. Quideau, D. Deffieux, C. Douat-Casassus, L. Pouységu, Plant polyphenols: chemical properties, biological activities, and synthesis, Angewandte Chemie International Edition 50 (2011) 586–621.

[133] C.G. Lim, Z.L. Fowler, T. Hueller, S. Schaffer, M.A.G. Koffas, High-yield resveratrol production in engineered *Escherichia coli*, Applied and Environmental Microbiology 77 (2011) 3451–3460.

[134] Y. Yang, Y. Lin, L. Li, R.J. Linhardt, Y. Yan, Regulating malonyl-CoA metabolism via synthetic antisense RNAs for enhanced biosynthesis of natural products, Metabolic Engineering 29 (2015) 217–226.

[135] O. Choi, C.-Z. Wu, S.Y. Kang, J.S. Ahn, T.-B. Uhm, Y.-S. Hong, Biosynthesis of plant-specific phenylpropanoids by construction of an artificial biosynthetic pathway in *Escherichia coli*, Journal of Industrial Microbiology & Biotechnology 38 (2011) 1657–1665.

[136] H. Morita, H. Noguchi, J. Schröder, I. Abe, Novel polyketides synthesized with a higher plant stilbene synthase, European Journal of Biochemistry 268 (2001) 3759–3766.

[137] I. Abe, T. Watanabe, H. Noguchi, Enzymatic formation of long-chain polyketide pyrones by plant type III polyketide synthases, Phytochemistry 65 (2004) 2447–2453.

[138] B. Hamberger, K. Hahlbrock, The 4-coumarate: CoA ligase gene family in *Arabidopsis thaliana* comprises one rare, sinapate-activating and three commonly occurring isoenzymes, Proceedings of the National Academy of Sciences of the United States of America 101 (2004) 2209–2214.

[139] E. Zhang, X. Guo, Z. Meng, J. Wang, J. Sun, X. Yao, et al., Construction, expression, and characterization of Arabidopsis thaliana 4CL and Arachis hypogaea RS fusion gene 4CL: RS in *Escherichia coli*, World Journal of Microbiology and Biotechnology 31 (2015) 1379–1385.

[140] A.C. Schroeder, S. Kumaran, L.M. Hicks, R.E. Cahoon, C. Halls, O. Yu, et al., Contributions of conserved serine and tyrosine residues to catalysis, ligand binding, and cofactor processing in the active site of tyrosine ammonia lyase, Phytochemistry 69 (2008) 1496–1506.

[141] T. Vannelli, W. Wei Qi, J. Sweigard, A.A. Gatenby, F.S. Sariaslani, Production of p-hydroxycinnamic acid from glucose in *Saccharomyces cerevisiae* and *Escherichia coli* by expression of heterologous genes from plants and fungi, Metabolic Engineering 9 (2007) 142–151.

[142] H. Zhang, G. Stephanopoulos, Engineering E. coli for caffeic acid biosynthesis from renewable sugars, Applied Microbiology and Biotechnology 97 (2013) 3333–3341.

[143] A.F. Fezea, H.N. Al-Zobaidy, M.F. Al-Quraishi, Total phenolic content, microbial content and sensory attributes evaluation of white soft cheese incorporated with mint (Mentha Spicata) leaf extract, IOSR Journal of Agriculture and Veterinary Sciences 10 (2017) 36–40.

[144] C.T.P. Deolindo, P.I. Monteiro, J.S. Santos, A.G. Cruz, M.C. da Silva, D. Granato, Phenolic-rich Petit Suisse cheese manufactured with organic Bordeaux grape juice, skin, and seed extract: technological, sensory, and functional properties, LWT 115 (2019) 108493.

[145] T. Ozdal, D.A. Sela, J. Xiao, D. Boyacioglu, F. Chen, E. Capanoglu, The reciprocal interactions between polyphenols and gut microbiota and effects on bioaccessibility, Nutrients 8 (2016) 78.

[146] G. Catalkaya, K. Venema, L. Lucini, G. Rocchetti, D. Delmas, M. Daglia, et al., Interaction of dietary polyphenols and gut microbiota: microbial metabolism of polyphenols, influence on the gut microbiota, and implications on host health, Food Frontiers 1 (2020) 109–133.

[147] F. Cardona, C. Andrés-Lacueva, S. Tulipani, F.J. Tinahones, M.I. Queipo-Ortuño, Benefits of polyphenols on gut microbiota and implications in human health, The Journal of Nutritional Biochemistry 24 (2013) 1415–1422.

[148] B. Mayo, L. Vázquez, A.B. Flórez, Equol: a bacterial metabolite from the daidzein isoflavone and its presumed beneficial health effects, Nutrients 11 (2019) 2231.

[149] A. González-Sarrías, J.C. Espín, F.A. Tomás-Barberán, Non-extractable polyphenols produce gut microbiota metabolites that persist in circulation and show anti-inflammatory and free radical-scavenging effects, Trends in Food Science and Technology 69 (2017) 281–288.

[150] K. Zhou, Strategies to promote abundance of *Akkermansia muciniphila*, an emerging probiotics in the gut, evidence from dietary intervention studies, Journal of Functional Foods 33 (2017) 194–201.

9

Bioprocess technologies for production of structured lipids as nutraceuticals

Suzana Ferreira-Dias[1], Natália Osório[2,3], Carla Tecelão[4]

[1]INSTITUTO SUPERIOR DE AGRONOMIA, LEAF, LINKING LANDSCAPE, ENVIRONMENT, AGRICULTURE AND FOOD, UNIVERSIDADE DE LISBOA, LISBON, PORTUGAL [2]INSTITUTO POLITÉCNICO DE SETÚBAL, ESCOLA SUPERIOR DE TECNOLOGIA DO BARREIRO, LAVRADIO, PORTUGAL [3]INSTITUTO SUPERIOR DE AGRONOMIA, CENTRO DE ESTUDOS FLORESTAIS, UNIVERSIDADE DE LISBOA, LISBON, PORTUGAL [4]POLITÉCNICO DE LEIRIA, ESCOLA SUPERIOR DE TURISMO E TECNOLOGIA DO MAR, MARE-MARINE AND ENVIRONMENTAL SCIENCES CENTRE, PENICHE, PORTUGAL

9.1 Introduction

9.1.1 Lipids as nutraceuticals

Lipids are important components of a balanced diet, since they provide energy, essential fatty acids, liposoluble vitamins (A, D, E, and K), and other micronutrients. Nowadays, lipids are not only considered as components to satisfy hunger, prevent diet-deficiency diseases, or to provide the essential nutrition for body maintenance and/or tissues growth and repair. They can have an important proactive role in health, preventing several aging-related chronic diseases, and promoting both physical and mental health and wellness. Thus, several lipids can be considered as functional foods and/or nutraceuticals [1].

The category of "Foods for Specified Health Use" (FOSHU) was first legislated in 1991, in Japan, after a decade on systematic studies on food functionality. These foods must demonstrate a positive influence in physiological functions. The concerted action on "Functional Food Science in Europe" (FUFOSE), created by the European Commission in the 1990s, defined functional food as "a food that beneficially affects one or more target functions in the body beyond adequate nutritional effects in a way that is relevant to either an improved state of health and well-being and/or reduction of risk of disease" [2]. Thus, functional foods are consumed as conventional foods, incorporated in a normal diet. However, this concept of "functional food" is sometimes in contrast with the concept of "nutraceutical." The concept of nutraceutical is less clear because it can be described differently. Nutraceuticals have been defined as the products obtained from foods presenting demonstrated physiological benefits and/or protection against chronic diseases, which are sold in medical forms (e.g., capsule, solution, or potion) and used in higher dosages than those obtained from a normal diet [1,3]. In Canada, nutraceuticals are sold in medical forms not associated with foods and referred as "natural health products" [4,5]. In MedicineNet, nutraceutical is defined as "a food or part of a food that allegedly provides medical or health benefits, including the prevention and treatment of disease. A nutraceutical may be a naturally nutrient-rich or medicinally active food, such as garlic or soybeans, or it may be a specific component of a food, such as the omega-3 fish oil that can be derived from salmon and other cold-water fish" [Shield Jr., W.C.; https://www.medicinenet.com/script/main/art.asp?articlekey = 9474].

Considering these definitions, the European Union stated that both functional foods and food supplements can be considered as nutraceuticals because they are obtained from natural sources [5]. Functional foods represent a market opportunity. Formulating foods for health is one of the top trends in the food industry. In addition to these new foods, functional foods can include traditional foods which have benefits for health that have been recognized [1].

9.1.2 Definition and types of structured lipids

In a broad sense, structured lipid (SL) is defined as tailormade fats and oils with improved functional, technological, and/or pharmaceutical properties. These modified lipids, which do not exist in nature, are obtained either chemically or enzymatically by changing the composition and/or positional distribution of fatty acids in the glycerol backbone of acylglycerols. The SL comprise restructured triacylglycerols (TAG) or phospholipids (SPL) and, more recently, partial acylglycerols (diacylglycerols, DAG, and monoacylglycerols, MAG) and phenolic lipids [6–8]. Over the past few years, several SLs have gained interest in the food industry, namely: (1) human milk fat substitutes (HMFS), (2) low-calorie TAG, (3) cocoa butter equivalents (CBE), (4) *trans*-free plastic fats, and (5) enriched TAG [6–14]. Some of their characteristics and functional properties are summarized in Table 9–1.

The increasing concern about healthy lifestyles has driven the food industry towards innovation, in order to meet consumer demands. Recently, oleogels (also called "physically structured oils") have gained interest as SLs, due to their versatility and cost-effective production [14–16]. Oleogels are gels in which the continuous phase is an oil structured by oleogelators (e.g., phytosterols, phospholipids, vegetable oils, and partial acylglycerols) to mimic solid-like properties. Oleogels may be produced to fulfill specific requirements, namely to incorporate specific fatty acids, to reduce saturated fat content, control cholesterol levels, or carry bioactive compounds to be absorbed in the gastrointestinal tract [16].

9.2 General aspects of lipids: definition, structure, and properties

The rationale of SL production is based on the chemical and physical properties of natural lipids and on the knowledge of how they are related to their nutritional and biological properties. Fats and oils may be attractive due to their appearance, aroma, flavor, and

Table 9–1 Main characteristics and functional properties of different types of structured lipids (see list of symbols at the end of the chapter).

Type of structured lipid	Composition	Functional properties
Human milk fat substitutes	1,3-Dioleoyl-2-palmitoylglycerol (OPO)	Increase calcium absorption; Prevent constipation
	TAG enriched with long-chain PUFA, mainly EPA, DHA, and ARA	Improve cognitive functions; Neurological system development; Immunological function; Visual acuity; Brain development
	TAG with high levels of medium-chain fatty acids (mostly C8:0, C10:0, and C12:0)	Increase lipid absorption; Rapid energy source
Low-calorie TAG	SLS—short-chain fatty acids at the external positions of glycerol backbone and long-chain fatty acid at the sn-2 positions	Prevent fat malabsorption; Control obesity
	MLM—medium-chain fatty acids (M) at the external positions and long-chain fatty acids (L) at the sn-2 positions	
	MMM type	
Cocoa butter equivalents	TAG with saturated fatty acids (C16:0 and C18:0), in the external positions, and a MUFA (C18:1) at position sn-2	Rheological and sensory properties of cocoa butter
$Trans$-free plastic fats	Low/zero $trans$ fatty acid contentFatty acid composition is maintained by interesterification	Prevent cardiovascular diseases; Improve cholesterol levels; Prevent type II diabetes
Structured phospholipids	Phospholipids with MCFA, n-3 PUFA (mainly DHA and EPA) and CLA at position sn-1 or sn2	Improve cognitive functions; Prevent cardiovascular diseases; Reduce inflammatory processes; Improve cholesterol levels
Partial acylglycerols	Monoacylglycerols and diacylglycerols obtained by lipase-catalyzed glycerolysis of TAG	Reduce obesity; Control total cholesterol, triacylglycerol, and glucose levels; Regulate appetite; Prevent fat malabsorption

texture, contributing to intensify and improve the organoleptic characteristics of the foods. The nutritional/physiological importance of lipids is based on their role as energetic molecules (37 kJ/g or 9 kcal/g triacylglycerol) and as a source of essential fatty acids, antioxidants, and vitamins. The physical, functional, nutritional, and organoleptic properties of a lipid are dependent, not only on its composition in fatty acids, but also on its positional distribution within the various TAG of oils and fats (sn-1,3 and sn-2) [17].

Lipids are a complex and chemically heterogeneous group of biomolecules, which have in common solubility in organic solvents, such as alcohols, ethers, and hydrocarbons, and high insolubility in water. Lipids can be divided into two main groups: the nonpolar lipids (acylglycerols, sterols, free fatty acids, hydrocarbons, alcohols, waxes, and steryl esters) and polar lipids (phosphoglycerides, glycosylglycerides, and sphingolipids) [18]. Lipids are also commonly distinguished based on their physical state at room temperature. They are called oils when they are liquid at room temperature (20°C), and fats when they are solid or pasty at 20°C [19,20].

Based on their composition, they can be classified as simple or complex lipids. Simple lipids contain carbon, hydrogen, and oxygen, yielding fatty acids and an alcohol upon saponification. Simple lipids can be divided into TAG and waxes. Animal and vegetable oils and fats contain more than 95% TAG. TAG are esters formed by a glycerol molecule to which three fatty acids are esterified at different positions. These positions are represented by a stereospecific numbering system (*sn*): *sn*-1 (first external position), *sn*-2 (middle position), and *sn*-3 (third external position) [21]. In addition, oils and fats also contain monoacylglycerols (MAG), diacylglycerols (DAG), and free fatty acids (FFA), in much smaller amounts. MAG and DAG are important compounds for their emulsifying capacity [20].

Complex lipids contain one or more additional elements, such as phosphorus, nitrogen, and sulfur, yielding fatty acids, alcohol, and other compounds by saponification. Complex lipids are classified as phospholipids or glycolipids. Phospholipids are composed of fatty acids and a phosphate group. Glycerol-based lipids called phosphoglycerides contain glycerol, two fatty acids, and a phosphate group. The phosphoglyceride structure contains a hydrophilic (polar) head, the phosphate unit, and two hydrophobic (nonpolar) fatty acid tails. The polar head can interact strongly with water, while the nonpolar tails interact strongly with organic solvents. These structures generate three or more different compounds when hydrolyzed [19]. Glycolipids differ from phospholipids since they possess a sugar group in place of the phosphate group. Their structure is again the polar head and dual tail arrangement in which the sugar is the hydrophilic unit.

Moreover, small amounts of nonacylglycerol components are found in all oils and fats. Some of these components (e.g., phosphatides, sterols, waxes, insoluble hydrocarbons, pigments, tocopherols, lactones, and methyl ketones) are partially or completely removed during the refining process to make oils and fats edible. The fatty acids in TAG differ in the length of the chain, generally from 4 to 24 carbon atoms, and in the number of double bonds, from 0 (saturated fatty acid) to 6 (1: monounsaturated; >1: polyunsaturated fatty acid, PUFA). Since the rotation around the double bond is severely restricted, two types of isomers appear: *cis* and *trans*. Normally in nature, fatty acids are in *cis* form, although they can also exist in *trans* form. Industrial processing operations, such as refining and hydrogenation, can cause the change of *cis* isomers to the *trans* form. *Trans* isomers are undesirable as their presence has been related to cardiovascular diseases by a mechanism that lowers the HDL-cholesterol [22]. One of the objectives of the food industry is to reduce this form to the lowest level in processed fats and oils.

Regarding the chain length, fatty acids can be classified as short-chain fatty acids, when they have between two to six carbon atoms, medium-chain fatty acids (MCFAs), when they have between eight and 12 carbon atoms, and long-chain fatty acids (LCFAs) if they have between 14 and 24 carbons in their constitution, regardless of the degree of unsaturation [6]. The most common fatty acids in oils and fats are palmitic (C16:0), palmitoleic (C16:1), stearic (C18:0), oleic (C18:1), and linoleic (C18:2) acids. All vegetable oils contain at least four major fatty acids, which theoretically can give rise to 40 different combinations in the three positions of the glycerol molecule, that is, 40 TAG with different chemical and physical properties [20]. However, in nature, fatty acid esterification to glycerol backbone in TAG does not occur at random. Vegetable oils and fats show a predominance of unsaturated fatty acids in the internal position of TAG, while saturated FA are mainly esterified at positions sn-1,3.

The physical properties, such as the melting point, the crystallization characteristics, the crystalline structure, and polymorphism of a fat, are very important parameters to predict and prevent undesirable changes during product processing and storage [23]. The melting point of a fat is the temperature at which, at atmospheric pressure, the melting of the lipid crystals begins [24]. It increases as the length of the fatty acid chain increases and is higher for saturated fatty acids than for unsaturated fatty acids with an equal number of carbon atoms [20].

TAG tend to crystallize in different ways, depending on the chemical composition of the fat phase and the cooling conditions that exist during crystallization. The distribution of FA in TAG will influence the physical properties of the fats [25]. In addition, it is also important to know its polymorphic behavior in relation to crystallization [23]. The polymorphic behavior is related to the way the crystals are organized, which may present greater or lesser stability. The three main forms that crystals can acquire are: α, β', and β, in increasing order of stability [26]. The polymorphic form α (crystals of the hexagonal system) has poorly ordered chains in space, is unstable, and has a low melting point. The α form appears in the initial phase of the crystallization process due to the sudden cooling of the fat, but quickly evolves into the β' and β forms. The β is the most stable form, with the highest melting point and latent heat. β crystals are of the triclinic system and due to their larger dimensions, they are responsible for a sandy and coarse texture of the crystallized fat. The β' form corresponds to orthorhombic twin-crystals responsible for a pleasant texture and mouthfeel of a fat [27].

9.3 The role of lipids in human nutrition

9.3.1 Functional/nutritional properties of lipids

Vegetable oils and fats, in adequate quantities and in a balanced diet, are essential for the human body. This is due to the need to consume some fatty acids that the organism does not produce in sufficient amounts for its needs. Omega-6 linoleic acid (C18:2 omega-6) and omega-3 linolenic acid (C18:3 omega-3) are considered essential fatty acids because they are not synthesized by the human body and they are absolutely necessary for human health. Thus, these FA must be included in the diet [9].

Animal organisms including humans can convert linoleic acid to arachidonic (C20:4 omega-6) and α-linolenic acid to longer chain omega-3 fatty acids such as eicosapentaenoic acid (EPA; C20:5 omega-3) and docosahexaenoic acid (DHA, C22:6 omega-3). EPA and DHA are omega-3 fatty acids found mainly in marine oils, with high benefits for human health. These fatty acids are important in the prevention and treatment of cardiovascular diseases, regulation of blood pressure, and the inflammation process [2]. DHA has proved to be very important in childhood development of the brain and nervous system and its normal functioning in adulthood [28,29].

Among the LCFAs, conjugated linoleic acid (CLA) and gamma-linolenic acid (GLA) also present important nutritional, physiological, and/or physical properties [9]. CLA, which is present in the milk and meat of ruminants, consists of a mixture of geometric and positional isomers of linoleic acid that contain conjugated double bonds in its carbon chain. Its intake has shown anticarcinogenic, antioxidant, antiatherosclerotic, antidiabetic, antiobesity, and improved immunity effects [30]. GLA is present in the oil seeds from plants such as *Borago officinalis* L. and *Oenothera biennis* L. It has recognized antitumoral effects and is a precursor to prostaglandin 1 and 15-hydroxy di-homo gamma linolenic acid, which have antiinflammatory and antithrombotic properties [31].

The importance shown by PUFA in human health has led to the establishment of recommended daily intake doses for these compounds [9]. Also, several studies have shown the importance of the omega-6 FA/omega-3 FA ratio in the diet. Western food is deficient in omega-3 FA and has an excess intake of omega-6 FA, considering the genetic pattern and sedentary lifestyle. This situation promotes an increase in cardiovascular diseases, cancers, autoimmune and antiinflammatory diseases, and obesity [32,33]. A MUFA-rich diet shows recognized health benefits namely antiatherogenic and antithrombotic potential, increases the HDL/LDL cholesterol ratio, decreases oxidized LDL and total cholesterol, and reduces platelet aggregation [34]. The replacement of saturated fats by oils rich in oleic acid has benefits in the prevention of cardiovascular diseases and improves insulin sensitivity [35]. Olive oil, containing 55%–83% oleic acid, has been shown to have anticancer effects [36] and to modulate the inflammatory response [37]. Monounsaturated oils can also be beneficial in terms of weight control strategy since they induce a reduction in postprandial triglyceridemic compared to the consumption of saturated fats [38].

Lipids also contribute to better absorption of fat-soluble vitamins (A, D, E, and K), and are essential for human growth and development. However, the excessive intake of lipids also presents negative aspects, causing obesity problems due to their high caloric value. In addition, there is evidence that high-fat diets may increase the risk of colon cancer and cardiovascular diseases [20]. Excessive intake of animal fats, consisting essentially of saturated fatty acids, may cause problems with cholesterol accumulation in the arteries and consequently coronary problems. The ingestion of fats of vegetable origin, rich in unsaturated fatty acids, and oils from vegetable or marine animal origin, rich in omega-3 PUFA, promotes a decrease in the level of cholesterol.

9.3.2 Absorption and lipid metabolism

The composition of fatty acids, their position in the acylglycerol backbone and their physical and biochemical properties result in different physiological properties, demonstrated in terms of absorption in the gastrointestinal tract and in the metabolic process [6]. These characteristics are directly related to coronary diseases, to energy balance, insulin sensitivity, and postprandial metabolism with a subsequent impact on human health [21]. The first step in the digestion of fats occurs in the stomach and is catalyzed by lingual or gastric lipase. The main digestion products of this gastric phase are diacylglycerols and free fatty acids, which facilitate the intestinal digestion phase [39]. In the duodenum region, the hydrolysis is catalyzed by the pancreatic lipase (*sn*-1,3 regioselective) generating free fatty acids, and *sn*-2 monoacylglycerols (*sn*-2 MAG), which form micelles with the bile salts [40]. When these micelles, also containing phospholipids, approach the apical side of the intestinal epithelial cells, they release their content and enable the absorption of nonpolar lipids in the microvilli membrane. The absorbed lipids are reesterified to reform TAG in the smooth endoplasmic reticulum [41]. The rate of hydrolysis reaction carried out by the pancreatic lipase depends on the chain length and degree of unsaturation of the FA present in the *sn*-1 and *sn*-3 positions [6].

In nature, vegetable oils and some animal fats present mono- and polyunsaturated fatty acids (MUFAs or PUFAs) mainly esterified at position *sn*-2 of TAG. This profile promotes the absorption of these FAs in the *sn*-2 position, some of them essential for humans, such as *sn*-2 MAG, by the mucosal membrane, in the form of micelles with bile salts, avoiding FA deficiency [42]. Short-chain fatty acids are more quickly absorbed in the stomach than other fatty acids, due to their high volatility, water-solubility, and low molecular weight. As such, they are suitable for controlling obesity due to their low caloric value: acetic acid (C2:0), 3.5 kcal/g; butyric acid (C4:0), 6.0 kcal/g; and caproic acid (C6:0), 7.5 kcal/g [43].

MCFAs can be absorbed in the stomach, after hydrolysis by the gastric lipase [44], and can also be solubilized in the aqueous phase of the intestinal content, where they weakly bind to albumin and are transported to the liver through the portal vein [45]. About 80%–100% of the MCFAs present in the entire portal flow is captured by the liver and the remaining portion goes through the bloodstream, becoming available to peripheral tissues. Unlike LCFAs, they are not significantly incorporated into lipoproteins (chylomicrons and VLDL—very low-density lipoproteins), allowing them to be absorbed directly into the bloodstream. The rate of absorption in the intestines of MCFAs is similar to that of glucose and these are quickly oxidized in the liver and used by the body. As they do not undergo significant reesterification in TAG, they do not accumulate in adipose tissue, which allows obesity control [46]. For these reasons, MCFAs have been used as a quick source of energy in some metabolic syndromes, such as pancreatic enzyme deficiency (cystic fibrosis) [47]. However, since MCFAs are saturated fatty acids, their ingestion promotes the increase of blood cholesterol and should not be exclusive in the diet [46].

The digestive bioavailability of medium-chain TAG is greater than that of long-chain TAG. The hydrolysis of medium-chain TAG starts in the stomach more quickly and its absorption is faster and more efficient [6].

The high hydrophobicity resulting from the long hydrocarbon chains prevents the absorption and direct transport of LCFAs. These fatty acids leave the intestine in the form of TAG, via the lymphatic route, after incorporation into chylomicrons (formed by TAG, phospholipids, cholesterol, and apoproteins) and are finally secreted into the bloodstream [48]. A fraction of these chylomicrons undergoes intravascular hydrolysis, releasing most of LCFAs to extrahepatic tissues, while the remaining fraction is transported to the liver. LCFAs reach this organ as fatty acids linked to albumin or in the form of TAG. All FAs use the two transport systems in varying proportions. The longer the carbon chain of FA, the more it is found in the lymph and less in the portal blood. In the lymph, LCFAs circulate as TAG associated with chylomicrons. In portal blood, FAs are linked to albumin [49].

9.4 Production of structured lipids

SLs can be produced either by chemical- or enzyme-catalyzed reactions, namely by acidolysis, interesterification (ester-interchange), transesterification (alcoholysis), or direct esterification [8]. Chemical-catalyzed reactions of oils and fats are currently carried out at high temperatures, under reduced pressure. Due to the lack of selectivity of chemical catalysts, the final products are contaminated by side-products with a subsequent decrease in product yield, the residual catalysts must be removed, and pollutant effluents are formed. Thus, products must be purified, increasing operation costs [50]. The replacement of chemical-catalyzed by enzyme-catalyzed processes, which are recognized as natural, is highly desirable. This is mainly due to the selectivity of biocatalysts and the mild reaction conditions used, resulting in highly pure products, higher yields, and environmentally friendly processes.

SLs production by enzyme-catalyzed processes has gained increased interest from the scientific community during the last 20 years. In fact, a search made in Scopus for "structured lipids and lipases" accounted for 517 documents up to the end of July 2021, showing the first publications in the late 1980s (accessed July 31, 2021).

9.4.1 Lipases and phospholipases

Lipases (acylglycerol acyl-hydrolases, EC 3.1.1.3.), phospholipases A1 (EC 3.1.1.32), and phospholipases A2 (EC 3.1.1.4), from microbial, animal, or plant origin, have been used to produce SLs [7,8,51]. Lipases and phospholipases are versatile biocatalysts that do not need cofactors and accept a wide variety of substrates. These enzymes are hydrolases that in aqueous media catalyze the hydrolysis of acylglycerols or phospholipids, respectively. Both lipases and phospholipases exhibit an interfacial activation kinetics [52–54]. It means that lipase and phospholipase activity is highly activated when the substrate concentration is higher than its critical micellar concentration and a lipid–water interface is formed (e.g., micelles, monomolecular or bimolecular layers of acylglycerols, or phospholipids).

These enzymes are also active in nonaqueous media at low water activity (a_w), where they can catalyze esterification, interesterification, alcoholysis, and acidolysis reactions, among others [55]. Particularly important is the selectivity exhibited by several lipases, either in aqueous or in organic media. They can be selective toward (1) the class of lipids (e.g., TAG, DAG, MAG, fatty acids methyl and ethyl esters); (2) the position of the fatty acids in acylglycerols (*sn*-1,3 regioselectivity; *sn*-2 regioselective lipases are not available from nature); (3) saturated, mono- or polyunsaturated fatty acids; (4) to an optical isomer; or (5) some combination of these situations [56]. The selectivity of each lipase can be affected by the water activity of the reaction medium and by the immobilization support used [57].

In the production of SLs, the use of an *sn*-1,3 regioselective lipase is of utmost importance to produce novel lipids with functional and/or nutraceutical properties that cannot be obtained by chemical catalysis. The *sn*-1,3 regioselective lipases will maintain the original fatty acids at the *sn*-2 position of TAG, since these lipases can only hydrolyze the *sn*-1,3 ester bonds. This is nutritionally desirable for faster absorption of the fatty acids at the internal position of the acylglycerols, in the form of *sn*-2 MAG. However, even using *sn*-1,3 regioselective lipases as biocatalyst, spontaneous acyl migration will occur, and undesirable TAG products will be obtained [7]. Thus, to minimize the extent of this undesirable side reaction, the following parameters must be controlled: reaction temperature, biocatalyst load, type of immobilization support, water content and water activity, solvent type, and reaction system [58]. When nonregioselective lipases are used, the obtained SLs are similar to those produced by chemical catalysis.

9.4.2 Biocatalyst immobilization

The commercial price of lipases and phospholipases is still higher than that of inorganic catalysts, which has been a constraint to the industrial scale-up of lipase-catalyzed processes in the food industry. The use of immobilized biocatalysts for SL production is especially important to decrease biocatalyst costs, by reusing it in successive batches or using it in continuous bioreactors. Cost reduction is particularly important to produce SLs such as interesterified fats with improved rheological properties or fats enriched in specific fatty acids, to be used as commodity fats. These SLs can only be industrially produced if the enzymatic processes are economically feasible.

The immobilized lipases must present both high specific activity and operational stability. In general, porous immobilization supports are preferred over nonporous supports since they allow the immobilization of higher amounts of enzyme molecules (at the surface and inside the porous matrix), intensifying the catalytic processes. Through the correct choice of the immobilization support, it is possible to create a microenvironment adequate to protect the enzyme and promote its activity and operational stability. Thus, the optimal support should foster the following conditions in the microenvironment: (1) high, but not inhibitory concentrations of substrates; (2) low product concentration to displace reaction equilibrium toward the synthesis; (3) low water activity, to favor the reactions of synthesis and avoid the reverse reaction of hydrolysis; and (4) protection of the biocatalyst against the toxicity effects of organic solvents and other inhibitory molecules (e.g., oxidation products, free fatty acids, pigments, phospholipids, and lipid polymers) [59,60].

However, during the immobilization process, stereochemical, and conformational modifications of the enzyme may occur, leading to a considerable decrease in the reaction rate. Also, internal diffusion and partition effects occur when porous supports are used. Mass transfer within solid particles, that is, the entrance of the substrates and the removal of the product away from the site of the reaction, only occur by molecular diffusion. Thus, the overall reaction rate can be significantly reduced if diffusion is slow, that is, under a diffusion-controlled reaction [61]. Partition effects between the bulk and the microenvironment of the enzyme can be estimated by applying the concept of partition coefficient between two immiscible liquid phases, K, adapted by Fukui et al. [62] to the system immobilization support/bulk medium, as follows:

$$K = \frac{C_{As}}{C_{Ab}} \qquad (9-1)$$

where C_{As} is the equilibrium concentration of component A in the support S (microenvironment) and C_{Ab} is the concentration of A in the bulk medium (macroenvironment), at equilibrium. The initial concentration of the component A in the reaction medium (C_{A0}) and its concentration in the reaction medium, C_{Ab}, after the addition of the support, the migration of A to the support until the concentration equilibrium is attained, can be easily quantified. Thus, considering the initial volume of the bulk medium, V_0, and the total volume of the system after addition of the support, V, the real volume of the immobilization support is $(V - V_0)$. Therefore, at equilibrium, K can be given by:

$$K = \left[\frac{C_{A0} - C_{Ab}}{C_{Ab}}\right] \times \left[\frac{V_0}{(V - V_0)}\right] \qquad (9-2)$$

If $K > 1$, the component A shows higher affinity for the support than for the bulk medium and tends to enter in it. When $K \ll 1$, the compound A has low affinity for the support and hardly enters in it. Thus, in an ideal biocatalytic system, substrates must have $K \gg 1$, while products must have $K \ll 1$.

Several organic and inorganic supports have been tested for lipase immobilization with quite encouraging results in terms of operational stability and retention of activity. Hydrophobic supports have been preferred for lipase immobilization because they promote the entrance of the hydrophobic substrates into the pores of the matrix. However, the composition of the microenvironment must be optimized for each reaction system to avoid high inhibitory substrate concentrations near the enzyme [63]. The interactions of the immobilization support with the enzyme and medium components are important aspects on the optimization of the microenvironment of the catalyst and, therefore, on the reaction rate and yield.

9.4.3 Reaction systems

The majority of SLs are obtained by lipase-catalyzed (1) acidolysis of a pure TAG, oil, or fat with FFA (omega-3 PUFA, MCFA, LCFA), (2) interesterification of blends of TAG, fats, and oils, or (3) interesterification of a TAG, fat, or oil with fatty acid ethyl or methyl esters [8]. Structured phospholipids (SPL) can be obtained by (1) acidolysis of phospholipids with omega-3 PUFA, MCFA, LCFA, or (2) interesterification of phospholipids with FA ethyl or methyl esters, using immobilized lipases or phospholipases A1 or A2 (PLA 1 and PLA2, respectively). PLA1 and PLA2 can act at sn-1 and sn-2 positions of phospholipids, respectively [51].

The use of ethyl esters over methyl esters as acyl donors is preferred due to the toxicity of methanol released from the reactions [64]. Also, ethyl esters, being more volatile than FFA, are more easily recovered from the reaction media by distillation, decreasing downstream processing costs. The mechanisms of lipase-catalyzed interesterification, acidolysis, and alcoholysis reactions consist of hydrolysis of ester bonds in acylglycerols followed by reesterification [65]. Therefore, the optimization of these reactions results from a balance between the rates of hydrolytic and esterification reactions, which are reversible reactions. In the presence of excess water (high a_w), the equilibrium is displaced toward the hydrolysis step. Conversely, when low water amounts are present, the global reaction is shifted toward reesterification, which is desirable to achieve higher product yield. The product yield under reaction equilibrium is decided by the substrate ratio [58,66].

The choice of the adequate lipase (or phospholipase) for the synthesis of a specific SL will also depend on the substrates used and on the reaction conditions. With oils as substrates, which are liquid at 20°C, the preferred system consists of a blend of substrates in appropriate amounts (solvent-free media) and a mesophilic lipase (or phospholipase) as biocatalyst. If a fat is used as substrate, two options are possible: to solubilize the fat in an organic solvent and use a lipase (or phospholipase) from a mesophilic organism (reaction in organic solvent medium); or to use higher reaction temperature to melt the substrates and use a thermophilic lipase (or phospholipase) as biocatalyst in a solvent-free medium.

The choice of an organic solvent to be used as a continuous phase of the reaction medium, must consider physicochemical (e.g., solubility of substrates/products, density, vapor pressure, melting and boiling points, viscosity, and superficial tension), biologic (toxicity for the biocatalyst), safety (toxicity for humans and flammability), logistic (availability, disposal of residues), and economic (cost) issues. Among them, the toxicity of the solvent for the biocatalyst is one of the most important parameters for the success of the biocatalysis.

The sensitivity to the nature of organic solvents varies with the lipase. Water is essential to the enzyme activity since it is responsible for the maintenance of native, catalytically active conformation of the enzyme, due to its role in the formation of hydrogen bonding and in van der Waals interactions. Zaks and Klibanov [67] demonstrated that the water monolayer bound to the enzyme corresponds to the essential water to maintain a lipase active in nearly anhydrous organic media. When a hydrophilic organic solvent is used, the water will be partitioned between the solvent and the enzyme, stripping the water monolayer around the enzyme, with subsequent loss of its catalytic activity. On the contrary, hydrophobic solvents are immiscible with water and will not strip the essential water from the enzymes. Thus, in enzyme-catalyzed processes in organic solvent, the interaction of the solvent with the enzyme-bound water is the main factor to consider for the success of the reaction [67].

Laane et al. [68] proposed the use of the Hansch parameter (log P) to assess solvent biocompatibility. Log P is related with solvent hydrophobicity. It is defined as the logarithm of the partition coefficient of the solvent in a standard biphasic system *n*-octanol/water. A sigmoid-shaped relationship between enzyme activity and log P value of the solvent has been observed. As a rule, biocompatible solvents have a log P higher than 4 (hydrophobic solvents); solvents with log P lower than 2 (hydrophilic solvents) will cause a strong inactivation of the biocatalyst, while the effect of solvents with a log P between 2 and 4 depends on the biocatalyst [68].

In the food industry, only food-grade organic solvents can be used. In this context, *n*-hexane is currently used in lipase-catalyzed reaction systems, despite presenting a log P value of 3.5. However, solvent-free systems are recommended for the food industry, due to solvent toxicity and negative effects on the environment and on enzyme activity and operational stability. In addition, the complexity of the system with organic solvents increases, as well as costs related with solvent acquisition and recovery and product purification.

9.4.4 Case studies on structured lipids production

In this section, several examples of SL production are presented: those where the original fatty acid composition is modified by incorporation of new fatty acids (e.g., dietetic low-calorie SL, HMFSs, and TAG enriched in specific fatty acids with health benefits) and fat blends where the original position of fatty acids is changed to reach specific rheological/functional properties.

9.4.4.1 Low-calorie triacylglycerols

Low-calorie TAG provide health and nutrition benefits by controlling fat malabsorption, obesity, and other metabolic disorders, due to their lower caloric density (~ 5 kcal/g) in comparison with conventional fats and oils (9 kcal/g). These SLs contain short- (S) or medium- (M) chain fatty acids, preferably esterified at external *sn*-1,3 positions, and long- (L) chain fatty acids esterified at the *sn*-2 position of the glycerol backbone [12,14]. Over the past few years,

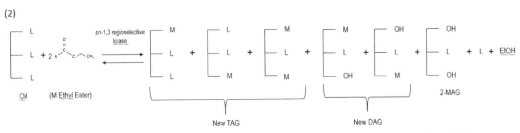

FIGURE 9–1 Schematic (1) acidolysis of a long-chain TAG (LLL) with medium-chain (M) fatty acids and (2) interesterification of a long-chain TAG (LLL) with medium-chain fatty acid ethyl ester, catalyzed by a sn-1,3 regioselective lipase (the L at position sn-2 is maintained) to produce low-calorie TAG (New TAG).

special interest has been devoted to MLM-type structured TAG. During digestion, medium-chain FA esterified at the sn-1,3 positions are released from TAG, by the action of pancreatic lipase, and transported to the liver where they are rapidly metabolized, providing a rapid source of energy [12,69]. The production of MLM SLs is usually carried out by (1) the acidolysis reaction between a TAG source rich in LCFAs (such as olive, olive pomace, grapeseed, avocado, soybean, and nut oils) and MCFAs, namely caprylic (C8:0) or capric (C10:0) acids or (2) interesterification of an oil with MCFA ethyl or methyl esters [12,69–74] (Fig. 9–1). Two-step reactions have also been implemented aiming at increasing structured TAG yield. A possible reaction scheme combines an alcoholysis, to get 2-MAG, followed by acidolysis between 2-MAG and MCFAs. Another approach uses esterification between glycerol and PUFA and the reaction product undergoes acidolysis with MCFAs to obtain MLM-type TAGs [12].

Since medium-chain FAs have melting points lower than 35°C, reactions can be carried out at near-room temperature, using mesophilic sn-1,3 regioselective lipases, in the absence of a solvent. Recent examples of low-calorie TAG produced either by acidolysis and interesterification reactions are presented in Table 9–2. Several SLs with low-calorie values have been commercialized, namely SALATRIM (Nabisco, United States), Olestra and Caprenin (both produced by Procter and Gambler, United States) [69].

9.4.4.2 Human milk fat substitutes

Human milk fat (HMF) is one of the main sources of nutrients and energy for the newborn, comprising TAG (98%–99%), phospholipids (0.26%–0.80%), sterols (0.25%–0.34%, mainly cholesterol), and trace amounts of partial acylglycerols (DAG and MAG) and free fatty acids

Table 9–2 Examples of MLM-type structured lipid production by batch acidolysis or interesterification (see list of symbols at the end of the chapter).

Reaction	Biocatalyst	Substrates	References
Acidolysis	ROL immobilized in MNP	Crude oils from spent coffee grounds and olive pomace + C10:0	[74]
	Lipozyme TL IM		[74]
	Novozym 435,	ARA-rich single cell oil from *Mortierella alpine* + capric acid	[73]
	Lipozyme 435		[73]
	Lipozyme TL IM		[73]
	Lipozyme RM IM		[73]
	NS 40086		[73]
	Lipozyme TL IM	Grapeseed oil + C8:0	[72]
	Lipozyme RM IM	Grapeseed oil + C10:0	[72]
	Novozym 435		[72]
	rROL in Amberlite IRA 96	Grapeseed oil + C8:0	[71]
	CPL self-immobilized in papaya latex	Grapeseed oil + C10:0	[71]
Interesterification	ROL immobilized in MNP	Crude spent coffee grounds oil + C10:0 ethyl ester	[74]
	Lipozyme TL IM	Crude olive pomace oil + C10:0 ethyl ester	[74]
	Lipozyme RM IM	Fish oil + C10:0 methyl ester	[75]
		Fish oil + medium-chain TAG	[75]

[76]. The FA composition of HMF is highly dependent on the mother's diet, on the time and frequency of breastfeeding, as well as on psychological and environmental factors that may affect breastfeeding practices. A total of 54 different fatty acids have been reported for HMF composition, with the following being the most abundant: palmitic (C16:0, 18%–25%), oleic (C18:1; 24%–39%), linoleic (C18:2 *n*-6; 8%–18%), linolenic (C18:3 *n*-3; 0.4%–2%), lauric (C12:0; 4%–14%), myristic (C14:0; 3%–12%), stearic (C18:0; 5%–8%), and capric (C10:0; 1, 5%–2.5%). Long-chain PUFA, such as arachidonic acid (C20:4 omega-6) and DHA (C22:6 omega-3), account for 0.1%–0.5% of total FA in mature milk [77].

The fatty acid distribution in HMF has a unique structure, characterized by the predominance of TAG containing unsaturated FAs in the *sn*-1 and *sn*-3 positions (c. 80%–90% of oleic acid) and with saturated FAs (about 60%–70% of palmitic acid) in the internal position, conversely to that observed in other natural fats. This structure is crucial for the efficient absorption of palmitic acid as monoacylpalmitate [78].

In infant formula, produced with vegetable oils or ruminant milk, palmitic acid and the remaining saturated FAs are predominantly esterified at the *sn*-1 and *sn*-3 positions of the TAG molecules. Saturated fatty acids, released by the action of the *sn*-1,3 regioselective lipase, may form insoluble calcium complexes, known as calcium soaps, which contribute to the poor absorption of calcium and FAs by children fed with infant formulas. This is also one of the causes of constipation in the first months of life. It is worth noting that, unlike free palmitic acid, monoacylglycerol *sn*-2-monopalmitate is efficiently absorbed [79–82].

The production of HMFS, to incorporate in infant formula, mimics the rather unique structure of HMF and has been extensively investigated over recent years [11,13]. The synthesis of HMFS may be attained by lipase-catalyzed acidolysis or interesterification reactions, performed batchwise or in continuous mode, either in solvent or solvent-free media (Fig. 9–2).

Both reaction systems require a TAG source rich in palmitic acid at sn-2 position, and sn-1,3 regioselective lipases as biocatalysts. Tripalmitin, palm stearin, palm oil, fractionated palm stearin, butterfat, and lard have been used as suitable substrates. Lard is a fat of low commercial value with a TAG structure very similar to that of human milk. However, it has a lower content of long-chain PUFA, namely linoleic and linolenic acids [83]. Nevertheless, ethical, religious, and cultural issues can be an obstacle to its use in the production of HMFS. In laboratory studies, tripalmitin is often used, but its implementation on a large scale is compromised owing to its high cost. Due to the high melting point of tripalmitin (66°C–68°C), and of the fats rich in palmitic acid at position sn-2, the production of HMFS in solvent-free media has to be performed at temperatures higher than the melting point of the reaction medium, using thermophilic sn-1,3 regioselective lipases. To produce HMFS at near room temperature, catalyzed by mesophilic lipases, the substrates must be solubilized in a biocompatible hydrophobic solvent (e.g., n-hexane).

In the acidolysis reaction, several substrates have been used as sources of FAs, namely vegetable oils (such as olive, sunflower, and hazelnut oils), MUFA, or omega-3 PUFA [83–86]. Blends of vegetable oils rich in long-chain FAs or methyl or ethyl esters (namely ethyl oleate) have been used as acyl donors in the interesterification reaction for HMFS production [66,87,88]. In order to obtain higher yields of structured TAG, two-step reactions have been carried out, using sn-1,3 regioselective lipases as catalyst. This multistep process comprises an alcoholysis to obtain

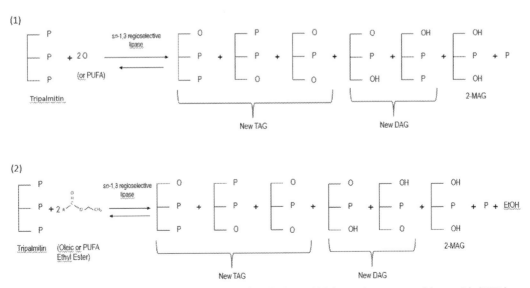

FIGURE 9-2 Schematic (1) acidolysis of tripalmitin (PPP) with oleic acid (O) or polyunsaturated fatty acids (PUFA) and (2) interesterification reaction of tripalmitin (PPP) with ethyl oleate or PUFA ethyl esters, catalyzed by a sn-1,3 regioselective lipase (the L at position sn-2 is maintained) to produce HMFS (new TAG).

Table 9-3 Examples of HMFS production by batch acidolysis or interesterification (see list of symbols at the end of the chapter).

Reaction	Biocatalyst	Substrates	References
Acidolysis	Lipozyme RM IM	Fungal oil (from *Mortierella alpina* ALK) + fractionated palm stearin + C18:1	[90]
	Novozym 435; Lipozyme 435; Lipozyme TL IM; Lipozyme RM IM	Microalgae oil from *Nannochloropsis oculata* rich in C16:0 at *sn*-2 position + FFA enriched in C18:2, SDA and DHA from the microalga *Isochrysis galbana*	[91]
	rROL; Lipozyme RM IM	PPP + FFA from camelina oil	[86]
	rROL in Accurel MP 1000; Novozym 435; Lipozyme TL IM; Lipozyme RM IM	Lard + omega-3 PUFA	[83]
Interesterification	Lipozyme RM IM, Lipozyme TL IM; NS 40086, DF Amano 15	Basa catfish oil (*Pangasisus haniltoa*) + coconut fat	[92]
	CpLip2 immobilized on Accurel MP 1000	PPP + ethyl oleate	[66]
	Lipozyme TL IM	Fractionated palm stearin + fish oil	[93]

sn-2-MAG rich in palmitic acid, followed by an esterification reaction between the purified *sn*-2-MAG and FFA [89]. Recent research on HMFS production is summarized in Table 9–3.

Nowadays, there are several HMFS produced by lipase-catalyzed reactions in the market. Betapol was the first product to reach the market, to mimic human milk fat for infant formula, developed by Bunge Loders Croklaan, using lipase-catalyzed processes (http://europe.bungeloders.com/applications/infant-nutrition/). More recently, other products such as Infat (Advanced Lipids, Sweden, Karlshamn), Alsoy (Nestlé), Cow & Gate Premium (Nutricia), and Bonamil (Wyeth Ayerst) are also commercialized [13].

9.4.4.3 Interesterified trans-free fat blends and triacylglycerols rich in specific long-chain fatty acids

The interesterification of fats and oils consists in the exchange of fatty acid moieties between the different TAG of oil or fat blends. This reaction is currently performed to obtain SLs with improved physical and functional properties (e.g., melting point, solid fat content, and crystallization pattern of fat blends) which is important for margarines and edible shortenings production [23]. In fact, the interesterified fat blends, obtained by chemical catalysis and currently used in the margarine industry, are SLs. In terms of physical properties, these interesterified fats may have a consistency ranging from creamy to hard and the texture must be smooth and not grainy.

By interesterification, ester-interchange occurs among TAG, without changing their original fatty acid composition (e.g., essential FAs and PUFA contents). Also, no *trans* fatty acids are produced, in contrast to what happens when the physical properties of fat blends are modified by hydrogenation [50]. The hydrogenation of vegetable oils destroys the unsaturated FAs and causes isomerization of nonhydrogenated *cis*-double bonds of unsaturated vegetable oil into their *trans* form with recognized negative impact on human health (c.f. 9.3.). However, when chemical catalysts are used (e.g., sodium, sodium methoxide), ester interchange occurs at random and does not respect the original structure of natural TAG.

Enzyme-catalyzed interesterification has emerged in recent years as a possible alternative to chemical interesterification, due to the benefits related to the use of biocatalysts (c.f. 9.4.1). When a nonregioselective lipase is used, the ester interchange occurs at random and the obtained product is similar to that obtained by chemical interesterification. When an *sn*-1,3 regioselective lipase is used, the *sn*-2 FAs are preserved, which has absorption benefits, in addition to the advantages related to the use of enzymatic processes. Fig. 9–3 shows the schematic interesterification catalyzed by these two types of lipases.

A wide variety of raw materials has been used for the production of these interesterified fat blends with the desirable plastic properties: low-cost vegetable fats rich in saturated fatty acids (e.g., palm fat and palm stearin), fats rich in lauric acid (C12:0) (e.g., palm kernel and coconut fats), and polyunsaturated vegetable oils (e.g., sunflower and soybean oils). In solvent-free systems, when the reaction media contain considerable amounts of fats with high melting point, the reaction temperature must be high enough to melt the fat blends and thermostable lipases must be used. Otherwise, an organic solvent must be used in the reaction medium.

The incorporation of omega-3 PUFA, especially EPA and DHA, essential FA, or other PUFA with recognized benefits in human health, in food products more readily available for consumption than marine fish oil, such as vegetable oils, dressings, and margarines, may be an interesting option for the human diet [6,32]. Thus, the use of vegetable and/or fish oils rich in PUFA as raw materials, will increase the functionality of the obtained interesterified fat blends. Therefore, SLs can be designed with different goals and approaches, namely nutrition, medical, nutraceutical use, or for functionality in foods.

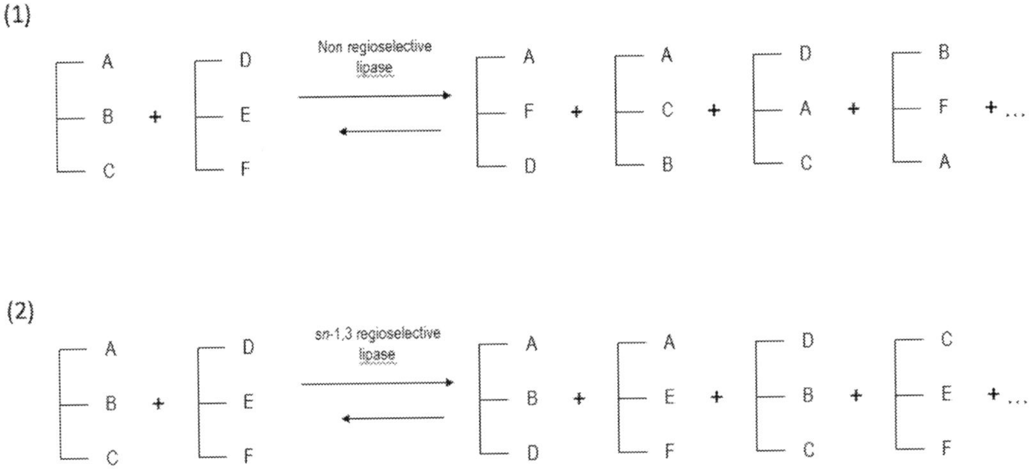

FIGURE 9–3 Schematic interesterification (ester interchange) between TAG with different fatty acid residues (A, ...F) catalyzed by (1) a nonregioselective lipase (the reesterification occurs at random) or (2) by a *sn*-1,3 regioselective lipase (the FAs at position *sn*-2 are maintained).

The supply of essential FAs and/or omega-3 PUFA can also be performed by replacing several FAs of natural oils and fats by these specific FAs. These novel TAG can be produced either by acidolysis with FFA or interesterification with those FA ethyl (methyl) esters, as reported for the production of HMFS or MLM dietetic TAG. Several studies have been carried in batch or continuous operation mode for the production of (1) low-*trans* or *trans*-free plastic fats by interesterification or (2) TAG rich in specific LCFAs, using commercial immobilized lipases such as Novozym 435 (*Candida antarctica* lipase) [94,95], Lipozyme RM IM (*Rhizomucor miehei* lipase) [31,96–98], and Lipozyme TL IM (*Thermomyces lanuginosus* lipase) [98–102]. In most of these examples, reaction equilibrium was attained in less than 3 h, which is comparable to the time needed for chemical interesterification.

However, these immobilized lipases are rather expensive when used to produce these commodity fats, which has been a constraint for industrial process implementation. Other lipases, such as the noncommercial *Candida parapsilosis* lipase/acyltransferase [103,104] and *Rhizopus oryzae* lipase immobilized in a hybrid silica-organic support [95] have been tested in solvent-free media, as alternatives to the high-cost commercial biocatalysts (Table 9–4).

There are already some examples of commercialized products obtained through enzymatic interesterification that meet new specific needs of the industry with different applications (e.g., production of margarines, shortenings, bakery, and pastry products). Some examples of these

Table 9–4 Examples of enzymatic synthesis of TAG rich in specific long-chain fatty acids and interesterified TAG blends, in solvent-free media, in batch or continuous bioreactors (see list of symbols at the end of the chapter).

Operation mode	Biocatalyst	Substrates	Product type	References
Batch	Lipozyme RM IM	Borage oil + Blend (50:50, wt.%) of PK and palm olein (1:9, substrate molar ratio)	TAG rich in GLA and C18:2	[31]
	Lipozyme TL IM	PS + PK + TAG rich in omega-3 PUFA (55:35:10, wt.%) and (45:45:10, wt.%).	SL rich in omega 3 PUFA for *trans*-free table margarines	[99]
	CPLip2	PS + PK + TAG rich in omega-3 PUFA (45:45:10, wt.%).	SL rich in omega 3 PUFA for *trans*-free table margarines	[103]
	Lipozyme TL IM	High-stearate soybean oil + stearidonic acid soybean oil (2:1, molar ratio)	SL rich in stearidonic acid for *trans*-free table margarines	[100]
	Lipozyme RM IM	Glycerol + sardine oil FAs	Omega-3 FA-rich oil	[96]
Continuous	Novozym 435	PS + soybean oil (55:45, wt.%)	SL for *trans*-free margarines	[94]
	Lipozyme TL IM	PS + PK + SO(55:25:20, wt.%);PS + PK + TAG rich in omega-3 PUFA(55:35:10, wt.%)	SL for *trans*-free margarines	[101]
	CpLip2	PS + PK + TAG rich in omega-3 PUFA (45:45:10, wt.%)	SL rich in omega 3 PUFA for *trans*-free table margarines	[104]
	Lipozyme RM IM	Rice bran oil + C8:0 (1:6, molar ratio)	Rice bran oil SL for *trans*-free plastic shortenings	[97]
	Lipozyme TL IM	Mustard oil + fish oil (2:1, molar ratio)	PUFA-rich mustard oil	[102]
	Lipozyme TL IM; Lipozyme RM IM	PS + PK oil + OO(45:30:25, wt.%)	*Trans*-free table margarines	[98]
	Novozym 435; ROL in organic-inorganic hybrid support	Milk fat + SBO (65:35, wt.%)	SL for *trans*-free margarines	[95]

commercial products include products from the Crokvitol line (Crokvitol Stand, Crokvitol Allround, Crokvitol Vitality) from Bunge Loders Croklaan (http://europe.bungeloders.com/applications/spreads/) and ADM's products (https://www.adm.com/products-services/food/oils).

In addition, it is possible to find on the market oils enriched in PUFA produced by biocatalysis. An example is Marinol D-40 from Stepan Lipid Nutrition (https://www.stepan.com/products-markets/product/MARINOLD40.html), which is a concentrate of natural fish oil with a high content of docosahexaenoic acid in glyceride form. This product can be used via tablet or capsule supplements, or incorporated in several types of foods (e.g., fruit sticks for children or convenient breakfast drinks for the elderly; fruit beverages and dairy drinks; margarines and dairy spreads; frozen desserts; fish cakes, sticks and fillets; soups, pastas, and pasta sauces).

9.4.5 Operational stability in batch and continuous bioreactors

Lipase-catalyzed processes are in general simple and require a lower investment than chemically catalyzed processes, since a lower number and less complex unit operations are needed. The implementation of an enzymatic process aimed at producing SL needs to be economically feasible. SL can be produced batchwise and the biocatalysts reused in successive batches or used in continuous bioreactors (e.g., packed-bed or fluidized-bed bioreactors), to reduce biocatalyst-related costs. Thus, low-cost biocatalysts presenting both high activity and operational stability must be used.

When compared to other enzymes, lipases are highly stable even under adverse conditions such as organic solvents and high temperatures. Nevertheless, like any enzymatic process, even in nonconventional media, the use of lipases is limited to a relatively narrow range of temperature, and inhibition by high concentrations of substrates and/or products, inactivation by heavy metals, lipid oxidation products, free fatty acids, and solvents, may occur. In fact, these major constraints will result in lower conversions and, therefore, in longer reaction times with low volumetric productivities, when compared with the chemical route. This may be a dramatic situation when inhibition effects and thermal deactivation occur simultaneously.

Several deactivation models have been fitted to biocatalysts during SLs production, either in consecutive batches or in continuous bioreactors. The most used are first-order deactivation model, and the series-type deactivation kinetics model proposed by Sadana [105] described by a parabolic profile. In some cases, the initial activity is maintained for a certain period, represented by a plateau, decreasing thereafter following a first-order deactivation model. This is known as a time-delay inactivation model [106] (Fig. 9−4).

The respective model equations and equations to estimate half-life times are given by Eqs. (9−3)−(9−6), respectively. The half-life time ($t_{1/2}$) of the biocatalyst corresponds to the operation time required for half the biocatalyst activity to be lost by deactivation.

First-order deactivation kinetics model:

$$A_t = Ae^{-k_d t} \qquad (9-3)$$

where A_t is the biocatalyst residual activity (%) at time t, A is a constant representing the initial activity of the enzyme before deactivation, and k_d is the deactivation rate constant, expressed in t^{-1}.

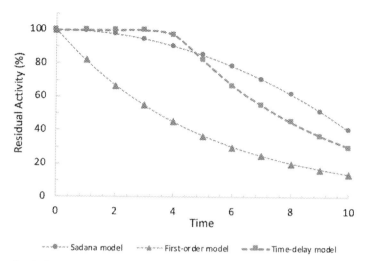

FIGURE 9–4 First-order, Sadana series-type, and time-delay first-order deactivation models.

When the biocatalyst follows a first-order deactivation profile, the half-life time is given by Eq. (9–4):

$$t_{1/2} = \frac{ln2}{k_d} \qquad (9-4)$$

Series-type deactivation kinetics model of Sadana:

$$A_t = 100 - 50 \times k_d t^2 \qquad (9-5)$$

The half-life time of a biocatalyst following a Sadana deactivation kinetics is given by the following equation:

$$t_{1/2} = k_d^{-1/2} \qquad (9-6)$$

If the biocatalyst is reused in successive batches, in the equation models, the time t is replaced by the number of batch, n.

Tables 9–5 to 9–7 show some examples of the operational stability tests of commercial and noncommercial immobilized biocatalyst used in batch or continuous production of SLs in solvent-free media. In the production of low-calorie TAG from olive oil, for the same biocatalyst, the use of caprylic (C8:0) or capric acid (C10:0) leads to a different stability and/or deactivation profile (Table 9–5). When commercial immobilized lipases were tested (Lipozyme RM IM, Lipozyme TL IM, and Novozym 435, from Novozymes), half-lives varied from 47 to 299 hours. With Lipozyme RM IM, in the presence of olive oil and capric acid, no deactivation was observed during 10 reuses of 23 hours each [107]. As alternatives to the expensive commercial immobilized lipases, several noncommercial lipases were immobilized and used in consecutive batches. When the noncommercial recombinant lipase from *Rhizopus oryzae* (rROL) was immobilized in different synthetic

Table 9-5 Batch operational stability of some biocatalysts used in the production of low-calorie SL (MLM) in solvent-free media (n.d., not determined; see list of symbols at the end of the chapter).

Reaction	Biocatalyst	Substrates	Deactivation model	Half-life time (h)	References
Acidolysis	Lipozyme RM IM	Olive oil + C8:0	First-order	299	[107]
	Lipozyme RM IM	Olive oil + C10:0	No deactivation	n.d.	[107]
	Lipozyme TL IM	Olive oil + C8:0	First-order	50.4	[107]
	Lipozyme TL IM	Olive oil + C10:0	First-order	47	[107]
	Novozym 435	Olive oil + C8:0	First-order	217	[107]
	Novozym 435	Olive oil + C10:0	First-order	225	[107]
	rROL in Eupergit C	Olive oil + C8:0	Time-delay	159	[108]
	rROL in Eupergit C	Olive oil + C10:0	Time-delay	136	[108]
	rROL in Eupergit C + rehydration	Olive oil + C8:0	First-order	39	[109]
	rROL in Eupergit C + rehydration	Olive oil + C10:0	First-order	54	[109]
	rROL in Lewatit VP OC 1600	Olive oil + C10:0	First-order	49	[109]
	rROL in Lewatit VP OC 1600 + rehydration	Olive oil + C10:0	Sadana series-type	234	[109]
	rROL in Amberlite IRA 96	Grapeseed oil + C8:0	First-order	166	[71]
	rROL in Amberlite IRA 96	Grapeseed oil + C10:0	First-order	118	[71]
	CPL	Grapeseed oil + C8:0	First-order	96	[71]
	CPL	Grapeseed oil + C10:0	First-order	81	[71]
	ROL in MNP	Crude olive pomace oil + C10:0	Sadana series-type	163	[74]
Interesterification	ROL in MNP	Crude olive pomace oil + C10:0 ethyl ester	Sadana series-type	220	[74]

Table 9-6 Batch operational stability of some biocatalysts used in the production of HMFS by acidolysis in solvent-free media (see list of symbols at the end of the chapter).

Biocatalyst	Substrates	Deactivation model	Half-life time (h)	References
Lipozyme RM IM	PPP + C18:1	No deactivation	n.d.	[110]
Lipozyme RM IM	PPP + omega-3 PUFA	Linear	276	[110]
Lipozyme TL IM	PPP + C18:1	Linear	154	[110]
Novozym 435	PPP + C18:1	Sadana series-type	253	[110]
Novozym 435	PPP + omega-3 PUFA	Sadana series-type	322	[110]
CPLip2	PPP + C18:1	First-order	35	[110]
CPLip2	PPP + omega-3 PUFA	Sadana series-type	127	[110]
Carica papaya Lipase	PPP + C18:1	Sadana series-type	80	[84]
rROL in Lewatit Accurel MP 1000	PPP + C18:1	Linear	35	[85]
rROL in Lewatit VP OC 1600	PPP + C18:1	Linear	64	[85]
rROL in Lewatit VP OC 1600 + rehydration	PPP + C18:1	Sadana series-type	202	[85]
rROL in Accurel MP 1000	Lard + omega-3 PUFA	First-order	112	[83]

supports and reused in batch, for the synthesis of MLM from olive oil or grapeseed oil, different deactivation profiles were observed and the half-life times varied from 39 to 234 hours [71,108,109]. Also, the low-cost self-immobilized *Carica papaya* lipase in the latex and the commercial lipase from *Rhizopus oryzae* immobilized in ferromagnetic nanoparticles showed high

Table 9–7 Operational stability of some biocatalysts used in the production of triacylglycerols modified in the original position of the fatty acids by interesterification in solvent-free media (see list of symbols at the end of the chapter).

Operation mode/ bioreactor type	Biocatalyst	Substrates	Deactivation model	Half-life time	References
Continuous fluidized-bed reactor	Novozym 435 (fresh and reused)	PS + SBO (55:45, wt.%).	Sadana series-type	17 days	[94]
Continuous packed-bed reactor	Lipozyme TL IM	PS + PK + SO (55: 25: 20, wt.%)	First-order	135 h	[101]
Continuous packed-bed reactor	Lipozyme TL IM	PS + PK + TAG rich in omega-3 PUFA (55:35:10, wt.%)	First-order	77 h	[101]
Continuous fluidized-bed reactor	CPLip2	PS + PK + "EPAX 4510 TG" (45:45:10, wt.%)	First-order	9 h	[104]
Batch	CPLip2	PS (45 wt.%) + PK (45 wt.%) + "EPAX 4510TG" (10 wt.%)	First-order	10 h18 h (water addition between batches)	[104]
Continuous packed-bed reactor	Lipozyme RM IM	PS + PK + OO (45:30:25, wt.%)	First-order	60 h	[98]
Continuous packed-bed reactor	Lipozyme TL IM	PS + PK + OO (45:30:25, wt.%)	First-order	88 h	[98]
Continuous fluidized-bed reactor	Novozym 435	Milk fat + SBO	No deactivation during 168 h operation	n.d.	[95]
Continuous fluidized-bed reactor	ROL in organic–inorganic hybrid support	Milk fat + SBO	Sadana series-type	190 h	[95]

batch operational stability in acidolysis of grapeseed oil [71] and acidolysis or interesterification of crude olive pomace oil [74], respectively.

In the production of HMFS in solvent-free media (Table 9–6), the same behavior was observed: in the system tripalmitin/oleic acid, no deactivation was observed for Lipozyme RM IM, in 10 consecutive reuses of 23 hours each. However, when oleic acid was replaced by a concentrate of omega-3 PUFA, a considerable change in the deactivation profile of Lipozyme RM IM was observed [110]. Also, rROL immobilized in different supports showed different inactivation profiles and stability in the acidolysis of tripalmitin with oleic acid or of lard with omega-3 PUFA, with half-lives from 35 to 202 hours [83,85].

Other examples of SL are the interesterified fat blends enriched in specific fatty acids or with improved functional properties for the food or pharmaceutical industries. Commercial immobilized lipases and noncommercial immobilized lipases were used as catalysts for the interesterification of different fat blends in the absence of solvent, either in consecutive batches or in continuous bioreactors (Table 9–7). Again, the operational stability and inactivation profile varied with the biocatalyst and, for the same biocatalyst, varied with reaction medium composition and operation conditions. The highest stability was observed for Novozym 435 with a half-life of 17 days in the interesterification of palm stearin with soybean oil in a continuous fluidized-bed reactor [94]. The same biocatalyst maintained its activity along 168 hours continuous operation in a fluidized-bed reactor for the interesterification of milk fat with soybean oil [95]. Using other commercial immobilized biocatalysts, half-life times varied from 60 to 135 hours. As alternative to these biocatalysts, *Rhizopus oryzae* lipase immobilized in an organic–inorganic hybrid support, showed excellent results in terms of operational stability with a half-life of 190 hours.

The results obtained with several noncommercial immobilized lipases, concerning their operational stability, are rather promising toward cost reduction and industrial implementation of enzymatic processes for SL production.

9.5 Conclusions and perspectives

Over the past few years, SLs have been recognized for their technological, functional, and health benefits. The use of lipases and phospholipases as catalysts for SL production offers several advantages in comparison with the chemical route, namely enzyme selectivity and mild operational conditions used. The *sn*-1,3 regioselectivity exhibited by some lipases and phospholipases allows for the production of SLs, which are impossible to obtain by chemical catalysis (e.g., low-calorie TAG, HMFS, and oils enriched in specific fatty acids in the positions *sn*-1,3). Nevertheless, the high cost of commercial enzymes remains a main constraint for industrial production of commodity fats (e.g., margarines and shortenings). The implementation of sustainable processes for SL production has been a challenge for the food industry. Promising results have been achieved in terms of yield and productivity of SLs, in solvent-free media, either in batch or continuous reactions. Also, high operational stability was observed with some biocatalysts in different reaction systems.

Several SLs obtained by lipase-catalyzed reactions are already in the market to be used incorporated in foods or sold as supplements. It is of utmost importance to search for novel low-cost enzymes with both high catalytic and operational stability, as well as to promote the use of cheap raw materials in the synthesis of SLs.

List of symbols

a_w	water activity
ARA	arachidonic acid (C20:4 omega-6)
CLA	conjugated linoleic acid
CPL	*Carica papaya* lipase
CpLip2	*Candida parapsilosis* lipase/acyltransferase
DAG	diacylglycerol(s)
DHA	docosahexaenoic acid (C22:6 omega-3)
EPA	eicosapentaenoic acid (C20:5 omega-3)
EPAX 4510 TG	TAG concentrate rich in EPA and DHA
FA	fatty acid(s)
FFA	free fatty acid(s)
GLA	gamma-linolenic acid
HMF	human milk fat
HMFS	human milk fat substitute
L	long-chain fatty acid(s)
LCFA	long-chain fatty acid(s)
Lipozyme RM IM	commercial immobilized lipase from *Rhizomucor miehei*

Lipozyme TL IM	commercial immobilized lipase from *Thermomyces lanuginosus*
M	medium-chain fatty acid(s)
MAG	monoacylglycerol(s)
MCFA	medium-chain fatty acid(s)
MLM	TAG containing medium-chain FAs at positions *sn*-1,3 and a long-chain FA at position *sn*-2
MMM	medium-chain TAG(s)
MNP	magnetic nanoparticles
MUFA	monounsaturated fatty acid(s)
Novozym 435	commercial immobilized lipases from *Candida antarctica*
O	oleic acid
OO	olive oil
PK	palm kernel fat
PL	phospholipid(s)
PO	palm oil
PPP	tripalmitin
PS	palm stearin
PUFA	polyunsaturated fatty acid(s)
ROL	commercial *Rhizopus oryzae* lipase
rROL	noncommercial recombinant lipase from *Rhizopus oryzae*
SBO	soybean oil
SDA	stearidonic acid (C18:4 omega-3)
SL	structured lipid(s)
SO	sunflower oil
SPL	structured phospholipid(s)
TAG	triacylglycerol(s)

References

[1] C.M. Hasler, The changing face of functional foods, Journal of the American College of Nutrition 19 (2000) 499S–506S.

[2] J. Howlett, Functional foods from science to health and claims, International Life Sciences Institute, ILSI Europe, Brussels, Belgium (2008) 36.

[3] K. Gul, A.K. Singh, R. Jabens, Nutraceuticals and functional foods: the foods for the future world, Critical Reviews in Food Science and Nutrition 56 (2016) 2617–2627.

[4] F. Shahidi, Functional foods: their role in health promotion and disease prevention, Journal of Food Science 69 (2004) 146–149.

[5] A.J. Stein, E. Rodríguez-Cerezo, Functional food in the European Union, JRC Scientific and Technical Report, EUR 23380 EN (2008) 76.

[6] H.T. Osborn, C.C. Akoh, Structured lipids – novel fats with medical, nutraceutical, and food applications, Comprehensive Reviews in Food Science and Food Safety 1 (2002) 93–120.

[7] B.H. Kim, C.C. Akoh, Recent research trends on the enzymatic synthesis of structured lipids, Journal of Food Science 80 (2015) C1713–C1724.

[8] S. Ferreira-Dias, N.M. Osório, J. Rodrigues, C. Tecelão, Structured lipids for foods, in: 1st edition, L. Melton, F. Shahidi, P. Varelis (Eds.), Encyclopedia of Food Chemistry, vol. 3, Elsevier, 2019.

[9] S. Ferreira-Dias, Enzymatic production of functional fats, in: A. Pandey, C. Larroche, C.R. Soccol, E. Gnansounou, P. Nigam (Eds.), Food Fermentation Biotechnology, 2, Asiatech Publishers Inc., New Deli, India, 2010, pp. 981–1014.

[10] S. Ferreira-Dias, The use of immobilised lipases in the Food Industry- Factors and facts, In: A. Koutinas, A. Pandey, C. Larroche (Eds.), Current Topics on Bioprocesses in Food Industry, Vol. II, chapter 8, pp. 105-131, Asiatech Publishers Inc., New Deli, India, 2008.

[11] S. Ferreira-Dias, C. Tecelão, Human milk fat substitutes: advances and constraints of enzyme-catalyzed production, Lipid Technology 26 (2014) 183−185.

[12] Q.D. Utama, A.B. Sitanggang, D.R. Adawiyah, P. Hariyadi, Lipase-catalyzed interesterification for the synthesis of medium-long-medium (MLM) structured lipids − a review, Food Technology & Biotechnology 57 (2019) 305−318.

[13] W. Wei, C. Sun, X. Wang, Q. Jin, X. Xu, C.C. Akoh, et al., Lipase-catalyzed synthesis of *sn-2* palmitate: a review, Engineering 6 (2020) 406−414.

[14] Y. Guo, Z. Cai, Y. Xie, A. Ma, H. Zhang, P. Rao, et al., Synthesis, physicochemical properties, and health aspects of structured lipids: a review, Comprehensive Reviews in Food Science and Food Safety 19 (2020) 759−800.

[15] H. Pehlivanoğlu, M. Demirci, O.S. Toker, N. Konar, S. Karasu, O. Sagdic, Oleogels, a promising structured oil for decreasing saturated fatty acid concentrations: Production and food-based applications, Critical Reviews in Food Science and Nutrition 58 (2018) 1330−1341.

[16] A. Puşcaş, V. Mureşan, C. Socaciu, S. Muste, Oleogels in food: a review of current and potential applications, Foods 9 (70) (2020) 27.

[17] J. Podmore, Application of modification techniques. Elsevier Applied Science in: R.J. Hamilton, A. Bhati (Eds.), Recent Advances in Chemistry and Technology of Fats and Oils, 1987, pp. 167−181.

[18] H.D. Belitz, W. Gorsch, Lipids, Food Chemistry, Springer-Verlag, 2009, pp. 158−247.

[19] C.C. Akoh, D.B. Min, Food lipids. Chemistry, Nutrition and Biotechnology, Marcel Dekker, New York, 2002, p. 1005.

[20] J. Giese, Fats, oils and fat replacers, Food Technology 50 (1996) 48−83.

[21] S. Berry, Triacylglycerol structure and interesterification of palmitic and stearic acid-rich fats: an overview and implications for cardiovascular disease, Nutrition Research Reviews 22 (2009) 3−17.

[22] D. Mozaffarian, R. Clarke, Quantitative effects on cardiovascular risk factors and coronary heart disease risk of replacing partially hydrogenated vegetable oils with other fats and oils, European Journal of Clinical Nutrition 63 (Suppl 2) (2009) S22−S33.

[23] P.J.M.W.L. Birker, F.B. Padley, Physical properties of fats and oils, Recent Advances in Chemistry and Technology of Fats and Oils, Elsevier Applied Science (1987) 1−11.

[24] E.G. Perkins, Modern analytical methodology for the evaluation of fats and oils, Oils and Fats in the Nineties, International Food Science Centre Proceedings (1992) 209−253.

[25] S. Bornaz, J. Fanni, M. Parmentier, Butter texture: the prevalent triglycerides, Journal of the American Oil Chemists' Society 70 (1993) 1075−1079.

[26] S. Metin, R.W. Hartel, Crystallization of fats and oils, in: F. Shahidi (Ed.), Bailey's Industrial Oil and Fat Products, 6, John Wiley & Sons, Inc, 2005.

[27] M.M. Chryson, Table spreads and shortenings, Bailey's Industrial Oil and Fat Products 3 (1985) 41−89.

[28] I.B. Helland, L. Smith, K. Saarem, O.D. Saugstad, S.A. Drevon, Maternal supplementation with very long chain n-3 fatty acids during pregnancy and lactation augments children's IQ at 4 years of age, Pediatrics 111 (2003) e39−e99.

[29] A. Valenzuela, Docosahexaenoic acid (DHA), an essential fatty acid for the proper functioning of neuronal cells: their role in mood disorders, Grasas y Aceites 60 (2009) 203−212.

[30] S.A.H. Goli, M. Kadivar, J. Keramat, M. Fazilati, Conjugated linoleic acid (CLA) production and lipase-catalyzed interesterification of purified CLA with canola oil, European Journal of Lipid Science and Technology 110 (2008) 400−404.

[31] S.E. Lumor, C.C. Akoh, Incorporation of gamma-linolenic and linoleic acids into a palm kernel oil/palm olein blend, European Journal of Lipid Science and Technology 107 (2005) 447–454.

[32] A.P. Simopoulos, An increase in the omega-6/omega-3 fatty acid ratio increases the risk for obesity, Nutrients 8 (2016) 128.

[33] J.J. DiNicolantonio, J.H. O'Keefe, Importance of maintaining a low omega−6/omega−3 ratio for reducing inflammation, Open Heart 5 (2018) e000946.

[34] S. Rajaram, E.H. Haddad, A. Mejia, J. Sabate, Walnuts and fatty fish influence different serum lipid fractions in normal to mildly hyperlipidemic individuals: a randomized controlled study, The American Journal of Clinical Nutrition 89 (2009) 1657S–1663S.

[35] X. Palomer, J.M. González-Clemente, F. Blanco-Vaca, D. Mauricio, Role of vitamin D in the pathogenesis of type 2 diabetes mellitus, Diabetes, Obesity and Metabolism 10 (2008) 185–197.

[36] T. Psaltopoulou, R.I. Kosti, D. Haidopoulos, M. Dimopoulos, D.B. Panagiotakos, Olive oil intake is inversely related to cancer prevalence: a systematic review and a meta-analysis of 13,800 patients and 23,340 controls in 19 observational studies, Lipids in Health and Disease 10 (2011) 127.

[37] H.G. Rodrigues, M.A.R. Vinolo, J. Magdalon, H. Fujiwara, H. Cavalcanti, D.M.H. Farsky, et al., Dietary free oleic and linoleic acid enhances neutrophil function and modulates the inflammatory response in rats, Lipids 45 (2010) 809–819.

[38] A. Palou, M.L. Bonet, Controlling lipogenesis and thermogenesis and the use of ergogenic aids for weight control, Novel Food Ingredients for Weight Control, Woodhead Publishing Series in Food Science, Technology and Nutrition (2007) 58–103.

[39] A.B. Thomson, M. Keelan, M.L. Garg, M.T. Clandinin, Intestinal aspects of lipid absorption: in review, Canadian Journal of Physiology and Pharmacology 67 (1989) 179–191.

[40] H. Carlier, A. Bernard, C. Caselli, Digestion and absorption of polyunsaturated fatty acids, Reproduction, Nutrition, Development 31 (1991) 475–500.

[41] O. Hernell, L. Bläckberg, Digestion of human milk lipids: physiologic significance of sn-2 monoacylglycerol hydrolysis by bile salt-stimulated lipase, Pediatric Research 16 (1982) 882–885.

[42] R.J. Jandacek, J.A. Whiteside, B.N. Holcombe, R.A. Volpenhein, J.D. Taulbee, The rapid hydrolysis and efficient absorption of triglycerides with octanoic acid in the 1 and 3 positions and long-chain fatty acid in the 2 position, American Journal of Clinical Nutrition 45 (1987) 940–945.

[43] C.C. Akoh, Fat replacers. IFT scientific status summary, Food Technology 52 (1998) 47–53.

[44] M.S. Christensen, C.E. Hoy, C.C. Becker, T.G. Redgrave, Intestinal absorption and lymphatic transport of eicosapentaenoic (EPA), docosahexaenoic (DHA), and decanoic acids: dependence on intramolecular triacylglycerol structure, American Journal of Clinical Nutrition, 61, 1995, pp. 56–61.

[45] E.A. Decker, The role of stereospecific saturated fatty acid positions on lipid nutrition, Nutrition Reviews 54 (1996) 108–110.

[46] A.E. Jeukendrup, W.H. Saris, A.J. Wagenmakers, Fat metabolism during exercise: a review. Part I: fatty acid mobilization and muscle metabolism, International Journal of Sports Medicine 19 (1998) 231–244.

[47] M.M. Jensen, M.S. Christensen, C.E. Høy, Intestinal absorption of octanoic, decanoic, and linoleic acids: effect of triglyceride structure, Annals of Nutrition and Metabolism 38 (1994) 104–116.

[48] M. Ramírez, L. Amate, A. Gil, Absorption and distribution of dietary fatty acids from different sources, Early Human Development 65 (SUPPL. 2) (2001) S95–S101.

[49] V.V. Colleone, Aplicações clínicas dos ácidos graxos de cadeia média, in: R. Curi, C. Pompéia, C.K. Miyasaka, J. Procopio (Eds.), Entendendo a Gordura: Os Ácidos Graxos, 2002, pp. 439–454. São Paulo, Manole, Brazil.

[50] M.D. Erickson, Interesterification, in: D.R. Erickson (Ed.), Practical Handbook of Soybean Processing and Utilization, AOCS Press and United State Soybean Board, Champaign, IL (USA), 1995, pp. 277–296.

[51] Lipases and Phospholipases. Methods and Protocols, in: G. Sandoval (Ed.), Springer Protocols, 2nd edition, Humana Press, 2018.

[52] R. Verger, "Interfacial activation" of lipases: facts and artifacts, Review, Trends in Biotechnology 15 (1997) 32–38.

[53] W. Cho, A.G. Tomasselli, R.L. Heinrikson, F.J. Kézdy, F.J., The chemical basis for interfacial activation of monomeric phospholipases A2. Autocatalytic derivatization of the enzyme by acyl transfer from substrate, Journal of Biological Chemistry 263 (1988) 11237–11241.

[54] S.A. Tatulian, Towards understanding interfacial activation of secretory phospholipase A2 (PLA$_2$): membrane surface properties and membrane-induced structural changes in the enzyme contribute synergistically to PLA$_2$ activation, Biophysical Journal 80 (2001) 789–800.

[55] R.J. Kazlauskas, U.T. Bornscheuer, Biotransformations with lipases, in: H.J. Rhem, G. Pihler, A. Stadler, P.J.W. Kelly (Eds.), Biotechnology, vol. 8, Wiley, New York, 1998, pp. 37–191.

[56] R.G. Jensen, D.R. Galluzzo, V.J. Bush, Selectivity is an important characteristic of lipases (acylglycerol hydrolases), Biocatalysis 3 (1990) 307–315.

[57] C.-H. Lee, K.L. Parkin, Effect of water activity and immobilization on fatty acid selectivity for esterification reactions mediated by lipases, Biotechnology and Bioengineering 75 (2001) 219–227.

[58] X. Xu, Engineering of enzymatic reactions and reactors for lipid modification and synthesis, European Journal of Lipid Science and Technology 105 (2003) 289–304.

[59] A.C. Correia, S. Ferreira-Dias, S., The effect of impurities of crude olive residue oil on the operational stability of the *Candida rugosa* lipase immobilized in polyurethane foams, in: A. Ballesteros, F.J. Plou, J.L. Iborra, P. Halling (Eds.), Stability and Stabilization of Biocatalysts, Elsevier, Amsterdam, 1998, pp. 71–76.

[60] X. Xu, C.-E. Høy, J. Adler-Nissen, Effects of lipid-borne compounds on the activity and stability of lipases in microaqueous systems for the lipase-catalyzed interesterification, in: A. Ballesteros, F.J. Plou, J.L. Iborra, P. Halling (Eds.), Stability and Stabilization of Biocatalysts, Elsevier, Amsterdam, 1998, pp. 441–446.

[61] P.M. Doran, Elsevier Science & Technology Books Bioprocess Engineering Principles, 1995.

[62] S. Fukui, A. Tanaka, T. Iida, Immobilization of biocatalysts for bioprocesses in organic solvent media, in: C. Laane, J. Tramper, M.D. Lilly (Eds.), Biocatalysis in Organic Media, Elsevier Science Publishers B.V, Amsterdam, 1987, pp. 21–41.

[63] P. Pires-Cabral, M.M.R. da Fonseca, S. Ferreira-Dias, Modelling the microenvironment of a lipase immobilised in polyurethane foams, Biocatalysis and Biotransformation 23 (2005) 363–373.

[64] G. Sandoval, L. Casas-Godoy, K. Bonet-Ragel, J. Rodrigues, S. Ferreira-Dias, F. Valero, Enzyme-catalyzed production of biodiesel as alternative to chemical-catalyzed processes: advantages and constraints, Current Biochemical Engineering 4 (2017) 109–141.

[65] X. Xu, Production of specific structured triacylglycerols by lipase-catalyzed reactions: a review, European Journal of Lipid Science and Technology 102 (2000) 287–303.

[66] C. Tecelão, V. Perrier, E. Dubreucq, S. Ferreira-Dias, Production of human milk fat substitutes by interesterification of tripalmitin with ethyl oleate catalyzed by *Candida parapsilosis* lipase/acyltransferase, Journal of the American Oil Chemists' Society 96 (2019) 77–787.

[67] A. Zaks, A. Klibanov, Enzyme-catalyzed processes in organic solvents, Proceedings of the.National Academy of Sciences of the United States of America 82 (1985) 3192–3196.

[68] C. Laane, S. Boeren, K. Vos, On optimizing organic solvents in multi-liquid-phase biocatalysis, Trends in Biotechnology 3 (1985) 251–252.

[69] Y.-Y. Lee, T.-K. Tang, O.-M. Lai, Health benefits, enzymatic production, and application of medium- and long-chain triacylglycerols (MLCT) in food industries: a review, Journal of Food Science 77 (2012) 137–144.

[70] A.L. Balieiro, N.M. Osório, A.S. Lima, C.M.F. Soares, F. Valero, S. Ferreira-Dias, Production of dietetic triacylglycerols from olive oil catalyzed by immobilized heterologous *Rhizopus oryzae* lipase, Chemical Engineering Transactions 64 (2018) 385–390.

[71] C.M. Costa, A. Canet, I. Rivera, N.M. Osório, G. Sandoval, F. Valero, et al., Production of MLM type structured lipids from grapeseed oil catalyzed by non commercial lipases, European Journal of Lipid Science and Technology 120 (2018) 1700320.

[72] N. Bassan, R.H. Rodrigues, R. Monti, C. Tecelão, S. Ferreira-Dias, A.V. Paula, Enzymatic modification of grapeseed (*Vitis vinifera* L.) oil aiming to obtain dietary triacylglycerols in a batch reactor, LWT—Food Science and Technology 99 (2019) 600–606.

[73] S.M. Abed, M. Elbandy, M.A. Abdel-Samie, A.H. Ali, S.A. Korma, A. Noman, et al., Screening of lipases for production of novel structured lipids from single cell oils, Process Biochemistry 91 (2020) 181–188.

[74] D.A. Mota, D. Rajan, G.C. Heinzl, N.M. Osório, J. Gominho, L.C. Krause, et al., Production of low-calorie structured lipids from spent coffee grounds or olive pomace crude oils catalyzed by immobilized lipase in magnetic nanoparticles, Bioresource Technology 307 (2020) 123223.

[75] M.M.C. Feltes, L.O. Pitol, J.F.G. Correia, R. Grimaldi, J.M. Block, J.L. Ninow, Incorporation of medium chain fatty acids into fish oil by chemical and enzymatic interesterification, Grasas y Aceites 60 (2009) 168–176.

[76] R.G. Jensen, The lipids in human milk, Progress in Lipid Research 35 (1996) 53–92.

[77] R. Yuhas, K. Pramuk, E.L. Lien, Human milk fatty acid composition from nine countries varies most in DHA, Lipids 41 (2006) 851–858.

[78] R.G. Jensen, Lipids in human milk, Lipids 34 (1999) 1243–1271.

[79] E. Lien, The role of fatty acid composition and positional distribution in fat absorption in infants, The Journal of Pediatrics 125 (1994) 62–68.

[80] W.M. Willis, R.W. Lencki, A.G. Marangoni, Lipid modification strategies in the production of nutritionally functional fats and oils, Critical Reviews in Food Science and Nutrition 38 (1998) 639–674.

[81] K.M. Linderborg, H.P.T. Kallio, Triacylglycerol fatty acid positional distribution and postprandial lipid metabolism, European Journal of Lipid Science and Technology 102 (2000) 287–303.

[82] A. López-López, A.I. Castellote-Bargalló, C. Campoy-Folgoso, M. Rivero-Urgèl, R. Tormo-Camicé, D. Infante-Pina, et al., The influence of palmitic acid triacylglyceride position on the fatty acid, calcium and magnesium contents of at term newborn faeces, Early Human Developement 65 (2001) S83–S94.

[83] T. Simões, F. Valero, C. Tecelão, S. Ferreira-Dias, Production of human milk fat substitutes catalyzed by a heterologous *Rhizopus oryzae* lipase and commercial lipases, Journal of the American Oil Chemists' Society 91 (2014) 411–419.

[84] C. Tecelão, M. Guillén, F. Valero, S. Ferreira-Dias, Immobilized heterologous *Rhizopus oryzae* lipase: a feasible biocatalyst for the production of human milk fat substitutes, Biochemical Engineering Journal 67 (2012) 104–110.

[85] C. Tecelão, I. Rivera, G. Sandoval, S. Ferreira-Dias, *Carica papaya* latex: a low-cost biocatalyst for human milk fat substitutes production, European Journal of Lipid Science and Technology 114 (2012) 266–276.

[86] A.R. Faustino, N.M. Osório, C. Tecelão, A. Canet, F. Valero, S. Ferreira-Dias, Camelina oil as a source of polyunsaturated fatty acids for the production of human milk fat substitutes catalyzed by a heterologous *Rhizopus oryzae* lipase, European Journal of Lipid Science and Technology 118 (2016) 532–544.

[87] A. Srivastava, C.C. Akoh, S.W. Chang, G.C. Lee, J.F. Shaw, *Candida rugosa* lipase LIP1-catalyzed transesterification to produce human milk fat substitute, Journal of Agricultural and Food Chemistry 54 (2006) 5175–5181.

[88] J.H. Lee, J.M. Son, C.C. Akoh, M.R. Kim, K.T. Lee, Optimized synthesis of 1,3-dioleoyl-2-palmitoylglycerol-rich triacylglycerol via interesterification catalyzed by a lipase from *Thermomyces lanuginosus*, New Biotechnology 27 (2010) 38–45.

[89] M.M. Soumanou, M. Pérignon, P. Villeneuve, Lipase-catalyzed interesterification reactions for human milk fat substitutes, European Journal of Lipid Science and Technology 115 (2013) 270–285.

[90] X. Wang, S. Zou, Z. Miu, Q. Jin, X. Wang, Enzymatic preparation of structured triacylglycerols with arachidonic and palmitic acids at the sn-2 position for infant formula use, Food Chemistry 283 (2019) 331–337.

[91] Y. He, C. Qiu, Z. Guo, J. Huang, M. Wang, B. Chen, Production of new human milk fat substitutes by enzymatic acidolysis of microalgae oils from *Nannochloropsis oculata* and *Isochrysis galbana*, Bioresource Technology 238 (2017) 129–138.

[92] T. Yuan, W. Wei, X. Wang, Q. Jin, Biosynthesis of structured lipids enriched with medium and long-chain triacylglycerols for human milk fat substitute, LWT – Food Science and Technology 128 (2020) 109255.

[93] M. Ghosh, A. Sengupta, D.K. Bhattacharyya, M. Ghosh, Preparation of human milk fat analogue by enzymatic interesterification reaction using palm stearin and fish oil, Journal of Food Science and Technology 53 (2016) 2017–2024.

[94] N.M. Osório, J.H. Gusmão, M.M.R. da Fonseca, S. Ferreira-Dias, Lipase-catalysed interesterification of palm stearin with soybean oil in a continuous fluidised-bed reactor, European Journal of Lipid Science and Technology 107 (2005) 455–463.

[95] A.V. Paula, G.F.M. Nunes, N.M. Osório, J.C. Santos, H.F. de Castro, S. Ferreira-Dias, Continuous enzymatic interesterification of milkfat with soybean oil produces a highly spreadable product rich in polyunsaturated fatty acids, European Journal of Lipid Science and Technology 117 (2015) 608–619.

[96] P. Bispo, I. Batista, R.J. Bernardino, N.M. Bandarra, Preparation of triacylglycerols rich in omega-3 fatty acids from sardine oil using a *Rhizomucor miehei lipase*: focus in the EPA/DHA ratio, Applied Biochemistry and Biotechnology 172 (2014) 1866–1881.

[97] B.H. Jennings, C.C. Akoh, Trans-free plastic shortenings prepared with palm stearin and rice bran oil structured lipid, Journal of the American Oil Chemists' Society 87 (2010) 411–417.

[98] F.A.S.M. Soares, N.M. Osório, R.C. da Silva, L.A. Gioielli, S. Ferreira-Dias, Batch and continuous lipase-catalyzed interesterification of blends containing olive oil for trans-free margarines, European Journal of Lipid Science and Technology 115 (2013) 413–428.

[99] N.M. Osório, M.H. Ribeiro, M.M.R. da Fonseca, S. Ferreira-Dias, Interesterification of fat blends rich in omega-3 polyunsaturated fatty acids catalysed by immobilized *Thermomyces lanuginosa* lipase under high pressure, Journal of Molecular Catalysis- B Enzymatic 52-53 (2008) 58–66.

[100] G. Pande, C.C. Akoh, R.L. Shewfelt, Production of trans-free margarine with stearidonic acid soybean and high stearate soybean oils-based structured lipid, Journal of Food Science 77 (2012) C1203–C1210.

[101] N.M. Osório, M.M.R. da Fonseca, S. Ferreira-Dias, Operational stability of *Thermomyces lanuginosa* lipase during fats interesterification in continuous-packed-bed reactors, European Journal of Lipid Science and Technology 108 (2006) 545–553.

[102] A. Sengupta, M. Ghosh, Hypolipidemic effect of mustard oil enriched with medium chain fatty acid and polyunsaturated fatty acid, Nutrition 27 (2011) 1183–1193.

[103] N.M. Osório, E. Dubreucq, M.M.R. da Fonseca, S. Ferreira-Dias, Lipase/acyltransferase-catalysed interesterification of fat blends containing omega-3 polyunsaturated fatty acids, European Journal of Lipid Science and Technology 111 (2009) 120–134.

[104] N.M. Osório, E. Dubreucq, M.M.R. da Fonseca, S. Ferreira-Dias, Operational stability of immobilised lipase/acyltransferase during interesterification of fat blends, European Journal of Lipid Science and Technology 111 (2009) 358–367.

[105] A. Sadana, A deactivation model for immobilized and soluble enzymes, Biotechnology Letters 2 (1980) 279–284.

[106] Z. Knezevic, N. Milosavic, D. Bezbradica, Z. Jakovljevic, R. Prodanovic, Immobilization of lipase from *Candida rugosa* on Eupergit® C supports by covalent attachment, Biochemical Engineering Journal 30 (2006) 269–278.

[107] P.A. Nunes, P. Pires-Cabral, S. Ferreira-Dias, Production of olive oil enriched with medium chain fatty acids catalysed by commercial immobilised lipases, Food Chemistry 127 (2011) 993–998.

[108] P.A. Nunes, P. Pires-Cabral, M. Guillén, F. Valero, D. Luna, S. Ferreira-Dias, et al., Production of MLM-Type structured lipids catalyzed by immobilized heterologous *Rhizopus oryzae* lipase, Journal of the American Oil Chemists' Society 88 (2011) 473–480.

[109] P.A. Nunes, P. Pires-Cabral, M. Guillén, F. Valero, S. Ferreira-Dias, Batch operational stability of immobilized heterologous *Rhizopus oryzae* lipase during acidolysis of virgin olive oil with medium-chain fatty acids, Biochemical Engineering Journal 67 (2012) 265–268.

[110] C. Tecelão, J. Silva, E. Dubreucq, M.H. Ribeiro, S. Ferreira-Dias, Production of human milk fat substitutes enriched in omega-3 polyunsaturated fatty acids using immobilized commercial lipases and *Candida parapsilosis* lipase/acyltransferase, Journal of Molecular Catalysis B- Enzymatic 65 (2010) 122–127.

10

Microbial fermentation for reduction of antinutritional factors

Ebenezer Jeyakumar, Rubina Lawrence

DEPARTMENT OF INDUSTRIAL MICROBIOLOGY, JACOB INSTITUTE OF BIOTECHNOLOGY AND BIOENGINEERING, SAM HIGGINBOTTOM UNIVERSITY OF AGRICULTURE, TECHNOLOGY AND SCIENCES, PRAYAGRAJ, INDIA

10.1 Introduction

Fermentation is among the ancient methods used for processing and preserving various food products that have existed since the Neolithic age. The traditional practices of fermentation vary considerably among regions with different cultural, religious, and ethnic practices. Bread and wine are classical examples of fermented foods that originated from great ancient civilizations. Initial scientific evidence of fermentation led to the establishment of microbial processes later on, leading to the role of enzymes. Thus, fermentation is a microbial process whereby the enzymes are involved in metabolizing organic substrates with the release of energy in the absence of an external electron acceptor.

Fermented foods, popular in various regions of the world, have been an integral part of several communities', culture and traditions [1]. Since the beginning of human civilization, fermented foods have been produced using various technologies from utilizing simple household tools to adopting recent food production system in industrial setups. The variation in the sensory attributes and quality of fermented food depends on environmental factors like temperature, rain, humidity, raw material, and fermentation time [2]. The inexpensive production technique using fermented food for efficient preservation with an increase in bioavailability of nutrients makes it an important component of the human diet globally [3]. Even though fermented foods' region-specific origin prevails, due to its nutritional and therapeutic significance, enormous emphasis is given to improving technologies for the large-scale production of these products. Limiting fermented food to bread and wine has now expanded to various other substrates like vegetables, milk, fish, meat products, and grains [4]. Therefore fermented foods can be classified based on the substrate (cereal, legume, fruit, and vegetable, milk, meat, fish) or type of fermentation (alcoholic and nonalcoholic). Enhancement of nutritional value in foods with respect to improved carbohydrate and protein digestibility, fiber solubilization, gluten degradation, vitamin biosynthesis, bioavailability of micronutrients, and degradation of antinutrient factors are the key benefits of

Table 10-1 Principal groups of global fermented food products [5].

Food origin	Food source	Fermented products
Plant	Cereals and Grains	Amazake, beer, bread, chunjang, doenjang, rice wine, malt and grain whisky, sourdough, sake, tamari, tempeh, tofu, miso, idli, dosa, vodka, vinegar
	Honey	Mead, metheglin
	Vegetables	Asinan, kimchi, pak dong, paw-tsay, sajur asin, sauerkraut
	Fruits	Atchara, burong mangga, brandy, cider, natade pina, vinegar, wine
	Tea	Kombucha, miang, pu-erh tea
Animal	Meat	Beef sticks, icelandic slátur, pepperoni, sausages, salami,
	Fish	Bagoong, jeotgal, garum, rakfisk, shidal, shiokara
	Dairy	Cheese, curd, filmjölk, kefir, kumis, skyr, yoghurt

fermented foods. Beneficial end-products formed due to fermentation are bioactive peptides, prebiotic oligosaccharides, and microbial metabolites. Simultaneously, toxic compounds, *viz.*, aflatoxin, cyanogens are significantly reduced during fermentation. Moreover, the improvement of flavors, textures, and aromas also makes the product enriched and palatable.

Natural or mixed culture fermentation with various genera of bacteria, yeast, and mold has been used to produce fermented foods globally (Table 10–1) [5]. The type of the fermented food is governed by one or more raw materials (cereals, legumes, and tubers) used for its preparation that contains a significant amount of antinutritional and toxic components, for example, tannins, phytates, saponins, lectins, oxalates, cyanogenic glycosides, and inhibitors of enzymes like trypsin, chymotrypsin, and α-amylase which interrupts normal physiological responses or causes diminished availability of certain nutrients [6,7]. Interference in the availability of minerals and digestion of proteins and carbohydrates in such foods is greatly affected. The present chapter describes the antinutritional factors (ANFs) and fermentative approach for its reduction.

10.2 Nutritive and antinutritive properties in food

Nutritional value in food is determined by the availability of nutrients, *viz.*, proteins, carbohydrates, vitamins, and minerals. Indigenous fermented foods serve as an important nutritional source for people in developing countries. They have been used to treat populations with caloric deficiencies and as an important source of the vitamins, proteins, and essential amino acids. Further, carbohydrate is the primary compound available in raw materials, which also serve as the energy source of fermenting microorganisms [8]. The nondigestible oligosaccharides, *viz.*, stachyose, raffinose, and verbascose, are utilized by microbes during fermentation, thus enhancing the food digestibility and preventing conditions like abdominal distension and flatulence [9]. Certain amino acids and organic acids like lactic and acetic acid are also produced during fermentation, which helps in preservation of fermented products.

Plant secondary metabolites that are highly bioactive and reduce nutrient utilization or intake are called ANFs. They limit the use of several plants since they are capable of precipitating the harmful effects in both man and animals. The quantity of harmful substances in plants varies greatly with respect to species, variety, and postharvest treatments applied, *viz*., soaking, drying, autoclaving, and seed germination. The dietary significance of proteins from pulses and their nutritive value is well known. However, the seeds also contain proteinaceous sources from water-soluble albumin class, along with lectins, protease inhibitors, and enzymes which also evokes physiological effects linked with essential human nutrition and have been accordingly classified as ANFs [10]. Several other nutrient-rich food products like cereals, pseudocereals also contain ANFs. The ANFs from these food products are broadly categorized as protein and nonprotein compounds. Commonly found protein ANFs include protease inhibitors (trypsin and chymotrypsin competitive inhibitors) and lectins. Alkaloids, phytic acid, phenolic compounds like saponins and tannins are the nonprotein ANFs found in food products (Table 10−2) [11,12].

10.2.1 Phytates

Phytic acid (myoinositol 1,2,2,4,5,6-hexakis di-hydrogen phosphate) is found primarily as a monovalent and divalent cationic salts in specific regions of legumes, cereal grains, roots, and tubers, where it accounts for 85% of total phosphorous. Being a strong chelator phytic acid forms complexes of protein and mineral-phytic acid which results in reduced protein and mineral bioavailability, and solubility [13]. It has a strong affinity toward minerals like calcium, magnesium, iron, copper, molybdenum, and zinc, thus impairing their absorption leading to mineral deficiencies. The health benefits of dietary phytates include lowering of

Table 10–2 Classification of antinutritional factors and their adverse effects [11,12].

Class	Antinutritional factors	Adverse effects
Protein	Protease inhibitors	Disruption of protein digestion and utilization
	Lectins	Malabsorption of digested end-products in small intestine
		Act as food allergens and hemagglutinins
		Disturbance in mineral metabolism
	Toxic amino acids	Cause hemolytic anemia
Lipid	Lipase inhibitors	Gastrointestinal disturbance
		Affects micro- and macronutrient absorption
Starch	Amylase inhibitors	Prevent absorption of dietary starch
Glycosides	Saponins	Causes reduced food intake, bioavailability of nutrients, protein digestibility
	Cyanogens	Potent respiratory inhibitor, hyperglycemia, cause shift from aerobic to anaerobic metabolism, inhibit absorption of nutrients
	Estrogens	Endocrine disruption
	Goitrogens	Interfere iodine uptake and affect thyroid function
Polyphenols	Flavonoids	Reduced iron and zinc absorption, induce DNA mutations
	Tannins	Inactivation of digestive enzymes, protein insolubility affecting utilization of vitamins and minerals
Alkaloids	Alkaloids	Acute gastrointestinal and neurological disturbances
Others	Phytates	Leads to mineral deficiencies caused by impaired absorption
	Oxalates	Affects peptic digestion, nutrient deficiencies, and irritation in the gut lining

blood glucose, regulate insulin secretion, reduce blood clot, cholesterol and triglycerides, prevent heart diseases, prevent renal stone formation, and helps in removing the traces of heavy metal ions [14,15].

10.2.2 Polyphenols

Polyphenols or polyhydroxyphenols are compounds with aromatic ring and several hydroxyl groups and often have functional groups beyond hydroxyl groups. These macromolecules can diffuse across cell membranes effortlessly to reach intracellular sites of action. These compounds are commonly found in fruits, vegetables, cereals, and beverages. Tannins and flavonoids are predominantly found polyphenols in foods (Fig. 10−1) [16]. Even though polyphenols have several beneficial health effects, the antinutritional properties of these compounds cannot be ruled out [17].

10.2.2.1 Tannins

These polyphenolic compounds (M.W. 500 to >3000 Da) possess astringent property and binds or precipitates protein along with organic compounds like alkaloids and amino acids. Even though tannins possess chemo-preventive activities against carcinogenesis and mutagenesis, they are involved in cancer formation, hepatotoxicity, or antinutritional activity. Tannins that have ANFs are broadly classified into two main groups, namely, condensed

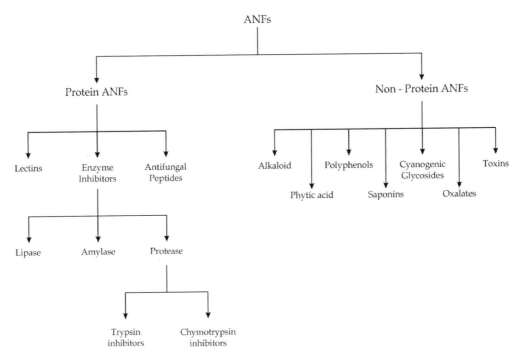

FIGURE 10−1 Classification of antinutritional factors (ANFs).

tannins (proanthocyanidins) and hydrolysable tannins (ellagitannins and gallotannins) [18,19]. Condensed tannins are the predominant class of polyphenols widely distributed in food grains and legumes. These are generally present in the external layers of seed coats and cereal grains or the testa of higher plants' legumes and seeds [20]. Food crops and legumes with a higher content of tannins include faba bean, lima bean, sorghum (up to 5% condensed tannins), sunflower seed meal (1.2%−2.7% chlorogenic acid), and rapeseed [21]. Naturally occurring tannins interact with the proteins to form a tannin−protein complex that causes inactivation of digestive enzymes and protein insolubility, affecting the utilization of vitamins and minerals. They also decrease protein digestibility resulting in reduced feed efficiency, decreased iron absorption and growth depression [21]. Damage to the mucosal lining of the gastrointestinal tract, increased excretion of proteins, and essential amino acids and alteration in the excretion of certain cations are some of the other harmful effects due to consumption of tannins. The increased antinutritional effect and decreased biological activity of tannins are directly proportional to its increase in molecular mass [22].

10.2.2.2 Flavonoids

Flavonoids (bioflavonoids) are polyphenolic compounds belonging to the class of plant secondary metabolites, the most common among them in the human diet being catechins. As compared to catechins, quercetin is found in lower amounts even though widely distributed in plant materials. Nonwheat cereals like Buckwheat (*Fagopyrum esculentum*), Amaranth (*Amaranthus* L.), Quinoa (*Chenopodium quinoa* Wild) are rich sources of flavonoids. Flavonoids can chelate metals such as iron and zinc [23] and reduce their absorption. Flavonoids in food are found to have strong inhibition of topoisomerase and cause DNA mutations in MLL gene, that is commonly found in neonatal acute leukemia [24,25].

10.2.3 Alkaloids

These are the smallest organic molecules commonly found in 15%−20% of vascular plants and one of the major groups of chemicals synthesized by plants from amino acids. Common potato tubers (*Solanum tuberosum*) contain two toxic glycoalkaloids: α-solanine and α-chaconine, which are also hemolytically active [26]. Even though the contents are low in general, consumption of potato with high levels of glycoalkaloids is often associated with acute poisoning causing gastrointestinal and neurological disturbances [27]. Since alkaloids can greatly affect the nervous system and disrupt the electrochemical transmission, it is considered to be an antinutrient factor. The most popular alkaloids like nicotine, cocaine, quinine, morphine, and solanine are found in plants such as tobacco, coca, chinchona, opium poppy, and solanum, respectively.

10.2.4 Saponins

Saponins found abundantly in plants (legumes and cereals) are a large family of structurally related compounds having one or more hydrophilic glycone moieties combined with lipophilic triterpenoid aglycone (Sapogenin). The soap-like behavior is attributed to the interaction of polar and nonpolar structural elements in their molecules hence called saponins. These are a heterogeneous group of naturally occurring secondary compounds that are nonvolatile surface-active compounds primarily found in the plant kingdom [16]. Triterpene saponins are more widely distributed in nature than steroidal types. The complex nature of saponins is responsible for various actions, including foaming, emulsifying properties, sweetness and bitterness, hemolytic properties, pharmacological and medicinal properties, antimicrobial, insecticidal, and molluscicidal activities [28]. Even though poorly absorbed, they form insoluble complexes with 3-β-hydroxysteroids, which further interacts and forms large mixed micelles with bile acids and cholesterol thus exhibiting a hypocholesterolemic effect. The triterpenoid saponin in legumes (Chickpeas), an effective calcium-activated potassium channel opener, can be explored for treatment of urological, respiratory, neurological, cardiovascular, and other disorders. Therefore saponin-rich foods are considered important in human diets as they can play a significant role in controlling plasma cholesterol, preventing peptic ulcers, osteoporosis, and reducing the risk of heart disease. However, saponins' bitterness is responsible for its limited use and contributes to its antinutritive factor [29]. It causes reduced food intake due to bitterness and throat-irritating activity. Further, it reduces the nutrient bioavailability, enzyme activity, affects protein digestibility by inhibiting various enzymes like trypsin and chymotrypsin [30], erythrocyte hemolysis, bloat, growth depression, inhibition of smooth muscle activity, and reduction in nutrient uptake. Further, saponins can also form insoluble complexes with calcium, iron, and zinc.

10.2.5 Cyanogenic glycosides/glucosides

Cyanogenic glycosides have amino acid-derived aglycones known as Cyanogens, widely distributed in beans and tubers (Cassava), which accounts for about 90% of the plant toxins. Cyanogenic glucosides are classified as phytoanticipins, distributed widely in the plant kingdom, occurring in more than 2500 plant species including Almonds (*Amygadalus communis*), Apricots (*Prunus armeniaca*), Apples (*Malus sylvestrius*), Cassava (*Manihot esculenta*), Linseed (*Linumusitatissmium*), Peaches (*Prunus persica*) Sorghums (*Sorghum* spp.), and White clover (*Trifolium repens*) [31]. Cyanogenic glycosides are commonly found in families like Fabaceae, Rosaceae, Leguminosae, Linaceae, and Compositae. Hydrolysis (acid or enzymatic) of toxigenic cyanogenic glycosides yields hydrocyanic acid (HCN) that inhibits cytochrome oxidase (terminal respiratory catalyst) in the mitochondria of cells by binding to Fe^{3+}/Fe^{2+} present in the enzyme causing a reduced oxygen utilization in the tissues thus acting as a potent respiratory inhibitor. Cyanide further increases the level of blood glucose and lactic acid along with decrease in adenosine triphosphate/adenosine diphosphate ratio causing a shift from aerobic to anaerobic metabolism. Several enzymatic systems are also inhibited by cyanide ions, affecting the growth by interfering with essential amino acids and absorption of nutrients. Cyanide also has an adverse effect on glycolysis and tricarboxylic acid cycle rate due to its glycogenolytic properties wherein it shunts glucose to pentose phosphate pathway [32]. Major symptoms of acute cyanide exposure are headache, bronchial constriction, weakness, acute toxicity, tropical ataxic neuropathy, and death. Several symptoms that can be noticed preceding death include weaker pulse rate, irregular breathing, dilation in pupils, salivation and frothing, muscular spasms, and bright red mucous membranes.

10.2.6 Oxalates

Oxalates are natural substances found in many foods that are poorly absorbed under nonfasting conditions in human beings. Calcium oxalates are present in varying concentrations in many kinds of edible plants among the different forms of oxalates. The strong bonds between oxalic acid and minerals, such as sodium, potassium, calcium, and magnesium, form complex oxalate salts. Oxalates of sodium and potassium are soluble, while that of calcium are insoluble. Insoluble calcium oxalates tend to form calcium oxalate crystals in the kidney or urinary tract [33]. Calcium bound oxalates in food limit the mineral availability for normal physiological and biochemical functions, including maintaining strong bones, teeth as a cofactor in enzymatic reaction, nerve impulse transmission, and blood clotting factor. It affects calcium and magnesium metabolism and reacts with protein-forming complexes having an adverse effect on peptic digestion. Intake of excess oxalates in food results in nutritional deficiencies and severe irritation in the gut lining.

10.2.7 Lectins

Lectins are proteins of nonimmune origin and glycoproteins that binds (noncovalently) reversibly and specifically to certain carbohydrates and can easily be disintegrated. Being ubiquitous, they have been found in microorganisms, plants, and animals with different functions, *viz.*, immune recognition of foreign carbohydrates, mediation of cell-to-cell interactions, and homeostatic regulation. More than 800 varieties of legume family possess 2%–10% of the total protein as lectins. Lectins prevent absorption of end-products of digestion in the small intestine. Lectins are also responsible for the protection of plants against insects and herbivory by other organisms along with humans due to their cytotoxic (anticancer), fungitoxic (antifungal), insecticidal, antiviral, and antinematode features. Lectins function both as allergens and as hemagglutinins and are present in 30% of the food, particularly in the whole grain diet [34]. Lectins are sugar-binding proteins that agglutinate cells or precipitate the glycoconjugates. The multivalent structure of lectins (hemagglutinins) determines the ability of lectin to agglutinate RBCs used for assay of blood types. Lectins when orally ingested are highly toxic to animals and humans, probably due to their ability to bind to the intestinal epithelial cells' carbohydrate moiety. Disruption of intestinal mucosa due to lectin permits entry of bacteria and their endotoxin causing bacteremia and toxemia, respectively [35]. Lectins can also have a systemic effect causing elevated protein catabolism and breakdown of accumulated fat and glycogen along with disturbance in mineral metabolism. The resulting phenomenon observed is reduced nutrient absorption from the digestive tract, endogenous loss of nitrogen and protein utilization, thus responsible for severe growth retardation and even death under extreme conditions [36,37]. A high concentration of lectins in food such as beans, cereals, seeds, nuts, and potatoes is harmful. They can cause nutritional deficiencies and allergic reactions if consumed in excess, particularly in raw form or undercooked conditions.

10.2.8 Enzyme inhibitors

10.2.8.1 Protease inhibitors

Protease inhibitors are found in nature being abundantly distributed in plants, animals, and microorganisms. In plants, a great variety of these inhibitors are found in cereals legumes and some fruits (pineapples, bananas, apples, and raisins) and vegetables (cabbage, cucumber, spinach, and tomatoes) [38]. Approximately 5%–10% of the soluble proteins found in barley, wheat, and grains are reported to be protease inhibitors [39]. Depending on the variety and physiological status of the plant and insect damage levels, the quantity of protease inhibitors has been found to vary. The mammalian protein digestion and nutrition are dependent on plant protease inhibitors, for example, carboxy protease, pepsin, serin proteases, trypsin, chymotrypsin, elastase, metalloproteases, carboxypeptidase A and B. Some plant protease inhibitors inhibit mammalian plasma serine proteases, plant sulfhydryl proteases, Kallikrein and plasmin, bromelain, papain, and ficin [38].

Plant protein protease inhibitors have little or no carbohydrate with a molecular weight ranging from 4000 to 80,000 Da. Bowman-Birk protease inhibitor found in soybeans has amino acids (71) with a molecular weight of 8000 Da, while Kunitz inhibitor from the same contains amino acids (198) having a molecular weight of 23,000 Da. Mostly larger protease inhibitors are polymeric in nature with a maximum of four subunits. A large degree of sequence homology in the amino acids, of these inhibitors have been observed within the same inhibitor as well as inhibitors from different plants [38].One or more reactive peptide bonds in these inhibitors interact with the active site of the corresponding enzyme. A protease inhibitor is capable of inhibiting two or more enzymes simultaneously or individually. The manner of interaction between the reactive sites of the inhibitor and the active site of the corresponding enzyme is similar to that between substrate and enzyme. However, the inhibitors form a stable complex with an enzyme that dissociates slowly, unlike the substrates.

10.2.8.2 Trypsin Inhibitor

Due to the importance of soya protein in animal and human nutrition, heat-labile trypsin inhibitors (TIs) have been studied in detail. Enhancement of soya beans' nutritional quality for rats, chickens, and mice due to heating and thus destroying heat-labile TIs was first realized in 1940. Moist heat treatment of soya flakes for increasing time period decreases the TI activity while the protein efficiency ratio increases [39]. However, TI activity cannot generally be considered a predictor for accessing the nutritive quality of soy protein preparations since the availability of sulfur amino acids and protein digestibility also has to be considered when determining the effect [40]. Amino acid-free diet supplemented with TI preparation results in decreased growth due to several mechanisms like pancreatic hypertrophy and hyperplasia (presumably due to overstimulation of exocrine pancreatic secretion) along with decreased intestinal proteolysis and digestion of dietary protein [41]. The digestion and absorption of dietary protein are reduced significantly by inhibiting the activity of pancreatic enzymes trypsin and chymotrypsin by the formation of indigestible complexes, even in the presence of digestible enzymes.

10.2.8.3 Lipase inhibitors

Food containing fats (triacylglycerol and phospholipids) is digested by lipase. Human lipid metabolism contains preduodenal (gastric and lingual) and extraduodenal (hepatic, pancreatic, endothelial, and lipoprotein) lipases. One-third of the ingested fat is digested by lingual lipase and approximately 10%–40% of dietary fat is digested by gastric lipase while 50%–70% of total dietary fat is hydrolyzed by the principle lipolytic enzyme, that is, pancreatic lipase (triacylglycerol acyl hydrolase), which efficiently digests the triglycerides. The macro- and micronutrient absorption in ingested food is greatly affected by the antinutrients that act as pancreatic lipase inhibitors [42]. Different phytoconstituents belonging to the class polysaccharides, terpenes, glycosides, saponins, polyphenols, alkaloids, and carotenoids which act on the pancreatic lipase are recognized as antinutrients. Saponins from soybeans, phytates from grains like polished rice, oilseeds, legumes, wheat bran, lectins from legumes and oilseeds (soybeans), TIs from soybeans, legumes (peas, beans), tannins from legumes (peas, beans), cereals (millet, sorghum), and polyphenols from extracts of citrus fruits, grape seeds, tea (oolong tea), peanut shells, apples are considered as antinutrients since they act as inhibitors of pancreatic lipase. Lipase inhibitors are also widely found in ginseng with saponin, extracts of peanut, green tea with epigallocatechin-3-gallate, theophylline, and theobromine, terpenes in *Salvia officinalis*, alkaloids with caffeine, glycoside in *Glycyrrhiza glabra*, procyanidin fraction of polyphenol extract, and soybean extracts [43,44]. Fermentation under controlled conditions can significantly reduce the effect of lipase inhibitors. Even though pancreatic lipase inhibitors are used as drugs in the treatment of obesity, several gastrointestinal disorders like flatulence, oily spooling, oily stools are experienced.

10.2.8.4 Amylase inhibitors

The postprandial blood glucose levels are managed by α-amylase as it cleaves starch during the digestive process. Therefore inhibition of α-amylase blocks glucose absorption by inhibiting carbohydrate-hydrolyzing enzymes present in the digestive tract. α-Amylase inhibitors are naturally occurring enzyme inhibitors of interest from the nutritional point of view as they can be used as a therapeutic agent in the treatment of diabetes and obesity since it modulates the postprandial blood glucose levels [45]. Phytochemicals in plants exert antidiabetic activity through inhibition of carbohydrate-hydrolyzing enzymes such as α-glucosidase and α-amylase. Fruit waste is a potential source of α-amylase inhibitors [46,47]. They prevent the absorption of dietary starches by the body, thus acting as starch blockers. The complex carbohydrate starch needs to be broken down by digestive enzyme amylase and other secondary enzymes before being absorbed into the body. The utilization of starch in the body is greatly reduced by proteinaceous inhibitors of α-amylase, which are widely distributed in legumes, cereals, and other plants, including pearl millet. Higher amounts of these inhibitors have been characterized from various common beans like red kidney bean, white kidney beans, black kidney beans, and pigeon peas. These inhibitors are heat-labile and show maximum activity in the pH range of 4.5–9.5 with hypoglycemic effects [48]. Further, they are effective in inhibiting the activity of bovine pancreatic amylase but fail to inhibit bacterial, fungal, and endogenous amylase.

10.2.9 Glucosinolates

Glucosinolates are generally known as goitrogens and are found exclusively in the plant kingdom. They interfere with iodine uptake and affect thyroid function. Vegetables belonging to the Cruciferae family including brocolli, cabbage, brussels sprouts, cauliflower, and kale are some of the goitrogen-rich foods [49]. Consumption of vegetables belonging to the genus *Brassica* eventually affects triiodothyronine (T3) and thyroxine (T4) levels by causing hypothyroidism [50]. Hydrolysis of glucosinolates by myrosinase (thioglucoside glucohydrolase), an internal component of seeds, converts the nontoxic intact glucosinolates to potentially toxic products that can be goitrogenic, hepatotoxic, and volatile with strong pungent nature.

10.2.10 Toxic amino acids

Toxic amino acids found commonly in *Lathyrus* and Broad Beans are nonproteinaceous in nature but reduce nutritious value and cause toxic effects. The most common toxic amino acid found in legumes is dihydroxyphenyl alanine, and responsible for hemolytic anemic due to faba bean consumption, a condition known as favism [29]. The seeds of legume sesbania and jack bean contain Canavanine, a toxic amino acid acting as arginine antagonist.

10.3 Neutralizing antinutritional factors in fermented foods

Major ANFs like phytates, saponins, lectins, goitrogens, protease inhibitors, antivitamins, estrogens, allergens, flatulence factor, and lysinoalanine are found in legumes [51]. Majority of these ANFs are heat-labile. Heat-stable ANFs like phytate and polyphenols cannot be removed by soaking and heating but can be effectively eliminated using either germination or fermentation [52].

Under controlled fermentative processes, enzymatic and nonenzymatic microbial reactions can result in hydrolysis and solubilization of macromolecules (proteins, cell wall polysaccharides). This process beneficially alters the macro- and microstructure of the substrate materials influencing retention, release, and absorption of nutrients and nonnutrients in the substrate material. Frequently encountered challenges in cereal foods using raw material for consumption include sensory acceptability, nutrient digestibility, bio-accessibility, and ANFs. These issues are effectively resolved by fermentation technology, enhancing the nutrient content and bioactive substances in the final product. Fermentation can also help in softening plant tissues by loosening and breakdown of cell walls inducing enzymatic degradation of macromolecules and ANFs like phytate and solubilization of minerals decomposition of carbohydrates and proteins [53–55]. Major and minor food components can be transformed using fermentative processes carried out by microbial and enzymatic conversions.

Among the functional and nonfunctional microorganisms that are present in traditional fermented foods, functional microbes play an important role in the enrichment of nutraceutical and health benefits of food including a decrease in ANFs and allergens (Fig. 10–2). The process of fermentation can enhance the bioavailability of nutrients in many foods. Microbial fermentations also lead to removing toxic or ANFs and preventing food allergies/intolerance in the gut. In the absence of α-galactosidase in humans' flatulence, producing oligosaccharides like stachyose, raffinose, ajugose, and verbascose undergoes microbial fermentation to produce carbon dioxide, hydrogen, and methane. For instance, black gram has 31%–76% of the total sugars that are oligosaccharides which is the major reason for its nutritional utilization [56]. The enzymatic action of yeast, bacteria, and mold or their combination in soybean materials during fermentation breaks down its constituents to yield desired texture and flavor along with the elimination of antinutritional substances. Destruction of ANFs such as nonstarch polysaccharides and phytates can be effectively achieved by microbial activity during fermentation [57–60] and release of encapsulated nutrients like starch, proteins, and minerals [61]. Higher coefficient of apparent total tract digestibility and coefficient of apparent ileal digestibility of

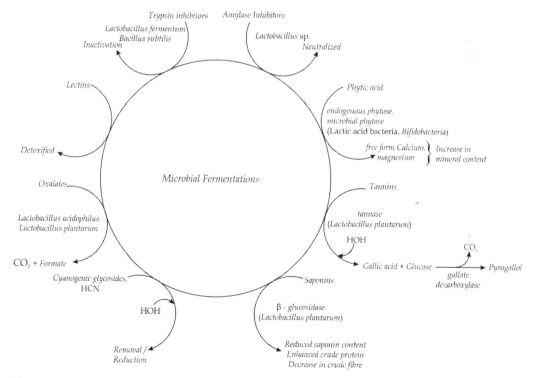

FIGURE 10–2 Conversion of toxic antinutritional factors in food by microbial fermentations.

nutrients and energy can be obtained as a result of fermentation since it causes an increase in available nutrients and reduction in indigestible components and ANFs [62].

10.3.1 Phytic acid

Fermentation significantly reduces phytic acid concentration in different foods, especially cereal-based food products. Phytic acid is effectively degraded by endogenous phytases and microbial phytases as a result of lower pH and longer fermentation time. The degradation of phytic acid during fermentation yields minerals, namely, calcium and magnesium, released from their complex substances into the free form [63,64]. Two phytases, namely, 3-phytase, characteristics of microorganisms, and 6-phytase, found in grains and seeds of higher plants, have been shown to catalyze phytate to inorganic ortho-phosphate and lower phosphoric esters of myoinositol and free myoinositol in some cases. The endogenous phytases reduce the phytate content during lactic acid fermentation of cereals, legumes, and tubers [20].

The process of fermentation has a reducing effect on the phytate content in white bread (100%), whole wheat bread (50%), Iranian flatbread (39%), Polish rye bread (80%), Polish white bread (64%), Arabic high bran bread (30%) [13,65–67], kenkey (31%), rabadi (27%) [68,69], tempeh (55%) [70,71], oncom (96%), dhokla (97%), khaman (58%), idli (40%) [72], and dawadawa (39%) [73]. Phytic acid content is effectively reduced during the fermentation process of cereal-based food product tarhana from 80% to 89% when cultured with *Lactobacillus sanfrancisco* and *Lactobacillus plantarum* [74,75]. Lower pH, longer fermentation time results in more intensive degradation of phytic acid and is attributed to endogenous phytases of the grain flour and microbial phytases [76,77]. An increase in mineral content (magnesium and calcium) is a consequence of phytic acid degradation. Kawal, which is a fermented product prepared from green leaves of *Cassia obtusilfolia* in Africa, is shown to have a dramatic improvement in the protein value of food due to fermentation. It is used as a meat replacer or meat extender. Reduction in the phytic acid content from 649.13 to 340.93 mg/100 gm in Kawal is due to the low pH of fermented dough which is considered to be optimum for phytase activity [78].

Lactic acid bacteria and *Bifidobacteria* are capable of producing phytase, which catalyzes the degradation of phytates. Phytates are reduced to a greater degree by these organisms in fermented soymilk. *Lactobacillus* sp. that secretes extracellular phytase, hydrolyzes phytates present in wheat to liberate 6 P moieties and inositol [57]. Since these phytates are present in complex forms with dietary nutrients like minerals, proteins, free amino acids, and starch, breakage of such complexes will enhance the bound nutrient availability [79]. Hence, an increase in starch and amino acid digestibility in fermented wheat diets is achieved. In addition, carbohydrate–protein complexes like glycoprotein and proteoglycans are also hydrolyzed by microbial enzymes during fermentation, enhancing amino acid digestibility. From the increase in the content of small size peptides (<20 kDa) in soybean meal, and increase in peptides (2–20 residues), nitrogen soluble trichloroacetic acid and free amino acid in rapeseed cake in the fermented product is a strong evidence for greater availability of amino acids [80].

10.3.2 Tannins

Heat treatment in the form of microwaving results in the reduction of 25.7% in total tannin concentration, thereby improving the nutritional characteristics of peas since tannins are considered as one of the ANFs [81]. Tannins are reduced prior to the fermentation step because of their localization in their outer layers or the raw ingredients' seed coats [13]. The polyphenol concentration in food grains can be effectively reduced by polyphenol oxidase released during microbial fermentation. Lactic acid fermentation reduces tannin levels and helps increase iron absorption with the exception of some high tannin cereals. *Lactobacillus plantarum* has the ability to degrade and detoxify harmful and antinutritional constituents in food containing tannins into simpler and harmless compounds [82]. *Lactobacillus plantarum* hydrolyzes tannins to gallic acid and glucose by tannase and gallic acid is further decarboxylated to pyrogallol in the presence of gallate decarboxylase. Fermentation during Kawal production was also found to decrease tannin content from 2.39% to 2.24% and polyphenols from 4.77% to 3.8% due to activation of enzymes like polyphenol oxidase [78].

10.3.3 Saponins

Saponins are not destroyed while cooking, but they are markedly reduced during fermentation [13]. However, Lactic acid bacteria and *Bifidobacteria* that secrets β-glucosidase are capable of splitting sugar side chains of steroids and triterpenoid saponins, lowering their water solubility. The catalytic action of β-glucosidase leads to a reduced content of saponin in fermented soymilk. An increase in the nutrient content with enhanced crude protein level and decrease in crude fiber content and reduction in saponin content is possible with the fermentation of trembesi leaves by *Aspergillus niger* and *Lactobacillus plantarum* [83].

10.3.4 Cyanogenic glycosides

Among the various factors responsible for eliminating HCN, fermentation by extraneous microflora has gained major significance in reducing its toxicity. The fermentation step in Gari and Lafun production from cassava was considered effective for hydrolysis of cyanogenic glycosides and HCN reduction subsequently [84]. The fermentation of grated cassava roots is efficient in the removal of cyanogen glucoside. Cassava pulp or dough fermented for 4–5 days decreases the total cyanide by 52%–63% [85]. Soaking and fermentation of bitter apricot kernels caused a reduction in cyanogen level by 70% [86]. Cyanide content from fermented cocoyam flour is reduced to 98.6%, while a reduction of 84.6% in fermented sorghum leaves [87]. Specific strain needs to be explored further for hydrolysis of cyanogenic glycosides

10.3.5 Oxalates

Food processing steps like soaking, cooking, boiling can significantly reduce the oxalate content. The process of fermentation eliminates up to 43% of oxalates in Dawadawa due to processes like soaking, dehulling, and washing steps prior to fermentation [13]. Fermentation of ameliorated must of *Garcinia xanthochymus* (Garcinia beverage) using native yeast (*Hanseniaspora* sp.) from the naturally fermented fruit results in complete reduction of oxalic acid [88]. Oral administration of *Bifidobacterium* or *Lactobacillus* sp. plays a significant role in luminal oxalate reduction, thereby decreasing the risk of urinary oxalate excretion in humans and animals [89–92]. The pH of the medium affects the two different forms of oxalates. At a pH below 6.0, there is a marked decrease in the deprotonated divalent oxalate ions with a corresponding reduced binding potential with divalent mineral cations forming insoluble oxalates [93]. Therefore reduction in pH due to fermentation enhances soluble oxalates utilized by anaerobic bacteria [94]. Some lactic acid bacteria, for example, *Lactobacillus plantarum*, *Lactobacillus acidophilus* cause degradation of oxalate and utilize it as a source of carbon and energy.

10.3.6 Lectins

Commonly used legumes such as lima beans, soybeans, black beans, lentils, kidney beans, and peas are source of lectins. The antinutritional effect of lectins in various beans and legumes depends upon the specific legume. Animals show toxic symptoms when fed with lectin fraction of black beans and kidney beans [95]. Lectins also have negative effect on the growth in animals when fed with raw soybean meal. Since lectins are heat-labile, traditional cooking before fermentation can have a detoxifying effect. The activity of lectin is often found in some of the fermented foods prepared in India with soaking and fermentation prior to steaming [13]. Regulated heat treatments, cooking or fermentation, can effectively remove lectins either partially or completely from cereal and legume-based foods.

10.3.7 Enzyme inhibitors

10.3.7.1 Protease inhibitors

Moist heat effectively decreases the activity of protease inhibitor thereby enhancing nutritional value of the plant protein. Baking is less effective as compared to other heat treatment methods used. The use of extrusion technology has its application in the legume industry for various purposes using temperature and high shear forces resulting in deformation of proteins. The technology has additional merit in legume processing for enhancement of nutritional value by reducing tannins, phytates, and improving protein digestibility up to twofold. In soya beans, extrusion also results in the inactivation of lipoxygenase, which catalyzes deoxygenation of some polyunsaturated fatty acids. Oxidized fatty acids produce undesirable and irreversible off-flavors and promote inflammation and hypersensitivity reactions [81]. The process of fermentation by bacteria and yeast results in the proteolytic activity, which increases the bioavailability of amino acids by degrading undesirable substances like proteinase inhibitors.

10.3.7.2 Trypsin inhibitors

TIs can be inactivated through various physical, chemical, and biological processes generally applicable to legumes seeds. The physical process includes thermal treatment, extrusion, ultrafiltration, ultrasound treatment, high hydrostatic pressure, instant controller pressure drop, radiation, and soaking [81]. Normal cooking procedures do not affect TI activity of potatoes, broad beans, cabbage, and eggs. Since these TIs are thermostable in nature, a higher temperature and time combination is generally applied, resulting in almost 95.6% of nutrient loss of vitamins such as riboflavin, niacin, thiamine, and pyridoxine in legumes seeds. Disruption of intermolecular bonds retaining the tertiary structure of TIs is promoted by the thermal treatment, which causes a change in the active site conformation. In soy-based products, heat treatments like spray-drying, canning, and sterilization also decrease TI activity and enhance protein digestibility. However, the high cost and negative impact of thermal treatment (above 80°C) on the environment and protein functionality has led to the development of alternate options [96–100].

Soaking is one of the simplest and most economical processes for TI inactivation influenced by pH, temperature, and soaking duration. This process provides hydration to the legume seeds and promotes the removal of water-soluble compounds along with ANFs in the soaking solution that is discarded [101]. The tertiary structure and stability of "TIs" are mainly dependent on the disulfide bonds, which can be disrupted using chemical treatments. TI inactivation is assisted using ammonium hydroxide, sodium hydroxide, or ammonium bicarbonate during heat treatment of soya beans since high and low pH results in unfavorable electrostatic interactions between amino acid residues causing conformational changes in the active site of the enzyme [81]. Acids have a delaying effect on TI inactivation since proteins are unavailable for thermal inactivation in the presence of acids. On the other hand, alkalis used during food processing can cause side effects influencing the nutrition of treated proteins. For example, the formation of lysinoalanine was reported to cause kidney damage in rats, racemization of amino acids resulting in decreased protein digestibility, and formation of toxic compounds [102,103]. Disruption of disulfide bonds for inactivation of "TIs" is also brought about by reducing agents like Sodium metabisulfite and tris (2-carboxiethyl)-phosphine in soybean. However, the latter being a carcinogen has restrictions regarding its application in food [81].

Germination and fermentation are two biotechnological approaches found effective for the inactivation of "TIs." Antinutritional compounds such as TIs are reduced by the process of germination whereby the proteases break down the cellular proteins and enzymes liberating free amino acids that are subsequently used for the autotrophic growth of seedlings. However, it greatly depends on the type of legume, germination conditions, and sprouting time [81]. Though germination is a medium efficiency process, it has the additional benefit of increasing the concentration and bioavailability of a variety of nutrients and improving certain characteristics of pulses. The effectiveness can be enhanced by combining germination with other strategies having similar effectiveness like gamma irradiation.

Inactivation of "TIs" is indicated in fermented soybean foods like soy sauce, meso, and *tempeh* when *Asperigillus oryzae, Lactobactillus plantarum, Lactobacillus acidophilus, Bacillus amyloliquefaceins,* and *Saccharomyces cerevisiae* are used for the fermentation process [104]. *Lactobacillus fermentum* is found effective in the fermentation of common beans and the inactivation of "TIs" [105]. However, *tempeh* produced from soybeans using *Rhizopus oligosporus* results in a slight increase of "TI" since they produce high amounts of carbohydrases that ferment sucrose, raffinose, and stachyose effectively [96]. Proteolytic enzymes released by *Bacillus subtilis* during fermentation processes degrade allergenic and antinutritional proteins such as "TIs" in soybean [106].

10.3.7.3 Amylase inhibitors

Nutrient-rich legumes contain high amounts of α-amylase inhibitors and other antinutrient factors. With the help of microbial fermentation, the nutrients that are locked in plant structures and cells as indigestible materials can be effectively released after the reduction in α-amylase inhibitors. Negligible levels of amylase inhibitor activity can be achieved in the fermentation of pearl millet with *Lactobacillus* sp. used alone or in combination with yeast [6]. Similarly, fermentation of sorghum causes complete neutralization of α-amylase inhibitors. Fermented products of rice like idli and dosa have no inhibitory effect on salivary amylase [29]. Food fermentation can be an effective approach for reduction of α-amylase inhibitors.

10.3.8 Glucosinolates

Antinutritional compounds present in Rapeseed meal especially high levels of glucosinolates are responsible for Goiter, leading to growth retardation, inactivation of digestive enzymes reduction in release rates of highly active peptides causing decreased nutrient utilization [107]. Microbial fermentation is considered as an alternative solution for removal of ANFs and improving nutritional value. The elevated level of peptide is due to hydrolytic activity of proteases' by fermentation of Rapeseed protein microorganisms [108].

10.4 Conclusions and perspectives

Nutrient-rich and biologically active food products with considerable health benefits also contain compounds with antinutritional properties with deleterious effects in digestibility, nutrient absorption, and human health. The commonly present antinutrients in plant materials, *viz.*, saponins, tannins, phytate, polyphenolic compounds, and protease inhibitors, cause a reduction in mineral absorption and affect protein digestibility and health issues, thereby interfering with the nutritional value of foods. Therefore various beneficial effects of fermentation should be considered when tackling the problem of ANFs in food. Microbial fermentations transform raw materials into organoleptically and biochemically useful products. It can also effectively detoxify, reduce immunoreactivity and allergic reactions, and improve the nutritional and functional properties as compared to the raw material. Recent approaches for addressing the issue include processing treatments like milling, soaking, germination, extrusion cooking, steam precooking, and fermentation. Various studies have confirmed fermentation and germination followed by fermentation to be effective methods in reducing the ANFs in food. However, there is a need for optimization of the processes for enhanced effectiveness concerning health and nutrition.

References

[1] T. Deak, Chapter 7: Yeasts in specific types of foods, e201 Handbook of Food Spoilage Yeasts, 2nd ed., CRC Press, Boca Raton, FL, 2007, p. 117.

[2] K.H. Steinkraus, Handbook of Indigenous Fermented Foods, Marcel Decker Inc, New York, 1996, p. 439.

[3] T.H. Gadaga, A.N. Mutukumira, J.A. Narvhus, S.B. Feresu, A review of traditional fermented foods and beverages of Zimbabwe, International Journal of Food Microbiology 53 (1999) 1–11.

[4] E.M. Selhub, A.C. Logan, A.C. Bested, Fermented foods, microbiota, and mental health: ancient practice meets nutritional psychiatry, Journal of Physiological Anthropology 33 (2014) 2. Available from: https://doi.org/10.1186/1880-6805-33-2.

[5] H. Xiang, D. Sun-Waterhouse, G.I.N. Waterhouse, C. Cui, Z. Ruan, Fermentation-enabled wellness foods: a fresh perspective, Food Science and Human Wellness 8 (2019) 203–243.

[6] A. Sharma, A.,C. Kapoor, Levels of antinutritional factors in pearl millet as affected by processing treatments and various types of fermentation, Plant Foods for Human Nutrition 49 (1996) 241–252.

[7] C.L. Ramos, R.F. Schwan, Technological and nutritional aspects of indigenous Latin America fermented foods, Current Opinion in Food Science 13 (2017) 97–102.

[8] C. Chavez-Lopez, A. Serio, C.D. Grande-Tovar, R. Cuervo-Mulet, J. Delgado-Ospina, A. Paparella, Traditional fermented foods and beverages from a microbiological and nutritional perspective: the Colombian heritage, Comprehensive Reviews in Food Science and Food Safety 13 (2014) 1031–1048.

[9] E.A. Shimelis, S.K. Rakshit, Antinutritional factors and in vitro protein digestibility of improved haircot bean (*Phaseolus vulgaris* L.) varieties grown in Ethiopia, International Journal of Food Science and Nutrition 56 (2005) 377–387.

[10] L.G. Saldanha, Summary of comments received in response to the Federal Register Notice Defining Bioactive Food Components, American Society of Nutrition response to the Federal Register notice of September 16, 2004. Federal Register 69 (179) (September 16, 2004) 55821–55822. <http://www.ods.od.nih.gov/Research-Food-Components-Initiatives.aspx>.

[11] I.E. Liener, Toxic Constituents of Plant Food Stuffs, second ed., Academic Press, New York and London, 1980, p. 502.

[12] J.P.F. D'Mello, Effects of antinutritional factors and mycotoxins on feed intake and on the morphology and function of the digestive system, in: R. Mosenthin, J. Zentek, T. Żebrowska (Eds.), Biology of Growing Animals, vol. 4, Elsevier Publications, 2006.

[13] N.R. Reddy, M.D. Pierson, S.K. Sathe, D.K. Salunkhe, Phytates in Cereals and Legumes, CRC Press, Boca Raton, FL, 1989.

[14] R. Selvam, Calcium oxalate stone disease: role of lipid peroxidation and antioxidants, Urological Research 30 (1) (2002) 35–47.

[15] A.M. Shamsuddin, Anti-cancer function of phytic acid, International Journal of Food Science and Technology 37 (7) (2002) 769–782.

[16] S. Khokhar, O.R.K. Apenten, Antinutritional factors in food legumes and effects of food processing, The Role of Food, Agriculture, Forestry and Fisheries in Human Nutrition, vol. 4, EOLSS Publications, 2003, pp. 82–116.

[17] J.M. Landete, Updated knowledge about polyphenols: functions, bioavailability, metabolism, and health, Critical Reviews in Food Science and Nutrition 52 (2012) 936–948.

[18] R. Kumar, M. Singh, Tannins: their adverse role in ruminant nutrition, Journal of Agriculture and Food Chemistry 32 (1984) 447–453.

[19] R. Munoz, B. de las Rivas, F. Lopez de Filipe, I. Reveron, L. Santamaria, M. Esteban-Torres, et al., Biotransformation of phenolics by *Lactobacillus plantarum* in fermented foods, Fermented Foods in Health and Disease Prevention, Elsevier Inc, 2017, pp. 63–83.

[20] N.R. Reddy, M.D. Pierson, Reduction in antinutritional and toxic components in plant foods by fermentation, Food Research International 27 (1994) 281–290.

[21] B.O. De Lumen, L.A. Salamat, Trypsin inhibitor in winged beans (*Psophocarpustetragnolobus*) and the possible role of tannins, Journal of Agricultural and Food Chemistry 28 (1980) 533–536.

[22] K.T. Chung, C.I. Wei, M.G. Johnson, Are tannins a double-edged sword in biology and health? Trends in Food Science and Technology 9 (1998) 168–175.

[23] M. Karamać, Chelation of Cu(II), Zn(II), and Fe(II) by Tannin constituents of selected edible nuts, International Journal of Molecular Sciences 10 (2009) 5485–5497.

[24] M.J. Thirman, H.J. Gill, R.C. Burnett, D. Mbangkollo, N.R. McCabe, H. Kobayashi, et al., Rearrangement of the MLL gene in acute lymphoblastic and acute myeloid leukemias with 11q23 chromosomal translocations, New England Journal of Medicine 329 (13) (1993) 909–914.

[25] R. Strick, P.L. Strissel, S. Borgers, S.L. Smith, J.D. Rowley, Dietary bioflavonoids induce cleavage in the MLL gene and may contribute to infant leukemia, Proceedings of National Academy of Sciences of the United States of America 9 (9) (2000) 4790–4795.

[26] G.Y. Koffi, M. Remaud-Simeon, A.E. Due, D. Combes, Isolation and chemoenzymatic treatment of glycoalkaloids from green, sprouting and rotting *Solanum tuberosum* potatoes for solanidine recovery, Food Chemistry 220 (2017) 257–265.

[27] V.A. Aletor, Allelochemicals in plant foods and feeding stuffs, part I. Nutritional, and physiopathological aspects in animal production, Veterinary and Human Toxicology 35 (1) (1993) 57–67.

[28] S.G. Sparge, M.E. Light, J. Van Staden, Biological activities and distribution of plant saponins, Journal of Ethnopharmacology 94 (2004) 219–243.

[29] V.R. Mohan, P.S. Tresina, E.D. Daffodil, Antinutritional factors in legume seeds: characteristics and determination, in: B. Caballero, P.M. Finglas, F. Toldrá (Eds.), Encyclopedia of Food and Health, Academic Press, 2016, pp. 211–220.

[30] W. Simee, Isolation and determination of antinutritional compounds from root to shells of peanut (*Arachis hypogae*), Journal of Dispersion Science and Technology 28 (2011) 341–347.

[31] M. Zagrobelny, S. Bak, B.L. Moller, Cyanogenesis in plants and arthropods, Phytochemistry 69 (2008) 1457–1468.

[32] B.O. Osuntoknun, Cassava diet and cyanide metabolism in Wister rats, British Journal of Nutrition 24 (1972) 797–805.

[33] S.C. Noonan, G.P. Savage, Oxalic acid and its effects on humans, Asia Pacific Journal of Clinical Nutrition 8 (1999) 64–74.

[34] S. Boehm, S. Huck, Presynaptic inhibition by concanavalin: are alphalatrotoxin receptors involved in action potential-dependent transmitter release? Journal of Neurochemistry 71 (2009) 2421–2430.

[35] K. Miyake, T. Tanaka, P.L. McNeil, Lectin-binding food poisoning: a new mechanism of protein toxicity, PLoS One 2 (8) (2007) e687.

[36] I.E. Liene, Protease inhibitors and lectins, in: A. Neuberger, T.H. Jukes (Eds.), Biochemistry of Nutrition, IA, vol. 27, University Park Press, Baltimore, MD, 1979.

[37] W.G. Jaffe, Hemagglutinins (lectins), in: I.E. Liener (Ed.), Toxic Constituents of Plant Foodstuffs, Academic Press, New York, 1980, pp. 73–102.

[38] M. Richardson, The protease inhibitors of plants and microorganisms, Phytochemistry 16 (1977) 159–169.

[39] J. Mikola, M. Kirsi, Differences between endospermal and embryonal trypsin inhibitors in barley, wheat and rye, Acta Chemica Scandinavica 26 (1972) 787–795.

[40] J.J. Rackis, J.E. McGhee, A.N. Booth, Biological threshold level of soybean trypsin inhibitors by rat bioassay, Cereal Chemistry 52 (1975) 85–92.

[41] J.J. Rackis, Physiological properties of soybean trypsin inhibitors and their relationship to pancreatic hypertrophy and growth inhibition of rats, Federation Proceedings 24 (1965) 1488–1493.

[42] R.B. Birari, K.K. Bhutani, Pancreatic lipase inhibitors from natural sources: Unexplored potential, Drug Discovery Today 12 (2007) 879–889.

[43] H. Sugiyama, Y. Akazome, T. Shoji, A. Yamaguchi, M. Yasue, T. Kanda, et al., Oligomeric procyanidins in apple polyphenol are main active components for inhibition of pancreatic lipase and triglyceride absorption, Journal of Agricultural and Food Chemistry 55 (11) (2007) 4604–4609.

[44] K. Satouchi, S. Matsushita, Purification and properties of a lipase inhibiting protein from soybean cotyledons, Agricultural and Biological Chemistry 40 (1976) 889–897. Available from: https://doi.org/10.1080/00021369.1976.10862147.

[45] H. Ali, P.J. Houghton, A. Soumyanath, α-Amylase inhibitory activity of some Malaysian plants used to treat diabetes; with particular reference to *Phyllanthus amarus*, Journal of Ethnopharmacology 107 (2006) 449–455.

[46] A. Aguilera-Carbo, J.S. Hernández, C. Augur, L.A. Prado-Barragan, E. Favela-Torres, C.N. Aguilar, Ellagic acid production from biodegradation of creosote bush ellagitannins by *Aspergillus niger* in solid state culture, Food and Bioprocess Technology 2 (2009) 208–212.

[47] P.P. McCue, K.A. Shetty, Model for the involvement of lignin degradation enzymes in phenolic antioxidant mobilization from whole soybean during solid-state bioprocessing by *Lentinus edodes*, Process Biochemistry 40 (2005) 1143–1150.

[48] J.J. Marshall, C.M. Lauda, Purification and properties of *Phaseolusvulgaris*, Journal of Biological Chemistry 250 (2007) 8030–8037.

[49] K.P. Latté, K.E. Appel, A. Lampen, Health benefits and possible risk of Broccoli—an overview, Food Chemistry and Toxicology 49 (12) (2011) 3287–3309.

[50] A. Abdul-Aziz, K.K. Kadhim, Efficacy of the cruciferous vegetable on the thyroid gland and the gonads in rabbits, Advances in Animal Veterinary Sciences 3 (3) (2015) 183–191.

[51] I.E. Liener, Factors affecting the nutritional quality of soy products, Journal of American Oil and Chemical Society 58 (1981) 406–415.

[52] K.R. Sridhar, S. Seena, Nutritional and antinutritional significance of four unconventional legumes of the genus *Canavalia*—a comparative study, Food Chemistry 99 (2006) 267–288.

[53] V.M. Koistinen, E. Nordlund, K. Kati, M. Ismo, P. Kaisa, H. Kati, et al., Effect of bioprocessing on the in vitro colonicmicrobial metabolism of phenolic acids from rye bran fortified breads, Journal of Agricultural and Food Chemistry 65 (9) (2017) 1854.

[54] P. Kaisa, F. Laura, K. Kati, Sourdough and cereal fermentation in a nutritional perspective, Food Microbiology 26 (7) (2009) 693–699.

[55] D. Lioger, L. Fanny, D. Christian, R. Christian, Sourdough fermentation of wheat fractions rich in fibres before their use in processed food, Journal of the Science of Food and Agriculture 87 (7) (2010) 1368–1373.

[56] K. Girigowda, T. Peterbaeur, V.H. Mulimani, Isolation and structural analysis of ajugose from *Vigna Mungo* L, Carbohydrate Research 341 (2006) 2156–2160.

[57] H.W. Lopez, A. Ouvry, E. Bervas, C. Guy, A. Messager, C. Demigne, et al., Strains of lactic acid bacteria isolated from sour doughs degrade phytic acid and improve calcium and magnesium solubility from whole wheat flour, Journal of Agricultural and Food Chemistry 48 (2000) 2281–2285.

[58] P.A. Vig, A. Walia, Beneficial effects of *Rhizopus oligosporus* fermentation on reduction of glucosinolates, fibre and phytic acid in rapeseed (*Brassicanapus*) meal, Bioresource Technology 78 (2001) 309–312.

[59] N. Canibe, B.B. Jensen, Fermented and nonfermented liquid feed to growing pigs: effect on aspects of gastrointestinal ecology and growth performance, Journal of Animal Science 81 (2003) 2019–2031.

[60] A. Skrede, S. Sahlstrøm, Ø. Ahlstrøm, K.H. Connor, G. Skrede, Effects of lactic acid fermentation and gamma irradiation of barley on antinutrient contents and nutrient digestibility in mink (*mustelavison*) with and without dietary enzyme supplement, Archives of Animal Nutrition 61 (2007) 211–221.

[61] M.R. Bedford, H. Schulze, Exogenous enzymes for pigs and poultry, Nutrition Research Reviews 11 (1998) 91–114.

[62] B. Koo, J.W. Kim, C.M. Nyachoti, Nutrient and energy digestibility and microbial metabolites in weaned pigs fed diets containing *Lactobacillus*-fermented wheat, Animal Feed Science and Technology 241 (2018) 27–37.

[63] I. Toufeili, C. Melki, S. Shadarevian, R.K. Robinson, Some nutritional and sensory properties of bulgur and whole wheatmeal kishk (a fermented milk-wheat mixture), Food Quality and Preference 10 (1999) 9–15.

[64] M.G. Albarracin, J. Rolando, S.R. Drago, Effect of soaking process on nutrient bio-accessibility and phytic acid content of brown rice cultivar, LWT—Food Science and Technology 53 (2013) 76–80.

[65] N. Khan, R. Zaman, M. Elahi, Effect of processing on the phytic acid content of wheat products, Journal of Agricultural and Food Chemistry 34 (1986) 1010–1012.

[66] S.M. Dagher, S. Shadarevian, W. Birbari, Preparation of high bran Arabic bread with low phytic acid content, Journal of Food Science 52 (1987) 1600–1603.

[67] M. Bartnik, J. Florysiak, Phytate hydrolysis during breadmaking in several sorts of Polish bread, Die Nuhrung 32 (1988) 37–42.

[68] B. Amoa, H.G. Muller, Studies on kenkey with particular reference to calcium and phytic acid, Cereal Chemistry 53 (1976) 365–375.

[69] N. Dhanker, B.M. Chauhan, Effect of temperature and fermentation time on phytic acid and polyphenol content of rabadi-a fermented pearl millet food, Journal of Food Science 52 (1987) 828–829.

[70] K.A. Buckle, Phytic acid changes in soybeans fermented by traditional inoculum and six strains of *Rhizopus oligosporus*, Journal of Applied Microbiology 58 (1985) 539–543.

[71] K.A. Sutardi, Buckle, Reduction in phytic acid levels in soybeans during tempeh production, storage, and frying, Journal of Food Science 50 (1985) 260–263.

[72] N.R. Reddy, M.D. Pierson, D.K. Salunkhe, Legume-based Fermented Foods, CRC Press, Boca Raton, FL, 1986.

[73] O.U. Eka, Effect of fermentation on the nutrient status of locust beans, Food Chemistry 5 (1980) 303–308.

[74] N. Bilgicli, A. Elgun, S. Turker, Effects of various phytase sources on phytic acid content, mineral extractability and protein digestibility of tarhana, Food Chemistry 98 (2006) 329–337.

[75] M. Magala, Z. Kohajdova, J. Karovicova, Degradation of phytic acid during fermentation of cereal substrates, Journal of Cereal Science 61 (2015) 94–96.

[76] M.A. Osman, Changes in sorghum enzyme inhibitors, phytic acid, tannins and *in vitro* protein digestibility occurring during Khamir (local bread) fermentation, Food Chemistry 88 (2004) 129–134.

[77] T. Zotta, A. Ricciardi, E. Parente, Enzymatic activities of lactic-acid bacteria isolated from Cornetto di Matera sourdoughs, International Journal of Food Microbiology 115 (2007) 165–172.

[78] M.Z. Algadi, N.E. Yousif, Anti-nutritional factors of green leaves of *Cassia obtusifolia* and Kawal, Journal Food Processing and Technology 6 (2015) 483. Available from: https://doi.org/10.4172/2157-7110.1000483.

[79] P.H. Selle, V. Ravindran, A. Caldwell, W.L. Bryden, Phytate and phytase: consequences for protein utilisation, Nutrition Research Reviews 13 (2007) 255–278.

[80] K.J. Hong, C.H. Lee, S.W. Kim, *Aspergillus oryzae* gb-107 fermentation improves nutritional quality of food soybeans and feed soybean meals, Journal of Medicinal Food 7 (2004) 430–435.

[81] S. Avilés-Gaxiola, C. Chuck-Hernández, S.O. Serna Saldívar, Inactivation methods of trypsin inhibitor in legumes: a review, Journal of Food Science 83 (1) (2018) 17–29. Available from: https://doi.org/10.1111/1750-3841.13985.

[82] N. Jiménez, M. Esteban-Torres, J.M. Mancheño, B. de las Rivas, R. Muñoz, Tannin degradation by a novel tannase enzyme present in some *Lactobacillus plantarum* strains, Applied and Environmental Microbiology 80 (2014) 2991–2997.

[83] A.K. Sariri, A.M.W. Mulyono, A.I.N. Tari, The utilization of microbes as a fermentation agent to reduce saponin in Trembesi leaves (*Sammaneasaman*), IOP Conference Series: Earth and Environmental Science 142 (1) (2018) 012041.

[84] A.T. Vasconcelos, D.R. Twiddy, A. Westby, P.J.A. Reilly, Detoxification of *Cassava duringgari* preparation, International Journal of Food Science Technology 25 (1990) 198–203.

[85] O.C.C. Kemdirimi, O.A. Chukwu, S.C. Anchinewhu, Effect of traditional processing of cassava on the cyanide content of gari and cassava flour, Plant Foods for Human Nutrition 48 (1995) 335–339.

[86] G. Tuncel, M.J.R. Nout, L. Brimer, Degradation of cyanogenic glycosides of bitter apricot seeds (*Prunus armeniaca*) by endogenous and added enzymes as affected by heat treatments and particle size, Food Chemistry 63 (1995) 65–69.

[87] I.F. Bolarinwa, M.O. Oke, S.A. Olaniyan, A.S. Ajala, A review of cyanogenic glycosides in edible plants, in: S. Sonia, L.L. Marcelo (Eds.), Toxicology—New Aspects to This Scientific Conundrum, IntechOpen, 2016. Available from: https://doi.org/10.5772/64886.

[88] K.A. Rai, A. Appaiah, Application of native yeast from Garcinia (*Garciniaxanthochumus*) for the preparation of fermented beverage: changes in biochemical and antioxidant properties, Food Bioscience 5 (2014) 101–107.

[89] J. Okombo, M. Liebman, Probiotic-induced reduction of gastrointestinal oxalate absorption in healthy subjects, Urological Research 38 (3) (2010) 169–178.

[90] C. Campieri, M. Campieri, V. Bertuzzi, et al., Reduction of oxaluria after an oral course of lactic acid bacteria at high concentration, Kidney International 60 (3) (2001) 1097–1105.

[91] J.C. Lieske, D.S. Goldfarb, C. de Simone, C. Regnier, Use of a probiotic to decrease enteric hyperoxaluria, Kidney International 68 (3) (2005) 1244–1249.

[92] C. Kwak, B.C. Jeong, J.H. Ku, et al., Prevention of nephrolithiasis by *Lactobacillus* in stone-forming rats: a preliminary study, Urological Research 34 (4) (2006) 265–270.

[93] T.S. Simpson, G.P. Savage, R. Sherlock, L.P. Vanhanen, Oxalate content of silver beet leaves (*Betavulgaris* var. cicla) at different stages of maturation and the effect of cooking with different milk sources, Journal of Agricultural and Food Chemistry 57 (2009) 10804–10808.

[94] Y. Wadamori, L. Vanhanen, G.P. Savage, Effect of Kimchi fermentation on Oxalate levels in silver beet (*Beta vulgaris* var. cicla), Foods 3 (2) (2014) 269–278.

[95] T. Shibamoto, L.F. Bjeldanes, Chapter 5—Natural toxins in plant foodstuffs, in: S. Takayuki, F.B. Leonard (Eds.), Food Science and Technology, Introduction to Food Toxicology, Academic Press, 1993, pp. 67–96. ISBN 9780080925776. Available from: https://doi.org/10.1016/B978-0-08-092577-6.50010-1.

[96] M. Egounlety, O.C. Aworh, Effect of soaking, dehulling, cooking and fermentation with *Rhizopus oligosporus* on the oligosaccharides, trypsin inhibitor, phytic acid and tannins of soybean (*Glycine max Merr.*), cowpea (*Vigna unguiculata L. Walp*) and groundbean (*Macrotylomageocarpa Harms*), Journal of Food Engineering 56 (2) (2003) 249–254.

[97] M. Barać, S. Stanojević, The effect of microwave roasting on soybean protein composition and components with trypsin inhibitor activity, Acta Alimentaria 34 (1) (2005) 23–31.

[98] F.P.P. Machado, J.H. Queiróz, M.G.A. Oliveira, N.D. Piovesan, M.C.G. Peluzio, N.M.B. Costa, et al., Effects of heating on protein quality of soybean flour devoid of Kunitz inhibitor and lectin, Food Chemistry 107 (2) (2008) 649–655.

[99] Y. Chen, Z. Xu, C. Zhang, X. Kong, Y. Hua, Heat-induced inactivation mechanisms of Kunitz trypsin inhibitor and Bowman-Birk inhibitor in soymilk processing, Food Chemistry 154 (2014) 108–116.

[100] H.W. Yang, C.K. Hsu, Y.F. Yang, Effect of thermal treatments on anti-nutritional factors and antioxidant capabilities in yellow soybeans and green-cotyledon small black soybeans, Journal of the Science of Food and Agriculture 94 (9) (2014) 1794–1801.

[101] T.A. El-Adawy, E.H. Rahma, A.A. El-Bedawy, T.Y. Sobihah, Effect of soaking process on nutritional quality and protein solubility of some legume seeds, Nahrung 44 (5) (2000) 339–343.

[102] M. Friedman, Chemistry, biochemistry, nutrition, and microbiology of lysinoalanine, lanthionine, and histidinoalanine in food and other proteins, Journal of Agricultural and Food Chemistry 47 (4) (1999) 1295–1319.

[103] G.S. Gilani, E. Sepehr, Protein digestibility and quality in products containing antinutritional factors are adversely affected by old age in rats, Journal of Nutrition 133 (1) (2003) 220–225.

[104] Y.L. Gao, C.S. Wang, Q.H. Zhu, G.Y. Qian, Optimization of solid-state fermentation with *Lactobacillus brevis* and *Aspergillus oryzae* for trypsin inhibitor degradation in soybean meal, Journal of Integrative Agriculture 12 (5) (2013) 869–876.

[105] Z. Barampama, R.E. Simard, Oligosaccharides, antinutritional factors, and protein digestibility of dry beans as affected by processing, Journal of Food Science 59 (4) (1994) 833–838.

[106] S. Seo, S. Cho, Changes in allergenic and antinutritional protein profiles of soybean meal during solid-state fermentation with *Bacillus subtilis*, LWT 70 (2016) 208–212. Available from: https://doi.org/10.1016/j.lwt.2016.02.035.

[107] C.Y. Shi, J. He, J.P. Wang, J. Yu, B. Yu, X.B. Mao, et al., Effects of *Aspergillus niger* fermented rapeseed meal on nutrient digestibility, growth performance and serum parameters in growing pigs, Animal Science Journal 87 (4) (2016) 557–563. <https://doi.org/10.1111/asj.12457.2016>.

[108] Y. Wang, J. Liu, F. Wei, X. Liu, C. Yi, Y. Zhang, Improvement of the nutritional value, sensory properties and bioavailability of rapeseed meal fermented with mixed microorganisms, LWT—Food Science and Technology 112 (2019) 108238.

11

Mycotoxins in foods: impact on health

Samuel Ayofemi Olalekan Adeyeye

DEPARTMENT OF FOOD TECHNOLOGY, HINDUSTAN INSTITUTE OF TECHNOLOGY AND SCIENCE, HINDUSTAN UNIVERSITY, CHENNAI, INDIA

11.1 Introduction

Mycotoxins are low-molecular-weight toxic secondary metabolites produced mainly by fungi. These toxins when ingested by humans or animals portend danger and result in serious side effects. Consumption of foods contaminated with mycotoxins could lead to vomiting, weight loss, tumors growth, and death depending on fatality. Several mycotoxins of importance have been detected in legumes, dried root, and tuber crops as well as in cereal grain crops. Researches have shown that mycotoxins are toxigenic to human organs such as blood, kidneys, skin, or central nervous system, and their side effects could be fatal to human or animal health. Some mycotoxins have been found to be carcinogenic while others are teratogenic [1–5].

Agricultural products may be contaminated either in the field before harvesting or after harvesting through poor cultural practices and when postharvest storage conditions are favorable. Although crops could be contaminated with mycotoxins when stored improperly, mycotoxins may also destroy crops in the field when they are infected with certain fungal diseases. In Africa and many tropical and subtropical countries, staple crops such as cereals (maize, rice, sorghum, and millet, etc.) and legumes (soybean, cowpea, groundnuts, etc.) are contaminated with mycotoxins produced by different fungi. The degree of contamination by mycotoxin is affected by several factors such as agronomic, sociological, and climatic [1].

An estimate of over 500 million people, according to the World Health Organization, is exposed to unsafe mycotoxin levels all over the world, with sub-Saharan Africa contributing the bulk. Unsafe level of mycotoxin exposure among infants leads to increased mortality and morbidity. It has been observed that contamination of foods with mycotoxins particularly in Africa has become a difficult obstacle in achieving 15 of the 17 Sustainable Development Goals of the United Nations [1].

Mycotoxicoses are common outcome of consumption of foods contaminated with mycotoxins by the people. Toxicity of mycotoxins could vary from acute to chronic, depending on prevailing factors such like the type of toxin involves, dosage, age, and susceptibility of consumers to mycotoxins based on their immune system. Data from prolong consumption of mycotoxin-contaminated foods revealed that mycotoxins could cause suppression in the immune system and risk of cancer in humans and animals [1–5].

About 25% of world's food crops produced is affected by mycotoxins, resulting in high annual economic losses in billions of dollars. Mycotoxins contamination of agricultural produce above the acceptable limits stipulated by the regulatory agencies could result to economic losses to both farmers and exporters due to rejection of produce at both domestic and international levels. Common economic losses due to mycotoxins contamination are loss of human and animal life, increased cost of health care, lower livestock production, crop losses, produce rejection, increased research, and regulatory costs associated with control and mitigation of mycotoxins [1–5].

In recent times, it has been a serious challenge to provide food of the right quality and nutrients free from mycotoxins particularly in Africa. Therefore this study examined mycotoxins in foods, impact on health and food security. This will assist in reducing postharvest losses, food contamination by fungi and mycotoxins as well as improving public health and consumer safety.

11.2 Mycotoxins in foods

Over 300 mycotoxins have been commonly identified, isolated, and characterized in agricultural produce [6] (Fig. 11-1). The major mycotoxins in agricultural produce are aflatoxins, ochratoxins, deoxynivalenol (DON), zearalenone, fumonisin, T-2 and T-2 like toxins (trichothecenes), and alternariol [6]. Aflatoxins are the most common important mycotoxins in agricultural produce used to feed humans and animals and are potent human and animal hepatocarcinogens causing cancer. They are metabolic products of *Aspergillus* species such as *Aspergillus flavus* and *Aspergillus parasiticus* that produce aflatoxins in maize and groundnuts under stress or drought in the field and also in stored produce [3,5,7]. Four common types of aflatoxins are B1, B2, G1, and G2 [4]. However, aflatoxin B1 (AFB1) is the most dangerous and toxic among the four aflatoxins. It is carcinogenic and has been known to cause liver cancer in animal models [1].

Ochratoxins are metabolites of *Penicillium* and *Aspergillus* species. There are three kinds of ochratoxins, namely, A, B, and C [2]. The three kinds of ochratoxins differ from each other, ochratoxin B is a nonchlorinated variant of ochratoxin A (OTA) while ochratoxin C is an ethyl ester variant of OTA [1–3,8]. *Aspergillus ochraceus* produces ochratoxins in beverages such as beer and wine. *Aspergillus carbonarius* produces ochratoxins in vine fruits, especially during juice making process. OTA is a potent carcinogen and nephrotoxin and implicated in the formation of tumors in the human urinary tract [1–3,8,9].

Citrinins are also metabolites of *Penicillium* and *Aspergillus* species. *Penicillium camemberti* produces mycotoxins in cheese while *Aspergillus oryzae* produces citrinins in sake, miso, and soy sauce. Citrinins are nephrotoxins and implicated in animal models and in yellow rice disease in Japan [1,2,10]. These have been identified, isolated, and implicated in

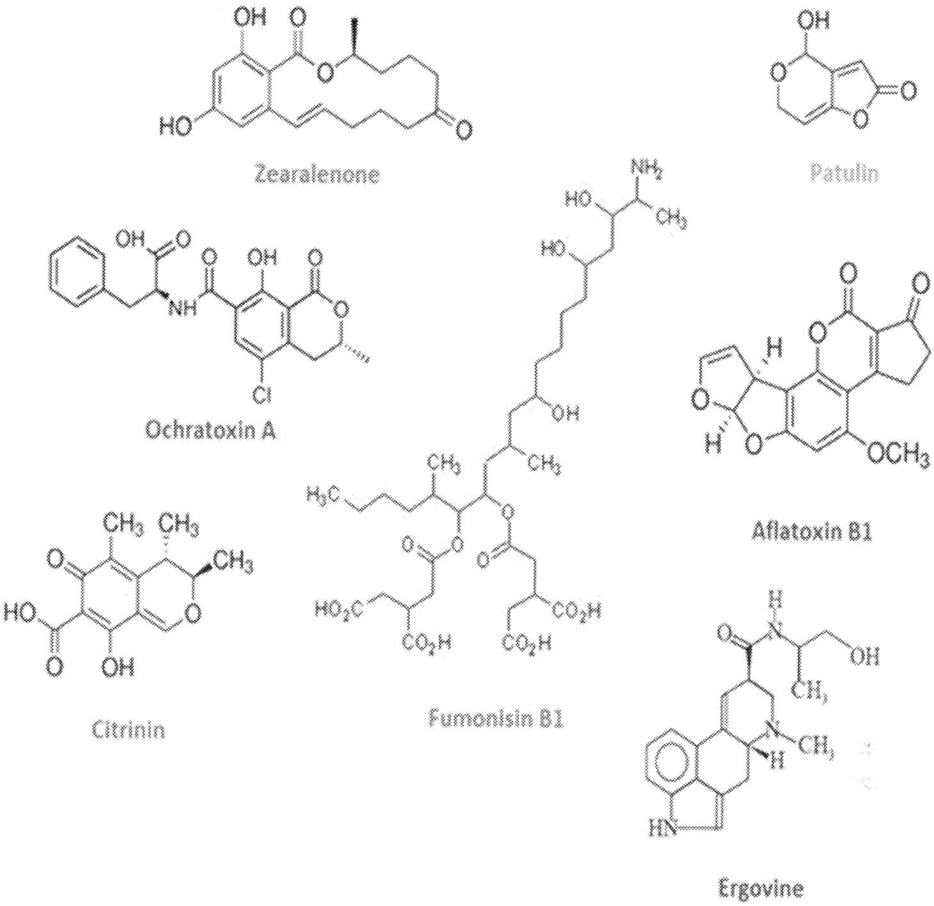

FIGURE 11–1 Molecular structure of some mycotoxins [13].

cereals such as barley, oats, rye, wheat, rice, and corn. Citrinin and OTA to act synergistically to depress RNA synthesis in murine kidneys but its effects on human health are yet to be established [1–3,9,10].

Ergots are alkaloids and are highly toxic compounds. They are toxins present in the sclerotia of species of *Claviceps*. Ergot sclerotia are found in cereals and the toxins could be consumed resulting in ergotism even when the flour is used for bread preparation. Good agricultural practices coupled with modern storage techniques eliminate ergotism as a human disease but it is still found in animal feed [1–3,9,10]. Patulin is a metabolic product of fungal species of *Penicillium*, *Aspergillus*, and *Paecilomyces* sp. *P. expansum* is a major contaminant of fruits and vegetables where its causes moldiness, specifically rotting of apples and figs [1,2,11,12]. However, fermentation is used to remove patulin and is eliminated in apple beverages as in cider. Although patulin is a noncarcinogenic mycotoxin, it is implicated in the destruction of the immune system of animals [1,2,11].

Fusarium mycotoxins are common in immature cereal grains such as maize, wheat, etc. Over 50 species of *Fusarium* species secret toxins of importance [1,2,14]. Common *Fusarium* mycotoxins are fumonisins, known to cause nervous disorders in horses. Fumonisins are potent carcinogens in rodents and animal models. Other *Fusarium* mycotoxins are trichothecenes, known to produce chronic and fatal toxicoses in animals and humans; and zearalenone. Some other *Fusarium* mycotoxins are beauvercin and enniatins, butenolide, equisetin, and fusarins [1,2,15]. Although fungi or molds infest agricultural produce in the field before harvest or after harvesting due to improper storage, certain crop diseases could also be responsible for the production of mycotoxins on the field. Factors that could boost fungal infestation in the field include excessive rain at the time of flowering, droughts during harvest and postharvest stages, and mold growth.

11.3 Mycotoxins in foods: impact on health

Mycotoxins are chemical substances produced by fungi, and their ingestion could lead to serious health implications for both animals and humans and result into great economic losses to the farmers, importers, and the nation. Several works have been carried out by researchers on health implications of ingesting mycotoxins by humans and animals. Adejumo and Adejoro [7] reported that aflatoxin could cause hepatocellular carcinoma or liver cancer. WHO [16] reported that aflatoxins are the third-leading cause of cancer in humans worldwide with staggering figure of 600,000 fresh cases annually. Adejumo and Adejoro [7], Hussein and Brussel [17], and Zain [18] also reported that aflatoxins can affect the kidney and the liver weights and also lower feed intake in livestock. They also linked aflatoxins to hepatitis and immune-suppression which lead to high death rate in livestock.

Mycotoxins have been linked to the outbreaks of several diseases in many parts of the world. In India, the outbreak of enteric ergotism was linked to ergot; also, the outbreak of vascular ergotism in Ethiopia was linked to ingestion of ergot [1,2]. The outbreak of DON mycotoxicosis in India and China has been linked to mycotoxin contamination of staple foods such as cereal and legume grains in those countries [1,2]. Most of the outbreaks have been linked to consumption of contaminated cereal grains and legumes grown under drought or stress conditions and stored under humid or tropical conditions that encourage the growth and proliferation of molds [2,19].

Aflatoxins are associated with several human diseases such as liver cancer, Reye's syndrome, Indian childhood cirrhosis, and chronic gastritis in many tropical, subtropical, and developing countries, with particular reference to Africa and Asia due to favorable weather conditions that encourage fungi growth and mycotoxins production [1,2,19,20].

Africa, particularly Kenya, recorded the most serious, largest, well-reported, and documented acute outbreaks of aflatoxicosis in human history in 2004, over 125 people were reported dead and 317 others hospitalized for consumption of maize contaminated with aflatoxin [1,2,21]. Kenya aflatoxicosis outbreak according to Lewis et al. (2005) [22] was attributed to the consumption of home-grown maize stored under moist conditions and

Table 11-1 Mycotoxins in staple grains and seeds.

Mycotoxin	Commodity	Fungal source(s)	Effects of ingestion	Reference
Deoxynivalenol/ nivalenol	Wheat, maize, barley	Fusarium graminearumFusarium crookwellenseFusarium culmorum	Human toxicoses in India, China, Japan, and Korea. Toxic to animals, especially pigs	Food and Agriculture Organization (FAO) [24]
Zearalenone	Maize, wheat	F. graminearumF. culmorumF. crookwellens	Identified by the International Agency for Research on Cancer (IARC) as a possible human carcinogen. Affects reproductive system in female pigs	FAO [24]
Ochratoxin A	Barley, wheat, and many other commodities	Aspergillus ochraceusPenicillium verrucosum	Suspected by IARC as human carcinogenCarcinogenic in laboratory animals and pigs	FAO [24]
Fumonisin B1	Maize	Fusarium moniliforme plus several less common species	Suspected by IARC as human carcinogen. Toxic to pigs and poultry. Cause of equine eucoencephalomalacia (ELEM), a fatal disease of horses	FAO [24]
Aflatoxin B1, B2Aflatoxin B1, B2, G1, G2	Maize, peanuts, and many other commoditiesMaize, peanuts	Aspergillus flavusAspergillus parasiticus	Aflatoxin B1, and naturally occurring mixtures of aflatoxins, identified as potent human carcinogens by IARC. Adverse effects in various animals, especially chickens	FAO [24]

Sources: Adapted from S.A.O. Adeyeye, Aflatoxigenic fungi and mycotoxins in food: a review, Critical Reviews in Food Science and Nutrition 60 (2019) 709–721 [1]; S.A.O. Adeyeye, Fungal mycotoxins in foods: a review, Cogent Food & Agriculture, 2 (2016) 1213127 [2].

contaminated with toxic AFB1 by people. Daily exposure of 50 mg/day AFB1was reported by Probst et al. [23] for individual during the outbreak. Aflatoxins cause other diseases (Table 11−1) such as esophageal cancer in southern Africa [24].

11.4 Economic implications of mycotoxins in foods

There are several economic implications due to contamination of agricultural produce particularly food and feed for man and animals by mycotoxins. In 1998 Food and Agriculture Organization (FAO) reported that global food trade in agricultural produce such as rice, wheat, barley, sorghum, corn, soybeans, groundnuts, and oilseeds was several millions of tons annually [1,2]. Contamination of the agricultural produce by mycotoxins may occur in the field before harvesting because of poor agricultural practices and or due to late harvesting or after harvesting due to poor postharvest handling practices. Poor and humid storage conditions also encourage fungal growth [1,2]. Large financial losses may be incurred due to contamination of staple foods by mycotoxins. Contamination due to mycotoxins could result to reduce productivity; losses of foreign exchange earnings through rejection of produce exported due to poor quality; high inspection costs, high sampling, and analysis of produce costs; high costs of compensation paid in case of claims; high costs of farmers' subsidies and high costs of research, training, and extension program that could enormous [1,2,25]. Reduced quality and quantity of grains cereals and legumes like wheat, maize, cowpea, and groundnuts as a result of contamination by mycotoxins could be staggering. Over 25% annual losses to the tune of one billion metric tons of agricultural produce worldwide has been reported by Schmale III and Munkvold [26].

11.5 Mitigation and control of mycotoxins in foods

Controlling and mitigating contamination of foods with mycotoxins as a food safety challenge need concerted efforts from all stakeholders in order to reduce the health, trade, income, and food security. The mitigation and control strategy may involve taking care of the entire food value chain—from farms to industries, traders, national storage systems, regulators, and relevant regulatory agencies. The strategy may need to consider several underlining factors such as the use of pre- and postharvest resistant crops, good agricultural practices, improved storage, bioremediation/control, and the study of economic losses and impact. Strengthening the public—private partnerships as well as harmonized regional standards and certification by various countries may go a long way to reduce the impact of mycotoxin contamination in foods. At household or family level traditional processing methods like hand-sorting, nixtamalization, and dietary improvement such as diversify diet may help while at industrial level, mechanical processing like optical-sorting, dry milling can reduce mycotoxin contamination of foods.

Currently, in many developing countries particularly in Africa, mycotoxin control measures are poorly handled and not easy to achieve the set objectives due to poor awareness among the people, poor implementation of food monitoring regulations, ineffective relevant regulatory agencies, and poor funding. Controlling and mitigating mycotoxins globally involves time and financial commitment from various nations to achieve desired results. Interventions from agencies such as FAO, Food and Drug Administration and Control, and Ministry of Health Officials may go a long way in making necessary impact.

The following steps could be taken to reduce mycotoxin contamination and mitigate its effects on human health.

11.5.1 Promoting and protecting human lives

In order to reduce the impact of mycotoxins on human health, government and its regulatory agencies should recognize mycotoxins as a major public health issue and control of mycotoxins should be prioritized. Various governments need to invest in developing comprehensive, sustainable, and effective interventions to control and mitigate mycotoxin contamination in foods. Governments at all levels national or regional should invest heavily and raise funds to develop practicable community- and technology-based control approaches. Government could achieve this through public–private partnerships in order to gather resources together to scale-up technologies and strategies for mycotoxin mitigation and control.

11.5.2 Development and application of innovative and resourceful technologies

Innovative and resourceful technologies could be developed and applied to share necessary information to the local farmers. Government at all levels could develop simple, cost-effective, practical mycotoxin control technologies at both pre- and postharvest stages. They may also develop good policies and develop active training programs that will help farmers and consumers. There is a need to develop and apply well-coordinated actions among farmers, governments, development partners, researchers, agricultural organizations, policymakers, health-sector, and other relevant stakeholders in order to achieve desirable results [1,27].

11.5.3 Adequate monitoring of mycotoxin management

Adequate and effective monitoring of mycotoxin management is a veritable way to achieve desired goal in mycotoxin management. Adequate and transparent monitoring of mycotoxin mitigation and control which will help in getting feedbacks from stakeholders such as farmers and rural households to open up to new improvement techniques [26,27].

11.5.4 Transparent and responsible data sharing

Data sharing is one of the ways to achieve good control and mitigation of mycotoxins contamination in foods. Large data are generated in various countries by relevant regulatory agencies and regional blocks on mycotoxin contamination of foods and sharing these data will assist in reducing the impact of mycotoxin contamination. There is also a need for collaboration between various regulatory agencies and sharing data that are continuously generated will boost mycotoxin control and mitigation in various countries involved. Developing good data repository system that promotes open, transparent, and responsible data sharing among nations and stakeholders will improve the fight against the mycotoxin threat [1,26,27].

11.5.5 Development and formulation of good policies and regulations

By developing and formulating good policies will assist in effective control and mitigation of mycotoxin contamination in foods. To develop and formulate effective policies and regulations for mycotoxins control, there is a need for quality and reliable data. Gathering high-quality data requires research and interpretation of information to get desirable outputs. Multidisciplinary collaborations are required from researchers to assist the policy-makers in taking informed decisions which may help in compliance and adherence to regulations made by the stakeholders [1,2].

Legislations on mycotoxins are very rare and nonexisting in some African countries. Legislations on control and intake of mycotoxins from food in many African countries are poorly coordinated because of poor policy planning and implementation. Although mycotoxins of importance like aflatoxins, ergot alkaloids, DON, and ochratoxins are recognized but legislations to promote and improve food safety and consumer health to control food contamination in Africa are nonexistent. In addition, Codex Alimentarius Commission (CAC) has developed international rules and regulations that could help to guide the tolerant amounts of mycotoxins, especially aflatoxins that are acceptable in agricultural produce [1,2,19,29]. Efforts have been made to mitigate and control mycotoxins in agricultural produce particularly those involved international trade. However, because of the food shortage and inadequacy in food supply among the people in Africa legislations for mycotoxin control in foods particularly staple food such cereal and legume grains may be difficult to achieve and implement [1,2,19,29].

Although CAC stipulated the afltoxin acceptable limits for major foodstuffs such as 0.05 mg/l of aflatoxin for milk and milk products, 4 to 5 mg/kg of aflatoxin for beans, 10 mg/kg of aflatoxin in nuts like peanuts and almonds, and 20 mg/kg of aflatoxin for cereals, the implementation could be difficult because of food shortages in developing countries particularly among African nations [1,2,7,21]. The European Union also stipulated limit for OTA in raw cereal to 0.5 mg/kg, with a tolerable weekly intake of 0.12 mg/kg, while maximum DON limits were set at 200 mg/kg for processed cereal-based foods and baby infants formula and foods for young children, and 1750 mg/kg for unprocessed wheat, oats, and maize [1,2,7,30]. Many African countries like Nigeria and Ghana have adopted the safety limits but implementing the regulations has been difficult to achieve.

11.6 Conclusions and perspectives

Food contamination by mycotoxins has continued to be a public health issue and the challenge has been due to poor storage conditions of agricultural produce that encourages fungal infestation and subsequent mycotoxin contamination of food in most African countries and other tropical countries. The current problems of food shortages, poor agricultural, and storage practices have exposed many people in the developing countries, particularly Africans to the consumption of mycotoxin-contaminated foods or agricultural produce which endanger the health and well-being of the people with attendance effects on public health and consumer safety. However, improving agricultural production practices, providing adequate and favorable storage facilities for agricultural produce will enhance food safety, consumer health, and availability of needed food for the consumers. Public education and enlightenment on the effects and impact of mycotoxins on economic and health need to be vigorously implemented. Proactive and effective implementation and enforcement of food safety legislations and regulations to control mycotoxin contamination of agricultural produce. Governments at various levels in the developing countries should establish better handling and storage facilities to reduce fungal infestation of foods and effects of mycotoxins in food consumers to promote consumer health and food security.

Effective management of mycotoxins, especially in agricultural produce that is prone to mycotoxin contamination and fungi growth along the food chain, the following steps could be taken.

1. There is a need for HACCP compliance with particular reference to good agronomic practices including transportation and distribution until the produce gets to the consumers' table;
2. Applying best postharvest practices including drying and storage of agricultural commodities and good pest control;
3. Adopting sustainable food safety procedures and farmers' education;
4. Proactive and effective implementation and enforcement of food safety legislations and regulations to control mycotoxin contamination of agricultural produce;
5. Implementing sustainable and reliable mycotoxins detection methods to protect the public and enhance international trade;
6. Creating mycotoxin awareness groups with officials of the Ministries of Agriculture and Natural Resources and Health that will take care of:
    a. Mycotoxins monitoring and surveillance in food and feeds;
    b. Providing mycotoxins information in foods and feeds to stakeholders and regulatory agencies for necessary actions;
    c. Providing relevant education and awareness on mycotoxins to the consumers;
    d. Providing training on regular basis on mycotoxin testing, and mitigation.

References

[1] S.A.O. Adeyeye, Aflatoxigenic fungi and mycotoxins in food: a review, Critical Reviews in Food Science and Nutrition 60 (2019) 709–721. Available from: https://doi.org/10.1080/10408398.2018.1548429.

[2] S.A.O. Adeyeye, Fungal mycotoxins in foods: a review, Cogent Food & Agriculture 2 (2016) 1213127.

[3] S. Ashiq, Natural occurrence of mycotoxins in food and feed: Pakistan perspective, Comprehensive Reviews in Food Science and Food Safety 14 (2015) 159–175. Available from: https://doi.org/10.1111/crf3.2015.14.

[4] Y.N. Yin, L.Y. Yan, J.H. Jiang, Z.H. Ma, Biological control of aflatoxin contamination of crops, Journal of Zhejiang University Science B 9 (10) (2008) 787–792. Available from: https://doi.org/10.1631/jzus.B0860003.

[5] M.L. Martins, H.M. Martins, F. Bernardo, Aflatoxins in spices marketed in Portugal, Food Additives and Contaminants 18 (4) (2001) 315–319. Available from: https://doi.org/10.1080/02652030120041.

[6] G.O. Adegoke, P. Letuma, Strategies for the prevention and reduction of mycotoxins in developing countries, mycotoxin and food safety in developing countries, IntechOpen, 2013. Hussaini Anthony Makun, Available from: https://doi.org/10.5772/52542.

[7] T.O. Adejumo, D.O. Adejoro, Incidence of aflatoxins, fumonisins, trichothecenes and ochratoxins in Nigerian foods and possible intervention strategies, Food Science and Quality Management 31 (2014) 127–146.

[8] P. Bayman, J.L. Baker, Ochratoxins: a global perspective, Mycopathologia 162 (3) (2006) 215–223. Available from: https://doi.org/10.1007/s11046-006-0055-4.

[9] P. Jeswal, D. Kumar, Mycobiota and natural incidence of aflatoxins, ochratoxin A, and citrinin in Indian spices confirmed by LC-MS/MS, International Journal of Microbiology 2015 (2015) 1. Available from: https://doi.org/10.1155/2015/242486.

[10] J.W. Bennett, M. Klich, Mycotoxins, Clinical Microbiology Reviews 16 (2003) 497–516. Available from: https://doi.org/10.1128/CMR.16.3.497-516.2003.

[11] M.O. Moss, Fungi, quality and safety issues in fresh fruits and vegetables, Journal of Applied Microbiology 104 (5) (2008) 1239–1243. Available from: https://doi.org/10.1111/j.1365-2672.2007.03705.x.

[12] M.W. Trucksess, P.M. Scott, Mycotoxins in botanicals and dried fruits: a review, Food Additives & Contaminants: Part A 25 (2008) 181–192. Available from: https://doi.org/10.1080/02652030701567459.

[13] T.H. Fernandes, J. Ferrao, V. Bell, T. Chabite, Mycotoxins, food and health, Journal of Nutrition Health Food Science 5 (7) (2017) 1–10. Available from: https://doi.org/10.15226/jnhfs2017.001118.

[14] A.W. Schaafsma, D.C. Hooker, Climatic models to predict occurrence of Fusarium toxins in wheat and maize, International Journal of Food Microbiology 119 (1–2) (2007) 116–125.

[15] A.E. Desjardins, R.H. Proctor, Molecular biology of fusarium mycotoxins, International Journal of Food Microbiology 119 (1–2) (2007) 47–50. Available from: https://doi.org/10.1016/j.ijfoodmicro.2007.07.024.

[16] WHO, Mycotoxins in African foods: Implications to Food Safety and Health. AFRO Food Safety Newsletter. World Health Organization Food Safety (FOS), July 2006, http://www.afro.who.int./des.

[17] H.S. Hussein, J.M. Brasel, Toxicity, metabolism, and impact of mycotoxins on humans and animals, Toxicology 167 (2) (2001) 101–134.

[18] E.M. Zain, Impact of mycotoxins on humans and animals, Journal of Saudi Chemical Society 15 (2) (2011) 129–144. Available from: https://doi.org/10.1016/j.js.

[19] C.F. Jelinek, A.E. Pohland, G.E. Wood, Worldwide occurrence of mycotoxins in foods and feeds (an update), Journal of the Association of Official Analytical Chemists 72 (1989) 223–230.

[20] M.S. Palmgren, A.W. Hayes, Aflatoxins in food, in: P. Krogh (Ed.), Mycotoxins in Food, Academic Press, Cambridge, MA, 1987, pp. 65–95.

[21] Council for Agricultural Science and Technology (CAST), Mycotoxins: risks in plant, animal, and human systems. Task Force Report, 139, Ames, IA, 2003.

[22] L. Lewis, M. Onsongo, H. Njapau, H. Schurz-Rogers, G. Luber, S. Kieszak, et al., Kenya Aflatoxicosis Investigation Group. Aflatoxin contamination of commercial maize products during an outbreak of acute aflatoxicosis in Eastern and Central Kenya, Environmental Health Perspectives 113 (12) (2005) 1763–1767, doi:10.1289/ehp.7998.

[23] C. Probst, H. Njapau, P.J. Cotty, Outbreak of an acute aflatoxicosis in Kenya in 2004: identification of the causal agent, Applied and Environmental Microbiology 73 (8) (2007) 2762–2764.

[24] FAO, Worldwide Regulations for Mycotoxins in Food and Feed in 2003, FAO, Rome, 2004, p. 81.

[25] B. Coulibaly, The problem of aflatoxin contamination of groundnut and groundnut products as seen by the African Groundnut Council, in: D. McDonald, V.K. Mehan (Eds.), Aflatoxin Contamination of Groundnuts, International Crops Research Institute for the Semi-Arid Tropics, Patancheru, 1989, pp. 47−55. Proceedings of the International Workshop.

[26] D.G. Schmale, III, G.P. Munkvold, Mycotoxins in crops: a threat to human and domestic animal health. Topics in Plant Pathology, Mycotoxins: Economic Impact. The American Phytopathological Society, 2018.

[27] R. Bandyopadhyay, A. Ortega-Beltran, A. Akande, P.J. Cotty, Biological control of aflatoxins in Africa: current status and potential challenges in the face of climate change, World Mycotoxin Journal 9 (2016) 771−789.

[28] C.N. Ezekie, A. Ortega-Beltran, R. Bandyopadhyay, The need for integrated approaches to address food safety risk: the case of mycotoxins in Africa. IFSC-1/19/TS1.5, in: The Future of Food Safety. FAO, WHO and African Union and World Trade Organization, 2019.

[29] E.M. Fox, B.J. Howlett, Secondary metabolism: regulation and role in fungal biology, Current Opinion in Microbiology 11 (6) (2008) 481−487. Available from: https://doi.org/10.1016/j.mib.2008.10.007.

[30] European Commission (EC), Commission regulation (EC) No 1881/2006 of 19 December 2006 setting maximum levels for certain contaminants in foodstuffs (text with EEA relevance), Official Journal of the European Union 364 (2006) 5−24.

12

Gut microbes: Role in production of nutraceuticals

Palanisamy Athiyaman Balakumaran[1], K. Divakar[2], Raveendran Sindhu[1], Ashok Pandey[3], Parameswaran Binod[1]

[1]MICROBIAL PROCESSES AND TECHNOLOGY DIVISION, CSIR-NATIONAL INSTITUTE FOR INTERDISCIPLINARY SCIENCE AND TECHNOLOGY (CSIR-NIIST), THIRUVANANTHAPURAM, INDIA [2]DEPARTMENT OF BIOTECHNOLOGY, SRI VENKATESWARA COLLEGE OF ENGINEERING, CHENNAI, INDIA [3]CENTRE FOR INNOVATION AND TRANSLATIONAL RESEARCH, CSIR-INDIAN INSTITUTE OF TOXICOLOGY RESEARCH (CSIR-IITR), LUCKNOW, INDIA

12.1 Introduction

The term "nutraceuticals" was first defined by Dr. Stephen De Felice in 1989, linking two words: nutrition and pharmaceutical. Nutraceuticals are the category of molecules originating from nutritional food products and having medicinal value or health benefits. They are nutritional supplements with health/therapeutic properties other than their nutritional value. Nutraceuticals include vitamins, minerals, antioxidants, amino acids, organic acids, fatty acids, plant polyphenols, probiotics, prebiotics, and synbiotics. Some common nutraceuticals which are widely used are Echinacea, green tea, glucosamine, omega-3, lutein, folic acid, and cod liver oil [1]. Among the various nutraceuticals known to date, prebiotics and probiotics have been widely studied in terms of basic research, products developed, and products available in the market [2,3]. Nutraceutical products have been reported to be an efficacious alternative to existing drugs for disorders related to oxidative stress, treatment of cancer, diabetes, osteoporosis, gastric, and Alzheimer, Parkinson, and other diseases [1]. A spectrum of nutraceuticals has been reported to play vital roles in the immune performance, immunity development, and decreased vulnerability to some disease conditions. They are known for their action as immune boosters by improving the immune system and its function. They have received significant attention owing to their prospective nutritional value, safety, and therapeutic properties. Nutraceutical diets can deliver biologically active compounds with health benefits in doses attainable and should be selected based on their other physiological properties, such as metabolic absorption. Compounds isolated from nonfood origins can also be described as nutraceuticals if they are known to be present in general food products. Recently, there has been a substantial increase in the number of human clinical trials carried

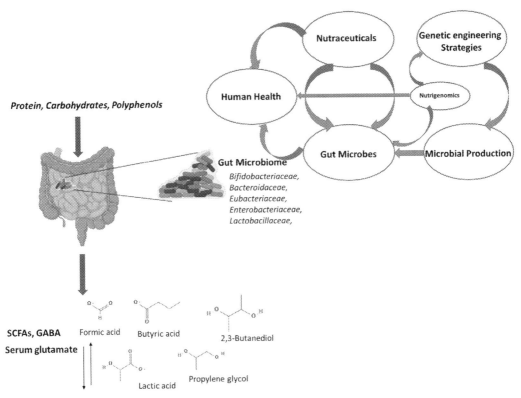

FIGURE 12-1 Overview of the role of gut microbes and nutraceuticals on human health.

out using nutraceuticals as potential therapeutic agents [4]. This chapter focuses on the role of the gut microbiome and nutraceuticals on human health, various nutraceuticals produced by gut microbes, recent developments in the microbial production of nutraceuticals, various metabolic engineering strategies applied for the production of nutraceuticals, and recent market trends and the economical value of nutraceuticals. An overview of the effect of gut microbes and nutraceuticals on human health is shown in Fig. 12–1.

12.2 Role of the gut microbiome in human health

The gut microbiome is purely the collection of microorganisms habituated in the gut of humans. The gut microbiome consists of millions of bacteria belonging to thousands of different species of bacteria, which acts as the key modulator of human health. The recent estimate on the bacterial population in the human gastrointestinal tract is over 10^{14} bacteria, this collective bacterial population is about 10 times greater than the total number of cells in the body and the total microbiome genome is approximately 100 times greater than that of the genome of cells present in the human body [5]. The Human Microbiome Project revealed that the intestinal microbiome is the most complex and stable, and later it was proved that the microbiome is subject to change within an individual and between individuals [6]. The microbial composition analysis revealed that even in monozygotic twins the variations in the gut microbiome are much greater than the differences in their own genomes [7]. Although numerous bacterial species are harbored by the human gut, to date very few species have been explored for their association with human health. *Firmicutes* are the most abundant phylum colonized in the gut, followed by *Bacteroidetes, Actinobacteria, Proteobacteria, Fusobacteria,* and some others belonging to the archaeal phyla. The most commonly reported bacterial species present in the gut microbiome include *Bacteroides, Peptococcus, Peptostreptococcus,* and *Bifidobacterium* [8,9]. The composition of microbes present in the human gut varies between individuals and is influenced by several factors, including age, genetics, diet, medication, and environmental/external factors [10]. For example, the gut microbiome of a healthy individual will be different from that of a diseased individual. The use of a medication or antibiotics will greatly influence the microbiome composition. The microbiome of a newborn infant is different from that of aged individuals. The dietary intake also impacts the gut microbiome composition. A key change in the composition of the microbiome which directly or indirectly affects human health is called dysbiosis. The state of dysbiosis is known to influence the immune system, cause several metabolic disorders, gastrointestinal disorders, obesity, inflammatory bowel disease, and other diseases including cancer and diabetes [11].

The close association of human health and the gut microbiome has been reported. The gut bacteria may inhibit the growth of several pathogenic bacteria, and also assist the immune system and maintain epithelial integrity [9]. Gut bacteria assist the host by producing vitamins, controlling gut motility and maturation, and also assist in the function of the central nervous system (CNS) [12]. Disruption to the epithelial integrity can allow bacterial toxins, unnecessary fats, and proteins to enter the bloodstream, leading to gastrointestinal problems. The gut bacteria are thought to be important for maintaining epithelial integrity by controlling tight junction permeability [12]. It has been reported that the gut bacterium *Lactobacillus plantarum* regulates the tight-junction proteins in epithelial cells and prevents chemical-induced disruption of the epithelial barrier [13].

It is well-known that obesity is strongly linked with diet and lifestyle; however, both diet and lifestyle can impact the composition of the gut microbiome. In vivo and human studies carried out recently have revealed that the gut microbiome has a significant role in obesity [14]. Reports suggest that the consumption of probiotic product yogurt can halt age-associated weight gain; this could be due to the beneficial effects of microbes present in the yogurt [15]. A polyphenol [epigallocatechin 3-O-(3-O-methyl) gallate] present in oolong tea was reported to act against obesity. The polyphenol showed changes in the composition of gut microbiota, and was found to increase the *Bacteroidetes* population and decrease the *Firmicutes* population in the intestinal microbiome of mouse models [16]. The increase in abundance of the gut bacteria *B. thetaiotaomicron* was reported to decrease obesity levels by altering the serum glutamate levels [16]. A recent study revealed that there was an improvement in lipid metabolism and changes in the gut microbiota structure when probiotic strains of *L. plantarum* were given to high-fat diet-fed mice [17]. An increase in the population of *Enterobacteriaceae* and *Streptococcus* sp. has been reported in patients with atherosclerotic cardiovascular disease [18].

12.3 Nutraceuticals and gut microbes

In recent years there has been growing interest in understanding the gut microbiota and their modulatory effect on human health through the use of probiotics, prebiotics, and synbiotics. Short-chain fatty acids (SCFAs) are rich sources of energy in humans and they are involved in metabolic signaling by acting as a substrate or source for lipogenesis, gluconeogenesis, and protein synthesis. The gut microbiota plays a key role in the degradation of amino acids for the formation of SCFAs such as acetate, propionate, and butyrate. Acetate and propionate were produced by the Gram-negative group of bacteria belonging to the *Bacteroidetes* family. Butyrate is generally produced by Gram-positive bacteria such as *Lachnospiraceae*, and *Ruminococcaceae* which belong to the *Firmicutes* family. Butyrate is also produced from lactic acid by lactate-utilizing bacteria. Some *Bacteroidetes* are saccharolytic bacteria that help in the conversion of nondigestible carbohydrates into oligosaccharides and then into monosaccharides by fermentative degradation. The fermentative action of these microbes results in the formation of CO_2, and the CO_2 produced then will be converted into acetate by acetogens [19]. Bacteria present in the large intestine, *Bacteroides* and *Proteobacteria*, ferment excess protein and convert them into ammonia, amines, phenols, sulfides, and SCFAs. A well-known gut microbe, *Escherichia coli*, produces indoles by metabolizing the amino acid tryptophan, while indoles acts as a signaling molecule for motility, biofilm formation, and antibiotic resistance of the bacterium [20,21]. The pH decreases from the ileum to the cecum and increases from the cecum to the rectum, and this change in pH across the different regions of the gastrointestinal tract is also regulated by controlling the concentration of SCFAs by gut microbiota, mainly *Roseburia* sp. and *Faecalibacterium prausnitzii*. The SCFAs play an important role in lipid, glucose, and cholesterol metabolism in various tissues; however, the concentration of SCFAs is altered/regulated by the composition of the gut microbiota [20]. The gut microbiota includes both pathogenic and nonpathogenic microflora, and the intake of nutraceuticals influences the balance between pathogenic and nonpathogenic bacteria. Also, probiotics are effective against diarrhea and other infections caused by *Clostridium* sp. Most of the probiotic formulations consist of one or more lactic acid bacteria, including *Lactobacillus* and *Bifidobacterium* sp. [1].

Plant polyphenols have been developed as a therapeutic agent, but the bottleneck in using polyphenols is their poor solubility and bioavailability. The polyphenol undergoes structural modification by the action of enzymes produced by gut bacteria; the derivatives of polyphenol are capable of being absorbed in the large intestine. Recent reports have revealed that the interaction of polyphenols with the gut microbiome, which leads to a modulatory effect in the composition of microbes, induces some metabolic changes with therapeutic effects [22]. Conversion of a less active isoflavone, Diadzein, to the very active form of its derivative, Equol, was catalyzed by the gut microbiota [23,24].

12.4 Nutrigenomics: Gut microbes and health benefits

It is well-known that micronutrients in foods are responsible for cellular growth and all metabolic reactions. Based on the literature and developments in the nutritional analysis field from the 18th century to date, it can be broadly classified under different eras: (1) analytical chemistry era (18th and 19th centuries), developments in how food was metabolized into energy and CO_2 by the body was accomplished; (2) biological era (19th century), developments in the science behind nutrition and its role in the prevention of several diseases including chronic diseases, cardiovascular diseases, neuro-disorders, and other metabolic disorders; (3) pre-genomic era (20th century), developments in understanding the relationship between the diet, body metabolism, and nutritional pathophysiology; and (4) postgenomic era (current period) or nutrigenomics era, the use of genomics, transcriptomics, proteomics, and metabolomics for a complete understanding of how diet/nutrition affect an individual's metabolism at the genome level [25].

Differential expressions of genes are required for different physiological functions, nutrition/a dietary component alters and modulates the expression of genes based on the genetic background. Nutrigenomics is the study of nutritional value using molecular biology tools to discover, analyze, and understand the individual's response or differences in responses between individuals or selected population groups, to a specific diet. It deals with the interaction of the human genome with nutrition, in other words understanding the changes in the genome with a change in dietary components and lifestyle. Later, the definition of nutrigenomics was redefined as a study of the relationship between nutrients and genotypic/phenotypic variation in the human physiological response using multiple omics techniques such as genomics, proteomics, and metabolomics. The area of nutrigenomics has been extended to include how the constituents of a particular diet affect the expression levels of various genes, which may have upregulatory or downregulatory effects, and seeks to elucidate how nutritional factors help in protecting the genome from damage.

Earlier reports on gene–nutrients interactions revealed the activation of nuclear receptors and the induction of transcriptional factors by vitamin A, vitamin D, and some fatty acids [26]. In recent reports, nutrigenomic analysis has revealed gene–nutrient interactions, dietary components' capacity to influence single-nucleotide polymorphisms (SNPs), and their capacity to bind to transcription factors. Further, interfering with gene–nutrient interactions reveals the interaction of transcription factors with other molecular elements leading to the binding and control of RNA polymerase [26–28]. Also, phytochemicals and antioxidant nutrients in the diet are known to boost DNA repair and decrease the oxidative damage of DNA [27]. Resveratrol (present in wine) and soy genistein have been reported to have indirect interference on the molecular signaling pathway elements such as factor-kappa B. The concentration of folic acid, a chemical compound that acts as a cofactor for enzymes mediating the biosynthesis of nucleotides, changes based on the intake of a folic acid-rich diet such as legumes, cereals, and citric acid-containing fruits [29]. Folate metabolism in the cell influences genetic integrity by balancing the amount of DNA and replication processes.

Reports published 10 years ago on nutrigenomics generally did not focus on change in the gut microbiome in response to the host's response to nutritive compounds. Due to limitations in the culturing techniques for studying gut microbes, the interaction between dietary nutrients, microbial composition, and the human genome is less explored. The human gut microbiota comprises a multifaceted network that has a significant impact on the host metabolism. However, recently, due to advancements in genomic techniques, the influence of nutrients on the microbial composition and population has been studied and is included in nutrigenomics [30]. The metagenomic analysis of gut microbiome gave way to an understanding of the interaction between nutrient intake, gut microbiome, the human genome, and health [31].

12.5 Developments in nutraceuticals production using gut microbes

Globally, there is an increase in trend for the nutraceuticals because of their biomedical applications. Due to the sudden increase in demand for nutraceuticals, there is a lag between supply and demand. Nevertheless, the increasing trend of nutraceutical demands cannot be satisfied by the production capacity of traditional nutraceutical manufacturers and industries. The extraction of nutraceuticals from plant and animal sources is limited because of the cost of plant/animal-based raw materials and their nonavailability in sufficient quantities to meet industry needs. Also, the process involved in improving the purity has become a bottleneck for their use as a source, mainly due to limitations in processing methods involved in the production and extraction of nutraceuticals from plant and animal sources. For the above reason, the scaling up of nutraceuticals production using plant and animal sources is unachievable to meet the current market demand [1,4,32]. Although nutraceutical molecules are chemical compounds which can be synthesized using chemical methods, this is limited to chemically synthesizing simple chemical molecules. However, many nutraceutical molecules are complex biochemical compounds that are difficult to synthesize and chemically unfavorable. Due to the above limitations in extraction methods and chemical synthesis, microbial production methods are viewed as an alternative approach. With the advancements in microbial production strategies and the ease of scale-up of production processes, nutraceutical production using microbial methods is attracting the interest of researchers and industrialists for their use as alternatives to traditional plant and animal sources. The other advantages include ease of optimization of fermentation conditions for production, using simple sugars for production which minimizes the downstream processing methods, and ease of interpretation of the metabolic pathways to enhance their production capacity. To date, the microbial production of a few nutraceuticals including polysaccharides, phytochemicals, and polyamino acids has been well studied and proven for industrial-scale production in two microbial strains: *E. coli* and *Saccharomyces cerevisiae* [33,34]. They are classified under the generally regarded as safe (GRAS) category of microorganisms. Hence, apart from producing a nutraceutical metabolite using these strains, the whole strain itself has been used as a nutraceutical in the form of probiotics [3,32,35].

Terpenoids have been accepted as a nutraceutical due to their health-complementary properties. Terpenoids (taxadiene and artemisinic acid) and tetraterpene carotenoids (β-carotene, lutein, lycopene, α-carotene) have been reported to be produced through microbial fermentation methods. Strains of *E. coli* and *Micrococcus luteus* have been reported to produce unusual

C_{50} carotenoids including sarcinaxanthin, decaprenoxanthin, and sarprenoxanthin [36]. Bacterial polysaccharides (alginate, dextrans, gellan, and xanthan) and fungal polysaccharides are known for their health benefits including antioxidant, hypoglycemic, hypocholesterolemic, and immunostimulating properties, and they are commercially produced using microbial fermentation methods. *Lactococcus* and *Streptococcus* have been reported for the production of exopolysaccharides (EPS)-based nutraceuticals and are widely used in dairy products [32,37,38]. Hyaluronic acid (HA), a polysaccharide generally extracted from animal tissue, has been reported to be produced using microbial production methods using *E. coli*, *Lactococcus lactis*, and *Streptomyces albulus* [32]. Prebiotics are another class of molecules with a modulatory effect on human health and the immune system based on their interaction with the gut microbiome. Well-known prebiotic nutraceuticals such as inulin, fructooligosaccharides (FOS), and galactooligosaccharides (GOS) are produced using *Lactobacillus* strains [1,3,39]. Although the microbial production platforms have the above-mentioned advantages, the major disadvantage is the lower efficiency of microbial enzymes compared to plant- and animal-derived enzymes, which may limit the productivity of target compounds.

12.6 Microbial strategies for the production of nutraceuticals

Many bacterial strains have been reported as safe and beneficial for the production of nutraceuticals. Not only probiotics like lactic acid bacteria but also other microbes such as *S. cerevisiae*, *Yarrowia lipolytica*, *Bacillus subtilis*, and *Spirulina* have been commercially exploited for the production of medically important nutraceuticals. Factors like conventional usage, health benefits, availability of genetic data, and safety impact the selection of strains suitable for the production of nutraceuticals.

12.7 Production of gamma-aminobutyric acid and hyaluronic acid by lactic acid bacteria

12.7.1 Gamma-aminobutyric acid

Gamma-aminobutyric acid (GABA) is a nonproteinogenic amino acid, which is well-known for conferring health benefits such as neurotransmission, protein synthesis, and secretion of hormones. Due to these beneficial factors, food-grade microbes like lactic acid bacteria have been extensively used for the production of GABA. Glutamate decarboxylase (GAD)-mediated decarboxylation of glutamate produces GABA. *Lactobacillus*, *Lactococcus*, and *Bifidobacterium* were observed to produce GABA at higher levels [40−42]. It was evident from earlier reports that the concentration of glutamate and acidic pH controlled the production of GABA [43]. Optimization of process parameters and fermentation through grading control and a shift in temperature from 37°C to 40°C effectively improved GABA production in *Streptococcus salivarius* [44]. The production of GABA in heterologous hosts like *E. coli* and *Corynebacterium glutamicum* (*C. glutamicum*) were perturbed due to the continuous maintenance of acidic conditions for GAD enzyme activity. Random mutagenesis to create a GadB mutant allowed for hassle-free production of GABA through the GAD enzyme active in a broad pH range. Nearly 39 g/L GABA was produced by *C. glutamicum* that expressed the GadB mutant [45–47].

12.7.2 Hyaluronic acid

HA is an EPS produced by different microbes. Microbial expolysaccharides like HA have immense applications in food (stabilization of dairy products) and pharmaceutical industries (anticancer, antiulcer). Glucuronic acid and N-acetyl glucosamine together constitute the HA molecule [48,49]. The synthesis of collagen and proliferation of fibroblasts is stimulated by HA. Therefore, HA is widely used not only in foods, but also in surgery and cosmetic industries [50,51]. The production of HA was initiated using the strain streptococci. However, due to pathogenic contamination issues, this strain was not continued with. Later, *L. lactis* with modified HA synthases produced almost 0.65 g/L of HA [52]. Coexpression of genes coding for the enzymes glucose pyrophosphorylase and glucose dehydrogenase improved the production of HA to 1.8 g/L [53].

12.8 Vitamin B12 and folate production by *Propionibacteria*

12.8.1 Vitamin B12

Propionibacteria are Gram-positive, anaerobic, rod-shaped bacteria that are widely used for the production of propionic acid, B vitamins, linoleic acids, and Swiss cheese [54]. Vitamin B12 is an important dietary supplement for humans and its deficiency leads to pernicious anemia [55]. Currently, the growing demand for vitamin B12 is supplied through microorganisms. Many microbes like *Streptomyces*, *Methanobacterium*, and *Klebsiella* are producers of vitamin B12. However, considering the presence of inbuilt pathways and safety, *Propionibacteria* are considered to be the microbe of choice for the production of vitamin B12 [56]. Fermentation strategies such as the addition of trace metals, dissolved oxygen concentration, the addition of precursors and analogs, genetic engineering strategies like random mutagenesis have augmented vitamin B12 production to 24 mg/L [57–60].

12.8.2 Folate

The synthesis of nucleotides is primarily mediated by folate or vitamin B9 compounds. A deficiency of folate levels in food leads to coronary malfunction, the proliferation of cancerous cells, and dysfunction of neurons [61]. Chemical methods for the synthesis of folate were adopted long ago. Considering positive factors like eco-friendliness and lower cost, the production strategy has now shifted to the microbial mode for synthesis of folate. As a GRAS organism, folate production using *Propionibacteria* was found to be higher compared to conventional production by *Streptococcus thermophilus* [62]. Starter cultures of *Propionibacteria* have been widely used for the production of folate-rich foods such as cheese and milk products. Propionic acid bacteria are also well-known producers of vitamin B2 (riboflavin), trehalose apart from folate and vitamin B12 [63]. *L. lactis* was found to be a good producer of folate. The expression of multiple folate pathway genes like GTP cyclohydrolase and glutamyl hydrolase under the nisin-controlled expressed system led to a threefold increase in folate production in *L. lactis* [62,64].

12.9 Nutraceuticals production by *Bacillus subtilis*

B. subtilis is well-known for its ability to produce industrially important enzymes and other bioproducts. It does not excrete toxins and is considered to be a safe organism for the production of commercially important compounds [65]. Almost 6.8 g/L HA was produced by *B. subtilis* harboring inducible operons coupled with HA synthesis pathway genes [66]. Likewise, riboflavin, glucosamine, and N-acetylglucosamine were also produced by *B. subtilis* [65,67]. Glucosamine acts as an important precursor for the synthesis of HA in addition to acting as a commercially important nutraceutical. It finds immense applications in the cosmetic, pharmaceutical, and food industries [68]. Glucosamine (GlcN) and N-acetylglucosamine (GlcNAc) were produced in *B. subtilis* by knocking out the genes responsible for degradation and overexpression of glmS and GNA1 [68]. GlcNAc increased to nearly 5.19 g/L as a result of the adopted pathway engineering strategy. Engineering the respiratory chain to substantially reduce the maintenance coefficient increased GlcNAc production to 20.6 g/L [69]. Similarly, the GlcNAc synthesis network was reorganized to further improve GlcNAc production to 31.7 g/L [70]. Thus, *B. subtilis* acts as a potential host for the production of nutraceuticals and other industrially important compounds.

12.10 Glutathione, carotenoids, and other nutraceuticals production by yeasts

Yeast species are known to be producers of vital nutraceuticals and functionally important compounds. *S. cerevisiae*, *Candida utilis*, and *Y. lipolytica* are considered to be industrially important yeasts. Among them, *S. cerevisiae* is known to be a GRAS organism with extensive applications in food industries, and beverages and alcohol production. The genome of *S. cerevisiae* has been fully sequenced and it is a popular eukaryotic model for the expression of heterologous genes [71].

An effective antioxidant, glutathione (GSH) is composed of three amino acids, namely glycine, cysteine, and glutamate. The cysteine contains a free sulfhydryl moiety that confers redox power to combat reactive oxygen species, carcinogens, and radiation [72]. Therefore GSH finds applications in the cosmetics, sports, and medical industries [17]. Limitations in conventional methods have shifted the method of GSH production to the fermentative mode. Random mutagenesis coupled with high cell density fermentation augmented GSH production in *S. cerevisiae* to 75 mg/L, with a 55% increase in biomass [17,73]. Amino acids like cysteine, glycine, glutamic acid, and serine supplemented to the culture medium improved the GSH levels to nearly 1.9 g/L [72]. Oxidative stress-induced accumulation of GSH by *C. utilis* proved to be a promising strategy to enhance production on a large scale [74].

Carotenoid pigments of different colors are known to confer protection from radiation, reactive oxygen species, and free radicals [75]. Carotenoids produced by various cold-adapted yeast strains have been reported [76,77]. β-Carotene is supplemented as an important antioxidant and coloring compound in food products [33]. A vitamin A precursor, β-carotene is omnipresent in animals and plants, and is synthesized naturally by different microbes [78]. *S. cerevisiae* lacks an endogenous pathway to synthesize β-carotene. However, heterologous expression of carotenogenic genes in *S. cerevisiae* improved the production of antioxidant. Verwaal reported *S. cerevisiae* accumulated nearly 6 mg/g dry weight of β-carotene [33]. Likewise, lycopene, a fat-soluble antioxidant is well-known for preventing the progression of cancer cells and confers several health benefits [79]. Abundantly present in grapes and watermelon, microbes like *Flavobacterium* sp., *Blakeslea* sp., *Streptomyces chrestomyceticus*, and *Phycomyces* sp. are known to synthesize lycopene [80]. Although *C. utilis* is known to accumulate carotenoid precursors, expression of the lycopene pathway gene augmented the lycopene production level to 8 mg/g dry cell weight [81].

α-Ketoglutaric acid (α-KG) is considered to be a key nutraceutical that finds immense applications in the pharmaceutical and food industries. Its role in the reduction of wrinkles and boosting endurance during exercise has widened its uses to the cosmeceuticals and sports industries [82–85]. Microbial synthesis was found to be a promising solution for the production of α-KG in large quantities. More precisely, the nonconventional yeast *Y. lipolytica* was found to be more effective for large-scale production [86]. Multiple substrate utilization and high product formation made *Y. lipolytica* an attractive choice for the production of not only α-KG but also organic acids and lipids [86]. The utilization of different carbon sources such as n-paraffin, glycerol, and ethanol increased α-KG production in *Y. lipolytica* to almost 195, 66, and 49 g/L, respectively [87,88]. The overexpression of key enzymes in the acetyl-coenzyme A pathway and rewiring of the pyruvate pathway substantially improved α-KG production with a concomitant decrease in by-product formation [89,90]. Polyunsaturated fatty acids (PUFAs) like docasahexanoic acid have also been successfully reported to be produced using *Y. lipolytica* [91]. Table 12–1 presents the strains used for the production of nutraceuticals.

12.11 Metabolic engineering of microbes for the production of nutraceuticals

12.11.1 Prebiotics

A nondigestible food component that enlivens the microbiota and confers health benefits to the host is commonly referred as a prebiotic compound. The microbiome increases gradually as it progresses from the stomach to the colon and is reported to be around 10^{12} CFU/mL [109]. More specifically, prebiotic compounds are known to stimulate lactic acid bacteria such as lactobacilli and *Bifidobacterium*. These beneficial microbes are known to boost the immune cells, prevent intolerance to lactose, inhibit the progression of pathogenic bacteria,

Table 12–1 Nutraceutical production by different strains.

Nutraceutical	Strain	References
EPS	Lactic acid bacteria	[92]
	Xanthomonas sp.	[93]
	Gluconobacter xylinus	[94]
	Leuconostoc mesenteroides	[95]
Hyaluronic acid	Bacillus subtilis	[96]
	Streptococcus zooepidemicus	[97]
GABA	C. glutamicum	[98]
Vitamins	Lactic acid bacteria	[99]
	Pseudomonas denitrificans	[100]
	Propionibacterium spp.	[101]
	Candida famata	[102]
Conjugated linoleic acid (CLA)	Lactobacillus plantarum	[51]
	Propionibacteria	[103]
EPA/DHA	Yarrowia lipolytica	[104]
	Isochyris galbana	[105]
ARA	Mortierella alpina	[106]
Carotenoids	Candida utilis	[107]
	Blakeslea trispora	[108]

and produce several B group vitamins [110]. Hence, a well-balanced diet can augment beneficial gut microbes in humans [111].

Glucooligosaccharides (GOS), FOS, N-acetylneuramic acid, inulin, and 2-fucosyllactose (2'-FL) are some of the well-known prebiotics with health benefits. *Lactobacillus gasseri* is known to produce the oligosaccharide inulin [112]. Similarly, from 400 g/L lactose, 177 g/L of GOS was produced by *Kluyveromyces lactis* through process optimization [113]. Similarly, *L. lactis* expressing β-galactosidase produced GOS from lactose [114]. 2-FL is considered to be a vital prebiotic in human milk and is recommended as a food additive for infants. Microbial routes such as the use of *E. coli* overexpressing fucosyltransferase have been successfully employed for large-scale production of 2'-FL [115,116]. Microbial polysaccharides, specifically EPSs, were reported to be produced in lactic acid bacteria and *Streptococcus* sp. through metabolic engineering approaches [38,117]. These EPS are known to demonstrate antimicrobial, anticancer, and immunomodulating activity. Supplementation of the amino acid L-lysine in the medium accelerated the production of scleroglucan by *Sclerotium rolfsii* [117]. *E. coli*, lactic acid bacteria, and *Streptomyces* sp. have been engineered to produce chondroitin and heparosan, which have important medical applications [118,119]. *L. lactis* expressing fructose diphosphatase enhanced EPS synthesis using fructose as substrate [120]. Glucose pyrophosphorylase and galactose epimerase significantly augmented EPS production by increasing the intracellular concentration of precursors like glucose and galactose [121].

One of the most important categories of essential fatty acids, connected by more than one double bond, is commonly referred to as PUFAs. Omega-3 and omega-6 fatty acids are the two major types of PUFA [122]. These fatty acids confer protective and health effects on the cardiovascular system and nervous system and prevent cancer [123,124]. α-Linolenic acid (ALA), eicosapentaenoic acid (EPA), and docosahexaenoic acid (DHA) are classical examples of omega-3 fatty acids. The demand for omega-3 fatty acids is expected to be more than 241,000 metric tons and hence there is a need to accelerate their production through microbial metabolic engineering strategies [125]. The anaerobic polyketide synthase pathway (PKS) and aerobic elongase pathway have been rewired to produce EPA [126]. Oleaginous yeasts like *Y. lipolytica* have been successfully engineered by the DuPont research team to synthesize nearly 57% EPA among the total fatty acids and nearly 15% of their dry cell weight [104]. Similarly, elimination of the β-oxidation pathway and enhanced carbon flux improved EPA production in *Y. lipolytica* Z5567 strain [127,128]. Fed-batch cultivation with nitrogen-fed and nitrogen-limited conditions produced EPA at 50% of their dry cell weight. New Harvest EPA and Verlasso salmon are two well-known commercial products using PUFA technology [129]. DHA has been produced on a large scale, predominantly using microalgae since microalgal strains accumulate DHA to nearly 40% by weight [130]. Yeasts and bacteria accumulate 6% or less of fatty acids [131,132].

The oleaginous yeast *Y. lipolytica* have been exploited to produce the essential fatty acid ALA. ALA biosynthesis typically occurs through the conversion of oleic acid and linoleic acid by the enzyme delta 14 desaturase [133]. Gene coding for delta 12/delta 15 desaturase enzymes from *R. kratochvilovae* was codon-optimized and expressed to improve the ALA content. Fermentation at 20°C improved the ALA productivity significantly to 1.4 g/L [134].

12.11.2 Mannitol and sorbitol production by lactic acid bacteria

Polyols like mannitol, sorbitol, and trehalose are alternatives to the commonly available sugars such as sucrose, fructose, and glucose. These polyols are low-calorie compounds, but with similar taste and sweetness [135,136]. Mannitol is reported to protect cells from extreme physical conditions such as drying and freezing by acting as an antioxidant [137,138]. Heterofermentative bacteria like *Leuconostoc mesenteroides* are known to ferment fructose for both energy generation and production of mannitol [139]. Knocking out the gene coding for lactate dehydrogenase (LDH) in both *L. plantarum* [140] and *L. lactis* [141] augmented mannitol production significantly. Likewise, *L. lactis* overexpressing mannitol-phosphate dehydrogenase (MPDH) and deficient in LDH showed a concomitant increase in mannitol production [142]. Heterologous expression of mannitol transporter from *Leuconostoc mesentroides* further improved the titers of mannitol in *L. lactis* [142]. Similarly, an *L. plantarum* strain deficient in MPDH and LDH but overexpressing DNA coding for sorbitol dehydrogenase enzyme was proved to effectively augment sorbitol production.

Another sucrose equivalent, low-calorie sugar known as tagatose is commercially available as a prebiotic and an antiplaque compound [143]. The lack of a biological route for the production of tagatose has mandated the use of arabinose coupled with an enzymatic isomerization process for the production of tagatose [144]. Lactic acid bacteria such as *L. lactis* alone contains the tagatose-6-phosphate pathway by utilizing lactose metabolism. The tagatose pathway genes, namely lacABCD, are found mainly in the lactose operon of *L. lactis*. Knocking the genes lacC/lacD or either one of the genes together with the nisin-controlled expression (NICE) system enhanced the accumulation of tagatose-6-phosphate by lactococcal cells [145]. Gene disruption occurs predominantly by a dual recombination process coupled with the incorporation of an erythromycin-resistance plasmid. Although tagatose-secreting *L. lactis* cells could be engineered, a process like dephosphorylation of phosphate-containing tagatose molecule could add value to the final compound produced.

A low-calorie sugar produced by different microbes and less easily digested by humans is trehalose. This is commonly referred to as dietetic sugar that suppresses plaque acidification, as is evident through dental studies in rats fed with trehalose [146]. Trehalose confers cellular protection under stressed environmental conditions by preventing protein denaturation [147–150]. In addition to protein protection, trehalose also averts aggregation of proteins and prevents illnesses such as Creutzfeld-Jakob disease [151]. Glucose-6-phosphate, glucose-1-phosphate, trehalose-6-phosphate, and UDP-glucose are known to be pathway intermediates for the synthesis of trehalose [152]. Variations in physical parameters during fermentation such as reduced temperature, increased osmolarity, and pH conditions are known to significantly improve trehalose production in *Propionibacterium* [153,154].

12.11.3 Polyphenols

Polyphenolic compounds like flavonoids, isoflavanoids, curcuminoids, and carotenoids fall under the category of phytochemicals. They are primarily compounds that confer protection to plants from environmental factors like infection and radiation [122]. In addition, these polyphenols also counter high blood pressure, cancer, metabolic disorders, and other diseases associated with the nervous system [155–157]. Phenylalanine and tyrosine act as starting precursors for the synthesis of different polyphenols.

Naringenin possesses the innate ability to scavenge free radicals, prevent cell apoptosis, and confers protection to the nervous system [157]. P-coumaric acid acts as a central precursor for the synthesis of naringenin. Phenylalanine ammonia lyase (PAL) converts phenylalanine to cinnamic acid, which is then converted to p-coumaric acid by cinnamate-4-hydroxylase [157]. Likewise, tyrosine ammonia-lyase (TAL) also directly synthesizes p-coumaric acid from tyrosine molecules [158]. P-coumaric acid is then converted as ester coumaroyl-CoA through the biocatalyst 4-coumarate: CoA ligase (4CL) [159]. Subsequently, the ester coumaroyl-CoA is condensed with malonyl CoA to form naringenin chalcone by the enzyme chalcone synthase. Finally, chalcone isomerase (CHI) converts naringenin chalcone into naringenin [157,160]. Due to the high cost of the supplementation of amino acids like phenylalanine and tyrosine, pathway engineering strategies have been successfully employed to augment naringenin production. Expression of TAL, PAL, 4CL, CHI, and CHS from various microbes in *S. cerevisiae* and *E. coli* produced nearly 113 and 100 mg/L of naringenin, respectively [161]. A coculture system in which *E. coli* consumes xylose to produce acetate and *S. cerevisiae* consumes acetate to produce naringenin was eightfold higher compared to a monoculture of bacteria or yeast [162]. Considered to be an agonist that overpowers the expression of a cyclooxygenase-2 enzyme known to promote cancer, resveratrol is considered to be a powerful nutraceutical. It also helps to protect from neurodegenerative and cardiac diseases [163,164]. Just like naringenin synthesis, resveratrol is synthesized from p-coumaric acid, but with a different enzyme known as stilbene synthase (STS). S. cerevisiae with mutants of different resveratrol pathway genes like TAL, CL, VST1, and other feedback-inhibition resistant DAHP synthase, chorismate mutase was reported to produce nearly 531 mg/L resveratrol by routing carbon sources toward the tyrosine pathway [165]. An increase in the supply of malonyl-CoA and phenylalanine and overexpression of cytochrome reductase and B5 enzyme enhanced de novo resveratrol production to nearly 812 mg/L in S. cerevisiae [166]. Likewise, an E. coli coculture technique in which one E. coli strain produced p-coumaric acid while another strain utilized the coumaric acid and converted it into resveratrol gave promising results [167]. Other flavonoids like quercetin and kaempferol have also been reported to be produced in microbes. These compounds confer health benefits such as anticancer, antiinflammatory, and protection against cardiotoxicity induced by doxorubicin [168]. Overexpression of flavanone-3-hydrolase and flavonol synthase in S. cerevisiae enhanced kaempferol production [169]. Similarly, afzelechin flavonoid was produced by splitting the pathway into the naringenin synthesis module and NADPH cofactor requiring module. Such a module incorporated in an E. coli coculture system drastically enhanced the production of afzelechin by 970-fold [170]. However, modeling of process parameters like pH, temperature, carbon/nitrogen sources, and inoculum size finally enhanced the production level to 40 mg/L.

12.11.4 Polyamino acids

Amino acids that are synthesized through an enzymatic process without the assistance of ribosomes are commonly referred as polyamino acids. Poly-γ-glutamic acids (γ-PGA), poly-ε-L-lysine (PL), and cyanophycin are potential nutraceutical candidates that belong to the polyamino acids [32]. Biodegradable polymer, PGA, is well-known to be produced by microbes like B. subtilis up to a titer of 50 g/L. Similarly, coculture of C. glutamicum and Bacillus subtilis synthesized nearly 33 g/L PGA after medium optimization [171]. Cyanobacteria and Acinetobacter sp. produce cyanophycin through optimal feeding of aspartic acid and arginine. Heterologous expression of cyanophycin synthetase in E. coli accelerated the productivity of cyanophycin to nearly 120 mg/g cell dry weight (CDW) [172]. Polymerization of the amino acid lysine by the enzyme poly lysine synthetase produces the PL molecule. The selection of appropriate carbon sources is known to augment PL production in Streptomyces sp. Use of glycerol and glucose enhanced PL production to nearly 35 g/L [173].

12.11.5 Carotenoids

Carotenoids are considered to be natural colorants and compounds with antioxidant properties [174]. They are molecules thought to prevent cardiac diseases, neuronal disorders, and tumor growth [80,175,176]. By 2022, the global carotenoid market is anticipated to be $2 billion [177]. To cater to this growing demand, industries and researchers are now focused on metabolic engineering strategies to boost productivity. Some of the important carotenoids such as astaxanthin, β-carotene, and lycopene are synthesized in non-carotenogenic microbes like *S. cerevisiae* and *E. coli* [35,86,178−182]. Isopentenyl pyrophosphate (IPP) and dimethylallyl pyrophosphate (DMAPP) are crucial precursors for carotenoid production and are synthesized through the mevalonic acid (MVA) pathway and methylerythritol 4-phosphate (MEP) pathway [183]. Expression of DXP synthase and IPP isomerase in *E. coli* synthesizes DMAPP and IPP, which are further converted to FPP by FPP synthase [183]. In eukaryotes like *S. cerevisiae*, lycopene is synthesized from FPP through the introduction of genes like *CrtE*, *CrtB*, and *CrtI* [174]. Lycopene was further converted to β-carotene through the action of lycopene cyclase. β-Carotene was further converted to astaxanthin by the enzyme β-carotene hydroxylase. An increase in carbon flux through the acetyl CoA pathway and mevalonate pathway improved the precursors and NADPH. Integration of the lycopene synthesis pathway (containing genes from *B. trispora* and *P. agglomerans*) further enhanced lycopene production [180]. Likewise, overexpression of the triacylglycerol (TAG) pathway genes like *FLD1* and *OLE1* significantly improved the lycopene levels to nearly 2.37 g/L. The gene copy number has proved to be a promising strategy to enhance lycopene production. Therefore *E. coli* clone having a high copy number of idi gene (Isopentenyl-diphosphate Delta-isomerase gene), synthesized nearly 1.44 g/L lycopene through a fed-batch fermentation strategy [129].

An *E. coli* strain engineered with geranyl diphosphate synthase (GPPS2) and isopentenyl pyrophosphate isomerase (FNI) gene was reported to produce 3.2 g/L β-carotene with glycerol as a carbon source in the fed-batch cultivation process [183]. An increase in the NADPH flux, overexpression of genes involved in MEP, TCA cycle, and PPP pathway synergistically augmented β-carotene production to 2.1 g/L in *E. coli* under controlled fermentation conditions [184]. It was reported that yeasts contain lipid droplets that stored carotenoids. Optimizing the expression of enzymes like carRP and carB and high gene copy numbers together led *Y. lipolytica* to produce 4 g/L β-carotene during the fed-batch fermentation process [86]. Heterologous expression of crt genes and ispDF in *E. coli* improved astaxanthin production to 433 mg/L during a fed-batch cultivation process [34]. Expression of CrtZ gene (origin: *A. aurantiacum*) and CrtW gene (gene source: *B. vesicularis*) in an astaxanthin-overproducing *S. cerevisiae* strain coupled with atmospheric and room temperature plasma mutagenesis, improved astaxanthin production to 218 mg/L [185]. Inhibition of squalene synthase expression, optimization of mevalonate pathway, and overexpression of astaxanthin biosynthetic genes enhanced the astaxanthin production to 10.4 mg/L [185]. Table 12−2 presents engineered strains that produce different nutraceuticals.

12.12 Market trends in the production of nutraceuticals

Nutraceuticals are gaining increasing attention around the world due to their therapeutic and nutritional properties. As per the World Health Organization (WHO) global health database, total spending by the United States was nearly 17% of GDP on healthcare [190]. This

Table 12-2 Production of nutraceuticals from engineered strains.

Strain	Product	Productivity (mg/L)	References
Yeast			
Y. lipolytica	Astaxanthin	5700	[182]
S. cerevisiae	Kaempferol	86	[186]
S. cerevisiae	Lycopene	3280	[187]
Bacteria			
E. coli	Naringenin	21160	[162]
E. coli	Resveratrol	2300	[188]
E. coli	Astaxanthin	385.04	[34]
C. glutamicum	GABA	8000	[189]

huge expenditure on nutraceuticals has made consumers and multinational companies rethink these nutritionally essential compounds. Based on their applications and chemical nature, nutraceuticals are classified as antioxidants, vitamins, essential fatty acids, prebiotics, and polyphenols [191,192]. As per the market research report, the nutraceutical market which was $231 billion in 2018 will reach $336.1 billion in the 2023 at a compounded annual growth rate (CAGR) of 7.8% [193]. Likewise, the functional beverages division and functional food division of the nutraceutical market are anticipated to reach $83.1 billion (by 2023) and $111 billion (by 2022) at CAGRs of 8.4% and 8.1%, respectively [193]. In Asian countries like India, as per the ASSOCHAM and market outlook of Indian nutraceuticals, the nutraceutical market is expected to grow threefold from 2018 to US $8.5 billion in 2022 at a CAGR of 16% [194]. The market for herbal supplements is also expected to grow to nearly US $1.7 billion. Nutraceuticals available in the market include the energy drink Rox (Rox America), Protinex (Pfizer Ltd.), and Snapple-a-day (Snapple Beverage group) [195].

12.13 Conclusions and perspectives

Nutraceuticals are compounds that enhance health and improve longevity in humans. Organ malfunctions or diseases can be treated with the aid of nutraceuticals. The constant growing demand for nutraceuticals has shifted the focus of the scientific community and industries toward more robust methodologies such as metabolic engineering compared to the traditional extraction process. Although many microbes have been exploited for nutraceutical production, those with GRAS status have attracted the attention of several industries. Selection of a strong promoter, transcription terminator, thermal inducers, and cap analogs to improve the translation, engineering of the respiratory chain, spatial modulation of enzymes involved in key pathways, and the coculturing technique could be some of the molecular engineering strategies to enhance nutraceutical production. Hence, the low production rate and other factors that limit the application of these microbes to industries could be bypassed through the application of systems and synthetic biology approaches.

Acknowledgment

Raveendran Sindhu acknowledges the Department of Science and Technology for sanctioning this project under the DST WOS-B scheme.

References

[1] H. Nasri, A. Baradaran, H. Shirzad, M.R. Kopaei, New concepts in nutraceuticals as alternative for pharmaceuticals, International Journal of Preventive Medicine 5 (2014) 1487–1499.

[2] S. Sugiharto, Role of nutraceuticals in gut health and growth performance of poultry, Journal of the Saudi Society of Agricultural Sciences 15 (2016) 99–111.

[3] E.M.M. Quigley, Nutraceuticals as modulators of gut microbiota: role in therapy, British Journal of Pharmacology 177 (2020) 1351–1362.

[4] A. González-Sarrías, M. Larrosa, M.T. García-Conesa, F.A. Tomás-Barberán, J.C. Espín, Nutraceuticals for older people: facts, fictions and gaps in knowledge, Maturitas 75 (2013) 313–334.

[5] R. Sender, S. Fuchs, R. Milo, Revised estimates for the number of human and bacteria cells in the body, PLOS Biology 14 (2016) e1002533.

[6] R.E. Hage, E. Hernandez-Sanabria, T.V. Van de, Emerging trends in 'smart probiotics': functional consideration for the development of novel health and industrial applications, Frontiers in Microbiology 8 (2017) 1889.

[7] J.A. Gilbert, M.J. Blaser, J.G. Caporaso, J.K. Jansson, S.V. Lynch, R. Knight, Current understanding of the human microbiome, Nature Medicine 24 (2018) 392–400.

[8] F. Guarner, J.R. Malagelada, Gut flora in health and disease, Lancet 361 (2003) 512–519.

[9] H. Wang, C.X. Wei, L. Min, L.Y. Zhu, Good or bad: gut bacteria in human health and diseases, Biotechnology & Biotechnological Equipment 32 (2018) 1075–1080.

[10] N. Hasan, H. Yang, Factors affecting the composition of the gut microbiota, and its modulation, Peer J 7 (2019) e7502.

[11] R.X. Ding, W.-R. Goh, R.-N. Wu, X.-Q. Yue, X. Luo, W.W.T. Khine, et al., Revisit gut microbiota and its impact on human health and disease, Journal of Food Drug Analalysis 27 (2019) 623–631.

[12] Q. Ma, C. Xing, W. Long, H.Y. Wang, Q. Liu, R.-F. Wang, Impact of microbiota on central nervous system and neurological diseases: the gut-brain axis, Journal of Neuroinflammation 16 (2019) 53.

[13] J. Karczewski, F.J. Troost, I. Konings, J. Dekker, M. Kleerebezem, R.-J.M. Brummer, et al., Regulation of human epithelial tight junction proteins by *Lactobacillus plantarum* in vivo and protective effects on the epithelial barrier, American Journal of Physiology-Gastrointestinal and Liver Physiology 298 (2010) G851–G859.

[14] R. Liu, J. Hong, X. Xu, Q. Feng, D. Zhang, Y. Gu, et al., Gut microbiome and serum metabolome alterations in obesity and after weight-loss intervention, Nature Medicine 23 (2017) 859–868.

[15] D. Mozaffarian, T. Hao, E.B. Rimm, W.C. Willett, F.B. Hu, Changes in diet and lifestyle and long-term weight gain in women and men, New England Journal of Medicine 364 (2011) 2392–2404.

[16] M. Cheng, X. Zhang, Y. Miao, J. Cao, Z. Wu, P. Weng, The modulatory effect of (-)-epigallocatechin 3-O-(3-O-methyl) gallate (EGCG3″Me) on intestinal microbiota of high fat diet-induced obesity mice model, Food Research International 92 (2017) 9–16.

[17] Y. Li, G. Wei, J. Chen, Glutathione: a review on biotechnological production, Applied Microbiology and Biotechnology 66 (2004) 233–242.

[18] Z. Jie, H. Xia, S.-L. Zhong, S.-L.,Q. Feng, S. Li, S. Liang, et al., The gut microbiome in atherosclerotic cardiovascular disease, Nature Communications 8 (2017) 845.

[19] T. Mohr, A. Infantes, L. Biebinger, P.-de Maayer, A. Neumann, Acetogenic fermentation from oxygen containing waste gas, Frontiers in Bioengineering and Biotechnology 7 (2019) 433.

[20] G.-D. Besten, K.V. Eunen, A.K. Groen, K. Venema, D.-J. Reijngoud, B.M. Bakker, The role of short-chain fatty acids in the interplay between diet, gut microbiota, and host energy metabolism, Journal of Lipid Research 54 (2013) 2325–2340.

[21] A. Woting, M. Blaut, The intestinal microbiota in metabolic disease, Nutrients 8 (2016) 202.

[22] A.K. Singh, C. Cabral, R. Kumar, R. Ganguly, R.H. Kumar, A. Gupta, et al., Beneficial effects of dietary polyphenols on gut microbiota and strategies to improve delivery efficiency, Nutrients. 1 (2019) 2216.

[23] B. Mayo, L. Vázquez, A.B. Flórez, Equol: a bacterial metabolite from the daidzein isoflavone and its presumed beneficial health effects, Nutrients 11 (2019) 2231.

[24] A.W.C. Man, N. Xia, A. Daiber, H. Li, The roles of gut microbiota and circadian rhythm in the cardiovascular protective effects of polyphenols, British Journal of Pharmacolog 177 (2020) 1278–1293.

[25] N.M.R. Sales, P.B. Pelegrini, M.C. Goersch, Nutrigenomics: definitions and advances of this new science, Journal of Nutrition and Metabolism (2014) e202759.

[26] S. Sirajudeen, I. Shah, A. Menhali, A narrative role of vitamin D and its receptor: with current evidence on the gastric tissues, International Journal of Molecular Sciences 20 (2019) 3832.

[27] M.N. Mead, Nutrigenomics the genome—food interface, Environmental Health Perspectives 115 (2007) A582–A589.

[28] M. Kussmann, P.J.V. Bladeren, The extended nutrigenomics— understanding the interplay between the genomes of food, gut microbes, and human host, Frontiers in Genetics 2 (2011) 1.

[29] E.B. Joseph, E.B., T. Tsunenobu, Folate-dependent purine nucleotide biosynthesis in humans, Advances in Nutrition 6 (2015) 564–571.

[30] D.A. Sela, D.A. Mills, The marriage of nutrigenomics with the microbiome: the case of infant-associated bifidobacteria and milk, American Journal of Clinical Nutrition 99 (2014) 697–703.

[31] C.A. Kolmeder, W.M.D. Vos, Gut health and the personal microbiome, in: K. Martin, P.J. Stover (Eds.), In Nutrigenomics and Proteomics in Health and Disease: Towards a Systems-Level Understanding of Gene–Diet Interactions, John Wiley & Sons Ltd, 2017, pp. 203–219.

[32] J. Wang, S. Guleria, M.A.G. Koffas, Y. Yan, Microbial production of value-added nutraceuticals, Current Opinion in Biotechnology 37 (2016) 97–104.

[33] R. Verwaal, J. Wang, J.P. Meijnen, H. Visser, G. Sandmann, J.A.V.D. Berg, et al., High-level production of beta-carotene in *Saccharomyces cerevisiae* by successive transformation with carotenogenic genes from *Xanthophyllomyces dendrorhous*, Applied Environmental Microbiology 73 (2007) 4342–4350.

[34] S.Y. Park, R.M. Binkley, W.J. Kim, M.H. Lee, S.Y. Lee, Metabolic engineering of *Escherichia coli* for high-level astaxanthin production with high productivity, Metabolic Enggineering 49 (2018) 105–115.

[35] A. Catinean, M.A. Neag, D.M. Muntean, I.C. Bocsan, A.D. Buzoianu, An overview on the interplay between nutraceuticals and gut microbiota, Peer J 3 (2018) 1.

[36] R. Netzer, M.H. Stafsnes, T. Andreassen, A. Goksøyr, P. Bruheim, T. Brautaset, Biosynthetic pathway for γ-cyclic *Sarcinaxanthin* in *Micrococcus luteus*: heterologous expression and evidence for diverse and multiple catalytic functions of C50 carotenoid cyclases, Journal of Bacteriology 192 (2010) 5688–5699.

[37] J.M. Laparra, Y. Sanz, Interactions of gut microbiota with functional food components and nutraceuticals, Pharmacology Research 61 (2010) 219–225.

[38] I. Giavasis, Bioactive fungal polysaccharides as potential functional ingredients in food and nutraceuticals, Current Opinion in Biotechnology 26 (2014) 162–173.

[39] H. Li, F. Liu, J. Lu, J. Shi, J. Guan, F. Yan, et al., Probiotic mixture of *Lactobacillus plantarum* strains improves lipid metabolism and gut microbiota structure in high fat diet-fed mice, Frontiers in Microbiology 11 (2020) 512.

[40] N. Komatsuzaki, J. Shima, S. Kawamoto, H. Momose, T. Kimura, Production of γ-aminobutyric acid (GABA) by *Lactobacillus paracasei* isolated from traditional fermented foods, Food Microbiology 22 (2005) 497–504.

[41] X. Lu, C. Xie, Z. Gu, Optimisation of fermentative parameters for GABA, Czech Journal Food Science 27 (2009) 433–442.

[42] K.B. Park, G.E. Ji, M.S. Park, S.H. Oh, Expression of rice glutamate decarboxylase in *Bifidobacterium longum* enhances γ-aminobutyric acid production, Biotechnology Letters 27 (2005) 1681–1684.

[43] H. Li, Y. Cao, Lactic acid bacterial cell factories for gammaaminobutyric acid, Amino Acids 39 (2010) 1107–1116.

[44] S.Y. Yang, F.X. Lu, Z.X. Lu, X.M. Bie, Y. Jiao, L.J. Sun, et al., Production of γ-aminobutyric acid by *Streptococcus salivarius subsp. thermophilus* Y2 under submerged fermentation, Amino Acids 34 (2008) 473–478.

[45] N.A.T. Ho, C.Y. Hou, W.H. Kim, T.J. Kang, Expanding the active pH range of *Escherichia coli* glutamate decarboxylase by breaking the cooperativeness, Journal of Bioscience Bioengineering 115 (2013) 154–158.

[46] Y. Soma, Y. Fujiwara, T. Nakagawa, K. Tsuruno, T. Hanai, Reconstruction of a metabolic regulatory network in *Escherichia coli* for purposeful switching from cell growth mode to production mode in direct GABA fermentation from glucose, Metabolic Engineering 43 (2017) 54–63.

[47] J.W. Choi, S.S. Yim, S.H. Lee, T.J. Kang, S.J. Park, K.J. Jeong, Enhanced production of gamma-aminobutyrate (GABA) in recombinant *Corynebacterium glutamicum* by expressing glutamate decarboxylase active in expanded pH range, Microbial Cell Factories 14 (2015) 21.

[48] A.D. Welman, I.S. Maddox, Exopolysaccharides from lactic acid bacteria: perspectives and challenges, Trends, Biotechnology 21 (2003) 269–274.

[49] P. Duboc, B. Mollet, Applications of exopolysaccharides in the dairy industry, International Dairy Journal 11 (2001) 759–768.

[50] P. Jin, Z. Kang, N. Zhang, G. Du, J. Chen, High-yield novel leech hyaluronidase to expedite the preparation of specific hyaluronan oligomers, Scientific Reports 4 (2014) 4471.

[51] L. Liu, Y. Liu, J. Li, G. Du, J. Chen, Microbial production of hyaluronic acid: current state, challenges, and perspectives, Microbial Cell Factories 10 (2011) 99.

[52] L.J. Chien, C.K. Lee, Hyaluronic acid production by recombinant *Lactococcus lactis*, Applied Microbiology and Biotechnology 77 (2007) 339–346.

[53] S.B. Prasad, G. Jayaraman, K. Ramachandran, Hyaluronic acid production is enhanced by the additional co-expression of UDP glucose pyrophosphorylase in *Lactococcus lactis*, Applied Microbiology and Biotechnology 86 (2010) 273–283.

[54] S. Pophaly, S. Tomar, S. De, R. Singh, Multifaceted attributes of dairy *propionibacteria*: a review, World Journal of Microbiology and Biotechnology 28 (2012) 3081–3095.

[55] J. Martens, H. Barg, M. Warren, D. Jahn, Microbial production of vitamin B12, Applied Microbiology and Biotechnology 58 (2002) 275–285.

[56] M.N. Ali, M.K. Mohd, Enhancement in vitamin B12 production by mutant strains of *Propionibacterium freudenreichii*, International Journal of Engineering Sciences 3 (2011) 4921–4925.

[57] N. Guan, X. Zhuge, J. Li, H.D. Shin, J. Wu, Z. Shi, et al., Engineering propionibacteria as versatile cell factories for the production of industrially important chemicals: advances, challenges, and prospects, Applied Microbiology and Biotechnology 99 (2015) 585–600.

[58] S. Marwaha, R. Sethi, J. Kennedy, Influence of 5, 6-dimethylbenzimidazole (DMB) on vitamin B12 biosynthesis by strains of *Propionibacterium*, Enzyme and Microbial Technology 5 (1983) 361–364.

[59] Y. Piao, M. Yamashita, N. Kawaraichi, R. Asegawa, H. Ono, Y. Murooka, Production of vitamin B12 in genetically engineered *Propionibacterium freudenreichii*, Journal of Bioscience and Bioengineering 98 (2004) 167–173.

[60] K. Ye, M. Shijo, S. Jin, K. Shimizu, Efficient production of vitamin B12 from propionic acid bacteria under periodic variation of dissolved oxygen concentration, Journal of Fermentation and Bioengineering 82 (1996) 484–491.

[61] Y. Thirupathaiah, C.S. Rani, M.S. Reddy, L.V. Rao, Effect of chemical and microbial vitamin B12 analogues on production of vitamin B12, World Journal of Microbiology and Biotechnology 28 (2012) 2267–2271.

[62] R.M. Pitkin, Folate and neural tube defects, American Journal of Clinical Nutrition 85 (2007) 285–288.

[63] J. Hugenholtz, J. Hunik, H. Santos, E. Smid, Nutraceutical production by propionibacteria, Le Lait 82 (2002) 103–112.

[64] S. Hugenschmidt, S.M. Schwenninger, C. Lacroix, Concurrent high production of natural folate and vitamin B12 using a co-culture process with *Lactobacillus plantarum* SM39 and *propionibacterium freudenreichii* DF13, Process Biochemistry 46 (2011) 1063–1070.

[65] W.F.H. Sybesma, M.J.C. Starrenburg, I. Mierau, M. Kleerebezem, W.M. de Vos, J. Hugenholtz, Production of B-vitamins in lactic acid bacteria by using metabolic engineering. In Abstracts of 102nd General Meeting American Society for Microbiology: May 19–23; Salt Lake City, USA: Washington DC: American Society for Microbiology; 2002, K-76.

[66] L. Liu, Y. Liu, H.-D. Shin, R.R. Chen, N.S. Wang, J. Li, et al., Developing *Bacillus spp.* as a cell factory for production of microbial enzymes and industrially important biochemicals in the context of systems and synthetic biology, Applied Microbiology and Biotechnology 97 (2013) 6113–6127.

[67] Y. Jia, J. Zhu, X. Chen, D. Tang, D. Su, W. Yao, et al., Metabolic engineering of *Bacillus subtilis* for the efficient biosynthesis of uniform hyaluronic acid with controlled molecular weights, Bioresour Technology 132 (2013) 427–431.

[68] L. Liu, Y. Liu, H.-d Shin, R. Chen, J. Li, G. Du, et al., Microbial production of glucosamine and N-acetyl-glucosamine: advances and perspectives, Applied Microbiology and Biotechnology 97 (2013) 6149–6158.

[69] Y. Liu, Y. Zhu, E. Ma, H.-D. Shin, J. Li, L. Liu, et al., Spatial modulation of key pathway enzymes by DNA-guided scaffold system and respiration chain engineering for improved N-acetylglucosamine production by *Bacillus subtilis*, Metabolic Engineering 24 (2014) 61–69.

[70] Y. Liu, Y. Zhu, J. Li, H.-D. Shin, R.R. Chen, G. Du, et al., Modular pathway engineering of *Bacillus subtilis* for improved N-acetylglucosamine production, Metabolic Engineering 23 (2014) 42–52.

[71] S. Ostergaard, L. Olsson, J. Nielsen, Metabolic engineering of *Saccharomyces cerevisiae*, Microbiology and Molecular Biology Reviews 64 (2000) 34–50.

[72] S. Wen, T. Zhang, T. Tan, Utilization of amino acids to enhance glutathione production in *Saccharomyces cerevisiae*, Enzyme Microbial Technology 35 (2004) 501–507.

[73] T. Zhang, S. Wen, T. Tan, Optimization of the medium for glutathione production in *Saccharomyces cerevisiae*, Process Biochemistry 42 (2007) 454–458.

[74] G. Liang, X. Liao, G. Du, J. Chen, A new strategy to enhance glutathione production by multiple H_2O_2-induced oxidative stresses in *Candida utilis*, Bioresour, Technology 100 (2009) 350–355.

[75] N.I. Krinsky, Antioxidant functions of carotenoids, Free Radical Biology and Medicine 7 (1989) 617–635.

[76] K. Chreptowicz, M. Jolanta, T. Jana, M. Mateusz, C. Milan, Carotenoid-producing yeasts: identification and characteristics of environmental isolates with a valuable extracellular enzymatic activity, Microorganisms 7 (2019) 653.

[77] J. Tkáčová, T. Klempová, M. Čertík, Kinetic study of growth, lipid and carotenoid formation in β-carotene producing *Rhodotorula glutinis*, Chemical Papers 72 (2018) 1193–1203.

[78] S.L. Wang, J.S. Sun, B.Z. Han, X.Z. Wu, Optimization of β-carotene production by *Rhodotorula glutinis* using high hydrostatic pressure and response surface methodology, Journal of Food Science 72 (2007) 325–329.

[79] J. Levy, E. Bosin, B. Feldman, Y. Giat, A. Miinster, M. Danilenko, et al., Lycopene is a more potent inhibitor of human cancer cell proliferation than either a-carotene or β-carotene, Nutrition and Cancer 24 (1995) 257–266.

[80] R. Ciriminna, A. Fidalgo, F. Meneguzzo, L.M. Ilharco, M. Pagliaro, Lycopene: emerging production methods and applications of a valued carotenoid, ACS Sustainable Chemistry & Engineering 4 (2016) 643–650.

[81] H. Shimada, K. Kondo, P.D. Fraser, Y. Miura, T. Saito, N. Misawa, Increased carotenoid production by the food yeast *Candida utilis* through metabolic engineering of the isoprenoid pathway, Applied and Environmental Microbiology 64 (1998) 2676–2680.

[82] B. Campbell, M. Roberts, C. Kerksick, C. Wilborn, B. Marcello, L. Taylor, et al., Pharmacokinetics, safety, and effects on exercise performance of L-arginine a-ketoglutarate in trained adult men, Nutrition 22 (2006) 872–881.

[83] U. Stottmeister, A. Aurich, H. Wilde, J. Andersch, S. Schmidt, D. Sicker, White biotechnology for green chemistry: fermentative 2-oxocarboxylic acids as novel building blocks for subsequent chemical syntheses, Journal of Industrial Microbiology and Biotechnology 32 (2005) 651–664.

[84] E.D. Son, G.H. Choi, H. Kim, B. Lee, I.S. Chang, J.S. Hwang, Alpha-ketoglutarate stimulates procollagen production in cultured human dermal fibroblasts, and decreases UVB-induced wrinkle formation following topical application on the dorsal skin of hairless mice, Biological and Pharmaceutical Bulletin 30 (2007) 1395–1399.

[85] L. Liu, N. Guan, J. Li, H.-d Shin, G. Du, J. Chen, Development of GRAS strains for nutraceutical production using systems and synthetic biology approaches: advances and prospects, Critical Reviews in Biotechnology 37 (2015) 139–150.

[86] S. Gao, Y. Tong, L. Zhu, M. Ge, Y. Zhang, D. Chen, et al., Iterative integration of multiple-copy pathway genes in *Yarrowia lipolytica* for heterologous beta-carotene production, Metabolic Enginering 41 (2017) 192–201.

[87] O. Chernyavskaya, N. Shishkanova, A. Il'chenko, T. Finogenova, Synthesis of a-ketoglutaric acid by *Yarrowia lipolytica* yeast grown on ethanol, Applied Microbiology and Biotechnology 53 (2000) 152–158.

[88] Z. Yu, G. Du, J. Zhou, J. Chen, Enhanced a-ketoglutaric acid production in *Yarrowia lipolytica* WSH-Z06 by an improved integrated fed–batch strategy, Bioresource Technology 114 (2012) 597–602.

[89] H. Guo, C. Madzak, G. Du, J. Zhou, J. Chen, Effects of pyruvate dehydrogenase subunits overexpression on the a-ketoglutarate production in *Yarrowia lipolytica* WSH-Z06, Applied Microbiology and Biotechnology 98 (2014) 7003–7012.

[90] J. Zhou, X. Yin, C. Madzak, G. Du, J. Chen, Enhanced a-ketoglutarate production in *Yarrowia lipolytica* WSH-Z06 by alteration of the acetyl-CoA metabolism, Journal of Biotechnology 161 (2012) 257–264.

[91] L.A. Horrocks, A.A. Farooqui, Docosahexaenoic acid in the diet: its importance in maintenance and restoration of neural membrane function, Prostaglandins, Leukotrienes & Essential Fatty Acids 70 (2004) 361–372.

[92] G. Caggianiello, M. Kleerebezem, A.S.G. Caggianiello, Exopolysaccharides produced by lactic acid bacteria: from health-promoting benefits to stress tolerance mechanisms, Applied Microbiology and Biotechnology 100 (2016) 3877–3886.

[93] S.K. Bhatia, N. Kumar, R.K. Bhatia, Stepwise bioprocess for exopolysaccharide production using potato starch as carbon source, 3 Biotech 5 (2014) 735–739.

[94] H. Kommann, P. Duboc, I. Marison, U. Stockar, Influence of nutritional factors on the nature, yield, and composition of exopolysaccharides produced by *Gluconacetobacter xylinus* I-2281, Applied and Environmental Microbiology 69 (2003) 6091–6098.

[95] C. Matsuzaki, A. Hayakawa, K. Matsumoto, T. Katoh, K. Yamamoto, K. Hisa, Exopolysaccharides produced by *Leuconostoc mesenteroides* strain NTM048 as an immunostimulant to enhance the mucosal

barrier and influence the systemic immune response, Journal of Agricultural and Food Chemistry 63 (2015) 7009–7015.

[96] A.W. Westbrook, X. Ren, J. Oh, M. Moo-Young, P. Chou, Metabolic engineering to enhance heterologous production of hyaluronic acid in *Bacillus subtilis*, Metabolic Engineering 47 (2018) 401–413.

[97] J.A. Vazquez, M. Montemayor, J. Fraguas, M.A. Murado, Hyaluronic acid production by *Streptococcus zooepidemicus* in marine by-products media from mussel processing wastewaters and tuna peptone viscera, Microbial Cell Factories 149 (2010) 46.

[98] K.A. Baritugo, H.T. Kim, Y. David, T.U. Khang, S.M. Hyun, K.H. Kaang, et al., Enhanced production of gamma-aminobutyrate (GABA) in recombinant *Corynebacterium glutamicum* strains from empty fruit bunch biosugar solution, Microbial Cell Factories 17 (2018) 129.

[99] K. Thakur, S.K. Tomar, S. De, Lactic acid bacteria as a cell factory for riboflavin production, Microbial Biotechnology 9 (2016) 441–451.

[100] W. Xia, W. Chen, W. Peng, K. Li, Industrial vitamin B12 production by *Pseudomonas denitrificans* using maltose syrup and corn steep liquor as the cost-effective fermentation substrates, Bioprocess and Biosystems Engineering 38 (2015) 1065–1073.

[101] K. Piwowarek, E. Lipinska, E. Hac-Szymanczuk, M. Kieliszek, I. Scibisz, *Propionibacterium spp.*—source of propionic acid, vitamin B12, and other metabolites important for the industry, Applied Microbiology and Biotechnology 102 (2018) 515–538.

[102] K. Dmytruk, O. Lyzak, V. Yatsyshyn, M. Kluz, V. Sibirny, C. Puchalaski, et al., Construction and fed-batch cultivation of *Candida famata* with enhanced riboflavin production, Journal of Biotechnology 172 (2014) 11–17.

[103] L. Wang, J. Lv, Z. Chu, Y. Cui, X. Ren, Production of conjugated linoleic acid by *Propionibacterium freudenreichii*, Food Chemistry 103 (2) (2007) 313–318.

[104] K. Gemperlein, D. Dietrich, M. Kohlstedt, G. Zipf, H.S. Bernauer, C. Wittmann, et al., Polyunsaturated fatty acid production by *Yarrowia lipolytica* employing designed myxobacterial PUFA synthases, Nature Communications 10 (2019) 4055.

[105] E.M. Grima, J.A.S. Perez, J.L.G. Sanchez, F.G. Camacho, D.L. Alonso, EPA from *Isochrysis galbana*. Growth conditions and productivity, Process Biochemistry 27 (5) (1992) 299–305.

[106] P.K. Bajpai, P. Bajpai, O.P. Ward, Production of arachidonic acid by *Mortierella alpina* ATCC 32222, Journal of Industrial Microbiolology 8 (3) (1991) 179–185.

[107] Y. Miura, K. Kondo, T. Saito, H. Shimada, P.D. Fraser, N. Misawa, Production of the carotenoids lycopene, β-carotene, and astaxanthin in the food yeast *Candida utilis*, Applied and Environmental Microbiology 64 (1998) 1226–1229.

[108] K. Nanou, T. Roukas, E. Papadakis, P. Kotzekidou, Carotene production from waste cooking oil by *Blakeslea trispora* in a bubble column reactor: the role of oxidative stress, Engineering in Life Sciences 17 (2017) 775–780.

[109] B. Kleessen, E. Bezirtzoglou, J. Mättö, Culture-based knowledge on biodiversity, development and stability of human gastrointestinal microflora, Microbial Ecology in Health and Disease 12 (2000) 53–63.

[110] G.R. Gibson, M.B. Roberfroid, Dietarymodulation of the colonic microbiota: introducing the concept of prebiotics, The Journal of Nutrition 125 (1995) 1401–1412.

[111] A. Gil, R. Rueda, Modulation of the intestinal microflora by specific dietary components, Microbial Ecology in Health and Disease 12 (2000) 31–39.

[112] M.A. Anwar, S. Kralj, A.V. Piqué, H. Leemhuis, M.J. van der Maarel, L. Dijkhuizen, Inulin and levan synthesis by probiotic *Lactobacillus gasseri strains*: characterization of three novel fructansucrase enzymes and their fructan products, Microbiology 156 (2010) 1264–1274.

[113] B. Rodriguez-Colinas, M.A. de Abreu, L. Fernandez-Arrojo, R. de Beer, A. Poveda, J. Jimenez-Barbero, et al., Production of galactooligosaccharides by the β-galactosidase from *Kluyveromyces lactis*: comparative

analysis of permeabilized cells versus soluble enzyme, Journal of Agriculture and Food Chemistry 59 (2011) 10477–10484.

[114] L. Yu, D. O'Sullivan, Production of galactooligosaccharides using a hyperthermophilic β-galactosidase in permeabilized whole cells of *Lactococcus lactis*, Journal of Dairy Science 97 (2014) 694–703.

[115] W.H. Lee, P. Pathanibul, J. Quarterman, J.H. Jo, N.S. Han, M.J. Miller, et al., Whole cell biosynthesis of a functional oligosaccharide, 20-fucosyllactose, using engineered *Escherichia coli*, Microbial Cell Factories 11 (2012) 48.

[116] F. Baumgartner, L. Seitz, G.A. Sprenger, C. Albermann, Construction of *Escherichia coli* strains with chromosomally integrated expression cassettes for the synthesis of 20-fucosyllactose, Microbial Cell Factories 12 (2013) 40.

[117] I. Giavasis, Production of microbial polysaccharides for use in food, microbial production of food ingredients, Enzymes and Nutraceuticals (2013) 413–468. Woodhead Publishing.

[118] W. He, L. Fu, G. Li, J.A. Jones, R.J. Linhardt, M. Koffas, Production of chondroitin in metabolically engineered *E. coli*, Metabolic Engineering 27 (2015) 92–100.

[119] C. Zhang, L. Liu, L. Teng, J. Chen, J. Liu, J. Li, et al., Metabolic engineering of *Escherichia coli* BL21 for biosynthesis of heparosan, a bioengineered heparin precursor, Metabolic Engineering 14 (2012) 521–527.

[120] P.J. Looijesteijn, I.C. Boels, M. Kleerebezem, J. Hugenholtz, Regulation of exopolysaccharide production by *Lactococcus lactis* subsp. *cremoris* by the sugar source, Applied and Environmental Microbiology 65 (1999) 5003–5008.

[121] I.C. Boels, A. Ramos, M. Kleerebezem, W.M. de Vos, Functional analysis of the *Lactococcus lactis* galU and galE genes and their impact on sugar nucleotide and exopolysaccharide biosynthesis, Applied and Environmental Microbiology 67 (2001) 3033–3040.

[122] S.F. Yuan, H.S. Alper, Metabolic engineering of microbial cell factories for production of nutraceuticals, Microbial Cell Factories 18 (2019) 46.

[123] L. Das, E. Bhaumik, U. Raychaudhuri, R. Chakraborty, Role of nutraceuticals in human health, Journal of Food Science and Technology 49 (2012) 173–183.

[124] A.H. Stark, M.A. Crawford, R. Reifen, Update on alpha-linolenic acid, Nutrition Reviews 66 (2008) 326–332.

[125] Experts I, Omega-3 Polyunsaturated Fatty Acids (PUFAs)—A Global Market Overview. <http://industry-experts.com/verticals/healthcare-and-pharma/omega-3-polyunsaturated-fatty-acids-pufas-a-global-market-overview>, 2014.

[126] Z. Xue, P.L. Sharpe, S.-P. Hong, N.S. Yadav, D. Xie, D.R. Short, et al., Production of omega-3 eicosapentaenoic acid by metabolic engineering of Yarrowia lipolytica, Nature Biotechnology 31 (2013) 734–740.

[127] D. Xie, E. Miller, B. Tyreus, E.N. Jackson, Q. Zhu, Sustainable production of omega-3 eicosapentaenoic acid by fermentation of metabolically engineered *Yarrowia lipolytica*, Quality Living Through Chemurgy and Green Chemistry (2016) 17–33. Springer.

[128] Q. Zhu, E.N. Jackson, Metabolic engineering of *Yarrowia lipolytica* for industrial applications, Current Opinion in Biotechnology 36 (2015) 65–72.

[129] F. Zhu, L. Lu, S. Fu, X. Zhong, M. Hu, Z. Deng, et al., Targeted engineering and scale up of lycopene overproduction in *Escherichia coli*, Process Biochemistry 50 (2015) 341–346.

[130] T.C. Adarme Vega, D.K.Y. Lim, M. Timmins, F. Vernen, Y. Li, P.M. Schenk, Microalgal biofactories: a promising approach towards sustainable omega-3 fatty acid production, Microbial Cell Factories 11 (2012) 96.

[131] H.G. Damude, D.J. Macool, S.K. Picataggio, J.J. Ragghianti, J.E. Seip, Z. Xue, et al., Docosahexaenoic acid producing strains of Yarrowia lipolytica, US Patent, US7, 550, 286, 2009.

[132] Y.-F. Peng, W.-C. Chen, K. Xiao, L. Xu, L. Wang, X. Wan, DHA production in Escherichia coli by expressing reconstituted key genes of polyketide synthase pathway from marine bacteria, PLoS One 11 (9) (2016) e0162861.

[133] J. Cui, S. He, X. Ji, L. Lin, Y. Wei, Q. Zhang, Identification and characterization of a novel bifunctional Δ12/Δ15-fatty acid desaturase gene from *Rhodosporidium kratochvilovae*, Biotechnology Letters 38 (2016) 1155–1164.

[134] L.T. Cordova, H.S. Alper, Production of α-linolenic acid in *Yarrowia lipolytica* using low-temperature fermentation, Applied Microbiology and Biotechnology 102 (2018) 8809–8816.

[135] B. Debord, C. Lefebvre, A.M. Guyot-Hermann, J. Hubert, R. Bouche, J.C. Guyot, Study of different forms of mannitol: comparative behaviour under compression, Drug Development and Industrial Pharmacy 13 (1987) 1533–1546.

[136] B.K. Dwivedi, Low Calorie and Special Dietary Foods, CRC Press, Inc, 1978.

[137] B. Shen, R.G. Jensen, H.J. Bohnert, Mannitol protects against oxidation by hydroxyl radicals, Plant Physiology 115 (1997) 527–532.

[138] B.J.O. Efiuvwevwere, L.G. Gorris, E.J. Smid, E.P.W. Kets, Mannitolenhanced survival of *Lactococcus lactis* subjected to drying, Applied Microbiology and Biotechnology 51 (1999) 100–104.

[139] W. Soetaert, K. Buchholz, E.J. Vandamme, Production of D-mannitol and D-lactic acid by fermentation with *Leuconostoc mesenteroides*, Agro Food Industry Hi-Tech 6 (1) (1995) 41–44.

[140] T. Ferain, A.N. Schanck, J. Delcour, 13C nuclear magnetic resonance analysis of glucose and citrate end products in an ldhL-ldhD double-knockout strain of *Lactobacillus plantarum*, Journal of Bacteriology 178 (1996) 7311–7315.

[141] A.R. Neves, A. Ramos, C. Shearman, M.J. Gasson, J.S. Almeida, H. Santos, Metabolic characterization of *Lactococcus lactis* deficient in lactate dehydrogenase using in vivo 13C-NMR, European Journal of Biochemistry 267 (2000) 3859–3868.

[142] H.W. Wisselink, R.A. Weusthuis, G. Eggink, J. Hugenholtz, G.J. Grobben, Mannitol production by lactic acid bacteria: a review, International Dairy Journal 12 (2002) 151–162.

[143] L.R. Zehner, D-tagatose as a low-calorie carbohydrate sweetener and bulking agent, European Patent Application EP0257626, 1988.

[144] P. Kim, S.H. Yoon, H.J. Roh, J.H. Choi, High production of D-tagatose, a potential sugar substitute, using immobilized L-arabinose isomerase, Biotechnology Progress 17 (2001) 208–210.

[145] R.J. Van Rooijen, S. van Schalkwijk, W.M. de Vos, Molecular cloning, characterization and nucleotide sequence of the tagatose-6-phosphate pathway gene cluster of the lactose operon of *Lactococcus lactis*, Journal of Biological Chemistry 266 (1991) 7176–7181.

[146] T. Neta, K. Takada, M. Hirasawa, Low-cariogenicity of trehalose as a substrate, Journal of Dentistry 28 (2000) 571–576.

[147] S. De Carlo, M. Adrian, P. Kalin, J.M. Mayer, J. Dubochet, Unexpected property of trehalose as observed by cryo-electron microscopy, Journal of Microscopy 196 (1999) 40–45.

[148] C.F. Felix, C.C. Moreira, M.S. Oliveira, M. Sola-Penna, J.R. Meyer-Fernandes, H.M. Scofano, et al., Protection against thermal denaturation by trehalose on the plasma membrane H+-ATPase from yeast. Synergetic effect between trehalose and phospholipid environment, European Journal of Biochemistry 266 (1999) 660–664.

[149] N. Guo, I. Puhlev, D.R. Brown, J. Mansbridge, F. Levine, Trehalose expression confers desiccation tolerance on human cells, Nature Biotechnology 18 (2000) 168–171.

[150] S.F. Wu, Y. Suzuki, A.K. Kitahara, H. Wada, Y. Nishimura, Skin flap storage with intracellular and extracellular solutions containing trehalose, Annals of Plastic Surgery 43 (1999) 289–294.

[151] M.A. Singer, S. Lindquist, Thermotolerance in *Saccharomyces cerevisiae*: the Yin and Yang of trehalose, Trends in Biotechnology 16 (1998) 460–468.

[152] C. Deborde, C. Corre, D.B. Rolin, L. Nadal, J.D. de Certaines, P. Boyaval, Trehalose biosynthesis in dairy *Propionibacterium*, Journal of Magnetic Resonance Analysis 2 (1996) 297–304.

[153] D.T. Welsh, R.H. Reed, R.A. Herbert, The role of trehalose in the osmoadaptation of *Escherichia coli* NCIB 9484: interaction of trehalose, K+ and glutamate during osmoadaptation in continuous culture, Journal of General Microbiology 137 (1991) 745–750.

[154] F. Cardoso, P. Gaspar, A. Ramos, J. Hugenholtz, H. Santos, Effect of environmental conditions on trehalose production by *Propionibacterium*. In Abstract Book of the 3rd International Symposium on Propionibacteria, July 8–11, Zurich, Switzerland: 41:P29, 2002.

[155] F. Shahidi, P. Ambigaipalan, Phenolics and polyphenolics in foods, beverages and spices: antioxidant activity and health effects—a review, Journal of Functional Foods 18 (2015) 820–897.

[156] M.A. Alam, N. Subhan, M.M. Rahman, S.J. Uddin, H.M. Reza, Effect of citrus flavonoids, naringin and naringenin, on metabolic syndrome and their mechanisms of actionS.D. anf Sarker Advances in Nutrition 5 (2014) 404–417.

[157] R.P. Pandey, P. Parajuli, M.A.G. Koffas, J.K. Sohng, Microbial production of natural and non-natural flavonoids: pathway engineering, directed evolution and systems/synthetic biology, Biotechnology Advances 34 (2016) 634–662.

[158] A. Rodriguez, Y. Chen, S. Khoomrung, E. Özdemir, I. Borodina, J. Nielsen, Comparison of the metabolic response to over-production of p-coumaric acid in two yeast strains, Metabolic Engineering 44 (2017) 265–272.

[159] C.M. Palmer, H.S. Alper, Expanding the chemical palette of industrial microbes: metabolic engineering for type III pks-derived polyketides, Biotechnology Journal 14 (2018) e1700463.

[160] E.I. Hwang, M. Kaneko, Y. Ohnishi, S. Horinouchi, Production of plant-specific flavanones by *Escherichia coli* containing an artificial gene cluster, Applied and Environmental Microbiology 69 (2003) 2699–2706.

[161] F. Koopman, J. Beekwilder, B. Crimi, A. van Houwelingen, R.D. Hall, D. Bosch, et al., De novo production of the flavonoid naringenin in engineered *Saccharomyces cerevisiae*, Microbial Cell Factories 11 (2012) 155.

[162] W. Zhang, H. Liu, X. Li, D. Liu, X.T. Dong, F.F. Li, et al., Production of naringenin from D-xylose with co-culture of *E. coli* and *S. cerevisiae*, Engineering in Life Sciences 17 (2017) 1021–1029.

[163] R. Nakata, S. Takahashi, H. Inoue, Recent advances in the study on resveratrol, Biological and Pharmaceutical Bulletin 35 (2012) 273–279.

[164] F. Zhang, J. Liu, J.S. Shi, Anti-inflammatory activities of resveratrol in the brain: role of resveratrol in microglial activation, European Journal of Pharmacology 636 (2010) 1–7.

[165] M. Li, K.R. Kildegaard, Y. Chen, A. Rodriguez, I. Borodina, J. Nielsen, De novo production of resveratrol from glucose or ethanol by engineered *Saccharomyces cerevisiae*, Metabolic Engineering 32 (2015) 1–11.

[166] M. Li, K. Schneider, M. Kristensen, I. Borodina, J. Nielsen, Engineering yeast for high-level production of stilbenoid antioxidants, Scientific Reports 6 (2016) 36827.

[167] J.M. Camacho-Zaragoza, G. Hernández-Chávez, F. Moreno-Avitia, R. Ramírez-Iñiguez, A. Martínez, E. Bolívar, et al., Engineering of a microbial coculture of *Escherichia coli* strains for the biosynthesis of resveratrol, Microbial Cell Factories 15 (2016) 163.

[168] J. Xiao, G.B. Sun, B. Sun, Y. Wu, L. He, X. Wang, et al., Kaempferol protects against doxorubicin-induced cardiotoxicity in vivo and in vitro, Toxicology 292 (2012) 53–62.

[169] A. Rodriguez, T. Strucko, S.G. Stahlhut, M. Kristensen, D.K. Svenssen, J. Forster, et al., Metabolic engineering of yeast for fermentative production of flavonoids, Bioresource Technology 245 (2017) 1645–1654.

[170] J.A. Jones, V.R. Vernacchio, A.L. Sinkoe, S.M. Collins, M.H.A. Ibrahim, D.M. Lachance, et al., Experimental and computational optimization of an *Escherichia coli* co-culture for the efficient production of flavonoids, Metabolic Engineering 35 (2016) 55–63.

[171] F. Shi, Z. Xu, P. Cen, Microbial production of natural poly amino acid, Science in China Series B 50 (2007) 291–303.

[172] W.C. Tseng, T.Y. Fang, C.Y. Cho, P.S. Chen, C.S. Tsai, Assessments of growth conditions on the production of cyanophycin by recombinant *Escherichia coli* strains expressing cyanophycin synthetase gene, Biotechnology Progress 28 (2012) 358–363.

[173] X.-S. Chen, X.-D. Ren, N. Dong, S. Li, F. Li, F.-L. Zhao, et al., Culture medium containing glucose and glycerol as a mixed carbon source improves e-poly-L-lysine production by *Streptomyces* sp. M-Z18, Bioprocess and Biosystems Engineering 35 (2012) 469–475.

[174] L.C. Mata-Gómez, J.C. Montañez, A. Méndez-Zavala, C.N. Aguilar, Biotechnological production of carotenoids by yeasts: an overview, Microbial Cell Factories 13 (2014) 12.

[175] S. Devaraj, S. Mathur, A. Basu, H.H. Aung, V.T. Vasu, S. Meyers, et al., A dose response study on the effects of purified lycopene supplementation on biomarkers of oxidative stress, Journal of the American College of Nutrition 27 (2008) 267–273.

[176] A.V. Rao, L.G. Rao, Carotenoids and human health, Pharmacology Research 55 (2007) 207–216.

[177] A. McWilliams, The global market for carotenoids. <https://www.bccresearch.com/market-research/food-and-beverage/the-global-market-for-carotenoids-fod025f.html>, 2018.

[178] H. Alper, K. Miyaoku, G. Stephanopoulos, Construction of lycopene-overproducing *E. coli* strains by combining systematic and combinatorial gene knockout targets, Nature Biotechnology 23 (2005) 612.

[179] W.R. Farmer, J.C. Liao, Improving lycopene production in *Escherichia coli* by engineering metabolic control, Nature Biotechnology 18 (2007) 533–537.

[180] T. Ma, B. Shi, Z. Ye, X. Li, M. Liu, Y. Chen, et al., Lipid engineering combined with systematic metabolic engineering of *Saccharomyces cerevisiae* for high-yield production of lycopene, Metabolic Engineering 52 (2018) 134–142.

[181] M. Larroude, E. Celinska, A. Back, S. Thomas, J.M. Nicaud, R.A. Ledesma-Amaro, Synthetic biology approach to transform *Yarrowia lipolytica* into a competitive biotechnological producer of beta-carotene, Biotechnology and Bioengineering 115 (2018) 464–472.

[182] K.R. Kildegaard, B. Adiego-Pérez, D. Doménech Belda, J.K. Khangura, C. Holk-enbrink, I. Borodina, Engineering of *Yarrowia lipolytica* for production of astaxanthin, Synthetic and Systems Biotechnology 2 (2017) 287–294.

[183] J. Yang, L. Guo, Biosynthesis of β-carotene in engineered *E. coli* using the MEP and MVA pathways, Microbial Cell Factories 13 (2014) 160.

[184] J. Zhao, Q. Li, T. Sun, X. Zhu, H. Xu, J. Tang, et al., Engineering central metabolic modules of *Escherichia coli* for improving β-carotene production, Metabolic Engineering 17 (2013) 42–50.

[185] J. Jin, Y. Wang, M. Yao, X. Gu, B. Li, H. Liu, et al., Astaxanthin overproduction in yeast by strain engineering and new gene target uncovering, Biotechnology for Biofuels 11 (2007) 230.

[186] X. Lyu, G. Zhao, K.R. Ng, R. Mark, W.N. Chen, Metabolic engineering of *Saccharomyces cerevisiae* for De Novo production of Kaempferol, Journal of Agriculture and Food Chemistry 67 (2019) 5596–5606.

[187] B. Shi, T. Ma, Z. Ye, X. Li, Y. Huang, Z. Zhou, et al., Systematic metabolic engineering of *Saccharomyces cerevisiae* for lycopene overproduction, Journal of Agriculture and Food Chemistry 67 (40) (2019) 11148–11157.

[188] C.G. Lim, Z.L. Fowler, T. Hueller, S. Schaffer, M.A.G. Koffas, High yield resveratrol production in engineered *Escherichia coli*, Applied and Environmental Microbiology 77 (2011) 3451–3460.

[189] J.M.P. Jorge, C. Leggewie, V.F. Wendisch, A new metabolic route for the production of gamma-aminobutyric acid by *Corynebacterium glutamicum* from glucose, Amino acids 48 (2016) 2519–2531.

[190] V. Sachdeva, A. Roy, N. Bharadvaja, Current prospects of nutraceuticals: a review, Current Pharmaceutical Biotechnology (2020). Available from: https://doi.org/10.2174/1389201021666200130113441.

[191] S.A. Hussain, N.R. Panjagari, R.R. Singh, G.R. Patil, Potential herbs and herbal nutraceuticals: food applications and their interactions with food components, Critical Reviews in Food Science and Nutrition 55 (2015) 94–122.

[192] H. Dureja, D. Kaushik, V. Kumar, Developments in nutraceuticals, Indian Journal of Pharmacology 35 (6) (2003) 363–372.

[193] https://www.bccresearch.com/market-research/food-and-beverage/nutraceuticals-global-markets.html.

[194] Indian Nutraceuticals Market Outlook: Vision 2022. Available from: <http://publication.assocham.tv/data/product-file/88%20Indian%20Nutraceutical%20Market%20Opportunities_Report_LR.pdf%20> (accessed 23.08.19).

[195] A. Sharma, P. Kumar, P. Sharma, B.A. Shrivastav, Comparative study of regulatory registration procedure of nutraceuticals in India, Canada and Australia, International Journal of Pharmaceutical Quality Assurance 4 (4) (2013) 61–66.

13

Bioprocessing of agri-food processing residues into nutraceuticals and bioproducts

Vinod Kumar[1,4], Sudesh K. Yadav[1], Anil K. Patel[2], Bhuwan B. Mishra[1], Vivek Ahluwalia[3], Lalitesh K. Thakur[3], Jitendra Kumar[3]

[1]CENTER OF INNOVATIVE AND APPLIED BIOPROCESSING, INDIA [2]DEPARTMENT OF CHEMICAL AND BIOLOGICAL ENGINEERING, KOREA UNIVERSITY, SEOUL, REPUBLIC OF KOREA [3]INSTITUTE OF PESTICIDE FORMULATION TECHNOLOGY, INDIA [4]DIVISION OF FERMENTATION & MICROBIAL BIOTECHNOLOGY, CSIR-INDIAN INSTITUTE OF INTEGRATIVE MEDICINE (IIIM), JAMMU, INDIA

13.1 Introduction

Since time immemorial, agriculture has been the primary source of livelihood for human beings. Agriculture holds an essential part in the economic development of a nation along with ensuring food security. It is also the largest sector in some Asian and African countries in terms of employment, policy development, and export earnings. In recent years, due to increase in establishment of agri-based industries, there is also threefold increase in agro-industrial waste production. The agricultural and agri-food industrial wastes such as sugarcane, rice, corn, wheat, tubers, etc. are some major sources of agro-food-based waste and generated in billions of tons of biomass per year, most accessible and renewable. These wastes are mostly dumped in the environment without proper treatment leading to land, water, soil, and air pollution. Consciousness deficiency and hesitancy to spend money on treatment are also reasons for the disposal of wastes into the surrounding environment (Fig. 13–1). Growing public awareness about the environmental problems caused by waste, changing governmental regulations, and economic considerations have led to use of these wastes for developing a process to derive chemicals and fuels from these as well as develop new materials [1,2].

Food waste is also a major global problem. Food processing and postconsumer wastes are the main stages for food waste generation. The waste is generated during usage of different techniques for the conversion of raw plant and animal materials to the consumable form for humans and animals. Discarded materials from the food industries such as fruit &

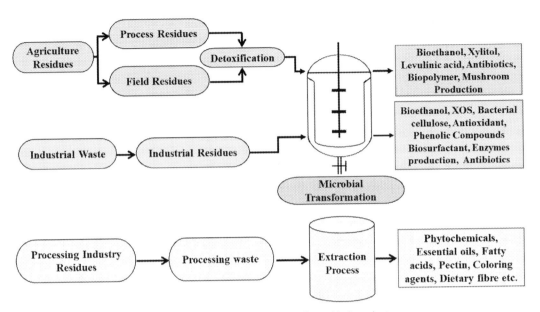

FIGURE 13-1 Systematic representation of agri-food residue to value-added products.

vegetable processing industries, meat & fish industries, winery & brewery industries, and milk processing plants are typical examples of food waste [3].

According to Food and Agriculture Organization estimation, the loss of food along the food supply chain is approximately 20% of the total world agricultural production and about one-third of the whole world food production for human consumption. The food loss by different edible commodities such as root crops, fruits and vegetables (40%–50%), fish (35%), cereals (30%), and oilseeds, meat, and dairy products (20%) is a large percentage of production [4]. Food waste is not recommended for direct consumption as animal feed as it leads to negative health effects, mainly due to the microbial toxins produced during the food decaying process. Food waste is also notorious for environmental health. Due to the above reasons, food waste must be disposed of through fermentation, composting, and landfilling. Changing lifestyle and urbanization are considered as the major factors for the increase in food waste generation which is anticipated to grow further [5–7].

Strategic valorization of organic waste is of concern as nonmanagement leads to severe environmental complications such as the greenhouse effect, soil pollution, and groundwater pollution. Effective management schemes for the sorting, treating, and valorizing of processing waste should be adopted. Agri-food waste has been used for the production of high added-value products such as bioactive peptides, organic acids, bioethanol, enzymes, etc. The lignocellulosic and starchy materials are degraded by microorganisms with the help of efficient enzymatic systems. Fruit processing wastes such as pomegranate peels, lemon peels, and green walnut husks have been utilized for the recovery of the compounds with health and pharmaceutical importance such as pectin, lipids, flavonoids, dietary fibers and/or bio-based products, biochemicals, biofuels, etc. [8–11].

13.2 Agri-food processing residues and valorization

The harvesting and processing of agricultural crops lead to generation of primary and secondary agricultural residues. Primary residues are also called field-based residues and are generated in the farm during the harvesting time. The waste generated during the processing stages is called secondary or processing-based residues. Examples of primary residues include sugarcane top, maize stalks, paddy straw, and coconut empty frond. Likewise, bagasse, paddy husk, coconut shell, maize cob, coconut husk, sawdust, and coir dust are some examples of secondary residues. Different crops produce a different type of primary and secondary residues [10,12–15]. Domestic and industrial sectors utilize agriculture residues for different applications. Primary residues to some extent are utilized as animal feed and composting. But, secondary residues are available at the processing sites which can be utilized by the bioprocessing industry for production of value-added products (Table 13–1).

Industrial growth in the agri-food sector has revolutionized with setup of food- and beverages-based industries. Industries based on commercial processing of fruits and vegetables have increased since the last decade. The by-products are produced by the food processing industry with high value of biochemical oxygen demand, chemical oxygen demand, and other suspended solids which lead to environmental and public health problems. This waste has been exploited for the invention of enormous value-added products. Like, in seed processing industries, a large amount of seed cake is left after obtaining the oil. Different type of seed processing units produces diversified seed cakes like canola oil cake, sunflower oil cake, coconut oil cake, sesame oil cake, mustard oil cake, palm kernel cake, soybean cake, and groundnut oil cake. These agro-industrial residues are comparatively economical, holding a good quantity of constituents that have an unrestricted potential to be consumed as alternative substrates for fermentation [37–40].

Valorization of wastes enables sustainable use of accessible natural resources. Processes and applications are two crucial features for considering the valorization of food processing residues. The agri-food processing residues contain innumerable valuable compounds from sugars to bioactive compounds. For example, the orange juice industry is one of major juices-producing industries that generate a high volume of orange peel as a by-product. Orange peel is discarded as a by-product which has been valorized to extract value-added essential oil [41]. Only pulp is used in avocado-based industry while seeds and peels signify a large amount of waste. Potential valuable ingredients for food, cosmetic, and other industrial fields are extracted from it. Nonfood applications of avocado as pulp oil are used for production of biodiesel. Recent studies proposed use of seed oil for the synthesis of liquid fuels, fuel additives, and charcoal. This represents a critical point in key strategy development against energy production plight and environmental pollution due to fossil fuels [42].

Table 13–1 Summary of some important products from agriculture food residue.

S. No.	Sources of residue	Types of waste generate	Name of products	Microorganism	Applications	Process	References
1.	Cereals and pulses	Wheat, rice, barley, maize, sorghum, millet, oat, sawdust, banana stalks, bahia grass, sugarcane molasses, and rye	Bioethanol	Saccharomyces cerevisiae	Biofuels	SmF and SSF	[16]
			Biobutanol	Clostridium species	Biofuels	SmF	[3,17,18]
			Xylitol	Candida Tropicalis	Food additive	SmF	[15]
			Levulinic acid		Fuel additives		
			Oxytetracycline		Antibiotic	SSF	[19]
			Poly-hydroxy butyrate (PHB)	Alcaligenes sp. NCIM 5085	Biopolymer	SmF	[20]
			Mushroom production	Pleurotus, oyster	Food	SSF	[21,22]
2.	Fruits and vegetables		Potato peel, banana peel, orange peel & pulp, cassava peel, and pomegranate peel	Bioethanol	Biofuels		
			XOS		Prebiotics	SmF	[23]
			Bacterial cellulose	Acetobacter pasteurianus strain RSV-4	Pharma industries	SmF	
			Antioxidant compounds, Phenolic compounds		Food additive	SSF	[24–26]
3.	Dairy	Whey, soya bean milk waste, and soy molasses	Biosurfactant	Pseudomonas aeruginosa SR17, P. aeruginosa ATCC 10145	Food additive	SmF	[27,28]
			Bacterial cellulose			SmF	
			Tagatose		Food additive	SmF	[29]
			Tempeh	Rhizopus strains		SSF	[16]
4.	Edible oil	Waste cooking oil Soybean oil cake, coconut oil cake, groundnut oil cake, and olive oil mill waste	Biodiesel		Biofuels		
			Biosurfactant	Bacillus subtilis, P. aeruginosa	Food additive	SmF	[30]
			Lipid	Penicillium expansum, Aspergillus sp. ATHUM 3482	Biofuels	SmF	[31]
			α-Amylase, xylanase, and β-glucosidase production	A. oryzae	Enzyme	SSF	[32,33]
			Lipase	Aspergillus ibericus	Enzyme		[34]
			Oxytetracycline	Streptomyces rimosus		SSF	[35]
			Rifamycin B	Amycolatopsis Mediterranean MTCC 14	Antibiotic	SSF	[36]

The pineapple waste has been used for production of bromelain, organic acids, and phenolic compounds. Pineapple waste contains insoluble fiber-rich fractions and sugars. Due to presence of sugars, it is used as a fermentable substrate or to produce enzymes [43].

Phenolic compounds are valuable assets for several industries like food and pharmaceutical industries, resins and adhesives, biofuels industries. Lignin-rich waste materials have been used for recovery of phenolic compounds. The lignin present in lignocellulosic processing residues then is valorized as an energy carrier [44]. Lipids have commercial importance because of their use in the renewable energy and food industry. Recent years have witnessed an increase in demand and consumption of edible oils. Meat processing industrial waste has been utilized for generation of lipids or transformed into bio-based fuels. Major key concerns for transformations are nature of waste and application estimation [45].

The industrial units based on cereals and pulses yield a huge amount of husk, bran, and germ during processing. Husk has been recycled to produce high-end products depending upon its origin. Rice husk is utilized as pet feed fiber, fertilizer, vermicompost, and fabrication of building material. The coconut husk is used for generation of second-generation bioethanol in addition to its household uses like making rope, broom, mat, tiles, and fishing net. The nutraceutical compounds like astaxanthin have been produced from the waste of shrimp processing industries. Marine animal waste is better source for some bioactive molecules like glycosaminoglycans than terrestrial organisms [7].

13.2.1 Extraction methods

There are various techniques for recovery of bioactive components from agro-industrial waste. The utilization technique for recovery depends on the specific bioactive components and agro-industrial waste. Solvent extraction, supercritical fluid extraction, subcritical water extraction, enzymes extraction, ultrasounds extraction, and microwaves extraction are mainly used extraction processes [46,47]. The appropriately sized plant matter is exposed to different organic solvents in the solvent extraction method. The solvent takes up compounds from raw material based on the solubility of the compound. Low processing cost and ease of operation are advantageous for this extraction process. It inherits some drawbacks like use of a toxic and a large amount of solvents [47–49].

A green extraction technique called supercritical fluid extraction is a very popular technique for natural compounds extraction from natural sources. The organic solvents are substituted with nonexplosive, nontoxic, and inexpensive supercritical carbon dioxide. This technique has better application with lipophilic compounds [50,51]. Subcritical water extraction is another green extraction method used for phenolic compounds where water is maintained at a temperature between 100°C & 374°C and pressure which is high ample with advantages of smaller extraction time, better extraction quality, and environment-friendly [52]. The polyphenolic compounds from grape pomace were extracted in good yield using this technique with optimal extraction temperature (140°C) and pressure (11.6 MPa) as compared to the conventional method [53]. Polyphenolic compounds are known for health benefits effects against chronic diseases like atherosclerosis, cancer, and cardiovascular disease.

The pomegranate seed residues were used as raw material for extraction of phenolic compounds. The high-performance liquid chromatography coupled with ABTS radical reaction system (HPLC-ABTS$\cdot^+$) was used to evaluate individual antioxidant capacities of each compound [54]. Recently, this technique was used to recover phenolics with notably higher yield and total phenolic contents from chestnut exocarps. The extracts exhibited better activity in comparison with extracts obtained by conventional methods [55].

Ultrasound-assisted extraction is an important extraction technique where ultrasound persuades superior dispersion of solvent into cellular materials. Ultrasound disrupts cell wall and increases mass transfer thus enabling release of bio-actives. Ultrasound frequency is a determining factor in the extraction process. Method with variables like methanol composition, temperature, ultrasound amplitude, cycle, solvent pH, and solvent-solid ratio was developed for anthocyanins and total phenolic compounds extraction from mulberries. Extraction temperature and solvent composition were found determining factors in the extraction process [56].

Microwave-assisted product extraction is a prevailing method that is a combination of microwave and conventional solvent extraction. This technique requires less solvent than traditional methods. The extraction time process with this technique is much lower with a higher extraction rate [57]. This technique was assessed for recovery/extraction of polysaccharides from Jamun (*Syzygium cumini*) fruit seeds waste. Four different combination factors comprising microwave power, pH, time, and solid to liquid ratio were considered. The yield of the polysaccharides recovered matched with the experimental predicted value suggesting that this procedure was good for the extraction process [58].

Enzyme-assisted extraction is regarded to be green, appreciative, and additional to a water-based extraction system. Different types of enzymes help in liberation of chemical constituents from agro-industrial waste. Organic solvents are not used in this process. Common enzymes used for extraction process are cellulase, β-glucosidase, xylanase, δ-amylase, β-gluconase, pectinase, and β-glucanase. [59].

13.2.2 Nutraceuticals and functional food from agri-food waste

Nutraceuticals are natural bioactive compounds that have health-nurturing, disease-halting, and/or medicinal values while functional foods are food products with boosted natural constituents with a defined physiological deterrent and/or health-boosting effect. It is estimated that vegetables and fruits have approximately 25%–30% nonedible portions in the form of skin, peelings, stems, shells, and seeds. This nonedible portion along with food and agricultural industries waste attracts researchers to recover nutraceuticals and functional foods from them. The by-products obtained are exceptional sources of proteins, lipids, fiber, and other bioactive compounds [60,61].

Grapes are appetizing fruit and used for production of wine. In fermentation process of grapes into wines, industries produce huge waste in the form of skin and seeds. The grape seed extract, grape seed oil, and floor have an unlimited perspective as a nutraceutical. Seeds contain vital nutrients like Vitamin E, flavonoids like catechin & epicatechin, fatty acids like linoleic acid, and oligomeric proanthocyanidin complexes [62]. Presently, grape seed is accessible as a dietary supplement in different forms. The grape seed oil contains important LDL-cholesterol-reducing constituent policosanols [63].

The apple industry produces apple pomace as waste after processing, which is a good source of pectin. It is estimated that apple pomace contributes about 14% of total pectin production around the world. Pectin has varied biological properties and used as a nutritional supplement. It helps in the reduction of blood cholesterol level and postprandial glycemic response. The food, cosmetics, and pharmaceutical industries use it for applications as a food stabilizer, gelling agent, and/or thickener [64]. Antidiabetic agent phlorizin with glucose transport inhibiting property is also found in apple industry by-products. Pectin oligosaccharides are produced by an enzymatic process with better prebiotic properties. Apple pomace is also a rich source of polyphenols [65,66].

Berries are rich source of nutraceuticals and consumed as fresh, frozen, or as processed products. Berry pomace and branches are rich source of polyphenols [67–69]. The oil obtained from berry seed is a good source of tocochromanols and phytosterols [70]. Citrus fruit waste is another by-product that is a good source of nutraceuticals and functional foods. Approximately 50% − 70% of wet weight of the processed fruit is generated as waste after processing. Due to presence of essential oil in the waste, it causes serious environmental problems by natural soil microflora growth inhibition. Citrus peel waste has been valorized for production of citrus pectin which is a nutraceutically valued compound. The commercial citrus pectin is modified chemically, enzymatically, or thermally to get modified citrus pectin. Modified citrus pectin is registered as a dietary supplement in the United States [64,71–73]. Also, the citrus waste contains phytochemicals such as phenolic acids, flavonoids and limonoids with antiinflammatory, anticancer, antibacterial, and antioxidant activities [74].

Plum pomace is an important agro-food industry waste. The oleic and linoleic acids are main components of plum seed oil. The plum seed oil is used for production of biodiesel as it has high ratio of unsaturated/saturated fatty acids (UFA/SFA) [75]. Solid-state fermentation of plum pomace with filamentous fungi increased recovery of polyphenols and lipids [76]. Pomegranate peel contains phytochemicals like phenolics, flavonoids, and complex polysaccharides. The sugar-free mono and oligomeric ellagitannins are present in abundant quantity in pomegranate peel. The ellagitannins are known for their antioxidant properties. The waste pomegranate seed contains oil with high content of conjugated fatty acids, dietary fibers, and high-quality proteins [77,78].

Natural coloring agents are in demand, and methods are being developed for the extraction of new natural coloring agents from biological sources. Betalains are red-violet (betacyanins) or yellow (betaxanthins) tyrosine-derived natural pigments. These compounds along with coloring capacity also possess antiviral, antioxidant, and antiinflammatory activities. In a recent study, beetroot pomace was estimated for betaxanthin, betacyanin, and phenolics extraction [79]. Phenolic acids like ferulic acid, vanillic acid, caffeic acid, and p-hydroxybenzoic acid were extracted from beetroot pomace. Similarly, beetroot pomace extraction led to extraction of three flavonoids namely catechin, epicatechin, and rutin [80].

Bioactive phytochemicals like sterols, tocopherols, and carotenes were recovered from tomato by-products. These phytochemicals have been utilized as natural antioxidants. Two different extraction methods were evaluated for pectin extraction from tomato waste. One method used conventional extraction while ultrasound-assisted extraction was performed in another method. A study summarized that pectin in ultrasound-mediated extraction exhibited better extraction quality; moreover, the process is eco-friendly [81,82]. Table 13–1 summarizes some important products from agriculture food residue.

13.2.3 Organic acids from agro-industrial wastes

Food waste is used as a valuable resource for an effective alternative waste valorization strategy for production of chemicals, materials, and fuels. Organic acids like succinic acid, fumaric acid, itaconic acid, lactic acid, gluconic acid, and xylonic acid are main building block chemicals that are produced from agri-food waste material.

Succinic acid is a naturally occurring vital dicarboxylic acid that is formed from biomass. It acts as an originator of several significant industrial chemicals like γ-butyrolactone, tetrahydrofuran, 1,4-butanediol, adipic acid, succinonitrile, succindiamide, 4,4-Bionolle. *Anaerobiospirillum succiniciproducens, Actinobacillus succinogenes, Mannheimia succiniciproducens,* and *Basfia succiniciproducens* are some microorganisms evaluated for the production of succinic acid from several carbon sources. Its production from natural succinate-producing strains rarely touched wanted objective for an achievable process. Efforts were made to develop recombinant *Escherichia coli* strains by metabolic engineering technique for strain improvement to enhance productivity [2,83]. Integrated upstream processing of succinic acid production from wheat waste or by-products was studied by fermentative production and chemical transformations [84]. Valorization of citrus peel waste was performed with *Actinobacillus succinogenes* with microbial fermentation for succinic acid production with good yield (6.13 g/L) [85]. Recently, the fruit and vegetable waste material was first hydrolyzed enzymatically to obtain a sugar mixture. Subsequent fermentation of this sugar mixture was performed by *A. succinogenes* to produce succinic acid (1.18 g per g of sugar) [86]. Fruit juices, syrups, and soft drinks contain high sugar content consisting mainly of glucose, fructose, and sucrose (50 – 1000 g/L). Valorization of by-product stream from the beverage industry was evaluated for production of succinic acid. *A. succinogenes* was used for this process and good results were reported [87].

United States Department of Energy has recognized fumaric acid as a platform chemical, mainly used as food acidulant and food additive. It is a vital intermediate in synthesis of products like L-malic acid and L-aspartic acid. Various microorganisms like bacteria (*Zymomonas Mobilis, Bacillus macerans, Thermoanaerobacter ethanolicus, Erwinia chrysanthemiet*), yeast (*Scheffersomyces stipitis, Brettanomyces, Pachysolen tannophilus, Candida utiliset*) and fungi (*Rhizopus nigricans, Rhizopus arrhizus, Rhizopus oryzae, Rhizopus Formosa, Penicillum griseofulvum, Aspergillus glaucus, Caldariomycels fumago*) are known for its microbial production [88]. *Rhizopus oryzae* has been employed for its production. The important characteristic of *R. oryzae* is that it is generally regarded as safe [2].

Itaconic acid is produced by microorganisms like *Aspergillus itaconicus*, *Aspergillus terreus*, *Ustilago zeae*, *Helicobasidium mompa*, *Candida* sp., and *Pseudozyma antartica* [89,90]. The cost of itaconic acid by glucose fermentation is not economically reasonable, and alternative sources are need of high priority. Two agro-industrial waste biomasses, wheat bran, and corn cobs were studied for production of itaconic acid industrially using liquid-state fermentation by *Aspergillus terreus* [91].

13.2.4 Volatile fatty acids production

Volatile fatty acids (VFAs) are treasured raw material which serves as initial raw molecules for bio-energy, biopolymers, reduced chemicals, and derivatives production. Among different VFAs, acetic acid has largest share with respect to market size which is followed by propionic acid & butyric acid. The production of VFAs can be made profitable by implementation of an efficient production process. The chemical transformation of petrochemicals is major route of synthesis, but recently biological transformation has also gained momentum. The organic waste streams act as raw materials for production of VFAs. The mixed culture engineering fermentation process is used for production of biofuels and valuable chemicals. For this biomass-based feedstock has been transformed [92]. The VFAs also have high potential as a renewable carbon source. VFAs are also used in different industries like pharmaceutical, food, and chemical industries [93].

Pretreatment is a very important stage before any bioprocess is functional. After efficient pretreatment, the substrate is ready for effective microbial activity. Low pH is a limiting factor for VFAs yield in bio-based production methods. Members of the bacteria *Acetobacter*, *Gluconacetobacter*, and *Gluconobacter* are known for commercial production of acetic acid by fermentation [94]. The chemical synthesis of butyric acid is preferred over other methods due to low production cost. The fermentation process is preferred over chemical synthesis as food or beverage industries cannot use chemically synthesized butyric acid. Bacteria like *Clostridium butyricum*, *Clostridium kluyveri*, *Clostridium beijerinckii*, *Clostridium barkeri*, *Clostridium acetobutylicum*, and *Eubacterium limosum* are commonly used for the production of butyric acid [95,96].

Although bio-based production process for propionic acid from renewable resources is not economical, yet researchers are searching for microorganisms that can enhance the biological yield. Some of the bacteria used for production of propionic acid are *Propionibacterium freudenreichii*, *Propionibacterium acidipropionici*, *Propionibacterium thoenii*, *Propionibacterium shermanii*, etc. The production of the VFAs depends on the pH, temperature, organic loading rate, and retention time [93,97].

Recovery is a very crucial step in any synthetic or biological process for the production of chemicals. If recovery of chemical from a process is poor, it will increase the overall cost of that molecule. The production cost increase will make the process economically nonfeasible. So it is also required to find out recovery processes with a prime aim of maximum VFAs recovery. The most common recovery methods used in VFAs recovery are adsorption, electrodialysis, solvent extraction, reverse osmosis, nanofiltration, and membrane contractor.

Studies suggested that pressure-driven reverse osmosis and high-voltage electrodialysis give good recovery along with high purity of the VFAs, but on other side, these processes are costly [93,98–100].

13.2.5 Polyunsaturated fatty acids

Polyunsaturated fatty acids (PUFAs) have attained substantial curiosity in the nutraceutical field due to various health benefits. Eicosapentaenoic acid (EPA), docosahexaenoic acid (DHA) and arachidonic acid are substantial PUFAs. The market is growing with high demand for DHA (DHA; 22:6, n-3)-rich oil. Fish oil is the main and limited source of PUFAs which is vulnerable to various organic chemical contaminations. The purification of PUFAs from fish oil is also an expensive step with pitiable oxidative stability. These lead to the exploration of alternative sustainable pathways for PUFAs production by the biotechnological route. Fish processing waste has been a good source of PUFAs, as consumption and waste generation are enormous worldwide [101–103]. Microorganisms like *Crypthecodinium cohnii Aurantiochytrium* sp., *Thraustochytrium* sp., *Skelotema costatum*, *Pythium* sp., *Mortiriella alpine*, *Chlorella stigmatophora* are known for their ability to produce PUFAs by fermentation process [104].

Worldwide research is going on for development of sustainable processes for industries that are more economical & environment-friendly. To make fish waste processing more economical and feasible, a process was developed involving one-step extraction and saponification cum enzymatic processing. It was anticipated that process will be more feasible in processing *Nile tilapia* waste for PUFA concentration. Experiments showed that high PUFAs yield (EPA + DHA content; 6.0% and 4.0%) was obtained using *Pseudomonas fluorescens* and *Thermomyces lanuginosa* lipases, respectively [105].

With growing population, food systems are impacted by production and environmental factors. Sustainable protein is one of the major requirements and insects are being evaluated as feed for livestock and/or aquaculture. Insects can play a very vital role in future with a dual role. They can be utilized for recycling the organic waste and themselves act as feed [106]. One important example is black soldier fly larvae (*Hermetia illucens*), which is also considered as a good protein source. These larvae are utilized for the recycling of organic waste into high-quality fat and protein-rich foods. Recently, it was demonstrated that feeding insects with fish by-products expressively raise their EPA + DHA and total PUFAs (n-3 levels) [107].

The oleo-chemical industry is also in process of sustainable growth phase with improvement in potential novel applications with shift from current linear approach to a circular economy. Life cycle assessment and process intensification options have been used in the fish oil processing industry for a sustainable circular economy [108].

13.2.6 Bioactive peptides

The peptides are definite protein fragments that may possess ability to improve human health. Protein fragments with a beneficial biological activity called bioactive peptides are endorsed with properties like antihypertensive, antioxidant, immunomodulatory, antiinflammatory, and hypocholesterolemic. These are enzymatically obtained from food protein. Studies have shown that enzyme-derived peptides & protein hydrolysates are biologically more active than their parent protein. It was suggested that peptide bond degeneration leads to release of potent peptides. Bioactive peptides have been produced using commercial enzymes through the hydrolysis process in different waste products like tomato by-products [109–111], olive seeds [112–114], prunus seeds [115], cherry seed [116], feather waste [117], meat industry waste [118], microalgae, fish, and marine processing waste [119–121]. The thermolysin, flavourzyme, alcalase, pepsin, and trypsin are main commercial enzymes used for the hydrolysis process [110,122]. As policymakers are shifting the human wellness goal from treatment of diseases to prevention of diseases, the bioactive peptides will play a crucial role in the disease prevention approach [123].

Recent research trends have shown exploration of marine sources like fish, oyster, macroalgae, giant squid, sea urchin, etc. as a source of bioactive peptides. Industrial marine waste had been explored for efficient recovery of the bioactive peptides [124]. The peptides have been explored for nonfood usage like manufacturing of adhesives, coatings, and polymers. The biggest benefit of application of peptides in nonfood items is that minor nonprotein impurities will not affect the final product. Also, antinutrients have no trouble as a threat to human and animal nutrition [125].

Tuna-based industry produces a large amount of waste in the form of tuna dark muscle. Experiments were performed aiming for valorization of the tuna waste. Orientase (OR) and protease XXIII (PR) enzymes were utilized to obtain hydrolyze waste and protein hydrolysate was chromatographed to obtain the peptides [126]. Higher added-value peptides were retrieved by the enzymatic hydrolysis followed by membrane fractionation process. Four novel peptides were isolated with antioxidant properties [127].

13.2.7 Fructooligosaccharides

The immune system stimulation and infection resistance are health benefits of fructooligosaccharides (FOS). These deliver very little energy to the body as they are seldom hydrolyzed by digestive enzymes. Different strains for production of FOS like *Aspergillus japonicus*, *Aspergillus ibericus*, *Aspergillus oryzae*, *Aspergillus niger*, *Aureobasidium pullulans*, *Rhizopus stolonifer*, and *Penicillium citreonigrum* have been reported. Recent review documented different agro-industrial wastes like sucrose rich solutions (sugarcane molasses, beet molasses), fruit peels, some bagasse, leaves (banana leaf, corn leaf, sugarcane leaf, etc.), pomaces (like apple pomace and grape pomace), and coffee processing by-products for production of FOS [128].

13.2.8 Bioethanol production

Overexploitation and limited amount of fossil fuels led researchers to look for sustainable energy sources, that is, bio-energy. Biomass is the most abundant natural, sustainable, economical, profitable, and viable source for commercial production of bio-energy. Bio-energy as bioethanol is produced from cellulosic agricultural waste. Sustainable fuels such as biodiesel and bio-hydrogen are derived from vegetable waste, fruit industries waste, coffee waste, sugarcane waste, corn waste, switchgrass, and algae. These sustainable fuels have properties that make them suitable for use as petroleum-based fuels [129].

The process of bioethanol production starts with feedstock preparation which includes steps like size reduction by milling, grinding, or chopping. The second major step is pretreatment (physiochemical or biological methods) which leads to release of free fermentable sugars. The fermentable sugars are freed by the hydrolysis process which is often carried out using acid or enzymes. The common enzymes used for this process are α- and β-amylase, glucoamylase, pullulanase, isoamylase, cellulases, and β-glucosidases. The fermentable sugars are fermented using microorganisms like *Saccharomyces cerevisiae*, *Escherichia coli*, *Zymomonas mobilis*, *Fusarium* species, etc. The last step is distillation using multistage distillation units leading to the production of bioethanol [10,130].

13.2.9 Bacterial cellulose

Bacterial cellulose (BC) has recently attained attraction of researchers from across the world due to its superior properties and nature which permit the wanted alterations. Due to its various beneficial properties, it has been explored for different roles in different industries for product development like medicine, pharmacy, food, chemistry, environment, engineering, and fashion. It has also been investigated for use in biomedical like manufacturing of artificial blood vessels, skin, cornea, cartilage, and bone. The biggest hurdle in commercial application of BC is high cost of media for its production. BC also has low productivity at an industrial scale. Because of high-cost researchers are exploring low-cost renewable carbon sources through bioprocess optimization. Agro-industrial waste products and by-products such as food, industrial, agricultural, and brewery wastes have been explored for commercial low-cost production of BC. Novel and competent BC-producing microbial strains have been identified and also genetically modified to enhance industrial BC production [131–133]. Studies have reported use of agriculture wastes (orange, apple, pineapple, fruit juice, etc.) and industrial by-products (corn steep liquor) for a novel, and nutritionally rich culture medium for the production of BC [134,135]. *Acetobacter xylinum* and *Acetobacter* sp. V6 were utilized for the production of BC by using low-cost substrates like orange pulp and cane molasses [136]. *Gluconacetobacter xylinus* was used to study effect of addition of vitamin C. The production touched 0.47 g/30 mL compared to 0.25 g/30 mL without vitamin C. It was proposed that decline in gluconic acid concentration influences the production. The influence of fructose media supplemented with 2% HNTs was evaluated for BC production and found increase in productivity [137,138]. Pineapple, pawpaw, and watermelon juice medium from agro-industries are rich in proteins, carbohydrates, and trace elements. These juices were evaluated as a medium for BC production by *Acinetobacter* sp. BAN1 and *Acetobacter pasteurianus* PW1. Results suggested that the Pawpaw juice medium is better in comparison to other media [139]. Recently, BC production was evaluated by using different agro-industrial waste as growth media like tomato juice, cane molasses, and orange pulp using *Acetobacter pasteurianus* RSV-4 (MTCC 25117) (Fig. 13–2) [23].

13.2.10 Mushroom production

The agro-industrial waste rich in organic compounds has been utilized as substrate for mushroom cultivation. Solid-state fermentation has proved to be very efficient process for mushroom cultivation and value-addition of agro-industrial residues. Mushroom cultivation is also considered as an efficient method for recovery of food protein from lignocellulosic waste material [140]. *Agrocybe aegerita*, *Volvariella volvacea*, *Agaricus bisporus*, *Pleurotus* spp., and *Lentinula edodes* are some popular species with high productivity in degradation of agri-food residues [141–143]. The cultivated mushroom *Lentinula edodes* with unique flavor, nutritional, and medicinal properties produces hydrolytic and oxidative enzymes for degradation of waste. The type of substrate has direct impact on the production as highest efficiency was achieved with sunflower seed hulls and sugarcane bagasse. Commercially significant edible mushrooms of *Pleurotus* species like *Pleurotus ostreatus*, *Pleurotus sajor-caju*, *Pleurotus pulmonarius*, etc. have numerous benefits including fast mycelial growth, cheap farming methods, and growth in different climatic conditions. Many researchers suggested that mushroom cultivation is an economically vital biotechnological process that practices a well-organized solid-state fermentation procedure of food protein recovery from agri-food waste [143].

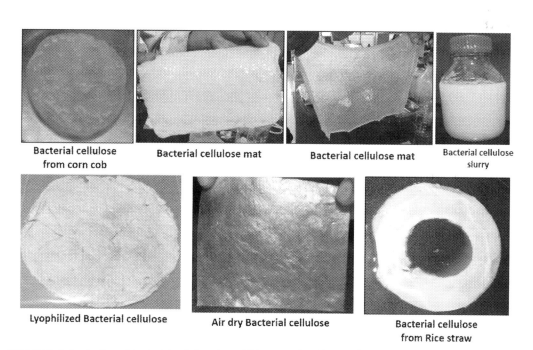

FIGURE 13–2 Production of bacterial cellulose from different agri-residues and its different forms.

13.3 Conclusions and perspectives

Industrial development brings socio-economic development for mankind; it also has some negative aspects. One of the major drawbacks of agro-food industrial development is generation of food wastes, which leads to environmental pollution. Recent alterations in consumer food habits and altering regulations regarding waste management policy or recycling norms by government agencies have drastically changed the waste management strategies of these industries. With high moisture, incineration is not a suitable and environmentally safe method to process large quantities of food waste. Agri-food waste is a rich source of carbohydrates, proteins, fatty acids, and several other inorganic minerals. Therefore, it can be potentially utilized by microbes as a sustainable source for producing value-added products. To date, few industrial developments have reported use of agri-food waste, and much research work have taken at the laboratory scale. A collaborative approach is need of the hour where public domain and government agencies along with industries can work synergistically for effective management of agro-industrial waste and their simultaneous conversion into value-added products.

Acknowledgments

Authors V.K., S.K.Y., and B.B.M. are thankful to CEO, CIAB, Mohali, and authors V.A. and L.K.T. are thankful to Director, Institute of Pesticide Formulation Technology, Gurugram for continuous support and encouragement.

References

[1] V.D. Katare, M.V. Madurwar, S. Raut, Agro-industrial waste as a cementitious binder for sustainable concrete: an overview, Sustainable Waste Management: Policies and Case Studies, Springer, 2020, pp. 683–702.

[2] V. Kumar, P. Binod, R. Sindhu, E. Gnansounou, V. Ahluwalia, Bioconversion of pentose sugars to value added chemicals and fuels: Recent trends, challenges and possibilities, Bioresource Technology 269 (2018) 443–451.

[3] F. Girotto, L. Alibardi, R. Cossu, Food waste generation and industrial uses: a review, Waste Management 45 (2015) 32–41.

[4] M. Shirzad, H.K.S. Panahi, B.B. Dashti, M.A. Rajaeifar, M. Aghbashlo, M.A. Tabatabaei, A comprehensive review on electricity generation and GHG emission reduction potentials through anaerobic digestion of agricultural and livestock/slaughterhouse wastes in Iran, Renewable and Sustainable Energy Reviews 111 (2019) 571–594.

[5] SAVE FOOD: Global initiative on food loss and waste reduction. Key facts on food loss and waste you should know, 2016.

[6] SAVE FOOD: Global initiative on food loss and waste reduction. Definitional Framework of Food Loss, Working Paper, 2014.

[7] P. Sharma, V.K. Gaur, S.-H. Kim, A. Pandey, Microbial strategies for bio-transforming food waste into resources, Bioresource Technology 299 (2020) 122580.

[8] M. Rosa, F.M. Souza, M. Figueiredo, J. Morais, S. Santaella, R. Leitão, Valorização de resíduos da agroindústria, II Simpósio Internacional sobre Gerenciamento de Resíduos Agropecuários e Agroindustriais–II SIGERA Foz do Iguaçu, PR 1 (98) (2011) 105.

Chapter 13 • Bioprocessing of agri-food processing residues into nutraceuticals and bioproducts

[9] J.V. Madeira Jr, F.J. Contesini, F. Calzado, M.V. Rubio, M.P. Zubieta, D.B. Lopes, Agro-industrial residues and microbial enzymes: an overview on the eco-friendly bioconversion into high value-added products, Biotechnology of Microbial Enzymes, Elsevier, 2017, pp. 475–511.

[10] V. Kumar, S.K. Yadav, J. Kumar, V. Ahluwalia, A critical review on current strategies and trends employed for removal of inhibitors and toxic materials generated during biomass pretreatment, Bioresource Technology 299 (2019) 122633.

[11] J. Banerjee, R. Singh, R. Vijayaraghavan, D. MacFarlane, A.F. Patti, A. Arora, Bioactives from fruit processing wastes: green approaches to valuable chemicals, Food Chemistry 225 (2017) 10–22.

[12] S. Balaman, Introduction to biomass—resources, production, harvesting, collection, and storage, Decision-Making for Biomass-Based Production Chains, Elsevier, 2019, pp. 1–23.

[13] B. Agarwal, V. Ahluwalia, A. Pandey, R.S. Sangwan, S. Elumalai, Sustainable production of chemicals and energy fuel precursors from lignocellulosic fractions, Biofuels, Springer, 2017, pp. 7–33.

[14] P. Kundu, S. Kumar, V. Ahluwalia, S.K. Kansal, S. Elumalai, Extraction of arabinoxylan from corncob through modified alkaline method to improve xylooligosaccharides synthesis, Bioresource Technology Reports 3 (2018) 51–58.

[15] V. Kumar, M. Krishania, P.P. Sandhu, V. Ahluwalia, E. Gnansounou, R.S. Sangwan, Efficient detoxification of corn cob hydrolysate with ion-exchange resins for enhanced xylitol production by *Candida tropicalis* MTCC 6192, Bioresource Technology 251 (2018) 416–419.

[16] E.U. Kiran, A.P. Trzcinski, W.J. Ng, Y. Liu, Bioconversion of food waste to energy: a review, Fuel 134 (2014) 389–399.

[17] H. Huang, V. Singh, N. Qureshi, Butanol production from food waste: a novel process for producing sustainable energy and reducing environmental pollution, Biotechnology for Biofuels 8 (2015) 147.

[18] V. Ujor, A.K. Bharathidasan, K. Cornish, T.C. Ezeji, Feasibility of producing butanol from industrial starchy food wastes, Applied Energy 136 (2014) 590–598.

[19] N. Ifudu, Indigenous resources for antibiotic productionAugust/September edition Expansion Today (Nigeria), AU Press Ltd, 1986pp. 27–32.

[20] A.D. Tripathi, T. Raj Joshi, S. Kumar Srivastava, K.K. Darani, S. Khade, J. Srivastava, Effect of nutritional supplements on bio-plastics (PHB) production utilizing sugar refinery waste with potential application in food packaging, Preparative Biochemistry and Biotechnology 49 (2019) 567–577.

[21] S. Jonathan, I. Fasidi, A. Ajayi, O. Adegeye, Biodegradation of Nigerian wood wastes by *Pleurotus tuber-regium* (Fries) Singer, Bioresource Technology 99 (2008) 807–811.

[22] S. Jonathan, B. Babalola, Utilization of agro-industrial waste for the cultivation of *Pleurotus tuber-regium* (Fries) Singer, a Nigerian Edible Mushroom. in: Proceedings of Tropentag Conference on Agricultural Development within the Rural-Urban Continuum, Stuttgart-Hohenheim, 2013.

[23] V. Kumar, D.K. Sharma, V. Bansal, D. Mehta, R.S. Sangwan, S.K. Yadav, Efficient and economic process for the production of bacterial cellulose from isolated strain of *Acetobacter pasteurianus* of RSV-4 bacterium, Bioresource Technology 275 (2019) 430–433.

[24] M.M. Rashad, A.E. Mahmoud, M.M. Ali, M.U. Nooman, A.S. Al-Kashef, Antioxidant and anticancer agents produced from pineapple waste by solid state fermentation, International Journal of Toxicological and Pharmacological Research 7 (2015) 287–296.

[25] S. Parashar, H. Sharma, M. Garg, Antimicrobial and antioxidant activities of fruits and vegetable peels: a review, Journal of Pharmacognosy and Phytochemistry 3 (2014).

[26] S. Singh, G. Immanuel, Extraction of antioxidants from fruit peels and its utilization in paneer, Journal of Food Processing & Technology 5 (2014) 1.

[27] R. Patowary, K. Patowary, M.C. Kalita, S. Deka, Utilization of paneer whey waste for cost-effective production of rhamnolipid biosurfactant, Applied Biochemistry and Biotechnology 180 (2016) 383–399.

[28] M.S. Rodrigues, F.S. Moreira, V.L. Cardoso, M.M. de Resende, Soy molasses as a fermentation substrate for the production of biosurfactant using *Pseudomonas aeruginosa* ATCC 10145, Environmental Science and Pollution Research 24 (2017) 18699–18709.

[29] G. Lim, Indigenous fermented foods in South East Asia, ASEAN Food Journal 6 (1991) 83–101.

[30] I.M. Ramírez, K. Tsaousi, M. Rudden, R. Marchant, E.J. Alameda, M.G. Román, et al., Rhamnolipid and surfactin production from olive oil mill waste as sole carbon source, Bioresource Technology 198 (2015) 231–236.

[31] S. Papanikolaou, A. Dimou, S. Fakas, P. Diamantopoulou, A. Philippoussis, M. Galiotou-Panayotou, et al., Biotechnological conversion of waste cooking olive oil into lipid-rich biomass using *Aspergillus* and *Penicillium* strains, Journal of Applied Microbiology 110 (2011) 1138–1150.

[32] P. Saharan, P.K. Sadh, J.S. Duhan, Comparative assessment of effect of fermentation on phenolics, flavanoids and free radical scavenging activity of commonly used cereals, Biocatalysis and Agricultural Biotechnology 12 (2017) 236–240.

[33] P.K. Sadh, P. Chawla, L. Bhandari, J.S. Duhan, Bio-enrichment of functional properties of peanut oil cakes by solid state fermentation using *Aspergillus oryzae*, Journal of Food Measurement and Characterization 12 (2018) 622–633.

[34] F. Oliveira, C.E. Souza, V.R. Peclat, J.M. Salgado, B.D. Ribeiro, M.A. Coelho, et al., Optimization of lipase production by *Aspergillus ibericus* from oil cakes and its application in esterification reactions, Food and Bioproducts Processing 102 (2017) 268–277.

[35] A.E. Asagbra, A.I. Sanni, O.B. Oyewole, Solid-state fermentation production of tetracycline by *Streptomyces* strains using some agricultural wastes as substrate, World Journal of Microbiology and Biotechnology 21 (2005) 107–114.

[36] B. Vastrad, S. Neelagund, Optimization of process parameters for rifamycin b production under solid state fermentation from *Amycolatopsis mediterranean* MTCC14, International Journal of Current Pharmaceutical Research 4 (2012) 101–108.

[37] S.G. Rudra, J. Nishad, N. Jakhar, C. Kaur, Food industry waste: mine of nutraceuticals, International Journal of Science Environmental Technology 4 (2015) 205–229.

[38] S. Ramachandran, S.K. Singh, C. Larroche, C.R. Soccol, A. Pandey, Oil cakes and their biotechnological applications—a review, Bioresource Technology 98 (2007) 2000–2009.

[39] P.K. Sadh, S. Duhan, J.S. Duhan, Agro-industrial wastes and their utilization using solid state fermentation: a review, Bioresources and Bioprocessing 5 (2018) 1.

[40] V. Kumar, P.P. Sandhu, V. Ahluwalia, B.B. Mishra, S.K. Yadav, Improved upstream processing for detoxification and recovery of xylitol produced from corncob, Bioresource Technology 291 (2019) 121931.

[41] M. Gavahian, Y.H. Chu, A. Mousavi Khaneghah, Recent advances in orange oil extraction: an opportunity for the valorisation of orange peel waste a review, International Journal of Food Science & Technology 54 (2019) 925–932.

[42] R. Colombo, A. Papetti, Avocado (*Persea americana* Mill.) by-products and their impact: from bioactive compounds to biomass energy and sorbent material for removing contaminants, a review, International Journal of Food Science & Technology 54 (2019) 943–951.

[43] A. Roda, M. Lambri, Food uses of pineapple waste and by-products: a review, International Journal of Food Science & Technology 54 (2019) 1009–1017.

[44] S. Kang, B. Li, J. Chang, J. Fan, Antioxidant abilities comparison of lignins with their hydrothermal liquefaction products, BioResources 6 (2011) 243–252.

[45] M. Déniel, G. Haarlemmer, A. Roubaud, E. Weiss-Hortala, J. Fages, Energy valorisation of food processing residues and model compounds by hydrothermal liquefaction, Renewable and Sustainable Energy Reviews 54 (2016) 1632–1652.

[46] T. Belwal, S.M. Ezzat, L. Rastrelli, I.D. Bhatt, M. Daglia, A. Baldi, Critical analysis of extraction techniques used for botanicals: Trends, priorities, industrial uses and optimization strategies, TrAC Trends in Analytical Chemistry 100 (2018) 82–102.

[47] K. Kumar, A.N. Yadav, V. Kumar, P. Vyas, H.S. Dhaliwal, Food waste: a potential bioresource for extraction of nutraceuticals and bioactive compounds, Bioresources and Bioprocessing 4 (2017) 18.

[48] C.Y. Gan, A.A. Latiff, Optimisation of the solvent extraction of bioactive compounds from *Parkia speciosa* pod using response surface methodology, Food Chemistry 124 (2011) 1277–1283.

[49] A. Ningombam, V. Ahluwalia, C. Srivastava, S. Walia, Growth inhibitory activity of Millettia pachycarpa (Bentham) extracts against Tobacco Caterpillar, Spodoptera litura (Fabricius) (Lepidoptera: Noctuidae), Archives of Phytopathology and Plant Protection 51 (2018) 550–559.

[50] L. Wang, C.L. Weller, Recent advances in extraction of nutraceuticals from plants, Trends in Food Science & Technology 17 (2006) 300–312.

[51] W.J. Lee, N. Suleiman, N.H.N. Hadzir, G.H. Chong, Supercritical fluids for the extraction of oleoresins and plant phenolics, Green Sustainable Process for Chemical and Environmental Engineering and Science, Elsevier, 2020, pp. 279–328.

[52] R. Gallego, M. Bueno, M. Herrero, Sub-and supercritical fluid extraction of bioactive compounds from plants, food-by-products, seaweeds and microalgae – An update, TrAC Trends in Analytical Chemistry (2019).

[53] B. Aliakbarian, A. Fathi, P. Perego, F. Dehghani, Extraction of antioxidants from winery wastes using subcritical water, The Journal of Supercritical Fluids 65 (2012) 18–24.

[54] L. He, X. Zhang, H. Xu, C. Xu, F. Yuan, Z. Knez, Subcritical water extraction of phenolic compounds from pomegranate (*Punica granatum* L.) seed residues and investigation into their antioxidant activities with HPLC–ABTS + assay, Food and Bioproducts Processing 90 (2012) 215–223.

[55] X. Liu, Y. Wang, J. Zhang, L. Yan, S. Liu, A.A. Taha, Subcritical water extraction of phenolic antioxidants with improved α-amylase and α-glucosidase inhibitory activities from exocarps of *Castanea mollissima* Blume, The Journal of Supercritical Fluids 158 (2020) 104747.

[56] E. Espada-Bellido, M. Ferreiro-González, C. Carrera, M. Palma, C.G. Barroso, G.F. Barbero, Optimization of the ultrasound-assisted extraction of anthocyanins and total phenolic compounds in mulberry (*Morus nigra*) pulp, Food Chemistry 219 (2017) 23–32.

[57] J. González-Rivera, A. Spepi, C. Ferrari, C. Duce, I. Longo, D. Falconieri, Novel configurations for a citrus waste based biorefinery: from solventless to simultaneous ultrasound and microwave assisted extraction, Green Chemistry 18 (2016) 6482–6492.

[58] N.A. Al-Dhabi, K. Ponmurugan, Microwave assisted extraction and characterization of polysaccharide from waste jamun fruit seeds, International Journal of Biological Macromolecules 152 (2020) 1157–1163.

[59] S.J. Marathe, S.B. Jadhav, S.B. Bankar, K.K. Dubey, R.S. Singhal, Improvements in the extraction of bioactive compounds by enzymes, Current Opinion in Food Science (2019).

[60] T. Vukasović, Functional foods in line with young consumers: challenges in the marketplace in Slovenia, Developing New Functional Food and Nutraceutical Products, Elsevier, 2017, pp. 391–405.

[61] T. Varzakas, P. Kandylis, D. Dimitrellou, C. Salamoura, G. Zakynthinos, C. Proestos, Innovative and fortified food: probiotics, prebiotics, gmos, and superfood, Preparation and Processing of Religious and Cultural Foods, Elsevier, 2018, pp. 67–129.

[62] S.X. Liu, White, E. Extraction and characterization of Proanthocyanidins from grape seeds, The Open Food Science Journal 6 (2012).

[63] S. Patel, Grape seeds: agro-industrial waste with vast functional food potential, Emerging Bioresources with Nutraceutical and Pharmaceutical Prospects, Springer, 2015, pp. 53–69.

[64] R. Ciriminna, A. Fidalgo, R. Delisi, L.M. Ilharco, M. Pagliaro, Pectin production and global market, Agro Food Industry Hi-Tech 27 (2016) 17–20.

[65] S. Rana, S. Gupta, A. Rana, S. Bhushan, Functional properties, phenolic constituents and antioxidant potential of industrial apple pomace for utilization as active food ingredient, Food Science and Human Wellness 4 (2015) 180–187.

[66] Z. Wang, B. Xu, H. Luo, K. Meng, Y. Wang, M. Liu, Production pectin oligosaccharides using *Humicola insolens* Y1-derived unusual pectate lyase, Journal of Bioscience and Bioengineering 129 (2020) 16–22.

[67] L. Klavins, J. Kviesis, I. Nakurte, M. Klavins, Berry press residues as a valuable source of polyphenolics: extraction optimisation and analysis, LWT 93 (2018) 583–591.

[68] P. Silva, S. Ferreira, F.M. Nunes, Elderberry (*Sambucus nigra* L.) by-products a source of anthocyanins and antioxidant polyphenols, Industrial Crops and Products 95 (2017) 227–234.

[69] V. Kitrytė, V. Kraujalienė, V. Šulniūtė, A. Pukalskas, P.R. Venskutonis, Chokeberry pomace valorization into food ingredients by enzyme-assisted extraction: process optimization and product characterization, Food and Bioproducts Processing 105 (2017) 36–50.

[70] S. Mildner-Szkudlarz, M. Różańska, A. Siger, P.è. Kowalczewski, M. Rudzińska, Changes in chemical composition and oxidative stability of cold-pressed oils obtained from by-product roasted berry seeds, LWT 111 (2019) 541–547.

[71] F. Naqash, F. Masoodi, S.A. Rather, S. Wani, A. Gani, Emerging concepts in the nutraceutical and functional properties of pectin—a review, Carbohydrate Polymers 168 (2017) 227–239.

[72] S. Ben-Othman, I. Jõudu, R. Bhat, Bioactives from agri-food wastes: present insights and future challenges, Molecules 25 (2020) 510.

[73] V.J. Morris, N.J. Belshaw, K.W. Waldron, E.G. Maxwell, The bioactivity of modified pectin fragments, Bioactive Carbohydrates and Dietary Fibre 1 (2013) 21–37.

[74] D.V. Dandekar, G. Jayaprakasha, B.S. Patil, Hydrotropic extraction of bioactive limonin from sour orange (*Citrus aurantium* L.) seeds, Food Chemistry 109 (2008) 515–520.

[75] P. Górnaś, M. Rudzińska, A. Soliven, Industrial by-products of plum *Prunus domestica* L. and *Prunus cerasifera* Ehrh. as potential biodiesel feedstock: impact of variety, Industrial Crops and Products 100 (2017) 77–84.

[76] F.V. Dulf, D.C. Vodnar, C. Socaciu, Effects of solid-state fermentation with two filamentous fungi on the total phenolic contents, flavonoids, antioxidant activities and lipid fractions of plum fruit (*Prunus domestica* L.) by-products, Food Chemistry 209 (2016) 27–36.

[77] S. Talekar, A.F. Patti, R. Singh, R. Vijayraghavan, A. Arora, From waste to wealth: high recovery of nutraceuticals from pomegranate seed waste using a green extraction process, Industrial Crops and Products 112 (2018) 790–802.

[78] S. Akhtar, T. Ismail, D. Fraternale, P. Sestili, Pomegranate peel and peel extracts: chemistry and food features, Food Chemistry 174 (2015) 417–425.

[79] R. Kushwaha, V. Kumar, G. Vyas, J. Kaur, Optimization of different variable for eco-friendly extraction of betalains and phytochemicals from beetroot pomace, Waste and Biomass Valorization 9 (2018) 1485–1494.

[80] J.J. Vulić, T.N. Ćebović, J.M. Čanadanović-Brunet, G.S. Ćetković, V.M. Čanadanović, S.M. Djilas, In vivo and in vitro antioxidant effects of beetroot pomace extracts, Journal of Functional Foods 6 (2014) 168–175.

[81] A.N. Grassino, M. Brnčić, D. Vikić-Topić, S. Roca, M. Dent, S.R. Brnčić, Ultrasound assisted extraction and characterization of pectin from tomato waste, Food Chemistry 198 (2016) 93–100.

[82] N. Kalogeropoulos, A. Chiou, A. Pyriochou, et al., Bioactive phytochemicals in industrial tomatoes and their processing byproducts, LWT—Food Science and Technology 49 (2012) 213–216.

[83] C. Pais, R. Franco-Duarte, P. Sampaio, J. Wildner, A. Carolas, D. Figueira, Production of dicarboxylic acid platform chemicals using yeasts: focus on succinic acid, Biotransformation of Agricultural Waste and By-Products, Elsevier, 2016, pp. 237–269.

[84] A.A. Koutinas, C. Du, C. Lin, C. Webb, Developments in cereal-based biorefineries, Advances in Biorefineries, Elsevier, 2014, pp. 303–334.

[85] M. Patsalou, K.K. Menikea, E. Makri, M.I. Vasquez, C. Drouza, M. Koutinas, Development of a citrus peel-based biorefinery strategy for the production of succinic acid, Journal of Cleaner Production 166 (2017) 706–716.

[86] W. Dessie, W. Zhang, F. Xin, W. Dong, M.M. Zhang, Succinic acid production from fruit and vegetable wastes hydrolyzed by on-site enzyme mixtures through solid state fermentation, Bioresource Technology 247 (2018) 1177–1180.

[87] M. Ferone, A. Ercole, F. Raganati, G. Olivieri, P. Salatino, A. Marzocchella, Efficient succinic acid production from high-sugar-content beverages by *Actinobacillus succinogenes*, Biotechnology Progress 35 (2019) e2863.

[88] F. Guo, M. Wu, Z. Dai, S. Zhang, W. Zhang, W. Dong, Current advances on biological production of fumaric acid, Biochemical Engineering Journal 153 (2020) 107397.

[89] T. Tabuchi, T. Sugisawa, T. Ishidori, T. Nakahara, J. Sugiyama, Itaconic acid fermentation by a yeast belonging to the genus *Candida*, Agricultural and Biological Chemistry 45 (1981) 475–479.

[90] W.E. Levinson, C.P. Kurtzman, T.M. Kuo, Production of itaconic acid by *Pseudozyma antarctica* NRRL Y-7808 under nitrogen-limited growth conditions, Enzyme and Microbial Technology 39 (2006) 824–827.

[91] A. Jiménez-Quero, E. Pollet, M. Zhao, et al., Itaconic and fumaric acid production from biomass hydrolysates by *Aspergillus* strains, Journal of Microbiology and Biotechnology 26 (2016) 1557–1565.

[92] S.K. Brar, S.J. Sarma, K. Pakshirajan, Platform Chemical Biorefinery: Future Green Chemistry, Elsevier, 2016.

[93] M. Atasoy, I. Owusu-Agyeman, E. Plaza, Z. Cetecioglu, Bio-based volatile fatty acid production and recovery from waste streams: current status and future challenges, Bioresource Technology 268 (2018) 773–786.

[94] P. Raspor, D. Goranovič, Biotechnological applications of acetic acid bacteria, Critical Reviews in Biotechnology 28 (2008) 101–124.

[95] J. Yin, X. Yu, Y. Zhang, D. Shen, M. Wang, Y. Long, et al., Enhancement of acidogenic fermentation for volatile fatty acid production from food waste: effect of redox potential and inoculum, Bioresource Technology 216 (2016) 996–1003.

[96] Z. Xu, L. Jiang, Butyric acid, Comprehensive Biotechnology, Elsevier (2011) 235–243. https://doi.org/10.1016/B978-0-444-64046-8.00162-2.

[97] G. Du, L. Liu, J. Chen, White biotechnology for organic acids, Industrial Biorefineries & White Biotechnology, Elsevier, 2015, pp. 409–444.

[98] L. Shi, Y. Hu, S. Xie, G. Wu, Z. Hu, X. Zhan, Recovery of nutrients and volatile fatty acids from pig manure hydrolysate using two-stage bipolar membrane electrodialysis, Chemical Engineering Journal 334 (2018) 134–142.

[99] S.S. Wadekar, R.D. Vidic, Influence of active layer on separation potentials of nanofiltration membranes for inorganic ions, Environmental Science & Technology 51 (2017) 5658–5665.

[100] A. Yaroshchuk, M.L. Bruening, An analytical solution of the solution-diffusion-electromigration equations reproduces trends in ion rejections during nanofiltration of mixed electrolytes, Journal of Membrane Science 523 (2017) 361–372.

[101] A.K. Rai, N. Bhaskar, V. Baskaran, Effect of feeding lipids recovered from fish processing waste by lactic acid fermentation and enzymatic hydrolysis on antioxidant and membrane bound enzymes in rats, Journal of Food Science and Technology 52 (2015) 3701–3710.

[102] A.K. Rai, H. Swapna, N. Bhaskar, P. Halami, N. Sachindra, Effect of fermentation ensilaging on recovery of oil from fresh water fish viscera, Enzyme and Microbial Technology 46 (2010) 9–13.

[103] A.K. Rai, H. Swapna, N. Bhaskar, V. Baskaran, Potential of seafood industry byproducts as sources of recoverable lipids: Fatty acid composition of meat and nonmeat component of selected Indian marine fishes, Journal of Food Biochemistry 36 (2012) 441–448.

[104] S. Abad, X. Turon, Valorization of biodiesel derived glycerol as a carbon source to obtain added-value metabolites: focus on polyunsaturated fatty acids, Biotechnology Advances 30 (2012) 733–741.

[105] K. Sangkharak, N. Paichid, T. Yunu, S. Klomklao, Improvement of extraction and concentration method for polyunsaturated fatty acid production from *Nile tilapia* processing waste, Biomass Conversion and Biorefinery (2020) 1–13.

[106] A. Van Huis, J. Van Itterbeeck, H. Klunder, E. Mertens, A. Halloran, G. Muir, et al., Edible Insects: Future Prospects for Food and Feed Security, Food and Agriculture Organization of the United Nations, 2013.

[107] F.G. Barroso, M.J. Sánchez-Muros, M.Á. Rincón, M. Rodriguez-Rodriguez, D. Fabrikov, E. Morote, et al., Production of n-3-rich insects by bioaccumulation of fishery waste, Journal of Food Composition and Analysis 82 (2019) 103237.

[108] R. Monsiváis-Alonso, S.S. Mansouri, A. Román-Martínez, Life cycle assessment of intensified processes towards circular economy: omega-3 production from waste fish oil, Chemical Engineering and Processing – Process Intensification 158 (2020) 108171.

[109] A. Moayedi, L. Mora, M.C. Aristoy, M. Safari, M. Hashemi, F. Toldrá, Peptidomic analysis of antioxidant and ACE-inhibitory peptides obtained from tomato waste proteins fermented using *Bacillus subtilis*, Food Chemistry 250 (2018) 180–187.

[110] A. Moayedi, M. Hashemi, M. Safari, Valorization of tomato waste proteins through production of antioxidant and antibacterial hydrolysates by proteolytic *Bacillus subtilis*: optimization of fermentation conditions, Journal of Food Science and Technology 53 (2016) 391–400.

[111] A. Moayedi, L. Mora, M.-C. Aristoy, M. Hashemi, M. Safari, F. Toldrá, ACE-inhibitory and antioxidant activities of peptide fragments obtained from tomato processing by-products fermented using *Bacillus subtilis*: effect of amino acid composition and peptides molecular mass distribution, Applied Biochemistry and Biotechnology 181 (2017) 48–64.

[112] R. Vásquez-Villanueva, L. Muñoz-Moreno, M.J. Carmena, M.L. Marina, M.C. García, In vitro antitumor and hypotensive activity of peptides from olive seeds, Journal of Functional Foods 42 (2018) 177–184.

[113] I.M. Prados, J.M. Orellana, M.L. Marina, M.C. García, Identification of peptides potentially responsible for *in vivo* hypolipidemic activity of a hydrolysate from olive seeds, Journal of Agricultural and Food Chemistry 68 (2020) 4237–4244.

[114] E. Hernández-Corroto, M.L. Marina, M.C. García, Multiple protective effect of peptides released from *Olea europaea* and *Prunus persica* seeds against oxidative damage and cancer cell proliferation, Food Research International 106 (2018) 458–467.

[115] P. Guo, Y. Qi, C. Zhu, Q. Wang, Purification and identification of antioxidant peptides from Chinese cherry (*Prunus pseudocerasus* Lindl.) seeds, Journal of Functional Foods 19 (2015) 394–403.

[116] M.C. García, J. Endermann, E. Gonzalez-Garcia, M.L. Marina, HPLC-Q-TOF-MS identification of antioxidant and antihypertensive peptides recovered from cherry (*Prunus cerasus* L.) subproducts, Journal of Agricultural and Food Chemistry 63 (2015) 1514–1520.

[117] H. Stiborova, B. Branska, T. Vesela, P. Lovecka, M. Stranska, J. Hajslova, et al., Transformation of raw feather waste into digestible peptides and amino acids, Journal of Chemical Technology & Biotechnology 91 (2016) 1629–1637.

[118] K. Ryder, A.E.-D. Bekhit, M. McConnell, A. Carne, Towards generation of bioactive peptides from meat industry waste proteins: generation of peptides using commercial microbial proteases, Food Chemistry 208 (2016) 42–50.

[119] P.A. Harnedy, R.J. FitzGerald, Bioactive peptides from marine processing waste and shellfish: a review, Journal of Functional Foods 4 (2012) 6–24.

[120] R. Ravallec-Plé, C. Charlot, C. Pires, V. Braga, I. Batista, A. Van Wormhoudt, et al., The presence of bioactive peptides in hydrolysates prepared from processing waste of sardine (*Sardina pilchardus*), Journal of the Science of Food and Agriculture 81 (2001) 1120–1125.

[121] P.A. Harnedy, R.J. Fitzgerald, Bioactive proteins and peptides from macroalgae, fish, shellfish and marine processing waste, Marine Proteins and Peptides: Biological Activities and Applications, Wiley, 2013, pp. 5–39.

[122] X. Yang, D. Teng, X. Wang, Q. Guan, R. Mao, Y. Hao, et al., Enhancement of nutritional and antioxidant properties of peanut meal by bio-modification with *Bacillus licheniformis*, Applied Biochemistry and Biotechnology 180 (2016) 1227–1242.

[123] Y. Mine, E. Li-Chan, B. Jiang, Biologically active food proteins and peptides in health: an overview, Bioactive Proteins and Peptides as Functional Foods and Nutraceuticals, Wiley-Blackwell, 2010, pp. 5–11.

[124] C.C. Udenigwe, I.D. Nwachukwu, R.Y. Yada, Advances on the production and application of peptides for promoting human health and food security, Global Food Security and Wellness, Springer, 2017, pp. 195–219.

[125] M. del Mar Contreras, A. Lama-Muñoz, J.M. Gutiérrez-Pérez, F. Espínola, M. Moya, E. Castro, Protein extraction from agri-food residues for integration in biorefinery: potential techniques and current status, Bioresource Technology 280 (2019) 459–477.

[126] K.-C. Hsu, Purification of antioxidative peptides prepared from enzymatic hydrolysates of tuna dark muscle by-product, Food Chemistry 122 (2010) 42–48.

[127] S. Saidi, M. Saoudi, R.B. Amar, Valorisation of tuna processing waste biomass: isolation, purification and characterisation of four novel antioxidant peptides from tuna by-product hydrolysate, Environmental Science and Pollution Research 25 (2018) 17383–17392.

[128] O. de la Rosa, A.C. Flores-Gallegos, D. Muñíz-Marquez, C. Nobre, J.C. Contreras-Esquivel, C.N. Aguilar, Fructooligosaccharides production from agro-wastes as alternative low-cost source, Trends in Food Science & Technology 91 (2019) 139–146.

[129] D.G. Panpatte, Y.K. Jhala, Agricultural waste: a suitable source for biofuel production, Prospects of Renewable Bioprocessing in Future Energy Systems, Springer, 2019, pp. 337–355.

[130] R. Nair, P.R. Lennartsson, M.J. Taherzadeh, Bioethanol production from agricultural and municipal wastes, Current Developments in Biotechnology and Bioengineering, Elsevier, 2017, pp. 157–190.

[131] Z. Hussain, W. Sajjad, T. Khan, F. Wahid, Production of bacterial cellulose from industrial wastes: a review, Cellulose 26 (2019) 2895–2911.

[132] M.U. Islam, M.W. Ullah, S. Khan, N. Shah, J.K. Park, Strategies for cost-effective and enhanced production of bacterial cellulose, International Journal of Biological Macromolecules 102 (2017) 1166–1173.

[133] A. Costa, M.A.V. Rocha, L. Sarubbo, Bacterial cellulose: an ecofriendly biotextile, International Journal of Textile and Fashion Technology 7 (2017) 11–26.

[134] A. Kurosumi, C. Sasaki, Y. Yamashita, Y. Nakamura, Utilization of various fruit juices as carbon source for production of bacterial cellulose by *Acetobacter xylinum* NBRC 13693, Carbohydrate Polymers 76 (2009) 333–335.

[135] F.P. Gomes, N.H. Silva, E. Trovatti, L.S. Serafim, M.F. Duarte, A.J. Silvestre, et al., Production of bacterial cellulose by *Gluconacetobacter sacchari* using dry olive mill residue, Biomass and Bioenergy 55 (2013) 205–211.

[136] S. Keshk, K. Sameshima, The utilization of sugar cane molasses with/without the presence of lignosulfonate for the production of bacterial cellulose, Applied Microbiology and Biotechnology 72 (2006) 291–296.

[137] S.M. Keshk, Vitamin C enhances bacterial cellulose production in *Gluconacetobacter xylinus*, Carbohydrate Polymers 99 (2014) 98–100.

[138] D. Tian, F. Shen, J. Hu, S. Renneckar, J.N. Saddler, Enhancing bacterial cellulose production via adding mesoporous halloysite nanotubes in the culture medium, Carbohydrate Polymers 198 (2018) 191–196.

[139] B.C. Adebayo-Tayo, M.O. Akintunde, J.F. Sanusi, Effect of different fruit juice media on bacterial cellulose production by *Acinetobacter* sp. BAN1 and *Acetobacter pasteurianus PW1*, Journal of Advances in Biology & Biotechnology 14 (2017) 1–9.

[140] S. Chiu, D. Moore, Threats to biodiversity caused by, Fungal Conservation: Issues and Solutions 111 (2001).

[141] A. Philippoussis, G. Zervakis, P. Diamantopoulou, Bioconversion of agricultural lignocellulosic wastes through the cultivation of the edible mushrooms *Agrocybe aegerita*, *Volvariella volvacea* and *Pleurotus* spp, World Journal of Microbiology and Biotechnology 17 (2001) 191–200.

[142] A. Philippoussis, P. Diamantopoulou, H. Euthimiadou, G.I. Zervakis, The composition and porosity of lignocellulosic substrates influence mycelium growth and respiration rates of *Lentinus edodes* (Berk.) Sing, International Journal of Medicinal Mushrooms 3 (2001) 198.

[143] A.N. Philippoussis, Production of mushrooms using agro-industrial residues as substrates, Biotechnology for Agro-industrial Residues Utilisation, Springer, 2009, pp. 163–196.

14

Genetically modified microorganisms for enhancing nutritional properties of food

Pardeep Kumar Bhardwaj[1], Kashmir Singh[2]

[1]INSTITUTE OF BIORESOURCES & SUSTAINABLE DEVELOPMENT, IMPHAL, INDIA
[2]DEPARTMENT OF BIOTECHNOLOGY, PANJAB UNIVERSITY, CHANDIGARH, INDIA

14.1 Introduction

Food biotechnology involves the application of various techniques to microorganisms, plants, and animals to improve the food quality, safety, ease of processing, and production. From years, humans have been involved in traditional processes used for making bread, beer, cheese, and various fermented products. To enhance the nutritional value, many characteristics of traditional foods have been identified like taste, flavor, vigor, and health benefits. These properties of foods are primarily obtained by fermentation using both genetically modified as well as wild-type microbes. The genetically modified organism (GMO) is an organism whose genetic material has been altered using recombinant techniques of modern biotechnology. The European Commission and Food and Agriculture organization defines GMO as a product "which does not occur naturally by mating or natural recombination" [1]. Food biotechnology, however, defines genetically modified foods as those produced from genetically modified microorganisms with enhanced nutritional properties, taste, flavor, shelf-life, and yield. From ancient time, mankind took advantage of natural variation by selecting microorganisms such as yeast, to produce domesticated variants better suited to specific environment as per the needs of humans. Due to many health benefits, production and consumption of fermented food products have been increased. Microbes are capable of growth on various substrates and able to induce their fermentation thus converting them into desired products with enhanced biological activity and improve the bioavailability of food products. For example, some potential enzymes isolated from microorganisms of naturally fermented soybean products which can help in fermentation processes for the production of bioactive compounds [2]. This also enhances the quality of food by increasing the bioavailability of nutrients, aroma, texture, and self-life of different food items [3]. Moreover, a number of biological activities of microbes are constantly being harnessed in production of fermented food products such as probiotic, antioxidant, antimicrobial properties, peptide production, nutritional properties etc. These serve as a crucial selection criteria for the starter culture which serves better in the production of functional foods. Fermented foods exploit the functional properties of microbes to promote various health benefits. The functional properties of these fermented foods can be increased using many biotechnological approaches to enhance their commercial value [4]. In view of the above examples, the present chapter is focused on various genetic approaches for the production of genetically modified microorganisms to enhance the nutritional properties of food.

14.2 Techniques used for genetic modification

During genetic modification, a gene is transferred from one organism to another using "cutting-copying-pasting" approach. For this, many bacterial enzymes are used which recognize, cut, and join DNA at specific locations and function as molecular scissors. Due to very less amount of genetic material, the selected gene is firstly amplified billions-fold and then transferred from one organism to another with the help of plasmids. In another technique, some plant cells are transformed using "Gene Gun" by bombarding small particles coated with the new DNA into the target cell.

14.3 Advantages of genetic modification

Genetic modification has potential for developing improved foods with improved flavor, texture of food, quality, chemical constituents, improved yield. For the development of improved food products, genetic modification can be targeted for improvement like yield, enhancement of micronutrients, removal of toxic components or allergens, increase in phytochemicals, and quality of macronutrients.

14.4 Need for genetically modified food products

In United States, fruits and vegetables are consumed very less than national dairy recommendations despite extensive outreach efforts, media campaigns for highlighting the relationship between health and different food choices [5]. Therefore, it is essential to develop foods enriched with vitamins and minerals with many health benefits. The fast-growing functional food market coupled many food products has motivated the biotechnology industry to develop genetically modified foods with enhanced nutrient levels. Different people have different perception about the genetically modified foods due to ethical, environmental, and safety issues. There are many food enzymes produced by genetically modified microorganisms via microbial fermentation [6]. The production of food enzymes can also be upscaled by standardizing optimal conditions for fermentation, either by using genetically modified microbes or by producing enzymes from recombinant technology. Microorganisms can easily be genetically modified for obtaining enzyme products with improved characteristics and higher yield. The use of genetically modified microorganisms is helpful in increasing the yield, improve enzyme production as well as limiting the unwanted metabolite production (such as mycotoxins), along with enzyme expression in organisms that normally would not produce enzyme of interest.

14.5 Microorganisms as source of food enzymes

The enzymes used in food manufacturing industry depict their big roles in governing their market value, which is expected to reach in the future to an estimated value of around $2.3 billion in 2021 [7,8]. Food enzymes are mainly obtained from microorganisms by fermentation process or by extraction from animals and plants. Microbes are preferred source over plants and animals for the production of enzymes. These enzymes can not only convert larger and complex chemical substrates into simple edible food products but also improve their overall nutritional value. Many fermented food products are produced by microorganisms by altering their metabolic activities. For example, lactic acid bacteria (LAB), found in milk, fruit juices, and in the intestinal mucosa of gut are important in the food processing due to lactic acid fermentation which prominently useful in production of dairy products, fermented vegetables, fermented meat, and sourdough bread. There are many bacteria like *Streptococcus*, *Lactobacillus*, *Pediococcus*, and *Leuconostoc* which are not only heterotrophic but also have complex beneficial prerequisites due to lack of a number of biosynthetic abilities. Most of these species present with differing prerequisites for growth as difference in requirement of amino acids and vitamins. In addition to microbes, yeast is being utilized in for the production of different fermented food products in beverage industry. Yeasts are eukaryotes possessing rich source of proteins and vitamin B. Yeasts are found either as dry powder or as compressed cakes and used in various types of food products particularly for baking and brewing foods, drinks, pharmaceuticals, and modern compounds. Examples are *Saccharomyces cerevisiae*, *Candida glabrata*, *Hansenula polymorpha*, etc. Similarly, molds are also important microbes used in many food products, such cheese, sausages, etc. Examples are *Rhizomucor miehei*, *Aspergillus niger*, *Endothia parasitica*, *A. nidulans*, etc. which are extensively utilized in the food processing industry.

14.6 Use of genetically modified microbes in food industry

14.6.1 Advantage of genetic modification of microorganisms

Modern biotechnology has made the utilization of microbes in food industry such as use of yeast in bread making and *Lactobacilli* for the production of various dairy items. Many enzymes used in food industry are obtained from resources like fungi (50%), bacteria (35%), and animals/plants (15%) [9]. There are some limitations of using the wild microorganisms for the production of these enzymes. Since wild microbes are typically unstable and unable to achieve the desired food quality during fermentation process because the conditions inside the fermenter inhibit the normal microbial growth or even challenge their survival.

Notably, in some cases, the metabolites or products from fermentation process are lethal to microbe itself and results in lower productivity and higher cost of production.

Many genetically modified microbes have the ability to resist the fermentation conditions. For example, cloning of gene encoding thermal resistance helped in the development of microbial stains that could withstand higher temperature inside the fermenter. Likewise, genetic engineering could generate steady and fast-growing clones of specific microorganisms. Hence, genetically modified microbes are able to produce desired food compounds when cultured on a suitable nutritive medium governed by defined conditions in the fermenter. A large number of such microorganisms are now used in different food industries.

14.6.2 Genetically modified microorganisms and fermented food products

Genetically modified microbes have been widely used not only in pharmaceutical, chemical, and other biotech industry but also in dairy industry. In many countries, most of the foods like yogurt, bread, cheese, butter, wine, kefir, and fermented meats as well as vegetables are synthesized by the microbes [1]. Few products in dairy industry are made via use of complex or wild microbial populations existing as crude products in nature such as unpasteurized cheeses, beers, etc. Some of these are produced by using genetically modified microbes such as cheese products which are produced from pasteurized milk. There is an extra advantage for genetically modified microbes over the wild microorganisms based on their organoleptic properties. For example, genetically modified microorganisms have been used to deliver numerous enzymes, organic acids, and biofuels in many chemical industries. Certain enzymes are used to catalyze the production of alcohol or organic acids in distillery. Many products like alcohol, beer, wine, amino acids, and vitamins are produced using genetically modified microorganisms by many pharmaceutical industries.

14.6.3 Genetically modified microorganisms and industrial food enzyme production

The food enzymes are produced mainly microbial fermentation using genetically modified microorganisms. The production of food enzymes can be up scaled by standardizing the process of fermentation using genetically modified microorganisms and producing recombinant enzymes [6]. At industrial level, microorganisms are preferred for the production of food enzymes in comparison with plants and animals. The main advantages are (1) less production time; (2) easy to extract and purify as they are produced extracellularly; (3) higher yield, activity, and stability; (4) easy to optimize. There are various proteins (or enzymes) with cloned genes in microorganisms for the production of recombinant enzymes. For examples, human insulin, somatostatin, hormones, hepatitis B immunization, superoxide dismutase enzyme, etc. [10,11]. Due to the ability of microbes to persist or pass through human as well as mucosa of animals, these can help to treat or prevent many diseases. For example, *Lactobacillus jensenii* which was engineered to make CD4 proteins which are used by the

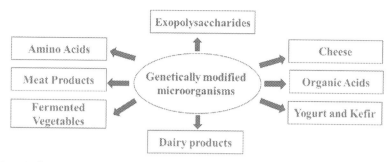

FIGURE 14–1 Genetically modified microorganisms and different food products for enhancing the nutritional properties.

HIV virus during infection in the vaginal mucosa in an attempt to infiltrate the lymphocyte cells [12].

14.6.4 Genetically modified microbes and metabolites production

The use of GMOs is better as it permits the combination of best required sources for the optimal metabolite production. Production of metabolites can be enhanced with in the cell using biotechnological techniques through metabolic engineering. Different targets for metabolic engineering can be categorized as: (1) extensive substrate range; (2) improved yield and productivity; (3) by-product elimination; (4) better processes; (5) advanced cell properties; (6) product range extensions such as in case of the heterologous protein production [13,14]. In genetically modified microorganisms, cell properties can be altered through the adjustment of biochemical processes or through introducing novel genes of a desired trait by using various biological systems. Therefore, genetic modifications are required to increase the product formation, stop the production of by-products, speed up the process, and develop the strains resistance to environmental stresses. Genetically modified microbes have been used for the production of antibiotics, enzymes, vitamins, fermented foods, drugs, sweeteners, flavoring agents in foods, etc. in various pharmaceutical, food, and brewery industries (Fig. 14–1).

14.7 Characteristics of genetically modified microorganisms

Different methods have been developed for the improvement of genetically modified microbes as well as manufacturing of various genetically modified food enzymes/products. The genetically modified microorganisms are produced via recombinant DNA technology, also known as "genetic engineering." Using recombinant DNA technology, a gene of interest coding for the desired character is inserted into the DNA of the nucleus, which translates the desired enzyme/protein. There are different methods for the genetic modification of microorganisms which involves conjugation, protoplast fusion, gene transformation, transduction, electroporation, and mutagenesis [15–17]. In metabolic engineering, diverse strategies have been utilized by specific microorganism for the production of secondary metabolites. In order to construct a recombinant strain, it is crucial to select a host organism which is cheaper as well as contains the required characteristics readily available for exploiting in order to produce and express the desired enzymes [6]. Later, it is important to determine the type of transformation methods to be used along with the expression vectors as well as the markers to be used to select the transformed strains. These parameters are important for screening an optimal genetically modified microorganism ready for industrial use. In recent years, new technologies such as CRISPR/Cas9 have emerged for genetic engineering. Here, the *Bacillus subtilis* strain WS5 was used for increasing the extracellular production of pullulanase enzyme where genes related to the protease production were disrupted in an attempt to protect the pullulanase which serves as the heterologous enzyme target of proteases [18]. Similarly, genes for undesirable characteristics, such as increased foam production during the fermentation process or excessive spore formation, can be targeted for disruption and can serve to further produce an industrially efficient enzymes [19]. In addition to production, strain modifications, enzyme modifications can also help to obtain an improved yield along with generation of better and more efficient enzyme properties. Some of these strategies include the following:

14.7.1 Selection of host organism

There are many expression systems for genetic modification but bacteria are simple and cheap host organism. However, simple proteins can be produced in bacterial system due to issue related to protein folding and posttranslational modifications. Examples are *Bacillus licheniformis, Bacillus amyloliquefaciens, E. coli* as well as *Bacillus subtilis*, etc. Besides, there are fungal host strains used for recombination with proper subcellular organizations and in turn allow correct folding of the proteins along with incorporation of various posttranslational modifications. Also, there is limited secretion of secondary proteins in case of a fungal host facilitating better downstream purification. Examples are *Aspergillus niger, Trichoderma reesei,* and *Aspergillus oryzae*.

14.7.2 Expression vectors and transformation process

For integration of modifications in host microorganisms, expression vectors are used which contains an origin of replication, a multiple cloning site, and a selection marker. Additionally, promoter and terminator regions are present in the expression system to control the expression of the inserted gene and are screened with the help of a selection marker.

14.7.3 Cloning of gene and heterologous expression

Recently with the advent of technology, the metabolite production has significantly improved with introduction of a wide plethora of microbial strains. This was accomplished with the transfer of an entire cluster of genes governing significant roles in biosynthetic pathway into a new host for the heterologous bioactive compound production. Such as in case of both tetrangulol and tetrangomycin produced in *Streptomyces rimosus* NRRL 3016 by transferring the entire cluster consisting of genes encoding the polyhydroxybutyrate pathway in *Ralstonia eutropha* into *S. cerevisiae*.

14.7.4 Engineering regulatory networks

The microbes can produce valuable products only under some stress conditions. However, for industrial purposes, there is a need for continuous metabolite production, which instigates the development of new engineering strategies for microbes in their up- and down-regulations. Various metabolite genes commonly exist together as clusters on the chromosomes in addition to other regulatory genes of their specific biological pathways, either as activators (positive) or as repressors (negative), thus varying the expression patterns accordingly for the gene cluster elements. For instance, constitutive expression of positive regulators of streptomyces antibiotic regulatory protein in *Streptomyces* strain has been demonstrated to enhance continuous production of actinorhodin.

14.7.5 Gene insertion and deletion

The intermediate metabolite genes from one organism can be manipulated to undergo deletion or insertion in another organism to get the product with desirable traits. For example, filamentous fungi known to produce a mycotoxin aflatoxin have also been utilized for the manufacturing of fermented foods. To override this issue, the mutant strain was constructed with aflatoxin containing cluster of genes as well as cyclopizonic acid genes were deleted [6].

14.7.6 Redirecting metabolic pathway

Most of the microbes are capable of producing different products along with other by-products. For instance, *E. coli* is known for producing products of little importance such as succinic acid along with other organic acids like acetic acid, ethanol, lactic acid, and formic acid in larger amounts during the process of anaerobic fermentation. So, in order to increase the amount of succinic acid in addition to lowering the production of other metabolic compounds, it is essential to redirect the metabolic pathways.

14.7.7 Stimulation by precursors

A number of stimulators have been employed to induce the quantity of a particular product in a biosynthetic pathway. Stimulators mainly inducers comprise of various amino acids involved in ergot alkaloid biosynthesis are tryptophan for dimethylallyl tryptophan synthase as well as leucine for bacitracin synthetase.

14.7.8 Genetic knockout of loci

These approaches in relation to "cell conditions" dynamically alter intracellular architectures as a response to various environmental cues thus altering heterologous product stability.

14.7.9 Clustered regularly interspaced short palindromic repeats-Cas9 (CRISPR-Cas9)

In recent years, new technological developments for introducing genetic modifications have emerged like "Clustered Regularly Interspaced Short Palindromic Repeat", "CRISPR-associated protein 9" and originated from bacterial and archaeal immune systems. In this technique, the Cas9 endonuclease incorporates double-stranded breaks in the DNA sequence of target gene. Specific donor sequence or mutations are introduced in the genomic DNA using endogenous repair pathways. This is carried out by the by the homologous repair system or by nonhomologous end-joining (NHEJ) DNA repair mechanism [6]. Contrarily to eukaryotes, bacteria do not have the NHEJ DNA repair system. Therefore, wild-type strains are not able to survive due to inability to repair the CRISPR/Cas9-mediated breaks. This capability in turn permits the CRISPR/Cas9 system to be utilized in selection of recombinant strains. This tool has been developed to engineer the formation of protease enzyme via disruption caused using CRISPR-mediated gene editing and thereby redirecting the increase of flux toward the substrates of ethanol acetyltransferase 1 in *Clostridium beijerinckii*.

14.7.10 Development of recombinant enzymes

Various techniques are involved in recombinant designing of enzymes based on the available protein sequence and 3D structure of the enzyme to target the desired amino acids for modifications. For example, a thermostable pullulanase was obtained by rational design in *Bacillus naganoensis*. This enzyme is involved in the hydrolysis of amylopectin to amylose in starch [20].

14.8 Products development using genetically modified microorganisms in food industry

LAB are considered as genetically safe. The lactic acid, formed as result of fermentation, enhances healthy bacteria of the gut to better proliferate and is linked to good gut health. The genetically modified strains in LAB include *Lactococcus*, *Lactobacillus*, *Bifidobacterium*, *Pediococcus*, and *Leuconostoc* with utility as flavor, texture, nutrition content, enhancer, as well as preservant for dairy and fermented products including milk, cheese, yogurt, butter, as well as pickled vegetables (Table 14–1).

Table 14–1 List of genetically modified microorganisms used for development of food products.

S. No.	Food product (s)	Genetically modified microorganisms	Targeted gene (s)/source/food quality	References
1	Cheese	E. coli, Kluyveromyces lactis, Aspergillus niger, and yeast	Chymosin from calf rennet	[21]
		Lactococcus lactis DN209	PepN, PepC, PepX, and PepI peptidases from L. helveticus strain	[22]
		Lactococcus lactis	Glutamate dehydrogenase from Peptoniphilus asaccharolyticus	[23]
2	Yogurt and Kefir	Saccharomyces cerevisiae, Saccharomyces boulardii, Kluyveromyces marxianus, Kluyveromyces lodderae, Kluyveromyces lactis, Yarrowia lipolytica, Candida inconspicua, Candida maris	Dairy products	[1]
3	Baking products	Saccharomyces cerevisiae, Saccharomyces exiguus, Issatchenkia orientalis, Pichia anomala, Pichia subpelliculosa, Candida humilis, Candida krusei, Candida milleri, and Torulaspora delbrueckii	From sourdough	[24]
3	Meat Products	Lactobacillus sake, L. curvatus, L. plantarum, Pediococcus pentosaceus, and P. acidilactici,	Taste and aroma of meat products	[25]
4	Fermented Vegetables	Leuconostoc mesenteroides, L. brevis, P. pentosaceus, and L. plantarum	Fermentation of cabbage, pickles, olives	[24]
		L. lactis subsp. lactis, L. sakei, L. curvatus, and L. fallax	Fermentation of cabbage	[26]
5	Dairy products	L. lactis biovar. Diacetylactis	α-Acetolactate decarboxylase for higher production of diacetyl (flavoring agent)	[27]
		Streptococcus thermophilus	GlyA gene encoding threonine aldolase for overproduction of acetaldehyde	[27]
		L. lactis	Pdc gene from Zygomonas mobilis for acetaldehyde production	[28]
9	Organic Acids	Lactobacillus species, E. coli and C. glutamicum	For the production of lactate, acetate, pyruvate, succinate	[29,30]
6	Vitamins	E. coli.	Biotin operon cluster gene for biotin production	[1]
		S. marcescens	S-2-aminoethylcysteine to enhance biotin production	
		C. ammoniagenes	Riboflavin biosynthesis genes for riboflavin production	
		Bacillus subtilis	ribC gene for riboflavin production	
		E. citreus	2,5-Diketo-D-gluconate reductase from Corynebacterium sp. for production of Vitamin C	
		L. lactis, L. gasseri, and L. reuteri	For the production of folic acid	

(Continued)

Table 14-1 (Continued)

S. No.	Food product(s)	Genetically modified microorganisms	Targeted gene (s)/source/food quality	References
7	Exopolysaccharides	*Lactococcus* and *Streptococcus*	Food grade exopolysaccharides such as xanthan and gurdlan	[1]
		S. thermophilus	*Phosphoglucomutase* and *UDP-glucose phosphorylase* for the increased production of exopolysaccharides	
8	Amino Acids	*Corynebacterium*, *Brevibacterium*, and *Serratia*	Production of different types of amino acids	[1]
		C. glutamicum	*Vitreoscilla hemoglobin* gene *vgb* for production of glutamic acid and glutamin	
		C. glutamicum	*xfp* gene from *Bifidobacterium animalis* for improved production of glutamate	
		C. glutamicum	Overexpression of pyruvate carboxylase or DAP dehydrogenase genes for the production of l-lysine	
		E. coli	Cloning of extra copy of *thrABC* operon for the production of threonine	
		C. glutamicum ssp. *lactofermentum*	*hom* (encoding homoserine dehydrogenase), *thrB* (encoding homoserine kinase), and *thrC* (encoding l-threonine synthase) genes for the production of l-threonine	
		Corynebacterium glutamicum VLA1	*ilvA* gene encoding threonine dehydratase for the production of l-valine	
		C. glutamicum	Threonine dehydratase gene *ilvA* from *E. coli* for the production of l-isoleucine	
		E. coli	Deletion of genes encoding a protein of the pyruvate dehydrogenase complex (*aceF*) and lactate dehydrogenase (*ldhA*), and replaced with the plasmid vector containing the *B. sphaericus* alanine dehydrogenase gene (*alaD*) for the production of l-alanine.	
		E. coli	Cloning of *aroF* and *pheA* genes for phenylalanine production	
		C. glutamicum	For l-Tyrosine overproduction	

14.9 Fermented food products and their health benefits

Fermentation in food sector has long evolved from primitive times. Presently, fermentation has long bypassed the households and is being commercialized at the industrial level thus captivating a huge mark in the market. This was possible due to their established roles in the maintenance of health and disease prevention. Fermented food products have been altered enzymatically by using microbes in a manner that enhances the taste and flavor. A number of potent health benefits associated with the fermented products have recently been under the radar of food industry. For example, health benefits of fermented foods like antihypertensive activity [31,32], blood glucose lowering benefits [32,33], and antidiarrheal and antithrombotic properties [32]. Fermented foods help in nutritional promotion and other health-related disease prevention by inducing changes in metabolism, inflammation, and oxidative stress [34,35]. Fermented pulses such as legumes and soybeans are consumed in Japan, China, Indonesia, and Korea as primary staple food crops and also exhibit antidiabetic effects. Fermented products might promote health benefits via compounds possessing enhanced bioactivity, and thus might help promote their synthesis during the fermentation process [36]. Fermented rice bran fractions, with *Saccharomyces boulardii*, depicted significant variations in their metabolite composition as compared with nonfermented ones [37] by decreasing the growth of lymphoma cells in humans. Fermented tea leaves using fungus, *Eurotium cristatum*, demonstrated an increased level of linoleamide, dodecanamide, epicatechin gallate and stearamide compounds in comparison with nonfermented green tea [38]. Blueberries were shown to lower the blood pressure in fermented probiotic strains of *Lactobacillus plantarum* [32]. Papaya preparation after fermentation shows strong antioxidant and immunomodulating potential. LAB strains associated with fermented foods demonstrate successful modulations in our immune system in action against the influenza virus [39]. *Lactobacillus plantarum* makes phenyl acids in trace portions along with stronger antifungal compounds. Fungal fermentation of polysaccharides separated from rice bran was used to demonstrate its therapeutic potential by activating the natural killer cells mediating anticancerous activity displayed via a nonspecific immune response [40].

14.10 Conclusions and perspectives

Genetically modified microorganisms have made progress in various food industries especially food enzymes. Due to their role in the enrichment of bioavailability of texture, nutrients, aroma, functional properties, and sensory qualities associated with genetically modified foods have been attracting people toward increased consumption. Genetically engineered microbes (involved in metabolic processes) might exhibit important roles in the production of several functional as well as food-based metabolites, comprising larger structures with difficulty in chemical synthesis. Globally, genetically modified foods are decoding many problems associated with people with an intent to provide sufficient food supplementation and other health benefits. Most of the microbes employed in the food industry are considered to be nonpathogenic. However, genetic manipulation of these microbes can also be associated with some human health and environmental complications. Hence, genetically modified microbes should be checked for assessing the risks involved either by using a "case-by-case" or "step-by-step" approach. The main intent behind of the risk measurement is to identify potential of adverse effects of genetically modified microbes on human health as well as various environmental issues. To combat complications as mentioned here, only acceptable genetically modified foods should be introduced into the market which have passed through rigorous risk assessments. In addition, the future strategies concerned with manipulation of genes are predicted to be rapidly evolving, cheaper, specific, and quantifiable and must promote multiplexing and their quantitative analysis in order to evoke an extensive and efficient perspective on the consumption of newer genetically modified food products.

References

[1] N. Mallikarjuna, K. Yellamma, Genetic and Metabolic Engineering of Microorganisms for the Production of Various Food Products: Recent Development in Applied Microbiology and Biochemistry, Elsevier Inc., 2019 (Chapter 13).

[2] A.K. Rai, S. Sanjukta, R. Chourasia, I. Bhat, P.K. Bhardwaj, D. Sahoo, Production of bioactive hydrolysate using protease, b-glucosidase and a-amylase of Bacillus spp. isolated from kinema, Bioresource Technology 235 (2017) 358–365.

[3] J.P. Tamang, D.-H. Shin, S.-J. Jung, S.-W. Chae, Functional properties of microorganisms in fermented foods, Frontiers in Microbiology 7 (2016) 578.

[4] R. Chourasia, M.M. Abedin, L.C. Phukon, D. Sahoo, S.P. Singh, A.K. Rai, Biotechnological approaches for the production of designer cheese with improved functionality, Comprehensive Review in Food Science and Food Safety 20 (2021) 960–979.

[5] G.J. Colson, W.F. Huffman, M.C. Rousu, Improving the nutrient content of food through genetic modification: evidence from experimental auctions on consumer acceptance, Journal of Agricultural and Resources Economics 36 (2) (2011) 343–364.

[6] M. Deckers, D. Deforce, M.A. Fraiture, N.H.C. Roosens, Genetically modified micro-organisms for industrial food enzyme production: an overview, Foods 9 (2020) 326. Available from: https://doi.org/10.3390/foods9030326.

[7] S. Raveendran, B. Parameswaran, S.B. Ummalyma, A. Abraham, A.K. Mathew, A. Madhavan, et al., Applications of microbial enzymes in food industry, Food Technology & Biotechnology 56 (2018) 16–30.

[8] R. Singh, M. Kumar, A. Mittal, P. Kumar, Microbial enzymes: industrial progress in 21st century, 3 Biotech 6 (2016) 1–15.

[9] P. Saranraj, M.A. Naidu, Microbial pectinases. a review, Global Journal of Traditional Medicinal Systems 3 (2014) 1–9.

[10] M. Rezaei, Zarkesh-Esfahani, Optimization of production of recombinant human growth hormone in Escherichia coli, Journal of Research in Medical Science 17 (7) (2012) 681–685.

[11] P.K. Bhardwaj, R. Sahoo, S. Kumar, P.S. Ahuja, A gene encoding autoclavable superoxide dismutase and its expression in E. coli, United States Patent (2011). 7,888,088.

[12] X. Liu, L.A. Lagenaur, D.A. Simpson, K.P. Essenmacher, C.L. Frazier-Parker, Y. Liu, et al., Engineered Vaginal Lactobacillus Strain for Mucosal Delivery of the Human Immunodeficiency Virus Inhibitor Cyanovirin-N, Antimicrobial Agents and Chemotherapy 50 (10) (2006) 3250–3259.

[13] N. Anesiadis, W.R. Cluerr, R. Mahadevan, Dynamic metabolic engineering for increasing bioprocess productivity, Metabolic Engineering 10 (2008) 255–266.

[14] R.R. Kumar, S. Prasad, Metabolic engineering of bacteria, Indian Journal of Microbiology 51 (3) (2011) 403–409.

[15] C. Hjort, Industrial Enzyme Production for Food Applications, Woodhead Publishing Limitedp, Cambridge, UK, 2007.

[16] J.L. Adrio, A.L. Demain, Genetic improvement of processes yielding microbial products, FEMS Microbiology Reviews 30 (2006) 187–214.

[17] R.S. Patnaik, S. Louie, V. Gavrilovic, K. Perry, W.P.C. Stemmer, C.M. Ryan, et al., Genome shuffling of Lactobacillus for improved acid tolerance, Nature Biotechnology 20 (2002) 707–712.

[18] K. Zhang, L. Su, J. Wu, Enhanced extracellular pullulanase production in *Bacillus subtilis* using protease-deficient strains and optimal feeding, Applied Microbiology and Biotechnology 102 (2018) 5089–5103.

[19] K. Zhang, X. Duan, J. Wu, Multigene disruption in undomesticated *Bacillus subtilis* ATCC 6051a using the CRISPR/Cas9 system, Scientific Reports 6 (2016) 1–11.

[20] M. Chang, X. Chu, J. Lv, Q. Li, J. Tian, N. Wu, Improving the thermostability of acidic pullulanase from *Bacillus naganoensis* by rational design, PLoS One 11 (2016) 1–12.

[21] J.S. Pai, Applications of microorganisms in food biotechnology, Indian Journal of Biotechnology 2 (2003) 382–386.

[22] V. Joutsjoki, S. Luoma, M. Tamminen, M. Kilpi, E. Johansen, A. Palva, Recombinant Lactococcus starters as a potential source of additional peptidolytic activity in cheese ripening, Journal of Applied Microbiology 92 (2002) 1159–1166.

[23] L. Rijnen, P. Courtin, J.-C. Gripon, M. Yvon, Expression of a heterologous glutamate dehydrogenase gene in Lactococcus lactis highly improves the conversion of amino acids to aroma compounds, Applied and Environmental Microbiology 66 (4) (2000) 1354–1359.

[24] A. Pulvirenti, L. Solieri, M. Gullo, L. De Vero, P. Giudici, Occurrence and dominance of yeast species in sourdough, Letters in Applied Microbiology 38 (2004) 113–117.

[25] D. Boyacioglu, D. Nilufer, E. Capanoglu, Flavor compounds in foods, in: F. Yildiz (Ed.), Advances in Food Biochemistry, CRC Press, Boca Raton, FL, 2010, pp. 301–302 (Chapter 9).

[26] V. Plengvidhya, F. Breidt Jr., L. Zhongjing, H.P. Fleming, DNA fingerprinting of lactic acid bacteria in sauerkraut fermentations, Applied and Environmental Microbiology 73 (2007) 7697–7702.

[27] A.C.S.D. Chaves, M. Fernandez, A.L.S. Lerayer, I. Mierau, M. Kleerebezem, J. Hugenholtz, Metabolic engineering of acetaldehyde production by *Streptococcus thermophilus*, Applied and Environmental Microbiology 68 (2002) 5656–5662.

[28] R.S. Bongers, M.H.N. Hoefnagel, M. Kleerebezem, High-level acetaldehyde production in Lactococcus lactis by metabolic engineering, Applied and Environmental Microbiology 71 (2005) 109–1113.

[29] V.F. Wendisch, M. Bott, B.J. Eikmanns, Metabolic engineering of *Escherichia coli* and *Corynebacterium glutamicum* for biotechnological production of organic acids and amino acids, Current Opinion in Microbiology 9 (2006) 268–274.

[30] R.P. John, D. Gangadharan, K. Madhavan Nampoothiri, Genome shuffling of Lactobacillus delbrueckii mutant and Bacillus amyloliquefaciens through protoplastic fusion for L-lactic acid production from starchy wastes, Bioresource Technology 99 (2008) 8008–8015.

[31] M. Koyama, S. Hattori, Y. Amano, M. Watanabe, K. Nakamura, Blood pressure-lowering peptides from neo-fermented buckwheat sprouts: a new approach to estimating ACE-inhibitory activity, PLoS One 9 (2014) e105802.

[32] I.L. Ahren, J. Xu, G. Önning, C. Olsson, S. Ahrne, G. Molin, Antihypertensive activity of blueberries fermented by Lactobacillus plantarum DSM 15313 and effects on the gut microbiota in healthy rats, Clinical Nutrition 34 (2014) 719–726.

[33] M.R. Oh, S.H. Park, S.Y. Kim, H.I. Back, M.G. Kim, J.Y. Jeon, et al., Postprandial glucose-lowering effects of fermented red ginseng in subjects with impaired fasting glucose or type 2 diabetes: a randomized, doubleblind, placebo-controlled clinical trial, BMC Complementary and Alternative Medicine 14 (2014) 237.

[34] C.L. Quave, A. Pieroni, Fermented foods for food security and food sovereignty in the Balkans: a case study of the Gorani people of Northeastern Albania, Journal of Ethnobiology 34 (2014) 28–43.

[35] J. Camps, A. Garcia-Heredia, Introduction: oxidation and inflammation, a molecular link between non-communicable diseases, Advances in Experimental Medicine and Biology 824 (2014) 1–4.

[36] S. Martins, S.I. Mussatto, G. Martinez-Avila, J. Montanez-Saenz, C.N. Aguilar, J.A. Teixeira, Bioactive phenolic compounds: production and extraction by solid-state fermentation. A review, Biotechnology Advances 29 (2011) 365–373.

[37] E.P. Ryan, A.L. Heuberger, T.L. Weir, B. Barnett, C.D. Broeckling, J.E. Prenni, Rice bran fermented with *Saccharomyces boulardii* generates novel metabolite profiles with bioactivity, Journal of Agricultural and Food Chemistry 59 (2011) 1862–1870.

[38] A.C. Keller, T.L. Weir, C.D. Broeckling, E.P. Ryan, Antibacterial activity and phytochemical profile of fermented Camellia sinensis (fuzhuan tea), Food Research International 53 (2013) 945–949.

[39] T. Kawashima, K. Hayashi, A. Kosaka, M. Kawashima, T. Igarashi, H. Tsutsui, et al., Lactobacillus plantarum strain YU from fermented foods activates Th1 and protective immune responses, International Immunopharmacology 11 (2011) 2017–2024.

[40] J.Y. Choi, D.J. Paik, D.Y. Kwon, Y. Park, Dietary supplementation with rice bran fermented with Lentinus edodes increases interferon-gamma activity without causing adverse effects: a randomized, double-blind, place bocontrolled, parallel-group study, Nutrition Journal 13 (2014) 35.

15

Exopolysaccharide producing microorganisms for functional food industry

Rwivoo Baruah, Kumari Rajshee, Prakash M. Halami

MICROBIOLOGY AND FERMENTATION TECHNOLOGY DEPARTMENT, CSIR – CENTRAL FOOD TECHNOLOGICAL RESEARCH INSTITUTE, MYSURU, 570020, INDIA

15.1 Introduction

The advent of functional foods in Japan during 1980s leads to the constant growth of the functional foods and beverages industry. In recent times, the consumer's desire for different functional foods stems from the fact that functional foods can provide various health benefits in a cost effective manner than conventional medicine. In many of functional foods, the value addition occurs through bacterial fermentation, which produces various bioactive compounds. These compounds can range from acids, enzymes, proteins, sugars, fatty acids, exopolysaccharides (EPS), and vitamins. These fermentative products are the basis of health-promoting properties and enriching the nutritive value in many functional foods.

Among the various microorganisms used to develop fermented functional foods, lactic acid bacteria (LAB) are the most important group. LAB have been regarded as generally regarded as safe (GRAS) by United States Food and Drug Administration (FDA) as well as qualified presumption of safety status according to the European Food Safety Agency [1]. LAB are very versatile group, which are natural inhabitant of several food matrices. Fermented foods involving LAB are rich in various health-promoting compounds as mentioned above. For over 1000 years, they have been used as starter cultures in various fermented foods like yogurt, cheese, bread, etc. [2,3]. Probiotics are the live microorganisms that confer health benefits on the host when administered in sufficient numbers and mainly comprise LAB and bifidobacterium [4]. They are generally isolated either from gut microbiota, raw food materials, or fermented foods. These probiotics are available mostly through dairy-based functional foods in the market or as food supplements in the form of capsules. Through enzymatic hydrolysis in fermented foods, LABs can produce prebiotics, which are defined as nondigestible compounds that can promote the growth of beneficial microorganisms in the gut resulting in various health benefits to the host [5].

Among the microbial metabolites in functional foods, EPS play a key role in maintaining the texture in many food products. EPS produced by microorganisms provides protection to adverse environmental conditions. In food matrixes, the property of EPS to modify the sensory and rheological properties depends on its concentration, structure, and strain of microorganism producing it. EPS also act as prebiotic, antioxidants, and have immunomodulatory and anticancer properties [6]. EPS in recent times have gained importance as a food hydrocolloid of biological origin. These properties of EPS are mostly exploited in dairy-based functional foods like yoghurt and in sourdough bread. Many traditional fermented foods contain EPS producing bacteria as seen in foods such as yoghurt, villi, kefir, kimchi, sourdough [7]. In recent time, EPS's use in functional foods through in situ EPS production or supplementation of EPS in food products has been reported. The introduction of EPS in the formulation of functional food helps improve the taste, texture, and health benefits of the product [8].

This chapter focuses on the various microbial genus and their importance in producing a wide variety of EPS used in the food industry. This chapter also focuses on the technological advantages of using EPS, their health-promoting properties, and functional foods that contain EPS.

15.2 Exopolysaccharides

EPS are biopolymers, which are composed of one or more repeating monosaccharide unit. Based on their structural composition, EPS are classified into homopolysaccharide (HoPS) and heteropolysaccharide (HePS). HoPS are those EPS, which are composed of single type of repeating monosaccharide unit [9]. EPS composed of repeating units of D-glucose is called as glucan. These HoPS are differentiated on the basis of linkages present between the repeating units, for example, α-glucan and β-glucan [10]. The α-glucan EPS is mostly produced by LABs and further subdivided on the basis of the major linkages present such as dextran (α-1,6), mutan (α-1.3), alteran (α-1,3 and α-1,6), and reuteran (α-1,4 and α-1,6). HePS are those EPS, which are composed of more than a single type of repeating monosaccharide unit. HePS contain various sugars such as galactose, glucose, and rhamnose and may also contain noncarbohydrate groups such as phosphate or acetyl groups (Fig. 15–1) [7].

The major producers of both HoPS and HePS are LAB, especially *Lactobacillus*, *Leuconostoc*, *Pediococcus*, *Weissella*, and *Streptococcus*. The mode of biosynthesis of HoPS and HePS greatly differs among these microorganisms. HoPS are produced extracellularly by the action of enzyme glycansucrase, a glucoside hydrolase (GH) such as glucansucrase belongs to GH70 family and fructansucrase belonging to GH68 family. Glycansucrase enzymes utilize sucrose as a substrate to convert into HoPS [12]. HePS are produced intracellularly by the action of a cascade of enzymes. The formation of HePS is governed by the action of an operon called EPS gene cluster, which is responsible for regulation, chain-length determination, polymerization, and export of HePS. HePS are produced from intermediates such as UDP (Uridine Diphosphate)-glucose, UDP-galactose, and dTDP (deoxyThymidine Diphosphate)-rhamnose [6,13]. The EPS gene cluster contains glycosyltranferase gene, responsible for HePS formation and any variability seen in this gene, resulting in differences in monosaccharides composition [14]. The great diversity of monosaccharides seen in HePS can range from two to nine different monosaccharides [15].

HoPS possesses higher molecular weight mostly $>10^6$ Da when compared to that of HePS ranging from 10^4 to 10^6 Da [10,16]. The high molecular weight of HoPS is correlated with their

Chapter 15 • Exopolysaccharide producing microorganisms for functional food industry 339

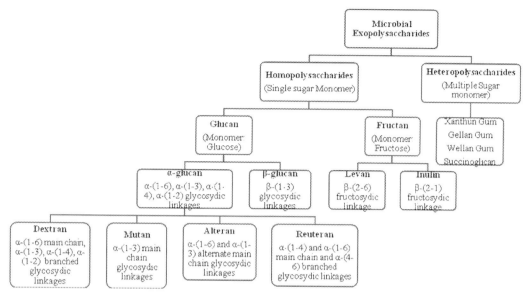

FIGURE 15–1 Classification of microbial exopolysaccharides [7,10,11].

properties as hydrocolloids and is currently widely used in the food industry. HePS, due to the highly variable monomer composition, can act as antioxidants, immunomodulators, and anticancer agents [6]. The yield of both types of EPS also varies, HoPS are produced in comparatively higher quantities (in grams per liters) as compared to HePS (in milligrams per liters).

15.3 Important exopolysaccharides producing bacterial genera

Among the great diversity of EPS producing bacteria, LAB are attributed to be one of the most prolific producers. However, besides LAB, several other bacterial genera produce unique EPS. Few of the EPS producing bacteria have been mentioned in Table 15–1. Following are the description of few important EPS producing genera that are utilized in food applications.

15.3.1 *Lactobacillus*

Lactobacillus is one of the most important producers of EPS among LAB known till date. *Lactobacillus* species are GRAS, making them food-grade organisms and the EPS produced by these cultures can be safely used in food products. Recently, the genera *Lactobacillus* was divided into 23 different genera on the basis various genome sequences; however, they will be collectively referred to as *Lactobacillus* in this chapter [31]. Among different species of *Lactobacillus*, well-known EPS producers are *La. casei*, *Lv. brevis*, *Lp. plantarum*, *Lt. curvatus*, *Lb. helveticus*, *Lb. acidophilus*, *La. rhamnosus*, *Lb. johnsonii*, *Lb. delbrueckii bulgaricus* [14]. *Lactobacillus* spp. produce both HoPS and HePS, though there is a large diversity of

Table 15–1 Exopolysaccharides (EPS) produced from different bacteria and their properties.

Microorganism	Type of EPS produced	Source of isolation	Functional properties	Reference
Fr. Sanfranciscensis LTH2590	Fructan	Sourdough	Improve sourdough texture	[17]
Lp. plantarum DM5	Dextran	Marcha (fermented food of Himalayan region)	Prebiotic in nature, antioxidant activity	[18]
Lb. kefiranofaciens	Kefiran	Kefir	Food hydrocolloid, antioxidant and antimicrobial activity	[19]
Lp. plantarum YW11	HePS	Tibitan-kefir	Viscosifier	[20]
W. confusa KR780676	Galactan	Idli-batter	Food hydrocolloid	[21]
W. confusa MD1	Mannan	Dosa batter	Antioxidant activity and antibiofilm activity	[22]
W. cibaria RBA12	Dextran	Pummello	Prebiotic	[23]
Leu. pseudomesenteroides YF32	Dextran	Soybean paste	Antimicrobial activity	[24]
Leuconostoc mesenteroides NTM048	Dextran, fructan	Green peas	Immunostimulant	[25]
P. pentosaceus CRAG3	Dextran	Cucumber	Anticancer activity	[26]
P. acidilactici M76	Glucan	Makgeolli, a Korean traditional rice wine	Antioxidant activity and antidiabetic effects	[27]
Oenococcus oeni	HePS, β-glucan	Wine		[28]
Xanthomonas campestris	Xanthan gum	Plant source	Food hydrocolloid, emulsifier, and stabilizer	[29]
S. paucimobilis	Gellan gum	Lake water	Stabilizer and a thickening agent	[30]

HePS produced. Generally, MRS medium or milk-based media (containing lactose) are used for the production of EPS. Some complex media used for EPS production usually hinder the purification process, making it difficult to remove contaminating carbohydrates [16].

Lactobacillus spp. produces both α- and β-glucan and some species also produce fructans. *Lm. fermentum, Lm. reuteri, Le. parabuchneri, Lt. sakei,* and *Le. hilgardi* are known producers of α- and β-glucan. *Lm. reuteri* is attributed to produce a unique α-glucan called reuteran having high molecular weight and contains both α-(1,4) and α-(1,6) linkages in main chain and α-(4,6) branched linkages [14]. *Fr. sanfranciscensis* LTH2590 and *Fr. sanfranciscensis* TMW 1.392 have been reported to produce fructan EPS during sourdough fermentation [17,32]. Dextran, aα-glucan, was produced by *Lb. acidophilus* ST76480.01 when sucrose is used as a carbon source, and it yielded a maximum of 4.24 mg/mL dextran and gave a maximum dextransucrase activity of 4.64 DSU/mL/h [33]. *Lp. plantarum* DM5 produced dextran of molecular weight of 1.11×10^6 Da and had 86.5% α-(1,6) linkages and 13.5% of α-(1,3) branched linkages [18]. Dextran from *Lp. plantarum* DM5 also possesses prebiotic potential by stimulating the growth of probiotic bacteria like *Lactobacillus* and Bifidobacteria [34].

Kefiran is an EPS found in kefir grains and is produced by several species of *Lactobacillus* which includes *Lb. kefiranofaciens, Lb. kefirgranum, Le. parakefir, Le. kefir,* and *Lb. delbrueckii* subsp. *bulgaricus*. Kefiran is HePS composed of glucose and galactose in a molar ratio of 1:1 and contains various linkages between these monomers. The linkages present in it include (1→6)-linked Glucose, (1→3)-linked Galactose, (1→4)-linked Galactose, (1→4)-linked Glucose, and (1→2,6)-linked galactose. Kefiran is used as a stabilizing and a gelling agent in food industry and has been reported to possess antioxidant and antimicrobial activity [19]. Great variability is seen in the composition of HePS produced by bacteria belonging to the genus *Lactobacillus*. The HePS-producing *Lactobacillus* strains include *Lp. plantarum* RJF4, *Lp. plantarum* YW11, *Lactiplantibacillus plantarum* SKT109, *Lp. paraplantarum* BGCG11, and *La. casei* LC2W. The HePS produced by *Lactobacillus* has several functional properties such as prebiotic, antioxidant, immunomodulatory, emulsifying, and stabilizing activities [6].

15.3.2 *Weissella*

Weissella are coccoid or short rod-shaped Gram-positive, catalase-negative, facultative anaerobic cells that belong to the phylum Firmicutes, class Bacilli, order Lactobacillales and family Leuconostocaceae [35,36]. Among the 19 know species of genus *Weissella* two species, that is, *W. confuse* and *W. cibaria*, are the most prolific producer of EPS. These two species have been reported to predominantly produce dextran, a HoPS composed of D-glucose linked with α-(1−6) linkages. The dextran produced by *W. confuse* and *W. cibaria* has characteristic features such as having high molecular weight and a relatively linear dextran backbone having very few α-(1−3) linkages. These features result in dextran from *Weissella* to be suitable for use in the food industry as hydrocolloids, stabilizers, and emulsifiers.

Beside dextran, *Weissella* have also been reported to produce other EPS like fructan a HoPS composed of fructose monomer, which are of two types levan (β-2−6 linked) or inulin (β-2−1 linked) [37]. Linear galactan containing α-(1,6)-linked galactose units was produced by *W. confusa* KR780676, isolated from an idli-batter [21]. Mannan type of EPS having a molecular weight of 2.9 kDa and having degradation temperature of 267.7°C was produced by *W. confusa* MD1 [22]. Dextran produced by genus *Weissella* has been used in bread making to improve the bread softness and increase self-life of the bread [38,39]. *W confuse* has been used to ferment sourdough, producing in situ dextran from added sucrose. Bread made with this sourdough showed improved specific volume [40]. Several gluten-free flours such as buckwheat, oat, quinoa, and teff were used to make sourdough using dextran from *Weissella* [41]. Dextran from *W. confusa* E-90392 was utilized to make a vegetable model using a fermenting carrot medium. The dextran formed during fermentation improved the texture, flavor, and smell [42].

Recently, prebiotic juice blends using dextran producing enzyme dextransucrase were developed utilizing the acceptor reactor, where maltose was used as an acceptor molecule [43]. Prebiotic isomalto-oligosaccharides from DP3-DP5 along with isomaltose (DP2) and leucrose (DP2) were synthesized in vitro using dextransucrase from *Weissella cibaria* RBA12 reducing the native sucrose content of the fruit juice [43]. Dextransucrase from *W. cibaria* 10M was used to synthesize isomalto-oligosaccharides in juice blend made using orange juice and malt extract. Isomalto-oligosaccharides (DP3) yielded 19.4 g/L in the juice blend having 36 g/L maltose and 19 g/L sucrose [44].

The use of *Weissella* as EPS producing starter culture in food products is still not popular due to reports of pathogenic traits. However, recent genomic studies of *Weissella* have revealed its potential to be used as a probiotic or a starter culture in food fermentations. Complete genome sequencing of *Weissella cibaria* CH2 isolated from cheese of western Himalaya revealed genes encoding for EPS production, gastrointestinal tolerance, and adhesion. Other functional traits such as cholesterol reduction, β-galactosidase production, vitamin, and essential amino acid biosynthesis were found by genomic analysis. *Weissella cibaria* CH2 lacked any transferable antibiotic resistance genes and virulence factors, which was confirmed using genome sequencing [45].

15.3.3 *Leuconostoc*

Leuconostoc are a genus of Gram-positive cocci, catalase-negative, facultative anaerobes that are arranged in pairs or chains. They are generally found in fresh plants, raw milk, and other chilled food products. *Leuconostoc* (L/Leu) are LAB that have GRAS status and is associated with fermentation in several foods products such as sausages, cereal products, fermented vegetables, and dairy products [46]. *L. mesenteroides* is generally associated with the production of dextran EPS. *L. mesenteroides* NRRL B-512F is used commercially for the production of dextran. The enzyme dextransucrase from *L. mesenteroides* NRRL B-512F is produced, purified, and used to synthesize dextran using sucrose as a substrate. The structure and function of both dextran and dextransucrase from *L. mesenteroides* NRRL B-512F were studied in several reports and therefore can be considered a model organism for EPS study [47].

Several strains of *L. mesenteroides* as well other *Leuconostoc* species have been associated with EPS production, where dextran being the major EPS produced. *L. mesenteroides* NRRL B-640 produces linear dextran with only α-(1−6) linkages with no branches, which was confirmed by 2D-NMR (2-dimensional Nuclear Magnetic Resonance) spectroscopy [48]. *L. pseudomesenteroides* YF32 isolated from soybean paste also produces linear dextran having α-(1−6) linked glucose and high degradation temperature of 307.62°C with a yield of 12 g/L [24]. *L. mesenteroides* strain NTM048 produced two types EPS having immunostimulatory properties. The major component of the two EPS was glucan in nature having 94% α-(1−6) linkages and 6% α-(1−3) branched linkages. The minor component was revealed to be a fructan composed of β-(2−6) and β-(2−1) linked fructose [25]. *L. kimchii* isolated from traditional fermented pulque a traditional Mexican alcoholic beverage produced two fractions of EPS one soluble and other cell-associated. The soluble fraction was identified as dextran having a majority of α-(1−6) linkages and a few α-(1−3) branched linkages. The cell-associated fraction was identified as levan having β-(2−6) linked fructose residues [49]. Besides producing dextran and levan, *Leuconostoc* have been associated with the production of HePS as seen in *Leuconostoc* sp. CFR 2181 isolated from traditional fermented milk dahi produces an EPS composed of glucose, rhamnose, and arabinose [50].

15.3.4 *Pediococcus*

Pediococcus are Gram-positive, catalase-negative, and facultative anaerobic LAB. They are homofermentative in nature only producing lactic acid and no CO_2 from glucose. They are unique in appearance as they form tetrads and do not form any chains that are typical to other LAB like *Leuconostoc* and *Streptococcus* [51]. *Pediococci* have been associated in several fermented foods such as in beer, cheese ripening, and starter culture in sausages [52].

Pediococcus has been associated with the production of both types of EPS, that is, HoPS and HePS. *P. pentosaceus* CRAG3 produces dextran with a molecular weight of 2.93×10^5 Da and highly branched dextran with 75% α-(1−6) linkages and 25% α-(1−3) linkages between glucose monomers. Dextran produced from *P. pentosaceus* CRAG3 showed anticancer activity as well as biocompatibility with human macrophage cells [26]. *P. acidilactici* M76 isolated from Makgeolli a Korean traditional rice wine produces glucan type EPS having a molecular weight of 67 kDa. This EPS also showed high antioxidant activity as well as antidiabetic effects by displaying protective effect on pancreatic RIN-m5F cell lines [53].

Beside the production of α-glucan such as dextran, *Pediococcus* have been reported to produce β-glucan which is rarely found from microbial sources. *P. parvulus* 2.6 produces 2-substituted (1,3)-β-D-glucan. It showed gut microbiota modulating activity as well as immunomodulatory activity by activating human macrophages [53]. In another study, β-glucan from *P. parvulus* 2.6 significantly influenced the survival of *Lp. plantarum*WCFS1 both under stimulated gastrointestinal condition and also in the fermented cereal-based food matrixes [54].

P. pentosaceus P 773 isolated from spoiled beer utilizes lactose as sole carbon, in the presence of sucrose it produces HePS composed of glucose and fructose in the ratio of 3:1 having a molecular mass of 2000 kDa. This HePS produced by *P. pentosaceus* P 773 was reported to have prebiotic potential by stimulating the growth of probiotic starter cultures like *Bifidobacterium lactis*, *Lb. acidophilus*, *Streptococcus thermophiles* [55]. *P. pentosaceus* M41 isolated from a marine source produces HePS composed of arabinose, mannose, glucose, and galactose in a molar ratio of 1.2:1.8:15.1:1.0 and having a molecular weight of 682.07 kDa. HePS from *P. pentosaceus* M41 showed antioxidant activity, α-amylase, and α-glucosidase inhibitory activity and antitumor activity [56].

15.3.5 Other lactic acid bacteria species

Besides the above-mentioned genera of LAB, there are other genera such as *Streptococcus*, *Lactococcus*, and *Oneococcus* reported to produce EPS. Some *Streptococcus* spp. like *S. mutans* produces EPS and are primarily associated with dental biofilms. EPS such as α-mutan is responsible for providing binding sites for microcolonies in the oral environment [57]. EPS from Streptococcal species like *S. thermophilus* has been associated with several diary food products. Capsular EPS from *S. thermophilus* Mr-1C was seen to improve moisture retention in low-fat mozzarella cheese. This capsular EPS was composed of D-galactose, L-rhamnose, and L-fucose in a molar ratio of 5:2:1 [58]. EPS from *S. thermophilus* strains, ST 285, and ST 1275 helped in improving the rheological properties of set-type yogurt. EPS production was growth associated, and the increase in the yoghurt's rheological properties was observed during prolonged cold storage [59]. EPS from *S. thermophilus* CRL1190 purified from fermented skim milk, displayed antioxidant activity as well as good flocculating and emulsifying activity [60].

Lactococcus spp. such as *Lc. lactis* strain NIZO B40 produces a phosphate containing HePS composed of galactose, glucose, and rhamnose, which is encoded by a 40 kb plasmid [61]. *Lc.*

lactis subsp. *cremoris* strain LC330 produces two HePS, one having a higher molecular weight of 1×10^6 and 10,000 Da. The larger HePS was composed of glucose, galactose, and glucosamine and the other HePS was composed of glucose, rhamnose, galactose, and glucosamine in an approximate ratio of 6:5:4:1 [62]. *Lc. lactis* Z-2 isolated from common carp produces EPS of molecular weight 18.65 kDa and is composed of rhamnose, xylose, mannose, glucose, and galactose at a molar percentage of 13.3%, 14.1%, 18.5%, 27.4%, and 26.7%, respectively. This EPS produced by *Lc. lactis* Z-2 displayed immunomodulatory and antioxidant activities [63].

Oenococcus oeni is a LAB found in wine fermentations. Several strains of *O. oeni* produce EPS both of soluble and capsular forms. HePS produced by these strains are mainly composed of glucose, galactose, and rhamnose. They have also been involved in β-glucan synthesis both in free and cell-associated forms. *O. oeni* strains have also been reported to produce both α-glucan and β-fructans from sucrose [28].

15.3.6 Non-lactic acid bacteria species

Besides the prevalence of EPS producing LAB, there are several non-LAB species that produce EPS of functional and economic importance. Among which xanthan gum a HePS produced by *Xanthomonas campestris* is composed of glucose, mannose, and glucuronic acid in a molar ratio of 2:2:1. Xanthan gum is nontoxic and safe to be used as food additives; hence, it has been approved by United States FDA and European Economic Community in 1984 and 1980, respectively. Xanthan gum is widely being used as a food hydrocolloid, emulsifier, and stabilizer in several food products [29].

Some species in *Bifidobacterium* genus are also capable of producing EPS. Bifidobacteria are beneficial human gut bacteria that reside in the distal part of the human large intestine. Several *Bifidobacteria* species such as *Bif. longum, Bif. breve, Bif. animalis, Bif. bifidum* have been reported to produce EPS [15]. These bifidobacterial EPS are HePS in nature and are mainly found in capsular form composed of glucose, galactose, and rhamnose. These EPS have been associated with immunomodulatory activity in the host as well as the ability to modify the composition of intestinal microbiota to benefit the host [64].

A strictly aerobic Gram-negative bacterium, *Sphingomonas paucimobilis* produces HePS gellan gum, composed of glucose, rhamnose, and glucuronic acid in a molar ratio of 3:1:1. Due to its superior rheological characteristics, gellan gum finds use in food, pharmaceuticals, and environmental bioremediation. In food products, it is used as stabilizer and a thickening agent and was approved to be used as a food additive by USFDA in 1992 [30].

Besides the EPS mentioned above, HePS diutan and welan are produced by members of genus *Sphingomonas*. Succinoglucan having glucose and galactose is produced by *Sinorhizobium meliloti*. *Azotobacter* spp. produce bacterial alginates and *Gluconobacter* spp. produce bacterial cellulose [65]. Curdlan a linear, (1→3)-β-glucan having (1→6) branched linkages is produced by some spp. of *Agrobacterium* and *Rhizobium* [66]. Pullulan is a EPS produced by fungus *Aureobasidium pullulans* and is composed of repeated maltotriose unit of glucose joined by α-(1−6) linkages. Pullulan due to its structural and physicochemical properties have found its use as blood plasma substitutes, food additive, cosmetic additives, and flocculants [67].

15.4 Fermented food containing exopolysaccharides

Several fermented foods utilize EPS producing strains, which mostly consists of LABs. EPS present in these fermented foods impart characteristic texture and flavor as well as impart several health-promoting benefits as well (Fig. 15–2).

15.4.1 Sour dough

Sour dough is a cereal-based fermented food product where the final product is sourdough bread. Sourdough bread have made for centuries but saw a loss of popularity with the rise of less time-consuming bread making with commercial yeast. Sour dough fermentation improves the texture, microbial safety, shelf-life, and nutritional value of the cereal matrix used [40]. EPS producing LABs found in sourdough are *Lactobacillus*, *Leuconostoc*, and *Weissella*, with HoPS production from sucrose [68]. As discussed in previous sections regarding the *in situ* EPS production in sour dough, EPS formation as a hydrocolloid can help develop gluten-free sourdough. Gluten-free flour (buckwheat, oat, quinoa, and teff) were used to prepare sourdough using dextran producing *W. cibaria* MG1. This bacterium produces dextran and oligosaccharides by utilizing sucrose in buckwheat, quinoa, and teff flour matrices. *cibaria* MG1 did not grow in oat flour [41]. Faba bean-wheat composite sour dough was developed using dextran producing strains *W. cibaria* Sj 1b and *Leu. pseudomesenteroides* DSM 20193. In situ fermentation after utilization of the supplemented sucrose resulted in dextran formation which lead to improvement of dough viscoelasticity, increased specific volume and reduced crumb hardness of final bread [69]. In situ dextran production by *W. confusa* A16 in wholegrain sorghum bread with 50% wheat flour yield dextran of 0.56% bread weight. The dextran

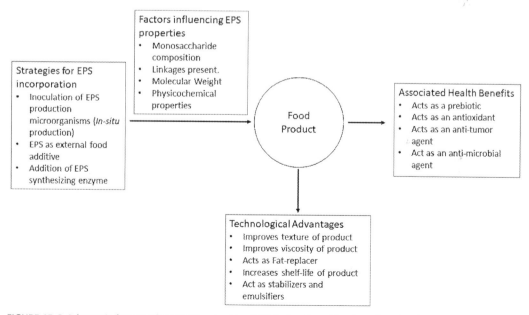

FIGURE 15–2 Schematic for use of exopolysaccharides (EPS) in functional food product development.

enriched bread showed considerable reduction of flavor intensity perception as compared to control bread without dextran. Dextran incorporation resulted in more cohesive, springy, and soft texture with significantly less perceived sourness and bitterness intensity [70].

15.4.2 Yoghurt

Yogurt is a fermented dairy product made using milk fermented with *Lb. delbrueckii bulgaricus* and *Str. thermophilus* in ratio 1:1 as starter culture. Both starter culture has been reported to produce EPS, where *Str. thermophiles* has higher EPS yield than *Lb. delbrueckii* subsp. *bulgaricus* [14]. Yoghurt made using non-EPS producing strains resulted in a less viscous product compared to EPS containing yoghurt. In yoghurt fermentation, due to low pH, micelles of milk protein casein form a network resulting in a gel-like structure. The whey is trapped by the casein gel matrix resulting in characteristic yoghurt texture. EPS also interacts with the casein and form bonds that are stronger than the bonds between the proteins. This interaction result in yogurt with EPS having higher viscosity [7]. Mixed culture of different EPS producing *Str. thermophilus* along with *Lb. delbrueckii* subsp. *bulgaricus* with negligible EPS production was used for yoghurt preparation. The yoghurt resulted in a different textural property such as mouth thickness, creaminess, and viscosity. Different EPS markers like free and capsular form in the yoghurt helped by distributing differently in the protein matrix and influenced yoghurt texture [71]. Yoghurt having different EPS producing bacteria along with the regular starter culture have also been developed. Probiotic low-fat yogurt was made using EPS producing *Bifidobacterium longum* subsp. *infantis* CCUG 52486 and yoghurt starter cultures. Probiotic yoghurt showed lower syneresis than control yoghurt and cell viability up to 28 days of storage [72]. HePS-producing *La. paracasei* was used as starter to prepare yoghurt that resulted higher viscosity. HePS from *La. paracasei* was composed of mannose, glucose, galactose, and glucuronic acid in a molar ratio of 0.87:0.92:1:0.24. HePS was found to inhibit the aggregation of casein in the protein matrix, thus enhancing its stability [73].

15.4.3 Kefir

Kefir is a popular Eastern European fermented milk-based drink. It made using a consortium of bacteria and yeast present in Kefir grains that are microbial aggregates in a polysaccharide matrix. The polysaccharide is a HePS "Kefiran" produced by *Lb. kefiranofaciens* at the center of the kefir grain produced only under specific conditions like the presence of ethanol and lactic acid. The ethanol produced is by the yeast and lactic acid is produced by LAB during fermentation. Kefiran comprises of 45% of the Kefir grains and are used to inoculate fresh milk for fermentation. Kefiran acts as a natural hydrocolloid enhancing texture and can be used as a food additive to increase viscosity [14]. The effect of kefiran in chemically acidified skim-milk gels on apparent viscosity and dynamic rheological properties was studied. Kefiran proved to be efficiently improving the skim-milk gels' viscoelasticity and showed the potential to be used as a natural thickening agent in dairy products [74]. Kefiran was used to form biodegradable edible film made using glycerol as plasticizer. Studies showed that the addition of glycerol improves the film's properties with a promise of its use in food technology applications [75].

15.5 Technological advantages of exopolysaccharides

EPS produced from microbial sources has several uses in various industries, ranging from dextran (cross-linked) for chromatographic separation to xanthan gum in oil recovery operations. Most EPS rely on their ability to act as an efficient hydrocolloid to influence food products' quality and texture. The functional properties of EPS depend upon their molecular weight, monomer composition, and type of branching present. An in-depth study on the physicochemical properties of EPS prior any planned applications are recommended. HoPS generally have larger molecular weight higher thickening capacity than HePS and hence are more prevalent in food applications.

Recently, some EPS have been used to replace synthetic hydrocolloids that are normally used in some food products. EPS as a naturally produced hydrocolloid can help to attain the safe label and consumer satisfaction of a product. The use of EPS as food additive to improve food texture was further explored by in situ EPS formation as seen in sourdough. EPS producing LAB are used to ferment the sourdough, they produce EPS in the dough matrix by utilizing natural sugars or supplemented sucrose [68]. There have been various reports of using different dough matrices such as wheat bran and rye bran, where the EPS formation was found higher than in wheat flour [76,77].

EPS have been used as fat replacers in food products like cheese, ice cream, or fermented meat products like sausages. EPS produced by LAB in cheese helps during the ripening of cheese, it helps in moisture retention and in turn improves cheese texture. *W. cibaria* MG1 and *Lm. reuteri* cc2 were used as adjunct culture in cheddar cheese manufacturing. The dextran produced by both strains increased the moisture retention of the cheese without any negative effects on its proteolysis [77]. HePS-producing strain *Lp. plantarum* TMW and HoPS-producing strains *Lt. sakei* TMW 1.411 and *Lt. curvatus* TMW 1.1928 were used to produces preadable raw fermented sausages (*Teewurst*). EPS was produced in situ and resulted in a maximum reduced fat content of 30%–33%. Sensory evaluations rated the HoPS containing sausages, softer, and more spreadable [78].

Many traditional fermented food products utilize the viscosifying properties of EPS from LABs as seen in Scandinavian fermented milk drinks like villi, taette, fil, and skyr. These products are fermented by HePS-producing LABs such as *Lc. lactic* subsp. *lactis* or *Lc. lactic* subsp. *cremoris* that are mesophilic in nature. The HePS interacts with the milk proteins to form gel, resulting in the product's characteristic consistency [7].

15.6 Health benefits of exopolysaccharides

EPS from microorganisms have shown potential to influence their host health and impact playing various functional roles. Consumption of these diverse dietary EPSs effects the host in different ways, for example, they serve as immune modulators, they also act as a prebiotic and also have shown to have antioxidant and antitumor properties. The Health benefits of EPSs are dependent on the structural and compositional variations of HePSs and HoPSs which pertains to different health effects on host health. In general, HoPSs are more related to modulation of beneficial gut microbiota growth (prebiotic), whereas HePSs are effectors of host function like immunity enhancers and antioxidant activity [8].

15.6.1 Exopolysaccharides as prebiotic

The EPS from *Str. thermophilus* SFi12 and SFi 39 have been shown to be capable to biological breakdown, whereas the EPS produced by, *Lt. sakei 0-1, Lc. lactis* ssp. *cremoris B40, Lb. helveticus Lh59*, and *Str. thermophilus SFi20* all remain intact during gastric transit [79]. Such variation is due to huge diverse polymers found to be produced from LAB. Dextran derived from *W. cibaria* RBA12 showed high resistance to hydrolysis by artificial gastric juice, intestinal fluid, and α-amylase, thus enhancing probiotic Bifidobacteria's growth *Lactobacillus* spp. [23]. Similarly, HePS produced by *La. rhamnosus* GD-11 which is mainly composed of mannose could also stimulate the growth of *Bifidobacteria* [80].

15.6.2 Exopolysaccharides as immunomodulator

Many EPS produced by different genera like *Bifidobacterium, Lactobacillus, Lactococcus*, and *Leuconostoc* were found to play a relevant role in their immunomodulation capabilities [81,82]. Some EPS of higher molecular weight have displayed the ability to induce the production of specific cytokines and induce immune cells' activation including splenocytes, dendritic cells, and macrophages. Neutrally charged EPS with a large size would have a suppressor effect while negatively charged EPS with small size molecules were observed to act as stimulators of immune cells [83]. Thus the immunomodulation effect of EPSs depends on the composition, size, and charge of the HoPS and HePS. Some examples of immunomodulation are the activation of macrophages in mice and cytokine production [interferon-γ and IL-1] by phosphorylated HePSs produced by *Lc. lactis* subsp. *cremoris* KVS20 [84]. Stimulation of macrophage activity has also been demonstrated in vitro for HePSs produced by *La. paracasei* and *Lp. plantarum*. Both EPSs stimulated proinflammatory cytokine and macrophage proliferation, release increased proportionately to increasing dosages of the EPSs [85]. In contrast to the above findings, EPS produced by *La. rhamnosus* RW-9595M demonstrated antiinflammatory or immunosuppressive properties [86]. Similarly, the immunosuppressive effect was reported of EPS produced by a strain of *Bif. longum* [87].

15.6.3 Exopolysaccharides as antioxidant

It is theorized that EPS attenuate oxidative stress through ROS (Reactive Oxygen Species), DPPH (2,2-diphenyl-1-picrylhydrazyl)-free radical, Mn^{2+} and superoxide anion, iron scavenging and hydrogen peroxide degradation, myeloperoxidase inhibition and erythrocyte hemolysis inhibition [88,89]. In recent studies, the antiproliferative nature of certain LAB EPS has been examined highlighting the potential effect as anticancer. However, these results are preliminary [90]. Oxidative stress and damage have played a crucial role in cell transformation and cancer pathogenesis. It is believed that the antiproliferative potential of EPS could be related to their antioxidant activity [90].

15.6.4 Exopolysaccharides as antitumor agent

Antitumor activity of EPSs can be reflected by the ability to inhibit in vitro proliferation of tumor cell lines such as HT-29, Caco-2, HepG-2, BGC-823, PANC-1, MCF7, and EAC [91]. According to Li et al., the antiproliferative potential of HePS fractions produced by *Lb.*

helveticus MB2-1 on BGC-823 gastric cancer cells in vitro has shown that the EPS fraction with the highest antiproliferative activity has the strongest free-radical scavenging activity [92]. According to Wang et al., cell-bound EPS (c-EPS) of *Lp. plantarum* 70810 displayed significant inhibition was observed against HepG-2, BGC-823 and especially HT-29 cell lines when compared to the control antitumor compound fluorouracil [91].

15.7 Conclusions and perspectives

EPS from various microorganisms have key role in the development of several functional foods. Novel functional food products can be developed with the incorporation of EPS either as an additive or by allowing its in situ production. The numerous functional properties of EPS ranging from hydrocolloids to prebiotic additives make them a key point of focus in food microbiology and food technology. The diversity seen in HePS can result in novel properties of EPS that would be exploited in the food industry. With the advancement of bioprocess engineering and metabolic engineering, the expected low yield of HePS could be improved to meet the industry demands. Advancement in molecular microbiology, synthetic biology, and metagenomics can lead to the discovery of new EPS from different sources. Future research on EPS should focus on understanding EPS biosynthesis in food matrices, key factors responsible for its yield, and exploring new ways to incorporate EPS as a safe and natural additive.

Acknowledgments

R.B. acknowledges the research associate fellowship provided by Council of Scientific and Industrial Research (CSIR). K.R. acknowledges Department of Biotechnology, Govt. of India for providing JRF fellowship. The authors acknowledge the contributions of Dr. SVN Vijayendra, Senior Principal Scientist, Dept. of Microbiology and Fermentation Technology for the critical comments and suggestions. Authors acknowledge kind approval of the Director, CSIR-CFTRI Mysuru for this work. This manuscript has been approved by planning, monitoring and coordination (PMC) section, CFTRI-CSIR vide approval number PMC/2020-21/64.

References

[1] M.G. Llamas-Arriba, A.M. Hernández-Alcántara, A. Yépez, R. Aznar, M.T. Dueñas, P. López, Functional and nutritious beverages produced by lactic acid bacteria, Nutrients in Beverages, Academic Press, 2019, pp. 419–465.

[2] J.P. Tamang, Health Benefits of Fermented Foods and Beverages, CRC Press, Taylor and Francis Group, New York, NY, 2015.

[3] M. Farhad, K. Kailasapathy, J. Tamang, Health aspects of fermented foods, in: J.P. Tamang, K. Kailasapathy (Eds.), Fermented Foods and Beverages of the World, CRC Press, Taylor and Francis Group, New York, NY, 2010, pp. 391–414.

[4] FAO/WHO: Report on joint FAO/WHO expert consultation on evaluation of health and nutritional properties of probiotics in food including powder milk with live lactic acid bacteria. Available from: <http://www.fao.org/es/ESN/Probio/probio.htm>, 2001.

[5] G.R. Gibson, M.B. Roberfroid, Dietary modulation of the human colonic microbiota: introducing the concept of prebiotics, The Journal of Nutrition 125 (6) (1995) 1401–1412.

[6] R. Baruah, D. Das, A. Goyal, Heteropolysaccharides from lactic acid bacteria: current trends and applications, Journal of Probiotics and Health 4 (141) (2016) 2.

[7] L. De Vuyst, B. Degeest, Heteropolysaccharides from lactic acid bacteria, FEMS Microbiology Reviews 23 (2) (1999) 153–177.

[8] K.M. Lynch, E. Zannini, A. Coffey, E.K. Arendt, Lactic acid bacteria exopolysaccharides in foods and beverages: isolation, properties, characterization, and health benefits, Annual Review of Food Science and Technology 9 (2018) 155–176.

[9] Y. Xu, Y. Cui, F. Yue, L. Liu, Y. Shan, B. Liu, et al., Exopolysaccharides produced by lactic acid bacteria and bifidobacteria: structures, physiochemical functions and applications in the food industry, Food Hydrocolloids 94 (2019) 475–499.

[10] M.L. Werning, S. Notararigo, M. Nácher, P. Fernández de Palencia, R. Aznar, P. López, Biosynthesis, purification and biotechnological use of exopolysaccharides produced by lactic acid bacteria, Food Additives (2012) 83–114.

[11] S. Patel, A. Majumder, A. Goyal, Potentials of exopolysaccharides from lactic acid bacteria, Indian Journal of Microbiology 52 (1) (2012) 3–12.

[12] L.D. Vuyst, F.D. Vin, Exopolysaccharides from lactic acid bacteria, Comprehensive Glycoscience 23 (2) (2007) 477–519.

[13] F. Mozzi, F. Vaningelgem, E.M. Hébert, R. Van der Meulen, M.R.F. Moreno, G.F. de Valdez, et al., Diversity of heteropolysaccharide-producing lactic acid bacterium strains and their biopolymers, Applied and Environmental Microbiology 72 (6) (2006) 4431–4435.

[14] S. Badel, T. Bernardi, P. Michaud, New perspectives for lactobacilli exopolysaccharides, Biotechnology Advances 29 (1) (2011) 54–66.

[15] N. Salazar, M. Gueimonde, C.G. de los Reyes-Gavilan, P. Ruas-Madiedo, Exopolysaccharides produced by lactic acid bacteria and bifidobacteria as fermentable substrates by the intestinal microbiota, Critical Reviews in Food Science and Nutrition 56 (9) (2016) 1440–1453.

[16] P. Ruas-Madiedo, N. Salazar, C.G. De los Reyes-Gavilan, Biosynthesis and chemical composition of exopolysaccharides produced by lactic acid bacteria, Bacterial Polysaccharides: Current Innovations and Future Trends, Caister Academic Press, 2009, pp. 279–310.

[17] M. Tieking, M.A. Ehrmann, R.F. Vogel, M.G. Gänzle, Molecular and functional characterization of a levansucrase from the sourdough isolate *Lactobacillus sanfranciscensis* TMW 1.392, Applied Microbiology and Biotechnology 66 (6) (2005) 655–663.

[18] D. Das, A. Goyal, Characterization and biocompatibility of glucan: a safe food additive from probiotic *Lactobacillus plantarum* DM5, Journal of the Science of Food and Agriculture 94 (4) (2014) 683–690.

[19] Z. Moradi, N. Kalanpour, Kefiran, a branched polysaccharide: preparation, properties and applications: a review, Carbohydrate Polymers 223 (2019) 115100.

[20] J. Wang, X. Zhao, Z. Tian, Y. Yang, Z. Yang, Characterization of an exopolysaccharide produced by *Lactobacillus plantarum* YW11 isolated from Tibet Kefir, Carbohydrate Polymers 125 (2015) 16–25.

[21] P.B. Devi, D. Kavitake, P.H. Shetty, Physico-chemical characterization of galactan exopolysaccharide produced by *Weissella confusa* KR780676, International Journal of Biological Macromolecules 93 (2016) 822–828.

[22] A.K. Lakra, L. Domdi, Y.M. Tilwani, V. Arul, Physicochemical and functional characterization of mannan exopolysaccharide from *Weissella confusa* MD1 with bioactivities, International Journal of Biological Macromolecules 143 (2020) 797–805.

[23] R. Baruah, N.H. Maina, K. Katina, R. Juvonen, A. Goyal, Functional food applications of dextran from *Weissella cibaria* RBA12 from pummelo (*Citrus maxima*), International Journal of Food Microbiology 242 (2017) 124–131.

[24] Y. Yang, F. Feng, Q. Zhou, F. Zhao, R. Du, Z. Zhou, et al., Isolation, purification and characterization of exopolysaccharide produced by *Leuconostoc pseudomesenteroides* YF32 from soybean paste, International Journal of Biological macromolecules 114 (2018) 529–535.

[25] C. Matsuzaki, C. Takagaki, Y. Tomabechi, L.S. Forsberg, C. Heiss, P. Azadi, et al., Structural characterization of the immunostimulatory exopolysaccharide produced by *Leuconostoc mesenteroides* strain NTM048, Carbohydrate Research 448 (2017) 95–102.

[26] R. Shukla, A. Goyal, Novel dextran from *Pediococcus pentosaceus* CRAG3 isolated from fermented cucumber with anti-cancer properties, International Journal of Biological Macromolecules 62 (2013) 352–357.

[27] Y.R. Song, D.Y. Jeong, Y.S. Cha, S.H. Baik, Exopolysaccharide produced by *Pediococcus acidilactici* M76 isolated from the Korean traditional rice wine, Makgeolli, Journal of Microbiology and Biotechnology 23 (5) (2013) 681–688.

[28] M. Dimopoulou, M. Vuillemin, H. Campbell-Sills, P.M. Lucas, P. Ballestra, C. Miot-Sertier, et al., Exopolysaccharide (EPS) synthesis by *Oenococcus oeni*: from genes to phenotypes, PLoS One 9 (6) (2014).

[29] F. García-Ochoa, V.E. Santos, J.A. Casas, E. Gómez, Xanthan gum: production, recovery, and properties, Biotechnology Advances 18 (7) (2000) 549–579.

[30] V.D. Prajapati, G.K. Jani, B.S. Zala, T.A. Khutliwala, An insight into the emerging exopolysaccharide gellan gum as a novel polymer, Carbohydrate Polymers 93 (2) (2013) 670–678.

[31] J. Zheng, S. Wittouck, E. Salvetti, C.M. Franz, H.M. Harris, P. Mattarelli, et al., A taxonomic note on the genus *Lactobacillus*: description of 23 novel genera, emended description of the genus *Lactobacillus* Beijerinck 1901, and union of *Lactobacillaceae* and *Leuconostocaceae*, International Journal of Systematic and Evolutionary Microbiology 70 (4) (2020) 2782–2858.

[32] M. Korakli, A. Rossmann, M.G. Gänzle, R.F. Vogel, Sucrose metabolism and exopolysaccharide production in wheat and rye sourdoughs by *Lactobacillus sanfranciscensis*, Journal of Agricultural and Food Chemistry 49 (11) (2001) 5194–5200.

[33] R.M. Abedin, A.M. El-Borai, M.A. Shall, S.A. El-Assar, Optimization and statistical evaluation of medium components affecting dextran and dextransucrase production by *Lactobacillus acidophilus* ST76480.01, Life Science Journal 10 (1) (2013) 1346–1353.

[34] D. Das, R. Baruah, A. Goyal, A food additive with prebiotic properties of an α-d-glucan from *Lactobacillus plantarum* DM5, International Journal of Biological Macromolecules 69 (2014) 20–26.

[35] M.D. Collins, J. Samelis, J. Metaxopoulos, S. Wallbanks, Taxonomic studies on some Leuconostoc-like organisms from fermented sausages: description of a new genus *Weissella* for the *Leuconostoc paramesenteroides* group of species, Journal of Applied Bacteriology 75 (6) (1993) 595–603.

[36] J. Björkroth, L.M. Dicks, A. Endo, The genus *Weissella*, Lactic Acid Bacteria: Biodiversity and Taxonomy (, Blackwell Science Publications, Oxford, 2014, pp. 417–428.

[37] S.K. Malang, N.H. Maina, C. Schwab, M. Tenkanen, C. Lacroix, Characterization of exopolysaccharide and ropy capsular polysaccharide formation by *Weissella*, Food Microbiology 46 (2015) 418–427.

[38] R. Di Cagno, M. De Angelis, A. Limitone, F. Minervini, P. Carnevali, A. Corsetti, et al., Glucan and fructan production by sourdough *Weissella cibaria* and *Lactobacillus plantarum*, Journal of Agricultural and Food Chemistry 54 (26) (2006) 9873–9881.

[39] G. Lacaze, M. Wick, S. Cappelle, Emerging fermentation technologies: development of novel sourdoughs, Food Microbiology 24 (2) (2007) 155–160.

[40] K. Katina, N.H. Maina, R. Juvonen, L. Flander, L. Johansson, L. Virkki, et al., In situ production and analysis of *Weissella confusa* dextran in wheat sourdough, Food Microbiology 26 (7) (2009) 734–743.

[41] A. Wolter, A.S. Hager, E. Zannini, S. Galle, M.G. Gänzle, D.M. Waters, et al., Evaluation of exopolysaccharide producing *Weissella cibaria* MG1 strain for the production of sourdough from various flours, Food Microbiology 37 (2014) 44–50.

[42] R. Juvonen, K. Honkapää, N.H. Maina, Q. Shi, K. Viljanen, H. Maaheimo, et al., The impact of fermentation with exopolysaccharide producing lactic acid bacteria on rheological, chemical and sensory properties of pureed carrots (*Daucus carota* L.), International Journal of Food Microbiology 207 (2015) 109–118.

[43] R. Baruah, B. Deka, A. Goyal, Purification and characterization of dextransucrase from *Weissella cibaria* RBA12 and its application in *in-vitro* synthesis of prebiotic oligosaccharides in mango and pineapple juices, LWT—Food Science and Technology 84 (2017) 449–456.

[44] P.M. Rolim, Y. Hu, M.G. Gänzle, Sensory analysis of juice blend containing isomalto-oligosaccharides produced by fermentation with *Weissella cibaria*, Food Research International 124 (2019) 86–92.

[45] M. Kumari, R. Kumar, D. Singh, S. Bhatt, M. Gupta, Physiological and genomic characterization of an exopolysaccharide-producing *Weissella cibaria* CH2 from cheese of the western Himalayas, Food Bioscience 35 (2020) 100570.

[46] J.C. Ogier, E. Casalta, C. Farrokh, A. Saïhi, Safety assessment of dairy microorganisms: the *Leuconostoc* genus, International Journal of Food Microbiology 126 (3) (2008) 286–290.

[47] M. Naessens, A.N. Cerdobbel, W. Soetaert, E.J. Vandamme, *Leuconostoc* dextransucrase and dextran: production, properties and applications, Journal of Chemical Technology and Biotechnology 80 (8) (2005) 845–860.

[48] R.K. Purama, P. Goswami, A.T. Khan, A. Goyal, Structural analysis and properties of dextran produced by *Leuconostoc mesenteroides* NRRL B-640, Carbohydrate Polymers 76 (1) (2009) 30–35.

[49] I. Torres-Rodríguez, M.E. Rodríguez-Alegría, A. Miranda-Molina, M. Giles-Gómez, R.C. Morales, A. López-Munguía, et al., Screening and characterization of extracellular polysaccharides produced by *Leuconostoc kimchii* isolated from traditional fermented pulque beverage, Springer Plus 3 (1) (2014) 583.

[50] S.V.N. Vijayendra, G. Palanivel, S. Mahadevamma, R.N. Tharanathan, Physico-chemical characterization of an exopolysaccharide produced by a non-ropy strain of *Leuconostoc* sp. CFR 2181 isolated from dahi, an Indian traditional lactic fermented milk product, Carbohydrate Polymers 72 (2) (2008) 300–307.

[51] W. Liu, H. Pang, H. Zhang, Y. Cai, Lactic Acid Bacteria: Fundamentals and Practice, vol. 10, Springer, Amsterdam, 2014, pp. 978–994.

[52] M. Raccach, Pediococci and biotechnology, CRC Critical Reviews in Microbiology 14 (4) (1987) 291–309.

[53] P.F. de Palencia, M.L. Werning, E. Sierra-Filardi, M.T. Dueñas, A. Irastorza, A.L. Corbí, et al., Probiotic properties of the 2-substituted (1, 3)-β-D-glucan-producing bacterium *Pediococcus parvulus* 2.6, Applied Environmental Microbiology 75 (14) (2009) 4887–4891.

[54] A. Pérez-Ramos, M.L. Mohedano, P. López, G. Spano, D. Fiocco, P. Russo, et al., *In-situ* β-glucan fortification of cereal-based matrices by *Pediococcus parvulus* 2.6: technological aspects and prebiotic potential, International Journal of Molecular Sciences 18 (7) (2017) 1588.

[55] P. Semjonovs, P. Zikmanis, Evaluation of novel lactose-positive and exopolysaccharide-producing strain of *Pediococcus pentosaceus* for fermented foods, European Food Research and Technology 227 (3) (2008) 851–856.

[56] M. Ayyash, B. Abu-Jdayil, A. Olaimat, G. Esposito, P. Itsaranuwat, T. Osaili, et al., Physicochemical, bioactive and rheological properties of an exopolysaccharide produced by a probiotic *Pediococcus pentosaceus* M41, Carbohydrate Polymers 229 (2020) 115462.

[57] H. Koo, J. Xiao, M.I. Klein, J.G. Jeon, Exopolysaccharides produced by *Streptococcus mutans* glucosyltransferases modulates the establishment of microcolonies within multispecies biofilms, Journal of Bacteriology 192 (12) (2010) 3024–3032.

[58] D. Low, J.A. Ahlgren, D. Horne, D.J. McMahon, C.J. Oberg, J.R. Broadbent, Role of *Streptococcus thermophilus* MR-1C capsular exopolysaccharide in cheese moisture retention, Applied Environmental Microbiology 64 (6) (1998) 2147–2151.

[59] U. Purwandari, N.P. Shah, T. Vasiljevic, Effects of exopolysaccharide-producing strains of *Streptococcus thermophilus* on technological and rheological properties of set-type yoghurt, International Dairy Journal 17 (11) (2007) 1344–1352.

[60] R.E. Lobo, M.I. Gómez, G.F. de Valdez, M.I. Torino, Physicochemical and antioxidant properties of a gastroprotective exopolysaccharide produced by *Streptococcus thermophilus* CRL1190, Food Hydrocolloids 96 (2019) 625–633.

[61] R.V. Kranenburg, J.D. Marugg, I.I. Van Swam, N.J. Willem, W.M. De Vos, Molecular characterization of the plasmid-encoded eps gene cluster essential for exopolysaccharide biosynthesis in *Lactococcus lactis*, Molecular Microbiology 24 (2) (1997) 387–397.

[62] V.M. Marshall, E.N. Cowie, R.S. Moreton, Analysis and production of two exopolysaccharides from *Lactococcus lactis* subsp. *cremoris* LC330, Journal of Dairy Research 62 (4) (1995) 621–628.

[63] J. Feng, Z. Cai, Y. Chen, H. Zhu, X. Chang, X. Wang, et al., Effects of an exopolysaccharide from *Lactococcus lactis* Z-2 on innate immune response, antioxidant activity, and disease resistance against *Aeromonas hydrophila* in *Cyprinus carpio* L, Fish and Shellfish Immunology 98 (2020) 324–333.

[64] N. Castro-Bravo, B. Sánchez, A. Margolles, P. Ruas-Madiedo, Biological activities and applications of bifidobacterial exopolysaccharides: from the bacteria and host perspective, The Bifidobacteria and Related Organisms, Academic Press, 2018, pp. 177–193.

[65] J. Schmid, V. Sieber, Fermentative production of microbial exopolysaccharides, Bioprocessing for Biomolecules Production, Wiley, 2019, pp. 145–166.

[66] M. Mcintosh, B.A. Stone, V.A. Stanisich, Curdlan and other bacterial (1→3)-β-D-glucans, Applied Microbiology and Biotechnology 68 (2) (2005) 163–173.

[67] K.C. Cheng, A. Demirci, J.M. Catchmark, Pullulan: biosynthesis, production, and applications, Applied Microbiology and Biotechnology 92 (1) (2011) 29.

[68] S. Galle, E.K. Arendt, Exopolysaccharides from sourdough lactic acid bacteria, Critical Reviews in Food Science and Nutrition 54 (7) (2014) 891–901.

[69] Y. Wang, P. Sorvali, A. Laitila, N.H. Maina, R. Coda, K. Katina, Dextran produced in situ as a tool to improve the quality of wheat-faba bean composite bread, Food Hydrocolloids 84 (2018) 396–405.

[70] Y. Wang, A. Trani, A. Knaapila, S. Hietala, R. Coda, K. Katina, et al., The effect of in situ produced dextran on flavour and texture perception of wholegrain sorghum bread, Food Hydrocolloids 106 (2020) 105913.

[71] D.M. Folkenberg, P. Dejmek, A. Skriver, R. Ipsen, Interactions between EPS-producing *Streptococcus thermophilus* strains in mixed yoghurt cultures, Journal of Dairy Research 73 (4) (2006) 385–393.

[72] P.H.P. Prasanna, A.S. Grandison, D. Charalampopoulos, Microbiological, chemical and rheological properties of low fat set yoghurt produced with exopolysaccharide (EPS) producing *Bifidobacterium* strains, Food Research International 51 (1) (2013) 15–22.

[73] X.W. Li, S. Lv, T.T. Shi, K. Liu, Q.M. Li, L.H. Pan, et al., Exopolysaccharides from yoghurt fermented by *Lactobacillus paracasei*: production, purification and its binding to sodium caseinate, Food Hydrocolloids 102 (2020) 105635.

[74] P.S. Rimada, A.G. Abraham, Kefiran improves rheological properties of glucono-δ-lactone induced skim milk gels, International Dairy Journal 16 (1) (2006) 33–39.

[75] M. Ghasemlou, F. Khodaiyan, A. Oromiehie, M.S. Yarmand, Development and characterisation of a new biodegradable edible film made from kefiran, an exopolysaccharide obtained from kefir grains, Food Chemistry 127 (4) (2011) 1496–1502.

[76] I. Kajala, J. Mäkelä, R. Coda, S. Shukla, Q. Shi, N.H. Maina, et al., Rye bran as fermentation matrix boosts in situ dextran production by *Weissella confusa* compared to wheat bran, Applied Microbiology and Biotechnology 100 (8) (2016) 3499–3510.

[77] K.M. Lynch, P.L. McSweeney, E.K. Arendt, T. Uniacke-Lowe, S. Galle, A. Coffey, Isolation and characterisation of exopolysaccharide-producing *Weissella* and *Lactobacillus* and their application as adjunct cultures in Cheddar cheese, International Dairy Journal 34 (1) (2014) 125–134.

[78] J. Hilbig, J. Gisder, R.M. Prechtl, K. Herrmann, J. Weiss, M. Loeffler, Influence of exopolysaccharide-producing lactic acid bacteria on the spreadability of fat-reduced raw fermented sausages (Teewurst), Food Hydrocolloids 93 (2019) 422–431.

[79] H.J. Ruijssenaars, F. Stingele, S. Hartmans, Biodegradability of food-associated extracellular polysaccharides, Current Microbiology 40 (3) (2000) 194–199.

[80] H. Sarikaya, B. Aslim, Z. Yuksekdag, Assessment of anti-biofilm activity and bifidogenic growth stimulator (BGS) effect of lyophilized exopolysaccharides (l-EPSs) from *Lactobacilli* strains, International Journal of Food Properties 20 (2) (2017) 362–371.

[81] M. Ciszek-Lenda, B. Nowak, M. Sróttek, A. Gamian, J. Marcinkiewicz, Immunoregulatory potential of exopolysaccharide from *Lactobacillus rhamnosus* KL37. Effects on the production of inflammatory mediators by mouse macrophages, International Journal of Experimental Pathology 92 (2011) 382–391.

[82] S. Fanning, L.J. Hall, D. van Sinderen, *Bifidobacterium breve* UCC2003 surface exopolysaccharide production is a beneficial trait mediating commensal-host interaction through immune modulation and pathogen protection, Gut Microbes 3 (2012) 420–425.

[83] C. Hidalgo-Cantabrana, P. López, M. Gueimonde, G. Clara, A. Suárez, A. Margolles, et al., Immune modulation capability of exopolysaccharides synthesised by lactic acid bacteria and bifidobacteria, Probiotics and Antimicrobial Proteins 4 (4) (2012) 227–237.

[84] H. Kitazawa, T. Itoh, Y. Tomioka, M. Mizugaki, T. Yamaguchi, Induction of IFN-γ and IL-1α production in macrophages stimulated with phosphopolysaccharide produced by *Lactococcus lactis* ssp. *cremoris*, International Journal Food Microbiology 31 (1996) 99–106.

[85] C.F. Liu, K.C. Tseng, S.S. Chiang, B.H. Lee, W.H. Hsu, T.M. Pan, Immunomodulatory and antioxidant potential of *Lactobacillus* exopolysaccharides, Journal of the Science of Food and Agriculture 91 (12) (2011) 2284–2291.

[86] C. Bleau, A. Monges, K. Rashidan, J.P. Laverdure, M. Lacroix, M.R. Van Calsteren, et al., Intermediate chains of exopolysaccharides from *Lactobacillus rhamnosus* RW-9595M increase IL-10 production by macrophages, Journal of Applied Microbiology 108 (2) (2010) 666–675.

[87] M.H. Wu, T.M. Pan, Y.J. Wu, S.J. Chang, M.S. Chang, C.Y. Hu, Exopolysaccharide activities from probiotic bifidobacterium: immunomodulatory effects (on J774A. 1 macrophages) and antimicrobial properties, International Journal of Food Microbiology 144 (1) (2010) 104–110.

[88] R. Xu, Q. Shen, X. Ding, W. Gao, P. Li, Chemical characterization and antioxidant activity of an exopolysaccharide fraction isolated from *Bifidobacterium animalis* RH, European Food Research and Technology 232 (2) (2011) 231–240.

[89] S.S. Choi, Y. Kim, K.S. Han, S. You, S. Oh, S.H. Kim, Effects of *Lactobacillus* strains on cancer cell proliferation and oxidative stress in vitro, Letters in Applied Microbiology 42 (5) (2006) 452–458.

[90] J. Mates, Effects of antioxidant enzymes in the molecular control of reactive oxygen species, Toxicology 153 (2000) 83–104.

[91] K. Wang, W. Li, X. Rui, X. Chen, M. Jiang, M. Dong, Characterization of a novel exopolysaccharide with antitumor activity from *Lactobacillus plantarum* 70810, International Journal of Biological Macromolecules 63 (2014) 133–139.

[92] W. Li, J. Ji, X. Chen, M. Jiang, X. Rui, M. Dong, Structural elucidation and antioxidant activities of exopolysaccharides from *Lactobacillus helveticus* MB2-1, Carbohydrate Polymers 102 (2014) 351–359.

Further reading

S. Galle, C. Schwab, E. Arendt, M. Gänzle, Exopolysaccharide-forming *Weissella* strains as starter cultures for sorghum and wheat sourdoughs, Journal of Agricultural and Food Chemistry 58 (9) (2010) 5834–5841.

R. Prete, M.K. Alam, G. Perpetuini, C. Perla, P. Pittia, A. Corsetti, Lactic Acid Bacteria Exopolysaccharides Producers: A Sustainable Tool for Functional Foods, Foods 10 (7) (2021) 1653, doi:10.3390/foods10071653.

16

Microbial metabolites beneficial in regulation of obesity

Khushboo[1], Kashyap Kumar Dubey[1,2]

[1]BIOPROCESS ENGINEERING LABORATORY, DEPARTMENT OF BIOTECHNOLOGY, CENTRAL UNIVERSITY OF HARYANA, MAHENDERGARH, INDIA [2]BIOPROCESS ENGINEERING LABORATORY, SCHOOL OF BIOTECHNOLOGY, JAWAHARLAL NEHRU UNIVERSITY, NEW DELHI, INDIA

16.1 Introduction

Obesity and other diseases associated with it are significant issues for public health at the global level. According to data provided by WHO in 2014, 13% of the adults were obese, out of which 11% were men and 15% were women. In addition to that, obesity is dominating day by day as it has been doubled from 1980 to 2014 [1]. Excessive accumulation of fat in adipose tissues due to high energy intake compared to energy expenditure leads to obesity. Weight loss at moderate rate along with an integrated treatment helps to resist the obesity. The integrated treatment includes intake of low-calorie diet and physical exercise and drug therapy to control hunger and appetite [2,3].

The environmental and genetic factors significantly alter gut microbiota's behavior, which leads to the development of metabolic disorders [4–10]. All the microorganisms living in gastrointestinal tract (GI) define gut microbiota. These microorganisms perform the process of digestion as well as vitamin synthesis along with other metabolic processes. The GI tract microbes help in the production of energy from diet and regulate the configuration of fatty acid tissues [4,11,12]. However, gut microbes' role behind the obesity is not clear due to the high diversity of gut microbes [7,13]. In spite of this, these microbes possess the ability to treat obesity and controlled diet and exercise [14–17]. Gut microbiota acts as a central regulator of the organism's diet as they convert the intake food into absorbable compounds. The members of gut microbiota also produce some vitamins as well as essential metabolites, which is not produced by the host. Alteration in the gut microbiome framework has a direct effect on the regulation of biochemical and metabolic pathways. Various metabolites produced by gut microbes maintain the host organism's immunological processes, homeostasis, and neurobiology [18–20]. In this condition, the storage capacity of adipose tissues to store triglycerides goes beyond the normal limit. As a result, the overflow of lipids occurs in blood circulation. That overflow lipid adjusted in nonadipose tissues, that is, liver and skeletal muscles [21]. Inflammation in adipose tissues triggers the production of pro-inflammatory adipokines [22]. Along with that, the gut microbiota is also responsible for obesity-related inflammation [23–25]. Increased availability of dietary fibers to microbes present in the colon region helps to modulate the gut microbiome for the prevention of metabolic disorders. These dietary fibers are converted into short-chain fatty acids (SCFAs), propionate, butyrate, acetate, succinate [26,27]. Researchers demonstrated that SCFAs and succinate help in the control of body weight, treatment of obesity-related inflammation, and insulin resistance [28–38].

16.2 Composition of gut microbiota

Trillions of microbes are there that reside in the human body for the whole life. At the time of pregnancy, the GI tract of an infant is completely free of microbes. The first exposure of the infant to the microbes is of maternal vaginal microbes, which takes place during the normal delivery [39], while the infants born through cesarean delivery get exposure to bacteria present at maternal skin with altered gut microbiota. Babies fed with breast milk possess different microbes in their gut, as compared to babies fed through formula-based milk. Solid food introduces another shift of microbes in the GI tract of babies [40]. That composition of the gut microbiome remains the same until old age. Bacteria are the gut microbiome's main members followed by viruses, fungi, and other microbes are also present [7]. Every individual possesses its unique microbiota. However, the gut microbiota is composed of four phyla, that is, Firmicutes, Bacteroidetes, Actinobacteria, and Proteobacteria [12]. The highest number of bacteria, that is, 10^{11} bacteria/g of intestinal content are present in large intestine. Approximately, 10^{12} bacteria are present in the mouth, whereas, 10^8–10^9 bacteria in Ileum [41]. Jejunum contains nearly 10^6 bacteria, while the stomachs possess the least number of bacteria (approximately 10^4). The next-generation sequencing technique has majorly contributed in the understanding of the gut microbes [42].

Various factors like diet, disease, genetic factors affect the composition of the gut microbiome. The use of antibiotics results in the reduction of bacterial diversity in the gut. Researchers demonstrated that vancomycin and gentamycin for the purpose of treatment result in major weight gain [43]. Recovery of normal gut microbiota after the treatment mainly depends on the type of antibiotic and its spectrum [44]. Studies suggest that board and wide spectrum antibiotics like clindamycin might have longer effects [45]. Disruption of normal microbial flora of gut promotes the drug resistance in virulent species via facilitating antibiotic resistance genes transfer. However, the main factor that decides the diversity of gut microbial flora is diet [46–50]. Studies suggest that diet change results in 57% alteration in gut microbiota while genetic variations account for 12% alterations only [51]. Probiotics and prebiotics are the main dietary supplements endorsed to control the diversity and activity of gut microbes. Probiotics are consisting of nonpathogenic microbes, while prebiotics are composed of fermented dietary fibers. Researchers demonstrated that *Lactobacillus reuteri* strain possesses the ability to decrease the level of low-density lipoprotein cholesterol in hypercholesterolemic individuals [52]. Dietary fibers of prebiotics significantly affect the bacterial diversity in the colon region to improve host individuals' health. Prebiotics like lactulose, inulin mainly target bifidobateria and lactobacilli, which themselves utilized as probiotics. Recent studies suggest the combined use of prebiotics and probiotics termed as synbiotics to reduce the obesity [53].

16.3 Obesity and lipid metabolism

Metabolism of lipid is essentially balanced to retain the homeostasis [54–56]. Disbalance in the metabolic process of lipid results in obesity, which in turn promotes various other diseases like diabetes, atherosclerosis, cardiovascular problems, etc. That's why the regulation of lipid metabolism could be a preventive measure for these diseases. Researchers are identifying the key enzymes involved in lipids' metabolic process as they might be a suitable target for the therapeutic purpose of obesity and other related diseases [57]. The best approach is to inhibit the digestion and absorption of nutrients from decreasing the uptake of energy via the GI tract [58]. Inhibition of dietary lipids digestion is a better approach as major unwanted calories are supplied through lipids mainly.

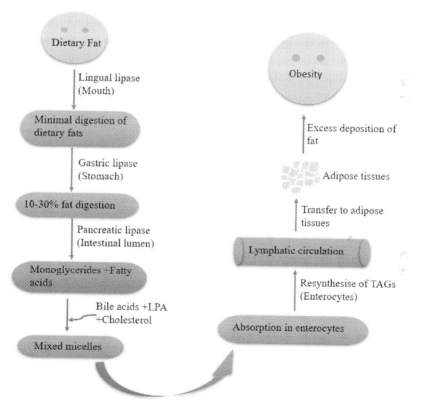

FIGURE 16–1 Metabolism of dietary lipids in human beings. *TAG*, Triacylglycerol; *LPA*, lysophosphatidic acid.

The digestion of lipids is performed by the enzymes termed as lipases (Fig. 16–1). Various types of lipases like lingual, gastric, pancreatic, hepatic lipases are there in human beings. Pancreatic lipase is the main player for the digestion of dietary lipids. Digestion of TAGs yields β-monoglycerides as well as long-chain fatty acids (saturated and polyunsaturated) as pancreatic lipase removes the α and α' positioned fatty acids [57]. Pancreatic lipase hydrolyzed 50%–70% of total lipids intake. The active site of pancreatic lipase in humans possesses His-263, Asp-176, and Ser-152 which form a triad. The activity of enzymes gets inhibited after chemical modification of serine residue present at 152 position, indicating the importance of Ser-152 to perform its normal function [59].

Other type of lipases also helps in digestion of lipids. Lingual lipase helps in digestion of 1/3rd dietary fats and is secreted by serous gland. Gastric lipase digests 10%–40% of fat taken in diet. Both these enzymes' activity reduces the impact of inhibition of fat absorption that was targeted via inhibiting the activity of pancreatic lipase.

16.4 Pancreatic lipase

Lipases are the enzymes involved in the digestion of triacylglycerol and phospholipases present in dietary fat. Pancreatic lipase is the exocrine enzyme produced by acinar cells of the pancreas. Pancreatic lipase is the part of pancreatic juice which plays an essential role in the digestion of lipids in the intestinal lumen. Hydrolysis of dietary fat is essential for their assimilation in the body [60,61]. The enzyme shows higher activity at pH 7–7.5 and got denatured below pH 3. Pancreatic lipase requires a cofactor, that is, colipase to perform its function. Colipase is a pancreatic protein, which helps in relieving the lipase-substrate complex. In addition to that, colipase stabilizes the active conformation of pancreatic lipase [62,63]. The catalytic site of the enzyme is conserved and is present at N-terminal. The enzyme loses its activity after the chemical alteration of serine 152 present at the edge of N/C terminal. The enzyme exhibits strong activity against water insoluble compounds, mainly triacylglycerols and binds nonspecifically to the nonpolar compounds. Pancreatic lipase possesses strong affection toward phospholipids and cholesterol [64].

Dietary fat consists of many lipid forms such as nonpolar lipids and polar phospholipids. Dietary lipids are converted into diacylglycerols (DAGs) and fatty acids by gastric lipase present in the stomach during digestion. DAGs and fatty acids promote the emulsification of dietary lipids in the stomach. In the small intestine, bile acids emulsify the lipids to enhance pancreatic lipase's lipolytic activity [64,65]. The action of pancreatic lipase over TAGs results in the formation of 2-monoacylglycerol and fatty acids for the uptake by enterocytes [66]. Pancreatic lipase is regioselective in nature as it hydrolyzes the ester bonds present at sn-1 and sn-3 positions only [67]. In today's lifestyle, obesity is the major health problem and is responsible for many other clinical disorders like diabetes mellitus, hyperlipidemia, and cardiovascular disease. Physical exercise and low-calorie diet are the best approach to losing weight, but drugs are recommended for fast results. Antiobesity drugs may act by reducing appetite or changing the rate of metabolism [68]. Antilipase inhibitor act on lipases and impedes their function [69].

16.5 Antilipid effect of Cineromycin B

Accumulation of lipids in adipose tissues results in the expansion of adipocytes in number and size via differentiation of preadipocytes to adipocytes [70]. 3T3-L1 cells are generally considered as preadipocytes to study the differentiation of adipocytes [71]. Researchers demonstrated that various genes and proteins are involved in the differentiation of fat cells [72]. Molecular study lying behind adipocytes' formation has identified some regulatory factors like members of the GATA binding protein family (GATA), and Krüppel-like factor (KLF) family. Members of GATA family acts as antiadipogenic factor by directly binding with PPARγ promoter [73] while members of KLF family are involved in cell proliferation and differentiation [74,75]. KLF2, 3, and 7 act as negative regulators as their overexpression inhibits adipocyte differentiation while KLF4, 5, 6, and 15 are considered positive regulators during differentiation [76–78].

Cineromycin B, an antiadipogenic compound isolated from *Streptomyces cinerochromogenes*, affects the expression of KLF family members. Further studies revealed that Cineromycin B mainly target KLF2 and 3 to inhibit the adipocytes differentiation. Matsuo et al. (2015) demonstrated that Cineromycin B affects the expression of KLF2 and 3 mRNA, that is, upregulating KLF2 and 3 to inhibit the differentiation of adipose tissues. Treatment with siRNA of KLF2 and 3 does not completely erase the inhibitory effect of Cineromycin B, suggesting the involvement of other factors in Cineromycin B's functioning. Thus Cineromycin B might be utilized for the development of antiobesity drug [79].

16.6 Gut–brain axis

Obesity and many other diseases associated with it are the major health problems worldwide [80]. As the prevalence of obesity is very high, new gut microbiota strategies have been investigated to control it. Behind the physic of any individual gut microbes are the deciding factors as they have full command over the metabolic process, appetite, adiposity, food reward signals [81,82]. In obese persons, gut microbes enhance the concentration of fermentation enzymes and nutrient transporters to extract more energy from diet compared to lean and thin individual's gut microbes [83]. The altered gut microbiome in obese individual could increase the permeability of intestinal membrane for the lipopolysaccharides (LPS) synthesized by the bacteria, that is, lead to endotoxaemia [23]. Absorption of LPS results in insulin resistance as it triggers the production of inflammatory cytokines via activating Toll-like receptor 4 in white adipose tissues [84].

Gut microbes are considered as the second brain because they could control our diet by releasing neurotransmitters like epinephrine, norepinephrine, dopamine, serotonin, etc. as per studies. The metabolites synthesized by gut microbes create an axis of communication between gut and brain to control the appetite [82]. These microbes can directly control the eating behavior of any individual by affecting the CNS via gut–brain axis. Metabolites synthesized by these microbes alter the secretion of glucagon-like peptide-1 (GLP-1), PYY, leptin, and many other gut hormones to act over neuroendocrine pathways [85]. Studies demonstrated that sulfate-reducing bacteria present in colon produces hydrogen sulfide (H_2S), which directly impacts the secretion of GLP-1. Changing the composition of the gut microbiome to increase the production of H_2S along with the secretion of GLP1 could be considered a novel approach to initiate the battle against obesity [86]. The metabolites of gut microbes could be taken in synthesized form or their production might be modulated by altering the microbes for the therapeutic purpose. SCFAs, the fermentation product of nondigestible carbohydrates, are effective metabolites against obesity [87].

Researchers are currently using probiotics, prebiotics, or nondigestible carbohydrates to feed gut microbes to alter the metabolite production and the modulation of gut microbiome [88]. Generally, fructo- and galacto-oligosaccharides are used as prebiotics but many research studies demonstrated that polyphenols as prebiotics might be beneficial to increase the growth of commensal bacteria in GI tract [89]. On the other hand, probiotics affect the production of microbial enzymes and metabolites to alter gut microbes' metabolic activity instead of their composition. Researchers have tried to identify new microbial species from the human gut microbiome to use them as next-generation probiotics for the treatment of obesity [90]. Environmental factors and the individual's lifestyle should also be considered during the modulation of gut microbes. Gut microbes could be considered as therapeutic target for the treatment of obesity as they maintain the homeostasis and appetite of any individual. The connection between gut–brain axis and microbiome should be decoded to introduce a better approach against obesity and its related diseases [91].

16.7 Short-chain fatty acids

Members of the gut microbiome produce various fatty acids named SCFAs with health benefits (Fig. 16–2). Metabolic compounds produced by gut microbes possess the ability to treat or control type 2 diabetes. SCFAs are the most studied metabolites that modulate the metabolic process of the host organism. Microbial fermentation of nondigestible carbohydrates leads to the production of SCFAs [92]. SCFAs can alter the level of gut peptides that are helpful in glucose metabolism and homeostasis [93,94]. *Bifidobacteria* present in GI tract generates conjugated linoleic acid, which plays an important role in altering fatty acid composition in fat depository organs like the liver [95]. In humans, acetate is the dominating form of SCFAs and plays mysterious role in remodeling the activity of 5′ AMP activated protein kinase along with infiltration of macrophages in adipose tissues [96]. Propionate is utilized for lipid synthesis to provide energy to the host [97]. Researchers demonstrated that butyrate and propionate abolish weight gain in high-fat diet (HFD)-induced obesity mice while acetate negatively affects diet intake in healthy mice [98,99]. The exact mechanism lying behind the effect of SCFAs remains unknown. Butyrate increases insulin sensitivity along with enhancement in the expenditure of energy in obese mice. Propionate reduces energy intake via increasing the release of postprandial plasma peptide YY (PYY). SCFAs perform tissue-specific function and control different processes to regulate obesity. Propionate

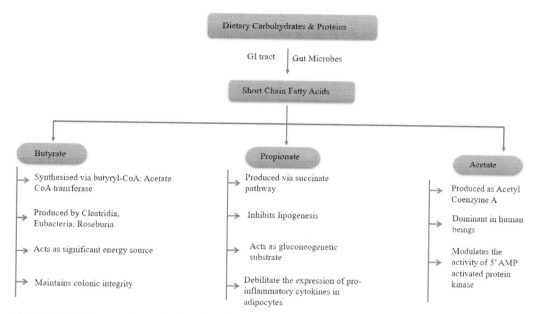

FIGURE 16–2 Metabolites of gut microbes beneficial in regulation of obesity.

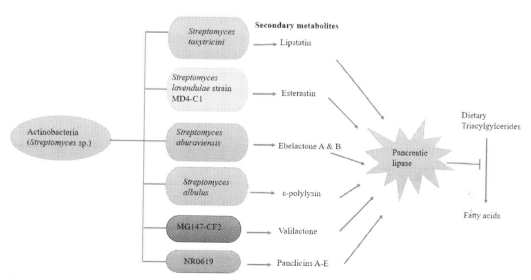

FIGURE 16–3 Secondary metabolites produced by Actinobacteria which showed inhibitory activity against pancreatic lipase.

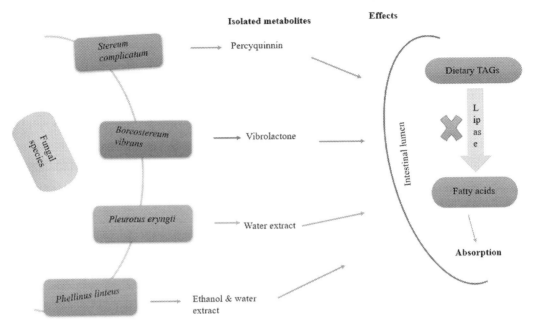

FIGURE 16-4 Fungal isolates as pancreatic lipase inhibitors.

and butyrate activate the gluconeogenesis in the intestine through gut-brain neural circuit to control the body weight and glucose intake [100]. Acetate regulates appetite regulatory neuropeptides' activity via stimulation of tricarboxylic acid cycle cycle to curtail the appetite [101,102] (Figs. 16-3 and 16-4).

Various studies indicated that G protein-coupled receptors, mainly GPR43 and 41, are involved in the effective functions of SCFAs. Interaction of SCFAs with GPR43 and 41 increases the level of GLP-1 as well as peptide YY (PYY) to reduce appetite. SCFAs produce 10% of the total energy utilized by a healthy individual. SCFAs control energy intake and metabolism along with improvement of host immunity [103–105].

16.8 New players for obesity treatment

Obesity is one of the major crises in human health due to high calorie diet intake compared to energy expenditure. Inhibition of nutrients digestion and absorption is one of the most effective approaches to control the obesity. Dietary lipids are the major contributors of unwanted calories. Thus fat digestion prohibition would be an effective approach to reduce fat absorption [106]. Pancreatic lipase plays an important role in the digestion of dietary lipids for their easy absorption in the small intestine. That's why pancreatic lipase is considered as the essential enzyme for the digestion of dietary fats. So, inhibition of pancreatic lipase will indirectly reduce the absorption of fats in small intestine. Researchers are trying their level best to introduce antiobesity drugs, but at present actinobacterium product termed as lipstatin is the only approved antiobesity drug in Europe as it inhibits the activity of pancreatic lipase to control the digestion of lipids [107]. Orlistat is the commercialized form of lipstatin, natural product of *Streptomyces toxytricini* [108]. Along with antiobesity activity, orlistat is also effective in preventing type 2 diabetes and treating cardiovascular disease [109,110]. Stemmed from of lipstatin, that is, tetrahydrolipstatin (THL) is amphiphilic in nature and binds with serine residue present at the active site of pancreatic lipase to inhibit its activity [111]. Similarly, various other compounds produced by fungi, bacteria, or marine algae are being tested for their ability to inhibit pancreatic lipase activity [112].

Along with *Streptomyces toxytricini* various other streptomyces like *S. albolongus*, *S. aburaviensis*, and *S. lavendulae* produce metabolites with inhibitory action on pancreatic lipase [59]. Complementary form of THL, that is, Panclicins which is produced by *Streptomyces* sp. NR 0619, possess strong inhibitory potential against lipase [113]. *S. aburaviensis* produces compounds named as ebelactones A and B (IC_{50} of 3 and 0.8 ng/mL, respectively) while the compound produced by *Boreostereum vibrans* is termed as vibralactone ($IC_{50} = 0.4$ μg/mL) are also the potent inhibitors of pancreatic lipase under in vitro conditions [114,115]. Some yeasts and fungi like *Candida antarctica*, *C. rugosa*, *Pseudomonas glumae*, and *Gestrichum candidum* are drawing the attention of researchers and pharmaceutical industries with their antilipase compounds. Research conducted in Slovenia demonstrated that the compounds produced by three fungal species, that is, *Laetiporus sulphureus*, *Tylopilus felleus*, *Hygrocybe conica*, possess very strong inhibitory effect against lipase, far more the Orlistat. Water extract obtained from *Pleurotus eryngii* also possesses the potential to inhibit lipase's function, thus helping prevent postprandial hyperlipidemia via regulating the absorption of fat in the intestine [116]. Water and ethanol extract of *Phellinus linteus* is another example of a lipase inhibitory compound with negative impact on obesity [117]. Monascus pigments produced by *Monascus* sp. also exhibit inhibitory action against pancreatic lipase after incorporation with unnatural amino acids.

IC_{50} of 0.14 μm of lipstatin is sufficient to inhibit the activity of pancreatic lipase irreversibly. Studies revealed that 50 mg/kg oral dose of lipstatin inhibit 80% lipases under ex vivo conditions after 2 hours of dose intake. β-Lactone ring is the most essential part for the inhibitory action of lipstatin as the compound loses its function after removing the β-lactone ring. Along with lipstatin various other β-lactone ring containing compounds like valilactone, percyquinnin, esterastin, panclicin A–E, and some non-β-lactone ring compounds like (*E*)-4-amino styryl acetate, caulerpenyne, ε-polylysine have been identified with lipase inhibition activity [118]. A streptomyces strain MG147-CF2 similar to *Streptomyces albolongus* produces a metabolite Valilactone inhibits pancreatic lipase with IC_{50} of 0.14 ng/mL, while another secondary metabolite esterastin isolated from *S. lavendulae* requires only IC_{50} of 0.9 ng/mL to stop the functioning of lipase. Both the metabolites, that is, valilactone and esterastin also possess the ability to inhibit the liver esterase activity with IC_{50} of 29 ng/mL and 50 μg/mL, respectively. In a similar manner, Basidiomycete *Stereum complicatum* produces the secondary metabolite with β-lactone ring named as percyquinnin, which inhibits the activity of lipase with IC_{50} of 2 μM [119]. Five types of panclicins, that is, A–E, are produced by NR

0619 strain of streptomyces. Out of these, A and B panclicins are alanine type while C, D, and E are glycine type. Panclicins also inhibit the activity of lipases irreversibly, like as orlistat but D and E panclicins are 2–3 folds more effective than orlistat.

16.9 Orlistat: Food and Drug Administration–approved antiobesity drug

Orlistat, also termed as THL, is synthesized from lipstatin, which lowers fat concentration in blood. At present, Roche medical company is producing orlistat under the trade name of Xenical. Orlistat works in the most effective and specific manner to inhibit the activity of pancreatic lipase. It is also involved in removal of endogenous cholesterol present in bile. The GI tract is the main site for lipase inhibitors' action over the enzyme involved in fat digestion. Hydrolysis of fat is the first step toward absorbing lipids in small intestine and their subsequent availability for energy metabolism. Lipase inhibitor inhibits the fat digestion and indirectly decreases the calories obtained by the metabolism of lipids. Research studies demonstrated that orlistat is able to inhibit about one-third on the dietary fat. Derivative of lipstatin might be utilized as dietary additive in very small concentration, that is, 0.1–100 mg in order to control the weight of animal. Utilization of the drug and water insoluble fibers increases the effectiveness of lipase inhibitors and reduce lipid digestion. Orlistat is as effective as other drugs like fenfluramine, sibutramine which acts on CNS to suppress appetite in order to lose body weight. The drugs acting on CNS put negative impact on CNS along with other toxic effects. This type of problem is not related to the utilization of lipase inhibitors. Lipase inhibitors works mainly to inhibit the absorption of fat without any side effects related to CNS. However, the lipase inhibitor causes fatty stool, oil leakage from anus, incontinence, vomiting, nausea, and GI distress [120].

16.9.1 Mechanism of action

As mentioned above, orlistat is prescribed as antiobesity drug under the brand name of Xenical or Alli. This drug irreversibly acts on pancreatic lipase as it is the saturated form of lipstatin. In addition to that orlistat also works against the activity of gastric lipase. These are the main enzymes responsible for the hydrolysis of triacylglycerols for the absorption of free fatty acids in the small intestine. Orlistat binds to the active site of pancreatic lipase to form a complex. The binding of orlistat induces alteration in the conformation of enzyme. This action of orlistat results in acylation of −OH group of the serine residue present at the active site of enzyme to inhibit the hydrolyzing function of lipase. Thus the undigested dietary fat will be excreted out directly in feces without any absorption in the small intestine, indirectly reducing the caloric intake. A research study demonstrated that orlistat also possesses the potential to inhibit the protein named as Niemann-Pick C1-like 1 involved in transportation of cholesterol [121]. In addition to that, orlistat is also able to inhibit the activity of upregulated fatty acid synthase (FAS) in tumor cells [122,123]. The inhibitory effect on FAS induces stress over endoplasmic reticulum resulting in control of tumor growth [124,125]. The drug is rarely absorbed in blood as it is of lipophilic nature. Studies described that 96% of ingested orlistat was excreted out through stool while only 1% was there in urine. The half-life of drug is 14–19 hours, indicating the drug's excretion without any change [126].

16.9.2 Side effects of orlistat

Some studies indicated that orlistat does not interfere with the normal functioning of GI tract. Orlistat also has no negative impact on the balance system of minerals and electrolytes. This drug did not interfere with the activity of other drugs taken by obese patients [127]. However, utilization of this drug put some adverse effect on the patient. The increased amount of fat in feces is the major side effect of orlistat as the study demonstrated that 14% of the patients were suffering from fatty stools while 4% of the patients were having incontinence fecal [128]. Prescription of natural fibers along with orlistat is beneficial to reduce the side effects related to GI tract [129]. Reduction in dietary fat content up to 15 g/meal might be effective to control oily stools and pomposity.

Fat soluble vitamins such as A, D, E, K, and β-carotene might be decreased in circulation of person treated with orlistat as hydrolysis of lipids is not occurring there. That's why multivitamin tablets are recommended along with orlistat. Absorption of some drugs like warfarin, thyroxine, and anticonvulsants is affected by use of orlistat. Patients prescribed with cyclosporine avoid taking orlistat as it reduces the concentration of cyclosporine in blood [130].

16.10 *Bacillus natto* probiotics

Probiotics are basically the living microorganisms possessing health benefits when supplemented in proper amount to the host [131]. Microorganisms like *Bifidobacterium, Bacillus natto, Lactobacillus plantarum,* and many others drawn the attention of researchers as probiotics to improve human health. *B. natto* is considered under potential probiotics as they are nonpathogenic to human beings and are producer of various antibiotics and enzymes and also exhibits antibacterial activity [132,133]. *B. natto* is resistant to acidic environment. This behavior helps it reach the intestine where it alters the composition of microbes and maintains the microecology of GI tract [134–137]. *B. natto* is among the 40 probiotics approved by US Food and Drug Administration [138]. Fermentation of soybeans is performed via *B. natto* to produce natto. Various studies demonstrated that the extract of *natto* exhibits hypolipidemic and antiobesity effect by inhibiting preadipocyte differentiation [139–141]. Naturally, soybeans are the rich source of proteins, isoflavones, unsaturated fatty acids, and many other nutrients. Production of *natto* through the fermentation process using *B. natto* diversifies soybean. The weight of an HFD mouse model reduced significantly after 8 weeks of feeding *natto*.

Same as to *B. natto*, *L. plantarum* is nonpathogenic and safe to reside in the GI tract. *L. plantarum* promotes the lipid oxidation and adversely affect the process involved in lipid synthesis to exhibits its antiobesity property [142–144]. *B. natto* is more effective than *L. plantarum* to reduce weight when administered in equal concentration. Both these microorganisms decrease the accumulation of visceral fat and decrease the expansion of adipose tissues. Alteration in visceral fat affects the presence of serum lipids as well as glucose metabolism. Various studies demonstrated that both these microorganisms reduce the concentration of triglycerides, cholesterol, and glucose in serum. However, *L. plantarum* was more effective than *B. natto* in reducing the glucose level while triglycerides and cholesterol level, *B. natto*, was more superior. Ingestion of *B. natto* in mice significantly increases the *Akkermansia muciniphila* in GI tract as this bacterium reduces the inflammation of the intestine and improves immunity. All these findings suggest that *B. natto* shows anti-obesity effect by reducing the level of triglycerides, glucose, and tumor necrosis factor-α in the blood serum.

B. natto also possesses the potential to improve the liver damage caused by lipid deposition in the form of triglycerides [145]. Supplementation of *B. natto* and *L. plantarum* reduces the fat droplets deposited in hepatocytes to return the liver in its original shape. *B. natto* controls lipid metabolism via regulating the activity of two genes, that is, Pparα and Srebp-1c involved in lipid metabolism. Tissues rich in mitochondria and performing β-oxidative activity is the center for the expression of Pparα [146]. Pparα promotes lipoprotein lipase synthesis and the dissociation of triglycerides to form lipoproteins and free amino acids [147]. Srebp-1c is mainly a membrane connexin present on the endoplasmic reticulum and performs fatty acid and glucose metabolism. Overexpression of this gene results in fat accumulation, and it is a transcriptional regulator for the genes involved in fat synthesis [148]. *B. natto* promotes the expression of Pparα to increase the synthesis of lipoprotein lipase while it inhibits the expression of Srebp-1c to control fat accumulation [149].

16.11 Inhibitory effect of fermented skim milk

A number of dairy products as fermented food are associated with huge health benefits. Various studies demonstrated the effect of lactobacilli fermented milk over obese mice, rats as well as human beings [150]. Bioactive peptides derived from milk are responsible for health benefits. Some strains of lactobacilli like *Lactobacillus curvatus*, *L. plantarum* reduces the accumulation of lipids and adipogenesis in 3T3-L1 cells [151,152]. *L. gasseri* LG2055 increases fat droplets' size to exhibit the antiobesity effect instead of their direct inhibitory action on pancreatic lipase. An increase in the size of fat droplets leads to a decrease in the interface area between oil and water, which in turn inhibits pancreatic lipase activity. Significantly, high fecal fat content was observed in the individuals supplemented with *L. gasseri* fermented milk [153].

The inhibitory effect of fermented milk over pancreatic lipase under in vitro conditions was studied [154]. However, the mechanism behind the lipase inhibitory effect of fermented milk is still unclear. Totally, 31 different strains of seven different species, that is, *Lactobacillus acidophilus* (4), *L. brevis* (2), *L. delbrueckii* (2), *L. helveticus* (13), *L. plantarum* (4), *Lactococcus lactis* (3), and *Pediococcus pentosaceus* (3) were used for this study to ferment the skimmed milk [154]. Fermentation process with each bacterial strain was performed under aerobic conditions at 30°C, 33°C, 37°C, and 42°C temperature. At the end of the fermentation process, pH of the medium was measured to analyze the strains' performance as they produce lactic acid as their by-product

[155]. The lipase inhibitory action of each sample was studied, and fractionation of the highest lipase inhibitory strains was performed to isolate the bioactive peptides to study their profile [156]. The fermentation culture of L. helveticus strains showed the lowest pH values at 42°C along with the increase in pH as the temperature decreased indicating the high metabolic activity. Out of 13 strains of L. helveticus, eight strains exhibited inhibition of pancreatic lipase up to 45%. SC8, SC44and SC45 strains of L. helveticus were registered for higher inhibition of pancreatic lipase, that is, 49.30%, 49.43%, and 49.75%, respectively.

These three samples of L. helveticus were subjected to fraction for protein isolation. Proteins of large size like casein were removed, and the remaining proteins were fractionated in three ranges of different sizes, that is, <10 kDa, 3–10 kDa, <3 kDa. Fractions of all three samples with <10 kDa exhibited slightly reduced inhibitory action against pancreatic lipase as compared to their respective supernatants without fractionation. The result indicates the involvement of small size proteins against pancreatic lipase activity as large proteins were eradicated through ultrafiltration. Possibly, these small size peptides might be produced through the hydrolysis of milk proteins performed by the proteolytic enzymes of microbes. The inhibitory action can also be observed in all three samples with <3 kDa fraction. However, the peptides with 3–10 kDa fraction not exhibited any inhibitory effect, suggesting the involvement of <3 kDa fraction in inhibiting pancreatic lipase. The protein profile of nonfermented skimmed milk (NFSM) produced two peaks. One with a large area could be of caseins, while the other with a smaller area could be globulin, albumin, and κ-casein [157]. In contrast, the peptide profile of all three samples of L. helveticus and their fractions exhibited two peaks with area smaller than that of NFSM, suggesting the hydrolysis of milk proteins in smaller peptides.

The fraction of <3 kDa peptide exhibiting inhibitory activity presented two peaks in its peptide profile. Out of two peaks, either one could be an active component, or both could be essential for inhibitory effect as the mechanism behind this beneficial effect needs to be elucidated [154].

16.12 Conclusions and perspectives

Gut microbes are drawing the attention of researchers toward their influential role in regulating body weight. Interaction between gut microbes and diet intake is becoming a part of the therapeutic process to treat many diseases and obesity. However, detailed knowledge about the connection between gut microbiota and their composition and diet intake is unknown. A number of secondary metabolites produced by different microbes provide a better opportunity to introduce novel clinical approaches to control obesity via inhibiting the digestion of dietary lipids. Some of the metabolites mentioned above possess the effective ability to control weight gain. However, more study is needed to define the effects, the minimum dose required, and the mechanism behind their actions, along with any possible side effects.

References

[1] World Health Organization, Global Status Report on Noncommunicable Diseases, 2014, World Health Organization, 2014.

[2] B. Thomas, J. Bishop, Manual of Dietetic Practice, John Wiley & Sons, 2013.

[3] E. Azzini, J. Giacometti, G.L. Russo, Antiobesity effects of anthocyanins in preclinical and clinical studies, Oxidative Medicine and Cellular Longevity 2017 (2017).

[4] A.J. Cox, N.P. West, A.W. Cripps, Obesity, inflammation, and the gut microbiota, The Lancet Diabetes & Endocrinology 3 (2015) 207–215.

[5] M. Rosenbaum, R. Knight, R.L. Leibel, The gut microbiota in human energy homeostasis and obesity, Trends in Endocrinology & Metabolism 26 (2015) 493–501.

[6] N. Tai, F.S. Wong, L. Wen, The role of gut microbiota in the development of type 1, type 2 diabetes mellitus and obesity, Reviews in Endocrine and Metabolic Disorders 16 (2015) 55–65.

[7] M. Villanueva-Millán, P. Perez-Matute, J. Oteo, Gut microbiota, a key player in health and disease, a review focused on obesity, Journal of Physiology and Biochemistry 71 (2015) 509–525.

[8] R.E. Ley, F. Bäckhed, P. Turnbaugh, C.A. Lozupone, R.D. Knight, J.I. Gordon, Obesity alters gut microbial ecology, Proceedings of the National Academy of Sciences 102 (2005) 11070–11075.

[9] E. Murphy, P. Cotter, S. Healy, T.M. Marques, O. O'sullivan, F. Fouhy, et al., Composition and energy harvesting capacity of the gut microbiota, relationship to diet, obesity and time in mouse models, Gut 59 (2010) 1635–1642.

[10] P.J. Turnbaugh, R.E. Ley, M.A. Mahowald, V. Magrini, E.R. Mardis, J.I. Gordon, An obesity-associated gut microbiome with increased capacity for energy harvest, Nature 444 (2006) 1027–1031.

[11] P.D. Cani, M. Osto, L. Geurts, A. Everard, Involvement of gut microbiota in the development of low-grade inflammation and type 2 diabetes associated with obesity, Gut Microbes 3 (2012) 279–288.

[12] S.J. Kallus, L.J. Brandt, The intestinal microbiota and obesity, Journal of Clinical Gastroenterology 46 (2012) 16–24.

[13] L.M. Cox, M.J. Blaser, Pathways in microbe-induced obesity, Cell Metabolism 17 (2013) 883–894.

[14] S. Park, J.-H. Bae, Probiotics for weight loss, a systematic review and meta-analysis, Nutrition Research 35 (2015) 566–575.

[15] A. Prados-Bo, S. Gómez-Martínez, E. Nova, A. Marcos, El papel de los probióticos en el manejo de la obesidad, Nutrición Hospitalaria 31 (2015) 10–18.

[16] E. Razmpoosh, M. Javadi, H.S. Ejtahed, P. Mirmiran, Probiotics as beneficial agents in the management of diabetes mellitus, a systematic review, Diabetes/Metabolism Research and Reviews 32 (2016) 143–168.

[17] O.A. Baothman, M.A. Zamzami, I. Taher, J. Abubaker, M. Abu-Farha, The role of gut microbiota in the development of obesity and diabetes, Lipids in Health and Disease 15 (2016) 1–8.

[18] J.K. Nicholson, E. Holmes, J. Kinross, R. Burcelin, G. Gibson, W. Jia, et al., Host-gut microbiota metabolic interactions, Science 336 (2012) 1262–1267.

[19] J.L. Sonnenburg, F. Bäckhed, Diet–microbiota interactions as moderators of human metabolism, Nature 535 (2016) 56–64.

[20] F. Brial, A. Le Lay, M.-E. Dumas, D. Gauguier, Implication of gut microbiota metabolites in cardiovascular and metabolic diseases, Cellular and Molecular Life Sciences 75 (2018) 3977–3990.

[21] G.I. Shulman, Ectopic fat in insulin resistance, dyslipidemia, and cardiometabolic disease, New England Journal of Medicine 371 (2014) 1131–1141.

[22] S.M. Reilly, A.R. Saltiel, Adapting to obesity with adipose tissue inflammation, Nature Reviews Endocrinology 13 (2017) 633.

[23] P.D. Cani, J. Amar, M.A. Iglesias, M. Poggi, C. Knauf, D. Bastelica, et al., Metabolic endotoxemia initiates obesity and insulin resistance, Diabetes 56 (2007) 1761–1772.

[24] P.D. Cani, R. Bibiloni, C. Knauf, A. Waget, A.M. Neyrinck, N.M. Delzenne, et al., Changes in gut microbiota control metabolic endotoxemia-induced inflammation in high-fat diet−induced obesity and diabetes in mice, Diabetes 57 (2008) 1470−1481.

[25] J. Henao-Mejia, E. Elinav, C. Jin, L. Hao, W.Z. Mehal, T. Strowig, et al., Inflammasome-mediated dysbiosis regulates progression of NAFLD and obesity, Nature 482 (2012) 179−185.

[26] E.E. Canfora, J.W. Jocken, E.E. Blaak, Short-chain fatty acids in control of body weight and insulin sensitivity, Nature Reviews Endocrinology 11 (2015) 577.

[27] A. Koh, F. De Vadder, P. Kovatcheva-Datchary, F. Bäckhed, From dietary fiber to host physiology, short-chain fatty acids as key bacterial metabolites, Cell 165 (2016) 1332−1345.

[28] L. Zhao, F. Zhang, X. Ding, G. Wu, Y.Y. Lam, X. Wang, et al., Gut bacteria selectively promoted by dietary fibers alleviate type 2 diabetes, Science 359 (2018) 1151−1156.

[29] F. De Vadder, P. Kovatcheva-Datchary, D. Goncalves, J. Vinera, C. Zitoun, A. Duchampt, et al., Microbiota-generated metabolites promote metabolic benefits via gut-brain neural circuits, Cell 156 (2014) 84−96.

[30] G. den Besten, A. Bleeker, A. Gerding, K. van Eunen, R. Havinga, T.H. van Dijk, et al., Short-chain fatty acids protect against high-fat diet−induced obesity via a PPARγ-dependent switch from lipogenesis to fat oxidation, Diabetes 64 (2015) 2398−2408.

[31] Y. Lu, C. Fan, P. Li, Y. Lu, X. Chang, K. Qi, Short chain fatty acids prevent high-fat-diet-induced obesity in mice by regulating G protein-coupled receptors and gut microbiota, Scientific Reports 6 (2016) 1−13.

[32] F. De Vadder, P. Kovatcheva-Datchary, C. Zitoun, A. Duchampt, F. Bäckhed, G. Mithieux, Microbiota-produced succinate improves glucose homeostasis via intestinal gluconeogenesis, Cell Metabolism 24 (2016) 151−157.

[33] M.P. Mollica, G.M. Raso, G. Cavaliere, G. Trinchese, C. De Filippo, S. Aceto, et al., Butyrate regulates liver mitochondrial function, efficiency, and dynamics in insulin-resistant obese mice, Diabetes 66 (2017) 1405−1418.

[34] Z. Li, C.-X. Yi, S. Katiraei, S. Kooijman, E. Zhou, C.K. Chung, et al., Butyrate reduces appetite and activates brown adipose tissue via the gut-brain neural circuit, Gut 67 (2018) 1269−1279.

[35] E.E. Canfora, C.M. van der Beek, J.W. Jocken, G.H. Goossens, J.J. Holst, S.W.O. Damink, et al., Colonic infusions of short-chain fatty acid mixtures promote energy metabolism in overweight/obese men: a randomized crossover trial, Scientific Reports 7 (2017) 1−12.

[36] E.S. Chambers, A. Viardot, A. Psichas, D.J. Morrison, K.G. Murphy, S.E. Zac-Varghese, et al., Effects of targeted delivery of propionate to the human colon on appetite regulation, body weight maintenance and adiposity in overweight adults, Gut 64 (2015) 1744−1754.

[37] E.S. Chambers, C.S. Byrne, K. Aspey, Y. Chen, S. Khan, D.J. Morrison, et al., Acute oral sodium propionate supplementation raises resting energy expenditure and lipid oxidation in fasted humans, Diabetes, Obesity and Metabolism 20 (2018) 1034−1039.

[38] E.E. Canfora, R.C. Meex, K. Venema, E.E. Blaak, Gut microbial metabolites in obesity, NAFLD and T2DM, Nature Reviews Endocrinology 15 (2019) 261−273.

[39] H. Makino, A. Kushiro, E. Ishikawa, H. Kubota, A. Gawad, T. Sakai, et al., Mother-to-infant transmission of intestinal bifidobacterial strains has an impact on the early development of vaginally delivered infant's microbiota, PLoS One 8 (2013) e78331.

[40] P.M. Munyaka, E. Khafipour, J.-E. Ghia, External influence of early childhood establishment of gut microbiota and subsequent health implications, Frontiers in Pediatrics 2 (2014) 109.

[41] H. Tlaskalová-Hogenová, R. Štěpánková, H. Kozáková, T. Hudcovic, L. Vannucci, L. Tučková, et al., The role of gut microbiota, commensal bacteria and the mucosal barrier in the pathogenesis of inflammatory and autoimmune diseases and cancer: contribution of germ-free and gnotobiotic animal models of human diseases, Cellular & Molecular Immunology 8 (2011) 110−120.

[42] B. Ji, J. Nielsen, From next-generation sequencing to systematic modeling of the gut microbiome, Frontiers in Genetics 6 (2015) 219.

[43] F. Thuny, H. Richet, J.-P. Casalta, E. Angelakis, G. Habib, D. Raoult, Vancomycin treatment of infective endocarditis is linked with recently acquired obesity, PLoS One 5 (2010) e9074.

[44] S.R. Modi, J.J. Collins, D.A. Relman, Antibiotics and the gut microbiota, The Journal of Clinical Investigation 124 (2014) 4212–4218.

[45] H.E. Jakobsson, C. Jernberg, A.F. Andersson, M. Sjölund-Karlsson, J.K. Jansson, L. Engstrand, Short-term antibiotic treatment has differing long-term impacts on the human throat and gut microbiome, PLoS One 5 (2010) e9836.

[46] J. Maukonen, M. Saarela, Human gut microbiota: does diet matter? Proceedings of the Nutrition Society 74 (2015) 23–36.

[47] G.D. Wu, F.D. Bushmanc, J.D. Lewis, Diet, the human gut microbiota, and IBD, Anaerobe 24 (2013) 117–120.

[48] M. Rothe, M. Blaut, Evolution of the gut microbiota and the influence of diet, Beneficial Microbes 4 (2013) 31–37.

[49] K.P. Scott, S.W. Gratz, P.O. Sheridan, H.J. Flint, S.H. Duncan, The influence of diet on the gut microbiota, Pharmacological Research 69 (2013) 52–60.

[50] S.F. Clarke, E.F. Murphy, K. Nilaweera, P.R. Ross, F. Shanahan, P.W. O'Toole, et al., The gut microbiota and its relationship to diet and obesity: new insights, Gut Microbes 3 (2012) 186–202.

[51] K. Brown, D. DeCoffe, E. Molcan, D.L. Gibson, Diet-induced dysbiosis of the intestinal microbiota and the effects on immunity and disease, Nutrients 4 (2012) 1095–1119.

[52] M. Jones, C. Martoni, S. Prakash, Cholesterol lowering and inhibition of sterol absorption by *Lactobacillus reuteri* NCIMB 30242: a randomized controlled trial, European Journal of Clinical Nutrition 66 (2012) 1234–1241.

[53] C. González de los Reyes-Gavilán, N. Delzenne, S. Gonzalez, M. Gueimonde Fernández, N. Salazar, Development of Functional Foods to Fight Against Obesity Opportunities for Probiotics and Prebiotics, November/December, 25 (6), Agro FOOD Industry Hi Tech, 2014.

[54] K. Hofbauer, Molecular pathways to obesity, International Journal of Obesity 26 (2002) S18–S27.

[55] R.A.K. Srivastava, N. Srivastava, Search for obesity drugs: targeting central and peripheral pathways, Current Medicinal Chemistry-Immunology, Endocrine & Metabolic Agents 4 (2004) 75–90.

[56] D.S. Weigle, Pharmacological therapy of obesity: past, present, and future, The Journal of Clinical Endocrinology & Metabolism 88 (2003) 2462–2469.

[57] Y. Shi, P. Burn, Lipid metabolic enzymes: emerging drug targets for the treatment of obesity, Nature Reviews Drug Discovery 3 (2004) 695–710.

[58] K.E. Foster-Schubert, D.E. Cummings, Emerging therapeutic strategies for obesity, Endocrine Reviews 27 (2006) 779–793.

[59] R.B. Birari, K.K. Bhutani, Pancreatic lipase inhibitors from natural sources: unexplored potential, Drug Discovery Today 12 (2007) 879–889.

[60] S. Esposito, M. Semeriva, P. Desnuelle, Effect of surface pressure on the hydrolysis of ester monolayers by pancreatic lipase, Biochimica et Biophysica Acta (BBA) – Enzymology 302 (1973) 293–304.

[61] J.W. Lagocki, J.H. Law, F.J. Kézdy, The kinetic study of enzyme action on substrate monolayers: pancreatic lipase reactions, Journal of Biological Chemistry 248 (1973) 580–587.

[62] M.E. Lowe, Molecular mechanisms of rat and human pancreatic triglyceride lipases, The Journal of Nutrition 127 (1997) 549–557.

[63] H. Brockman, Kinetic behavior of the pancreatic lipase-colipase-lipid system, Biochimie 82 (2000) 987–995.

[64] P. Tso, Gastrointestinal digestion and absorption of lipid, Advances in Lipid Research 21 (1985) 143–186.

[65] J.B. Watkins, Lipid digestion and absorption, Pediatrics 75 (1985) 151–156.

[66] C.T. Phan, P. Tso, Intestinal lipid absorption and transport, Frontiers in Bioscience 6 (2001) D299–D319.

[67] C.-W. Ko, J. Qu, D.D. Black, P. Tso, Regulation of intestinal lipid metabolism: current concepts and relevance to disease, Nature Reviews Gastroenterology & Hepatology 17 (2020) 169–183.

[68] F. Bietrix, D. Yan, M. Nauze, C. Rolland, J. Bertrand-Michel, C. Coméra, et al., Accelerated lipid absorption in mice overexpressing intestinal SR-BI, Journal of Biological Chemistry 281 (2006) 7214–7219.

[69] M. Mukherjee, Human digestive and metabolic lipases—a brief review, Journal of Molecular Catalysis B: Enzymatic 22 (2003) 369–376.

[70] D. Haslam, W. James, Obesity [J], Lancet 366 (2005) 197–1209.

[71] C.S. Rubin, A. Hirsch, C. Fung, O.M. Rosen, Development of hormone receptors and hormonal responsiveness in vitro, Insulin receptors and insulin sensitivity in the preadipocyte and adipocyte forms of 3T3-LI cells, Journal of Biological Chemistry 253 (20) (1978) 75070–75078.

[72] E.D. Rosen, O.A. MacDougald, Adipocyte differentiation from the inside out, Nature Reviews Molecular Cell Biology 7 (2006) 885–896.

[73] Q. Tong, G. Dalgin, H. Xu, C.-N. Ting, J.M. Leiden, G.S. Hotamisligil, Function of GATA transcription factors in preadipocyte-adipocyte transition, Science 290 (2000) 134–138.

[74] R. Pearson, J. Fleetwood, S. Eaton, M. Crossley, S. Bao, Krüppel-like transcription factors: a functional family, The International Journal of Biochemistry & Cell Biology 40 (2008) 1996–2001.

[75] G. Suske, E. Bruford, S. Philipsen, Mammalian SP/KLF transcription factors: bring in the family, Genomics 85 (2005) 551–556.

[76] N. Sue, B.H. Jack, S.A. Eaton, R.C. Pearson, A.P. Funnell, J. Turner, et al., Targeted disruption of the basic Krüppel-like factor gene, Klf3 reveals a role in adipogenesis, Molecular and Cellular Biology 28 (2008) 3967–3978.

[77] K. Birsoy, Z. Chen, J. Friedman, Transcriptional regulation of adipogenesis by KLF4, Cell Metabolism 7 (2008) 339–347.

[78] D. Li, S. Yea, S. Li, Z. Chen, G. Narla, M. Banck, et al., Krüppel-like factor-6 promotes preadipocyte differentiation through histone deacetylase 3-dependent repression of DLK1, Journal of Biological Chemistry 280 (2005) 26941–26952.

[79] H. Matsuo, Y. Kondo, T. Kawasaki, N. Imamura, Cineromycin B isolated from Streptomyces cinerochromogenes inhibits adipocyte differentiation of 3T3-L1 cells via Krüppel-like factors 2 and 3, Life Sciences 135 (2015) 35–42.

[80] H.-S. Ejtahed, P. Angoorani, S. Hasani-Ranjbar, S.-D. Siadat, N. Ghasemi, B. Larijani, et al., Adaptation of human gut microbiota to bariatric surgeries in morbidly obese patients: A systematic review, Microbial Pathogenesis 116 (2018) 13–21.

[81] H.-S. Ejtahed, A.-R. Soroush, P. Angoorani, B. Larijani, S. Hasani-Ranjbar, Gut microbiota as a target in the pathogenesis of metabolic disorders: a new approach to novel therapeutic agents, Hormone and Metabolic Research 48 (2016) 349–358.

[82] C. Torres-Fuentes, H. Schellekens, T.G. Dinan, J.F. Cryan, The microbiota–gut–brain axis in obesity, The Lancet Gastroenterology & Hepatology 2 (2017) 747–756.

[83] R. Krajmalnik-Brown, Z.E. Ilhan, D.W. Kang, J.K. DiBaise, Effects of gut microbes on nutrient absorption and energy regulation, Nutrition in Clinical Practice 27 (2012) 201–214.

[84] S. Moran-Ramos, B.E. López-Contreras, S. Canizales-Quinteros, Gut microbiota in obesity and metabolic abnormalities: a matter of composition or functionality? Archives of Medical Research 48 (2017) 735–753.

[85] A. Everard, P.D. Cani, Gut microbiota and GLP-1, Reviews in Endocrine and Metabolic Disorders 15 (2014) 189–196.

[86] J. Pichette, N. Fynn-Sackey, J. Gagnon, Hydrogen sulfide and sulfate prebiotic stimulates the secretion of GLP-1 and improves glycemia in male mice, Endocrinology 158 (2017) 3416–3425.

[87] M. van de Wouw, M. Boehme, J.M. Lyte, N. Wiley, C. Strain, O. O'Sullivan, et al., Short-chain fatty acids: microbial metabolites that alleviate stress-induced brain–gut axis alterations, The Journal of Physiology 596 (2018) 4923–4944.

[88] P. Hemarajata, J. Versalovic, Effects of probiotics on gut microbiota: mechanisms of intestinal immunomodulation and neuromodulation, Therapeutic Advances in Gastroenterology 6 (2013) 39–51.

[89] L.B. Bindels, N.M. Delzenne, P.D. Cani, J. Walter, Towards a more comprehensive concept for prebiotics, Nature Reviews Gastroenterology & Hepatology 12 (2015) 303.

[90] P.W. O'Toole, J.R. Marchesi, C. Hill, Next-generation probiotics: the spectrum from probiotics to live biotherapeutics, Nature Microbiology 2 (2017) 1–6.

[91] H.-S. Ejtahed, S. Hasani-Ranjbar, Neuromodulatory effect of microbiome on gut-brain axis; new target for obesity drugs, Journal of Diabetes & Metabolic Disorders 18 (2019) 263–265.

[92] S. Fukuda, H. Toh, K. Hase, K. Oshima, Y. Nakanishi, K. Yoshimura, et al., Bifidobacteria can protect from enteropathogenic infection through production of acetate, Nature 469 (2011) 543–547.

[93] Z. Gao, J. Yin, J. Zhang, R.E. Ward, R.J. Martin, M. Lefevre, et al., Butyrate improves insulin sensitivity and increases energy expenditure in mice, Diabetes 58 (2009) 1509–1517.

[94] H.V. Lin, A. Frassetto, E.J. Kowalik Jr, A.R. Nawrocki, M.M. Lu, J.R. Kosinski, et al., Butyrate and propionate protect against diet-induced obesity and regulate gut hormones via free fatty acid receptor 3-independent mechanisms, PLoS One 7 (2012) e35240.

[95] E.F. O'Shea, P.D. Cotter, C. Stanton, R.P. Ross, C. Hill, Production of bioactive substances by intestinal bacteria as a basis for explaining probiotic mechanisms: bacteriocins and conjugated linoleic acid, International Journal of Food Microbiology 152 (2012) 189–205.

[96] B. Carvalho, D. Guadagnini, D. Tsukumo, A. Schenka, P. Latuf-Filho, J. Vassallo, et al., Modulation of gut microbiota by antibiotics improves insulin signalling in high-fat fed mice, Diabetologia 55 (2012) 2823–2834.

[97] N.M. Delzenne, P.D. Cani, A. Everard, A.M. Neyrinck, L.B. Bindels, Gut microorganisms as promising targets for the management of type 2 diabetes, Diabetologia 58 (2015) 2206–2217.

[98] H.J. Binder, Role of colonic short-chain fatty acid transport in diarrhea, Annual Review of Physiology 72 (2010) 297–313.

[99] A.D. Sabatino, R. Morera, R. Ciccocioppo, P. Cazzola, S. Gotti, F. Tinozzi, et al., Oral butyrate for mildly to moderately active Crohn's disease, Alimentary Pharmacology & Therapeutics 22 (2005) 789–794.

[100] P. Kovatcheva-Datchary, A. Nilsson, R. Akrami, Y.S. Lee, F. De Vadder, T. Arora, et al., Dietary fiber-induced improvement in glucose metabolism is associated with increased abundance of *Prevotella*, Cell Metabolism 22 (2015) 971–982.

[101] G. Frost, M.L. Sleeth, M. Sahuri-Arisoylu, B. Lizarbe, S. Cerdan, L. Brody, et al., The short-chain fatty acid acetate reduces appetite via a central homeostatic mechanism, Nature Communications 5 (2014) 1–11.

[102] M. Kasubuchi, S. Hasegawa, T. Hiramatsu, A. Ichimura, I. Kimura, Dietary gut microbial metabolites, short-chain fatty acids, and host metabolic regulation, Nutrients 7 (2015) 2839–2849.

[103] C.L. Boulangé, A.L. Neves, J. Chilloux, J.K. Nicholson, M.-E. Dumas, Impact of the gut microbiota on inflammation, obesity, and metabolic disease, Genome Medicine 8 (2016) 1–12.

[104] S. Duranti, C. Ferrario, D. Van Sinderen, M. Ventura, F. Turroni, Obesity and microbiota: an example of an intricate relationship, Genes & nutrition 12 (2017) 1–15.

[105] J. Kałużna-Czaplińska, P. Gątarek, M.S. Chartrand, M. Dadar, G. Bjørklund, Is there a relationship between intestinal microbiota, dietary compounds, and obesity, Trends in Food Science & Technology 70 (2017) 105–113.

[106] G.A. Bray, D.H. Ryan, Drug treatment of the overweight patient, Gastroenterology 132 (2007) 2239–2252.

[107] K.S. McClendon, D.M. Riche, G.I. Uwaifo, Orlistat: current status in clinical therapeutics, Expert Opinion on Drug Safety 8 (2009) 727–744.

[108] E. Weibel, P. Hadvary, E. Hochuli, E. Kupfer, H. Lengsfeld, Lipstatin, an inhibitor of pancreatic lipase, produced by *Streptomyces toxytricini*, The Journal of Antibiotics 40 (1987) 1081–1085.

[109] S.B. Heymsfield, K.R. Segal, J. Hauptman, C.P. Lucas, M.N. Boldrin, A. Rissanen, et al., Effects of weight loss with orlistat on glucose tolerance and progression to type 2 diabetes in obese adults, Archives of Internal Medicine 160 (2000) 1321–1326.

[110] J.S. Torgerson, J. Hauptman, M.N. Boldrin, L. Sjöström, XENical in the prevention of diabetes in obese subjects, XENDOS study: a randomized study of orlistat as an adjunct to lifestyle changes for the prevention of type 2 diabetes in obese patients, Diabetes Care 27 (2004) 155–161.

[111] A. Jawed, G. Singh, S. Kohli, A. Sumera, S. Haque, R. Prasad, et al., Therapeutic role of lipases and lipase inhibitors derived from natural resources for remedies against metabolic disorders and lifestyle diseases, South African Journal of Botany 120 (2019) 25–32.

[112] P. Slanc, B. Doljak, A. Mlinarič, B. Štrukelj, Screening of wood damaging fungi and macrofungi for inhibitors of pancreatic lipase, Phytotherapy Research: An International Journal Devoted to Pharmacological and Toxicological Evaluation of Natural Product Derivatives 18 (2004) 758–762.

[113] K. Yoshinari, M. Aoki, T. Ohtsuka, N. Nakayama, Y. Itezono, M. Mutoh, et al., Panclicins, novel pancreatic lipase inhibitors, The Journal of Antibiotics 47 (1994) 1376–1384.

[114] Y. Nonaka, H. Ohtaki, E. Ohtsuka, T. Kocha, T. Fukuda, T. Takeuchi, et al., Effects of ebelactone B, a lipase inhibitor, on intestinal fat absorption in the rat, Journal of Enzyme Inhibition 10 (1995) 57–63.

[115] D.-Z. Liu, F. Wang, T.-G. Liao, J.-G. Tang, W. Steglich, H.-J. Zhu, et al., Vibralactone: a lipase inhibitor with an unusual fused β-lactone produced by cultures of the basidiomycete *Boreostereum vibrans*, Organic Letters 8 (2006) 5749–5752.

[116] T. Mizutani, S. Inatomi, A. Inazu, E. Kawahara, Hypolipidemic effect of *Pleurotus eryngii* extract in fat-loaded mice, Journal of Nutritional Science and Vitaminology 56 (2010) 48–53.

[117] F.I. GarzaMilagro-Yoldi, N. Boque, J. Campión-Zabalza, J.A. Martinez, Natural inhibitors of pancreatic lipase as new players in obesity treatment, Planta Medica 77 (8) (2011) 773–785.

[118] N.A. Lunagariya, N.K. Patel, S.C. Jagtap, K.K. Bhutani, Inhibitors of pancreatic lipase: state of the art and clinical perspectives, EXCLI Journal 13 (2014) 897.

[119] C. Hopmann, M. Kurz, G. Mueller, L. Toti, Percyquinnin, a process for its production and its use as a pharmaceutical, Google Patents (2003).

[120] M.A. Al-Omar, A. Al-Suwailem, A. Al-Tamimi, M. Al-Suhibani, Safety and mechanism of action of orlistat, tetrahydrolipstatin as the first local antiobesity drug, Journal of Applied Sciences Research 2 (2006) 205–208.

[121] S. Alqahtani, H. Qosa, B. Primeaux, A. Kaddoumi, Orlistat limits cholesterol intestinal absorption by Niemann-pick C1-like 1, NPC1L1 inhibition, European Journal of Pharmacology 762 (2015) 263–269.

[122] F.P. Kuhajda, Fatty acid synthase and cancer: new application of an old pathway, Cancer Research 66 (2006) 5977–5980.

[123] S.J. Kridel, F. Axelrod, N. Rozenkrantz, J.W. Smith, Orlistat is a novel inhibitor of fatty acid synthase with antitumor activity, Cancer Research 64 (2004) 2070–2075.

[124] J.L. Little, F.B. Wheeler, D.R. Fels, C. Koumenis, S.J. Kridel, Inhibition of fatty acid synthase induces endoplasmic reticulum stress in tumor cells, Cancer Research 67 (2007) 1262–1269.

[125] C.D. Browne, E.J. Hindmarsh, J.W. Smith, Inhibition of endothelial cell proliferation and angiogenesis by orlistat, a fatty acid synthase inhibitor, The FASEB Journal 20 (2006) 2027–2035.

[126] X. Qi, Review of the Clinical Effect of Orlistat, in: IOP Conference Series: Materials Science and Engineering, vol. 301, IOP Publishing, 2018, p. 012063.

[127] R. Guerciolini, Mode of action of orlistat, International Journal of Obesity and Related Metabolic Disorders: Journal of the International Association for the Study of Obesity 21 (1997) S12–S23.

[128] J. Wilding, Orlistat: should we worry about liver inflammation? British Medical Journal Publishing Group (2013).

[129] H. Cavaliere, I. Floriano, G. Medeiros-Neto, Gastrointestinal side effects of orlistat may be prevented by concomitant prescription of natural fibers, psyllium mucilloid, International Journal of Obesity 25 (2001) 1095–1099.

[130] P. Errasti, I. Garcia, J. Lavilla, B. Ballester, J. Manrique, Reduction in blood cyclosporine concentration by orlistat in two renal transplant patients, Transplantation Proceedings 34 (2002) 137–139.

[131] R.B. Salah, I. Trabelsi, K. Hamden, H. Chouayekh, S. Bejar, *Lactobacillus plantarum* TN8 exhibits protective effects on lipid, hepatic and renal profiles in obese rat, Anaerobe 23 (2013) 55–61.

[132] X. Jiang, H. Ding, Q. Liu, Y. Wei, Y. Zhang, Y. Wang, et al., Effects of peanut meal extracts fermented by *Bacillus natto* on the growth performance, learning and memory skills and gut microbiota modulation in mice, British Journal of Nutrition 123 (2020) 383–393.

[133] Q.-Y. Wang, Q.-L. Lin, K. Peng, J.-Z. Cao, C. Yang, D. Xu, Surfactin variants from *Bacillus subtilis natto* CSUF5 and their antifungal properties against *Aspergillus niger*, Journal of Biobased Materials and Bioenergy 11 (2017) 210–215.

[134] Y. Li, Q. Xu, Z. Huang, L. Lv, X. Liu, C. Yin, et al., Effect of *Bacillus subtilis* CGMCC 1.1086 on the growth performance and intestinal microbiota of broilers, Journal of Applied Microbiology 120 (2016) 195–204.

[135] Z. Li, H. Jin, S.Y. Oh, G.E. Ji, Anti-obese effects of two Lactobacilli and two Bifidobacteria on ICR mice fed on a high fat diet, Biochemical and Biophysical Research Communications 480 (2016) 222–227.

[136] D. Song, H. Kang, J. Wang, H. Peng, D. Bu, Effect of feeding *Bacillus subtilis natto* on hindgut fermentation and microbiota of holstein dairy cows, Asian-Australasian Journal of Animal Sciences 27 (2014) 495.

[137] H.L. Zhang, W.S. Li, D.N. Xu, W.W. Zheng, Y. Liu, J. Chen, et al., Mucosa-reparing and microbiota-balancing therapeutic effect of *Bacillus subtilis* alleviates dextrate sulfate sodium-induced ulcerative colitis in mice, Experimental and Therapeutic Medicine 12 (2016) 2554–2562.

[138] X.-H. Cao, Z.-Y. Liao, C.-L. Wang, W.-Y. Yang, M.-F. Lu, Evaluation of a lipopeptide biosurfactant from *Bacillus natto* TK-1 as a potential source of anti-adhesive, antimicrobial and antitumor activities, Brazilian Journal of Microbiology 40 (2009) 373–379.

[139] J.W. Hwang, H.J. Do, O.Y. Kim, J.H. Chung, J.-Y. Lee, Y.S. Park, et al., Fermented soy bean extract suppresses differentiation of 3T3-L1 preadipocytes and facilitates its glucose utilization, Journal of Functional Foods 15 (2015) 516–524.

[140] J. Kim, J.N. Choi, J.H. Choi, Y.S. Cha, M.J. Muthaiya, C.H. Lee, Effect of fermented soybean product, Cheonggukjang intake on metabolic parameters in mice fed a high-fat diet, Molecular Nutrition & Food Research 57 (2013) 1886–1891.

[141] M. Kushida, R. Okouchi, Y. Iwagaki, M. Asano, M.X. Du, K. Yamamoto, et al., Fermented soybean suppresses visceral fat accumulation in mice, Molecular Nutrition & Food Research 62 (2018) 1701054.

[142] C. Hill, F. Guarner, G. Reid, G.R. Gibson, D.J. Merenstein, B. Pot, et al., Expert consensus document: the International Scientific Association for Probiotics and Prebiotics consensus statement on the scope

and appropriate use of the term probiotic, Nature Reviews Gastroenterology & Hepatology 11 (2014) 506.

[143] S. Park, Y. Ji, H.-Y. Jung, H. Park, J. Kang, S.-H. Choi, et al., *Lactobacillus plantarum* HAC01 regulates gut microbiota and adipose tissue accumulation in a diet-induced obesity murine model, Applied Microbiology and Biotechnology 101 (2017) 1605–1614.

[144] M.C. De Vries, E.E. Vaughan, M. Kleerebezem, W.M. de Vos, *Lactobacillus plantarum*—survival, functional and potential probiotic properties in the human intestinal tract, International Dairy Journal 16 (2006) 1018–1028.

[145] R.A. van der Heijden, F. Sheedfar, M.C. Morrison, P.P. Hommelberg, D. Kor, N.J. Kloosterhuis, et al., High-fat diet induced obesity primes inflammation in adipose tissue prior to liver in C57BL/6j mice, Aging (Albany, NY) 7 (2015) 256.

[146] P.-M. Chao, M.-F. Yang, Y.-N. Tseng, K.-M. Chang, K.-S. Lu, C.-J. Huang, Peroxisome proliferation in liver of rats fed oxidized frying oil, Journal of Nutritional Science and Vitaminology 51 (2005) 361–368.

[147] M. Panadero, C. Bocos, E. Herrera, Relationship between lipoprotein lipase and peroxisome proliferator-activated receptor-α expression in rat liver during development, Journal of Physiology and Biochemistry 62 (2006) 189.

[148] C. Zhang, J. Hu, L. Sheng, M. Yuan, Y. Wu, L. Chen, et al., Ellagic acid ameliorates AKT-driven hepatic steatosis in mice by suppressing de novo lipogenesis via the AKT/SREBP-1/FASN pathway, Food & Function 10 (2019) 3410–3420.

[149] P. Wang, X. Gao, Y. Li, S. Wang, J. Yu, Y. Wei, *Bacillus natto* regulates gut microbiota and adipose tissue accumulation in a high-fat diet mouse model of obesity, Journal of Functional Foods 68 (2020) 103923.

[150] J. Aguilar-Toalá, L. Santiago-López, C. Peres, C. Peres, H. Garcia, B. Vallejo-Cordoba, et al., Assessment of multifunctional activity of bioactive peptides derived from fermented milk by specific *Lactobacillus plantarum* strains, Journal of Dairy Science 100 (2017) 65–75.

[151] Y. Moon, J. Soh, J. Yu, H. Sohn, Y. Cha, S. Oh, Intracellular lipid accumulation inhibitory effect of W Eissella Koreensis OK 1-6 isolated from K Imchi on differentiating adipocyte, Journal of Applied Microbiology 113 (2012) 652–658.

[152] D.-Y. Park, Y.-T. Ahn, C.-S. Huh, S.-M. Jeon, M.-S. Choi, The inhibitory effect of *Lactobacillus plantarum* KY1032 cell extract on the adipogenesis of 3T3-L1 cells, Journal of Medicinal Food 14 (2011) 670–675.

[153] A. Ogawa, T. Kobayashi, F. Sakai, Y. Kadooka, Y. Kawasaki, *Lactobacillus gasseri* SBT2055 suppresses fatty acid release through enlargement of fat emulsion size in vitro and promotes fecal fat excretion in healthy Japanese subjects, Lipids in Health and Disease 14 (2015) 1–10.

[154] A.M. Gil-Rodríguez, T.P. Beresford, Lipase inhibitory activity of skim milk fermented with different strains of lactic acid bacteria, Journal of Functional Foods 60 (2019) 103413.

[155] M.G. Gaenzle, Lactic metabolism revisited: metabolism of lactic acid bacteria in food fermentations and food spoilage, Current Opinion in Food Science 2 (2015) 106–117.

[156] J. Giacometti, A. Buretić-Tomljanović, Peptidomics as a tool for characterizing bioactive milk peptides, Food Chemistry 230 (2017) 91–98.

[157] D. Vincent, V. Ezernieks, A. Elkins, N. Nguyen, P.J. Moate, B.G. Cocks, et al., Milk bottom-up proteomics: method optimization, Frontiers in Genetics 6 (2016) 360.

17

Potential bovine colostrum for human and animal therapy

Maria Giovana Binder Pagnoncelli[1], Fernanda Guilherme do Prado[2], Juliane Mayara Casarim Machado[3], Andreia Anschau[3], Carlos Ricardo Soccol[2]

[1]DEPARTMENT OF CHEMISTRY AND BIOLOGY, FEDERAL UNIVERSITY OF TECHNOLOGY—PARANÁ (UTFPR), CURITIBA, BRAZIL [2]BIOPROCESS ENGINEERING AND BIOTECHNOLOGY DEPARTMENT, FEDERAL UNIVERSITY OF PARANÁ (UFPR), CURITIBA, BRAZIL [3]BIOPROCESS ENGINEERING AND BIOTECHNOLOGY DEPARTMENT, FEDERAL UNIVERSITY OF TECHNOLOGY—PARANÁ (UTFPR), CURITIBA, BRAZIL

17.1 Introduction

Colostrum, known as a premilk fluid secreted in the mammary glands of all mammals species during the first few days after parturition. This lacteal secretion is the first natural source of nutrients for the growth and development of newborns. Colostrum is food with high nutritional value and contains the same constituents found in milk, however, in different concentrations. Besides that, colostrum contains a lot of biologically active molecules, which are transmitted to the newborn. Then it is essential to encourage the first feed or the supply of colostrum right after birth, because from two hours after delivery, the secretion of gastric juices begins, which reduces the absorption of nutrients and other bioactive substances [1–3].

Colostrum consumption is important both from a nutritional point of view and in the acquisition of growth and immunity factors in the neonatal period. The bioactive components contribute to the growth, development, maturation, and integrity of the neonatal gastrointestinal tract (GI). These components have antimicrobial and endotoxins-neutralizing effects throughout the GI, fighting gut inflammation, and also help to repair cases of tissue lesions. Besides that, the immunoglobulins present in colostrum could get across the intestinal barrier and promote systemic effects, contributing to initial immunological defense [4–8]. This passive immunity mechanism is primordial to increase the likelihood of the newborns surviving due to their immature immune system. In human, immunoglobulin transmission occurs via the placenta, while in some other species the maternal immunoglobulins do not cross the placenta and the newborn is dependent on the intestinal absorption of the colostrum immunoglobulins to protect against infection, for example, horses, pigs, cows, and goats [9,10]. In large systems of animal production, it is necessary to have management strategies to reduce neonatal mortality and morbidity and consequently economic losses.

Bovine colostrum (BC) is normally produced in a higher volume than the calf can consume during the first 5 days and it accounts for approximately 0.5% of a cow's annual milk output. The colostrum is not allowed to be mixed with normal milk, as it has an unpleasant taste [11]. The benefits of colostrum have been discussed for many decades, and many possibilities for use in human nutrition have already been reported [3,8,12–16]. However, large amounts of BC are still being discarded and disposed of in the environment due to the poor knowledge about how to process the colostrum. Nowadays, because of the bioactive molecules, the use of BC as a health food supplement for humans and animals has raised great interest in the market [12]. BC is not species specific, so it can be beneficial for humans and other mammals [8]. The use of BC as a human and animal therapy will be detailed in this chapter.

17.2 Bovine colostrum composition

BC consists of the first cow's milk after delivery and, like human colostrum, it contains much higher levels of protein, antimicrobial factors, immunoregulatory factors, and trophic factors than mature milk (e.g., immunoglobulins, lactoferrin, lysozyme, lactoperoxidase, osteopontin, transforming growth factor (GF), insulin-like GFs, epidermal GF) [17].

Although there are some studies about the composition of the colostrum, there is no consensus by authors concerned to it. The composition, physical-chemical and nutritional properties, is influenced by a number of intrinsic and extrinsic factors, such as farrowing, number of lactations, genetics, race, climate, dry period of management (applicable to ruminants), pre- and postpartum food management, energy balance, among other factors. These factors directly influence the BC's final product [10,18]. BC is homologous to human colostrum, although immune factors are present in higher concentrations. Table 17–1 shows the composition of macronutrients and bioactive components in BCF compared to values in mature milk.

In addition to chemical attributes, the physical properties between BC and mature milk are also mentioned. For example, the pH observed in colostrum, is slightly acid, with values ranging from 6.0 to 6.5 on average, there is no agreement on the reason for the change in colostral pH, some authors credit this change due to the permeability of the membranes, which causes higher protein concentration, and by the flow of blood at the end of pregnancy. The color of the colostrum also differs from milk, caused by the higher concentration of fat molecules and carotenoids, which can cause the emulsion to acquire more yellowish colors. The presence of blood may also contribute to the change in color [11].

BC is rich in nutrients, and growth and immune factors (Fig. 17–1). The GF is responsible to promote muscle growth (IGF-I and IGF-II). BC stimulates to repair inflammation process at the site, stimulates the gastrointestinal growth and repair, inhibits the acid secretion, stimulates mucosal restitution after injury, and increases the gastric mucin concentration.

Table 17–1 Macronutrient composition and bioactive factors in bovine and human colostrum.

	Bovine colostrum	Human colostrum
Protein (g/L)	60–135[a]	11–32[c,f]
Casein (g/L)	26[b]	3.0–5.6[c,f]
Whey (g/L)	35–119[a]	4.3–11.1[c,f]
α-Lactalbumin (g/L)	2.04[c]	2.56[c]
β-Lactoglobulin (g/L)	14.3[c]	None[c]
Lactoferrin (g/L)	1.0–2.0[c]	5.0–7.0[c]
Immunoglobulins (g/L)	20–150[c]	1.14–20[c]
Lactoperoxidase (mg/L)	11–45[c]	5.17[c]
Osteopontin (mg/L)	Not determined[c]	1493.4[c]
Lysozyme (mg/L)	0.14–0.7[c]	270–430[c]
Superoxide dismutase (U/mL)	0.06–2.88[c]	18.7–22.5[c]
Platelet-activating factor-acetylhydroxylase (μg/L)	None[c]	0.95–1.19[c]
Alkalinephosphatase (μkat/L)	6.84[h]	1.79[c]
Transforming growth factor-β (μg/L)	150–1150[c]	1366[c]
Insulin-like growth-I (μg/L)	49–2000[c]	29–49[c]
Insulin-like growth-II (μg/L)	400–600[c]	10.5[c]
Epidermal growth factor (μg/L)	4–324.2[c]	35–438[c]
Lactose (g/L)	18.9–32[a]	44–59[d,g]
Fat (g/L)	50–80[a]	20–29[e]

[a][20].
[b][21].
[c][22].
[d][23].
[e][24].
[f][25].
[g][19].
[h][26].

The platelet derived growth factor is a potent mitogen for fibroblasts and arterial smooth muscle, while the GH stimulates the neonate to release GH from the pituitary gland cell [4,12,27].

The protein content in colostrum tends to be higher on the first day after parturition and dramatically decreases as lactation progresses. Most of it is responsible for the immune factors (immunomodulatory and antimicrobial properties), such as immunoglobulins, lactoferrin, lysozymes, peroxidases, and cytokines. The high protein concentration in colostrum is supposed to increase passive immunity and reduce the incidence of diarrhea rate in calves by the local effects against pathogens.

Besides antimicrobial peptides, BC contains proteins such as lactoferrin, which earlier studies in vitro have attested its antimicrobial activity against gram-positive and negative bacteria, fungi, and viruses; rotavirus, enterovirus, and adenovirus showed positive responses. Lactoferrin exhibits antagonistic properties against pathogens such as *Escherichia coli, Salmonella typhimurium, Shigella dysenteria, Listeria monocytogens, Streptococcus*

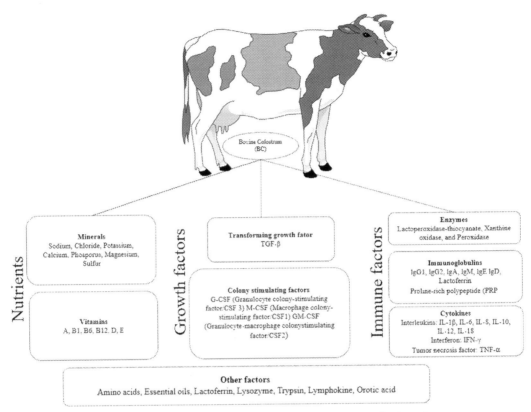

FIGURE 17–1 Diversity of nutrients, growth factors, and immune factors present in bovine colostrum.

mutans, Bacillus stearothermophilus, and *Bacillus subtilis* and is active against the herpes simplex virus type I (HSV-I), the human immunodeficiency virus I (HIV-I) and human cytomegalovirus [17]. The supplementation of low birth weight, bovine lactoferrin showed positive results in the reduction in initial sepsis [3,6,10,28].

Immunoglobulin concentration is very high, and it is divided into three main classes, type G (immunoglobins [IgG]), type A (IgA), and type M (IgM), which constitute about 35% of the protein composition and can reach values of 30–20 g/L, with IgG1 comprising more than 75% of Immunoglobulins, followed by IgM, IgA, and IgG2 [29]. The Ig present in higher concentration is the type G (IgG), being the main antibody, which can present viability after transit through the digestive system and can replace the need for immunoglobulin type A. IgG has particular importance to the newborn, whose intestine in the first hours after parturition allows the transit and better absorption of large molecules, thus providing passive immunity [3,10].

Cytokines are a diverse group of proteins, peptides, or glycoproteins, which, even in low concentrations, have biological effects. The main function of cytokines is the immunomodulation of the body, such as tumor necrosis and interferons. Despite its presence in colostrum, there are no reports of it in milk [11,30]. In general, the enzymatic content and catalytic activity of colostrum is higher than in milk, such as lactoperoxidase and lysozymes. Both milk and colostrum contain a number of enzyme inhibitors at rates that are higher initially than during lactation. Enzyme inhibitors are believed to play an important role in the mechanism of absorption of immune components by the calf, protecting immunoglobulins from cleavage. The concentration of inhibitors decreases dramatically during the first three days after calving when they reach stable levels [11].

There are few studies on lipid profile in the literature. The vast majority of reports are concentrated on the content of immunoglobulins and proteins. Lipids, similarly to other colostrum constituents, are of great importance to newborn development [11]. Lipids provide energy for thermoregulation in body temperature, aid in glucose homeostasis, and are responsible to carry liposoluble nutrient triglycerides and bioactive molecules to neonates. In addition, some fatty acids are beneficial not only for their nutritional properties but also for their positive effects on health [31].

In contrast to protein, the lipid content tends to remain constant at all stages of lactation; however, it may present some variations according to the cow's age, that is, primiparous cows present fatty acids with variation in the number of carbonaceous compared to multiparous cows. Colostrum has high concentration of cholesterol which aims to meet the needs of the newborn, as it is an essential component for their development and an integral part of cell membranes where it affects the content of other lipids, specifically sphingolipids [31,32]. In addition, cholesterol is a precursor of steroids and is responsible for the standardization and development of the central nervous system.

Phospholipid values have shown no differences in relation to the number of births; however, the concentrations found in the first 24 hours changed significantly with the advance of lactation. Among the phospholipid classes, sphingolamyelin has presented higher concentration which may be related to the newborn needs due to their function on the membrane structure, participating in the transduction of biological signals through the membrane, increasing the protection against gastrointestinal bacterial infections [31,32]. Therefore, the colostrum secreted in the first 24 hours after parturition contains a different lipid profile from the one observed as the lactation progresses, acquiring some characteristics found in mature milk (from 5/7 days after parturition).

The most concentrated carbohydrate in milk is lactose, which is responsible for 50% of the osmotic pressure of milk. However, in colostrum, there is a low lactose content, as it presents higher viscosity due to the absence of the lactose osmoregulator [11,33]. In addition to lactose, milk and colostrum contain traces of other sugars, including glucose, fructose, glucosamine, galactosamine, N-acetylneuraminic acid, and oligosaccharides, ranging from three to 10 monosaccharides covalently bound by glycosidic bonds.

Nucleotide concentration in colostrum is very low, but it reaches its maximum concentration after 24–48 hours after birth, followed by a gradual decrease as lactation continues until the third week when levels stabilize. Nucleosides play a major biochemical role, being the precursors of nucleic acids, and consequently enhance immune system responses. Nucleosides are present in low concentrations in colostrum and are stable 3 weeks after delivery [11]. BC has aroused interest from the scientific community in order to elaborate studies and protocols for its use in different phases and moments of mammalian and human life, due to its composition and physical characteristics.

17.3 Passive immunity mechanism in human and different animal species

The antibodies transfer from mother to child is known as passive immunity. This mechanism is essential for the newborn health, which does not have an adequate immune system, with active antibodies, presenting a large predisposition to develop pathogenicity, which in severe cases can cause death [7]. In mammals, the permeability of the placenta to macromolecules is inversely proportional to the number of tissue layers between the maternal and fetal circulation. Thus species with placental constitution with higher tissue increment do not present satisfactory and/or null antibody transfer during the gestational phase [2].

Mammals can be classified into classes, according to the permeability of the immunoglobulins. In humans and rabbits, for example, the acquisition of maternal antibodies occurs mainly during the gestational phase through placental translocation [10,34]. Unlike this group, most ruminant mammals present such complexity in the constitution of the placenta. This placenta is the syndescorial type, it is constituted of five layers of tissue, and during the gestational phase does not translocate defense cells to the fetus, this way passive immunity is acquired in the postnatal period, with the ingestion of colostrum indispensable to provide maternal antibodies to the newborn [2]. The transfer of passive immunity in mammals, especially ruminants, is intrinsically dependent on colostrum and its formation, and the ingestion and absorption of immunoglobulins by the intestinal epithelium. Maternal ability, a poor udder formation, or difficulty of the newborn in ingesting colostrum, can cause damage to the immune system development and economic losses to the producer [2,3].

Immunoglobulins represent the mother's history of exposure to adverse factors and the response of her immune system. This mechanism represents great benefits to newborns, which are still developing the immune system, who are susceptible to pathogen attacks [7,35]. As premature newborns do not have immunoglobulins at adequate levels they are more likely to develop diseases, increasing the mortality rate due to complications in the first days of life, especially when an adequate passive immunization does not occur, which is provided via colostrum. As the newborns' immune system presents high immaturity, the necessary protection is provided by intaking IgA, as it provides the necessary protection especially to the gastrointestinal and respiratory mucosa, which are more susceptible to exogenous bacterial attacks, as they may adhere to the mucosa at parturition [7]. Soon after parturition, there is no satisfactory colonization in the gastrointestinal system yet. IgG can supply the IgA deficiency and is effective in preventing gastrointestinal diseases, including diarrhea. It prevents pathogenic microorganisms from increasing their rate of colonization in mucous membranes [34].

Because of its composition, especially the high concentration of immunoglobulins, BC has been the focus of studies on product enrichment, such as hyperimmune milk, and food supplementation, which has had a positive impact in groups with a deficient immune system, such as patients with chronic diseases, premature birth defects, and postsurgical use [36,37]. In a study with dogs, which received an increased diet with BC, it was observed an increased level of fecal IgA, those animals have developed stable intestinal microbiota and better immune response to canine kinomose vaccine, compared to the control group, and no side effects caused by supplementation with BC were reported [36].

The deficiency or failure in the immunoglobulins supply to a newborn can cause several metabolic dysfunctions to the individual such as low immunological responses to resulting diseases since its birth. BC is produced in larger quantities than the needs as a result it can be used as a vector for acquiring passive immunity to other animal species and humans. Due to its readiness to acquire and low cost, colostrum can be offered to undernourished children and vulnerable people in developing countries and countries of extreme poverty, assisting in the development of the immune system and providing indispensable nutrients for a good metabolic performance.

17.4 Bovine colostrum application for human therapy

For some years, researchers have highlighted the enormous importance of BC in the trade of functional therapeutic products that have a beneficial impact on human health, since, concerning immunologically active milk proteins, human and bovine milk are quite similar. Its nutraceutical uses in maintaining health are an expanding niche. Produced immediately after delivery, the "first milk" is a fluid rich in nutrients such as proteins, carbohydrates, fats, vitamins, minerals, and significantly richer in biologically active peptides [38,39].

BC is homologous to human colostrum; however, due to the fast growth of calves, the colostrum produced by cows has higher levels of nutritional and immune factors [40]. The protective effects of BC are associated with bioactive constituents which have immunomodulatory and antimicrobial properties, such as immunoglobulins, lactoferrin, lactoperoxidase, lysozyme, β-GFs, glycoprotein, and glycolipids [8,41,42]. Bovine IgG is present in concentrations 100 times higher in colostrum than in mature milk and bind to a variety of human intestinal and respiratory bacteria—as well as viral pathogens and some allergies. Unlike maternal immunoglobulins in humans, IgG does not cross the placenta. Thus the newborn calf depends on intestinal immunoglobulin absorption to acquire passive immunity after birth, which explains the higher levels of IgG in BC [9,27,43].

In humans, the use of IgGs has been studied since the 1970s to ensure its efficiency as an immunotherapeutic ingredient, it is important that a significant amount can survive the passage from the stomach to the small intestine or even throughout the GI in order to keep their structures intact. Some studies have suggested that up to 10%–20% of immunoglobulins orally ingested will survive to the gastrointestinal passage in babies and adults, especially in babies as they have a higher gastric pH and lower levels of GI proteolysis [44,45].

IgGs are able to recognize and make connections with various bacterial pathogens associated with the GI, including *Yersinia enterocolitica, Campylobacter jejuni, E. coli, Klebsiella pneumoniae, Serratia marescens, Salmonella typhimurium, Staphylococcus, Streptococcus, Cryptosporidium,*

Helicobacter, *E. coli* EHEC O157:H7, *Pseudomonas*, Rotavirus, and respiratory pathogens such as human respiratory syncytial virus (RSV), influenza virus, and *Streptococcus pneumoniae* [46–48].

Colostrum is obtained from nonimmunized samples, and all the preparation was carried out using the usual industry techniques, the whey being spray dried. After ingestion, IgGs find and bind to respiratory pathogens, leading to immune modulation of the tonsils that make up Waldeyer's ring—the first immune defense line of the aero-digestive system which protects the entry from the external environment agents to the digestive and respiratory tubes [48]. The protective immunity against respiratory pathogens (influenza and RSV) is mediated by IgG and IgA (human immunoglobulins). Increased IgG levels in neonates are inversely associated with an increased prevalence of RSV infections; thus, the use of BC together with breastfeeding reduces the incidence and severity of these types of infections [49–51].

In the small intestine, IgGs exclude pathogens, preventing adhesion to epithelial surfaces and can promote the uptake of immune complexes of IgG with pathogens via Fc receptors (a protein found on the surface of cells that contributes to the protection of the immune system), which results in the regulatory immune responses and IgA induction, contributing to defense against pathogenic bacteria [52]. In the colon, IgGs prevent the leakage of LPS (endotoxin that causes responses by immune systems), modifying the composition of the microbiota and the production of short-chain fatty acids, still preventing the adhesion of pathogens [41,53].

The main role of immunoglobulins on mucous surfaces is binding to pathogens to prevent their entry into the body, a process called immune exclusion. Fig. 17–2 shows the proposed effects of BC.

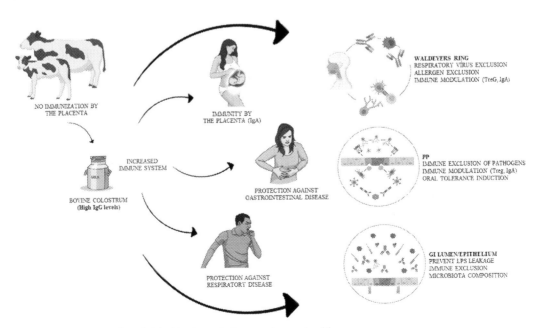

FIGURE 17–2 The main beneficial of bovine colostrum on human health.

Other components of BC are related to the innate immune system. Lactoferrin, for example, is a glycoprotein with antibacterial and antiviral properties, binding to lipopolysaccharides and regulating growth. Lactoperoxidase consists of an antibacterial enzyme capable of inhibiting bacterial metabolism, being toxic to a variety of gram-positive and gram-negative bacteria. The lytic lysozyme enzyme acts on the innate immune system, attacking the peptidoglycan cell constituents found in gram-positive bacteria, leading to bacterial lysis [8,27].

Patients with acquired immunodeficiency syndrome (AIDS) or patients infected with HIV are severely immunocompromised as a result of the absence of CD4 + T cells. With the GI widely exposed, they are not able to resist infections, becoming highly prone to diarrhea induced by pathogens such as Cryptosporidium, Amoebae, and Campylobacter [54,55].

The use of colostrum in these patients comes from reports that describe the positive effect of IgGs. Studies generally show the use of colostrum preparations as a dry lyophilized powder. Thus the particles allow a slower intestinal transport time than the liquid, in order to obtain the ideal exposure by the active components. Studies indicate that the use of preparations enriched with more than 65% of bovine IgG strongly reduces the severity and the occurrence of diarrhea associated with HIV. BC contains GFs, whose main forms IGF-1 and TGF-β2, are identical to human forms. These promote recovery of the mucosa and intestinal integrity in patients with severe diarrhea, with reduced scores of fatigue, weight gain, and CD4 + T cell count [54,56–58].

Some fats found in BC have anticancer properties, such as conjugated linolenic acid (CLA). It acts by stimulating the production of lymphokines and interleukins 2 by 32% and 29%, respectively, and by increasing certain levels of immunoglobins. CLA inhibits carcinogens by inducing apoptosis through a mechanism that involves inhibiting the synthesis of eicosanoid inflammatory mediators and reducing immunosuppressive substances such as leukotrienes and prostaglandins [59,60].

For athletes, BC is seen as an important ally. Research has shown that BC helps in increasing strength, endurance, lean muscle mass, burning body fat, immune function, reduces recovery time, and accelerates wound healing [61]. As a "health-promoting" product, BC is freeze-dried in a powder with a total protein content of up to 80%. Insulin-like growth factors (IGF-I and IGF-II) and growth hormone (GH) are found at high levels in colostrum, and IGF-I is the only natural hormone capable of promoting muscle growth by itself. Such hormone induces protein synthesis, which leads to an increase in lean muscle mass without a corresponding increase in adipose tissue (fat) and increases the uptake of glucose in the blood, thus facilitating the transport of glucose to the muscles that maintain levels high energy levels [62–64].

Another potential use of BC consists of IgG preparations to control bacterial populations in the formation of dental caries, one of the most common bacterial infections in humans. The pathogen *Streptococcus mutants* are associated as an organism that causes human tooth decay [65] and adheres to tooth surfaces by independent and sucrose-dependent mechanisms. When these factors are inhibited, the teeth are protected from colonization. The use of pasteurized BC preparations, which is usually performed at 72°C for 15 seconds, contains more than 60% immunoglobulin, of which 80% is IgG1, that inhibits the adhesion of

Streptococcus mutants and causes the aggregation of suspended bacteria, inhibiting the recolonization of saliva and protecting teeth against dental caries [66–68].

17.5 Bovine colostrum application for animal therapy

Due to its marked biological activity, no side effects, or drug interactions, BC is considered to be a therapeutic and nutraceutical effective for enhancing immune function in a wide range of animal species, such as calves, goats, pigs, pets, horses, and fishes [2,69–75]. BC immunoglobulins are not species specific, so other species can benefit from it, too. BC used in diets of supplementation can influence immune function, improving protective immune and gut responses to various stress factors, including infections [36].

The BC immunoglobulins are the best line of defense against invading pathogens in the newborn dairy calf, which are absorbed in the small intestine. Management strategies in order to reduce neonatal morbidity and mortality and consequently economic losses are required in large systems of cattle production [76]. Research has indicated that dairy farms often do not feed calves according to the best practice recommended, despite legislation, and industry advisory efforts [77]. Feeding newborn Holstein heifer calves with heat-treated colostrum could significantly increase apparent efficiency of absorption; therefore, IgG concentration in animals' blood serum [78]; decrease fecal scores, diarrhea, and pneumonia incidence [79]; stimulate the renewal of intestinal epithelium in the first life days [76].

Cattle colostrum contains immunoglobulin concentrations between five and ten times that found in serum [80]. Colostral whey contains even higher concentrations of immunoglobulins than colostrum and can be readily extracted by incubation with rennin followed by centrifugation [81]. A few researches on Bovine Tuberculosis (TB) or Johne's disease using colostrum or colostral whey are directed to parameterize antibody assays and wish to draw attention to the potential of these samples to increase diagnostic test sensitivities. Current bovine TB and Johne's disease research may be able to incorporate the collection of colostrum without major additional expense.

The relevance of colostrum in the competing horse area lies in several important features. Colostrum supplementation on these animal bodies can be an effective way to improve muscular strength, increase endurance and capacity, and increase lean body mass [82]. Respiratory disease in young race-horses has taken a significant economic toll in terms of missed training days, and often missed races as well [83]. In humans, the athletic performance enhancement has been attributed to an improvement in the anabolic cytokine IGF-1. BC supplementation did not increase serum IGF-1 concentrations in horses, and therefore, any effects of BC supplementation cannot be directly attributed to changes in serum IGF-1 [84].

During the first 24–48 hours of life (the period of passive immunity acquisition), the ruminants GI has special characteristics which allow the immunoglobulin uptake from colostrum [85]. In young goats, the colostrum can transmit some pathogens, especially caprine arthritis-encephalitis viruses, causing health problems and consequent losses in animal performance with high rates of mortality [86]. To prevent this problem and to offer adequate passive immunity to small ruminants, alternative IgG supplementation methods have been evaluated, such as banks of BC (refrigerated or frozen) and artificial colostrum [2,34,72]. The supplementation of lyophilized bovine colostrum (LBC) has been used to newborn small ruminants which is considered a promising alternative management [87–89].

Studies have shown that the enteric histology of young goats is not affected when fed with LBC, suggesting the use as an alternative IgG source to goat colostrum for young goats [89]. Absorption of higher proteins from LBC and the increase in maturity cell in the muscle and enteric tissues have also been observed in the first hours of newborns [70]. The higher is the absorption of bovine first milk secretion in the jejunum epithelium of the goat newborns, the lower is the number of goblet cells, highlighting a reaction of the intestinal epithelium with nonrecognized substances from the LBC, increasing secretion [90].

In preterm piglets (a model for preterm infants), BC improves absorptive and digestive functions, increases immunity, dampens inflammation, and protects against necrotizing enterocolitis (NEC) relative to composition [91–94]. It is known that colostrum administered as minimal enteral nutrition may prevent the inflammatory cascade leading to NEC lesions in preterm piglets [95], while human milk may reduce NEC in preterm infants compared with milk formula [96]. However, the ability of colostrum to regenerate an already committed gut that was exposed to a short period of formula feeding is unknown [91]. There is a need to know to which extent mother's own milk, or a possible substitute bioactive product like BC, may help to suppress the pro-inflammatory state of the immature intestine resulting from a few days of total parenteral nutrition followed by a period of formula feeding. Species-specific immune development may be important, but if BC protects against systemic infections in preterm piglets, it may be relevant to provide a colostrum supplement to the mother's own milk in the first weeks subsequential of the preterm birth in piglets [91].

BC is used as the initial feeding, preparing the immune and digestive systems in preterm infants when maternal milk is not available immediately after birth [97]. Colostrum supplementation also prevented brain barrier disruption and ameliorated the neuroinflammatory response during sepsis [91]. Used as a fortifier to donor human milk (DM), BC is better than formula-based fortifiers (FFs) as bacterial defense mechanisms, supporting nutrient absorption and gut function in preterm neonates [93]. Rasmussen et al. [92] suggest BC as an important supplement to mother's own milk (MM) and DM, stimulating enteral nutrient supply and gut maturation in preterm piglets.

It is not known yet if longer supplementation using BC as a fortifier to DM and MM would improve NEC, sepsis resistance, and growth. Sun et al. [94] provided preclinical evidence for a dietary-dependent response to the nutrient fortification of DM to preterm neonates and the results have suggested that BC is superior to the tested FFs in promoting nutrient metabolism, body growth, and intestinal maturation. It remains to be established whether BC supplementation may benefit sepsis-sensitive preterm piglets, particularly those with no access to their own mother's milk during the first days and weeks after preterm birth.

Perturbations in cerebral and systemic metabolism and the impact of BC, during bloodstream infection were investigated in a preterm piglet model [98]. Although few studies have evaluated the effect of BC on cytokine production [99], researches on the specific metabolic impact of BC intervention in infants are still lacking. The effect of BC supplementation on energy metabolism and significant results were identified on plasma levels of methionine, valine, and leucine, as well as the content of methionine in cerebrospinal fluid [98]. These effects reflect direct differences in amino acid absorption and quantity in BC and the parenteral formulation. However, it remains unknown if different levels of amino acids exert some biological effect.

Heat treatment, which includes spray drying and pasteurization, increases shelf life, reduces bacterial load, and facilitates the handling of BC products, but, on the other hand, the BC may also have affect on its bioactivity due to loss or reduction in bioactive factors [100]. Pasteurized, spray-dried BC reduces NEC and inflammation and increases gut function in preterm piglets [101]. Spray-dried BC does not markedly damage colostrum milk proteins, whereas the pasteurization can result in some protein aggregation and reduced concentration of bioactive factors.

Another recent scientific topic of interest is the effect of BC as a supplement incorporated in the fish diet [102]. The biological molecules which are present in colostrum can affect fish health and gut physiology [70,103]. The consumption of BC by fish could provide intestinal morphometrical alteration, as a consequence of the increased activity of absorption processes [104], changes in the pattern of goblet cells present in enteric mucosa [70], the indicators of cellular activity in muscle, liver and intestine tissue [105], and absence of implications in intestinal enzymatic activity [106].

The inclusion of LBC in the fish diet is an innovative idea that can elicit a protective effect in the aquatic organism [103]. Studies of LBC incorporated in fish diets have been shown to prompt enteric morphometrical alterations and changes in goblet cell density in enteric mucosa, with no implications in intestinal enzymatic activity [70,104,107]. According to the same authors, there was the possibility for this lacteal secretion to be used as a partial substitute for dietary protein commonly given in aquafeeds, complementing other protein sources.

Rodrigues et al. [108] and Pauletti et al. [105] observed the effect of LBC in the diet of *Pseudoplatystoma fasciatum* in the intestinal characteristics. Diets with LBC were nutritionally suitable for Pacu and Dourado species indicating its use as a partial substitute of protein diet [109]. The consumption of a diet containing LBC increases the superoxide dismutase activity in the gut of juveniles Pacu *Piaractus mesopotamicus*, indicating a protective action of colostrums [103]. Juvenile Dourado *Salminus brasiliensis* presents lower muscle RNA and RNA/DNA and higher TP/RNA when supplemented with LBC [110].

Bovine lactoferrin can repress the stress response in *Siberian sturgeon* [111]. In *Oreochromis niloticus* (Nile tilapia), the supplementation of bovine lysozyme in the dietary has increased resistance to diseases and nonspecific immunity [112]. Attenuation of fish stress and better performance rates are potential strategies to minimize the rate of mortality. The issue development and redox balance of juvenile *Piaractus mesopotamicus* submitted to high stocking density and diets containing LBC have shown that LBC could improve the protection of the enteric tissue from blood antioxidant and superoxide radicals capacity [113]. It seems that BC can be used in high concentration as a nutraceutical for fish, with a positive effect on tissue development and in redox balance.

17.6 Conclusions and perspectives

Unlike mature milk, colostrum has a higher concentration of substances and compounds, which are fundamental for good nutritional and physiological development of neonates. A large amount of BC is produced on dairy farms and in many countries, it is discarded due to the lack of knowledge of its beneficial properties. Some bioactive molecules which are present in the BC can promote health benefits for human and other animals, such as antimicrobial and immunomodulatory properties. As BC contains a high content of bio compounds, especially immunoglobulins, several clinical tests are being conducted to attest and prove the efficiency of the use of colostrum as an alternative therapy, for treatment and curative measurement in some diseases, including some pathologies caused by bacteria, especially those which have already developed resistance to antibiotics currently available on the market, and the use as food supplementation.

Due to its marked biological activity with no side effects or drug interactions, BC is considered a therapeutic and nutraceutical product in order to enhance immune function in a wide range of animal species. Therefore, a lot of products and processes have been developed to attend to human and animal health, besides that BC could be considered a new resource for circular product design e recycling, increasing the circular bioeconomy. However, the largest challenge is to effectively separate and purify the active molecules which are present in BC and there is still a need to establish processing procedures to ensure that the molecules do not lose their biological activity.

References

[1] A.W. Scammell, Production and uses of colostrum, Australian Journal of Dairy Technology 56 (2001) 74–82.

[2] D.B. Moretti, P. Pauletti, L. Kindlein, R. Machado-Neto, Enteric cell proliferation in newborn lambs fed bovine and ovine colostrum, Livestock Science 127 (2010) 262–266.

[3] E.G.S.O. Silva, A.H.N. Rangel, L. Mürmam, M.F. Bezerra, J.P.F. Oliveira, Bovine colostrum: benefits of its use in human food, Food Science and Technology 39 (2019) 355–362.

[4] B.R. Thapa, Therapeutic potentials of bovine colostrums, The Indian Journal of Pediatrics 72 (2005) 849–852.

[5] K. Stelwagen, E. Carpenter, B. Haigh, A. Hodgkinson, T.T. Wheeler, Immune components of bovine colostrum and milk, Journal of Animal Science 87 (2009) 3–9.

[6] H.S. Buttar, S.M. Bagwe, S.K. Bhullar, G. Kaur, Health benefits of bovine colostrum in children and adults, in: R.R. Watson, R.J. Collier, V.R. Preedy (Eds.), Dairy Health and Disease Across the Lifespan, Elsevier, London, 2017, pp. 3–20.

[7] E. Mizelman, W. Duff, S. Kontulainen, P.D. Chilibeck, The health benefits of bovine colostrum, in: R.R. Watson, R.J. Collier, V.R. Preedy (Eds.), Nutrients in Dairy and Their Implications for Health and Disease, Elsevier, London, 2017, pp. 51–60.

[8] M. Rathe, K. Müller, P.T. Sangild, S. Husby, Clinical applications of bovine colostrum Therapy: A Systematic Review, Nutrition Reviews 72 (2014) 237–254.

[9] W.L. Hurley, P.K. Theil, Perspectives on immunoglobulins in colostrum and milk, Nutrients 3 (2011) 442–474.

[10] M.G.B. Pagnoncelli, G.V. de Melo Pereira, M.J. Fernandes, V.O.A. Tanobe, C.R. Soccol, Milk immunoglobulins and their implications for health promotion, 2017, in: R.R. Watson, R.J. Collier, V.R. Preedy (Eds.), Nutrients in Dairy and Their Implications for Health and Disease, Elsevier, London, 2017, pp. 87–96.

[11] B.A. McGrath, P.F. Fox, P.L.H. McSweeney, A.L. Kelly, Composition and properties of bovine colostrum: a review, Dairy Science and Technology 96 (2015) 133–158.

[12] R.J. Playford, C.E. Macdonald, W.S. Johnson, Colostrum and milk-derived peptide growth factors for the treatment of gastrointestinal disorders, National Library of Medice 72 (2000) 5–14.

[13] C.K. Fenger, T. Tobin, P.J. Casey, E.A. Roualdes, J.L. Langemeier, R. Cowles, et al., Enhanced bovine colostrum supplementation shortens the duration of respiratory disease in thoroughbred yearlings, Journal of Equine Veterinary Science 42 (2016) 77–81.

[14] K. Saad, M.G.M. Abo-Elela, K.A.A. El-Baseer, A.E. Ahmed, F.A. Ahmad, M.S.K. Tawfeek, et al., Effects of bovine colostrum on recurrent respiratory tract infections and diarrhea in children, Medicine 95 (37) (2016) e4560.

[15] A.M. Ahnfeldt, N. Hyldig, Y. Li, S.S. Kappel, L. Aunsholdt, P.T. Sangild, et al., FortiColos—a multicentre study using bovine colostrum as a fortifier to human milk in very preterm infants: study protocol for a randomised controlled pilot trial, Trials 20 (2019) 279.

[16] J. Li, Y.W. Xu, J.J. Jiang, Q.K. Song, Bovine colostrum and product intervention associated with relief of childhood infectious diarrhea, Scientific Reports 9 (2019) 3093.

[17] V. Tripathi, B. Vashishtha, Bioactive compounds of colostrum and its application bioactive compounds of colostrum, Food Reviews International 22 (2006) 225–244.

[18] FAO, Food and Agriculture Organization, Milk and Dairy Products in Human Nutrition (2013).

[19] O. Ballard, A.L. Morrow, Human milk composition: nutrients and bioactive factors, Pediatric Clinics of North America 60 (2013) 49–74.

[20] A.M.A. El-Fattah, F.H.R.A. Rabo, S.M. EL-Dieb, H.A. El-Kashef, Changes in composition of colostrum of Egyptian buffaloes and Holstein cows, BMC Veterinary Research 8 (2012) 19.

[21] H.J. Korhonen, Bioactive milk proteins and peptides: from science to functional applications, Australian Journal of Dairy Technology 64 (2009) 16–25.

[22] D.E.W. Chatterton, D.N. Nguyen, S.B. Bering, P.T. Sangild, Anti-inflammatory mechanisms of bioactive milk proteins in the intestine of newborns, The International Journal of Biochemistry & Cell Biology 45 (2013) 1730–1747.

[23] I. Espinosa-Martos, A. Montilla, A.G. Segura, D. Escuder, G. Bustos, C. Pallás, et al., Bacteriological, biochemical, and immunological modifications in human colostrum after holder pasteurisation, Journal of Pediatric Gastroenterology and Nutrition 56 (2013) 560–568.

[24] R.G. Jensen, Lipids in human milk, Lipids 34 (1999) 1243–1271.

[25] C. Boyce, M. Watson, G. Lazidis, S. Reeve, K. Dods, K. Simmer, et al., Preterm human milk composition: a systematic literature review, British Journal of Nutrition 116 (2016) 1033–1045.

[26] I.A. Zanker, H.M. Hammon, J.W. Blum, Activities of γ-glutamyltransferase, alkaline phosphatase and aspartate-aminotransferase in colostrum, milk and blood plasma of calves fed first colostrum at 0–2, 6–7, 12–13 and 24–25 h after birth, Journal of Veterinary Medicine Series A 48 (2001) 179–185.

[27] R. Pakkanen, J. Aalto, Growth factors and antimicrobial factors of bovine colostrum, International Dairy Journal 7 (1997) 285–297.

[28] A. Chae, A. Aitchison, A.S. Day, J.I. Keenan, Bovine colostrum demonstrates anti-inflammatory and antibacterial activity in in vitro models of intestinal inflammation and infection, Journal of Functional Foods 28 (2017) 293–298.

[29] S. Bagwe, L.J.P. Tharappel, G. Kaur, H.S. Buttar, Bovine colostrum: an emerging nutraceutical, Journal of Complementary and Integrative Medicine 12 (2015) 175–185.

[30] P. Sacerdote, F. Mussano, S. Franchi, A.E. Panerai, G. Bussolati, S. Carossa, et al., Biological components in a standardized derivative of bovine colostrum, Journal of Dairy Science 96 (2013) 1745–1754.

[31] X. Zou, Z. Guo, Q. Jin, J. Huang, L. Cheong, X. Xu, et al., Composition and microstructure of colostrum and mature bovine milk fat globule membrane, Food Chemistry 185 (2015) 362–370.

[32] A. Coroian, S. Erler, C.T. Matea, V. Mirsane, C. Rãducu, C. Bele, et al., Seasonal changes of buffalo colostrum: physicochemical parameters, fatty acids and cholesterol variation, Chemistry Central Journal 7 (2013) 40.

[33] G.T. Bleck, M.B. Wheeler, L.B. Hansen, D.J. Miller, Lactose synthase components in milk: concentrations of alpha-lactalbumin and beta 1,4-galactosyltransferase in milk of cows from several breeds at various stages of lactation, Reproduction in Domestic Animals 44 (2009) 241–247.

[34] D.B. Moretti, L. Kindlein, P. Pauletti, R. MacHado-Neto, IgG absorption by Santa Ines lambs fed Holstein bovine colostrum or Santa Ines ovine colostrum, Animal 4 (6) (2010) 933–937.

[35] B. Resch, N. Hofer, S. Kurath-Koller, Is there enough evidence for the use of immunoglobulins in either prevention or treatment of bacterial infection in preterm infants? Journal of Neonatal Nursing 21 (2015) 88–92.

[36] E. Satyaraj, A. Reynolds, R. Pelker, J. Labuda, P. Zhang, P. Sun, Supplementation of diets with bovine colostrum influences immune function in dogs, British Journal of Nutrition 110 (2013) 2216–2221.

[37] A.S. Cross, H.J. Karreman, L. Zhang, Z. Rosenberg, S.M. Opal, A. Lees, Immunization of cows with novel core glycolipid vaccine induces anti-endotoxin antibodies in bovine colostrum, Vaccine 32 (2014) 6107–6114.

[38] S. Donovan, J. Odle, Growth factors in milk as mediators of infant development, Annual Review of Nutrition 14 (1994) 147–167.

[39] N.W. Solomons, Modulation of the immune system and the response against pathogens with bovine colostrum concentrates, European Journal of Clinical Nutrition 56 (2002) S24–S28.

[40] J. Steele, J. Sponseller, D. Schmidt, O. Cohen, S. Tzipori, Hyperimmune bovine colostrum for treatment of GI infections: a review and update on *Clostridium difficile*, Human Vaccines & Immunotherapeutics 9 (2013) 1565–1568.

[41] M.H. Labbok, D. Clark, A.S. Goldman, Breastfeeding: maintaining an irreplaceable immunological resource, Nature Reviews Immunology 4 (2004) 565–572.

[42] G. Loss, M. Depner, L.H. Ulfman, R.J.J. Van Neerven, V. Kaulek, A.J. Hose, et al., Consumption of unprocessed cow's milk protects infants from common respiratory infections, Asthma and Lower Airway Disease 135 (2015) 56–62.

[43] V.S. Jasion, B.P. Burnett, Survival and digestibility of orally-administered immunoglobulin preparations containing IgG through the gastrointestinal tract in humans, Nutrition Journal 14 (2015) 22.

[44] H. Hilpert, H. Brüssow, C. Mietens, J. Sidoti, L. Lerner, H. Werchau, Use of bovine milk concentrate containing antibody to rotavirus to treat rotavirus gastroenteritis in infants, The Journal of Infectious Diseases 156 (1987) 158–166.

[45] N. Roos, S. Mahé, R. Benamouzig, J. Sick, J. Rautureau, D. Tomé, 15N-labeled immunoglobulins from bovine colostrum are partially resistant to digestion in human intestine, The Journal of Nutrition 125 (1995) 1238–1244.

[46] J.A. Rump, R. Arndt, A. Arnold, C. Bendick, H. Dichtelm, M. Franke, et al., Treatment of diarrhoea in human immunodeficiency virus-infected patients with immunoglobulins from bovine colostrum, The Clinical Investigator 70 (1992) 588–594.

[47] G.S. Kelly, Bovine colostrums a review of clinical uses, Alternative Medicine Review 8 (2003) 378–394.

[48] G. Hartog, S. Jacobino, L. Bont, L. Cox, L.H. Ulfman, J.H.W. Leusen, et al., Specificity and effector functions of human RSV-specific IgG from bovine milk, PLoS One 9 (2014) e112047.

[49] R.J. Cox, K.A. Brokstad, P. Ogra, Influenza virus: immunity and vaccination strategies. Comparison of the immune response to inactivated and live, attenuated influenza vaccines, Scandinavian Journal of Immunology 59 (2004) 1–15.

[50] J.C. Rebecca, K.A. Brokstad, M.A. Zuckerman, J.M. Wood, L.R. Haaheim, J.S. Oxford, An early humoral immune response in peripheral blood following parenteral inactivated influenza vaccination, Vaccine 12 (1994) 993–999.

[51] E.E. Walsh, A.R. Falsey, Humoral and mucosal immunity in protection from natural respiratory syncytial virus infection in adults, The Journal of Infectious Disease 190 (2004) 373–378.

[52] P. Brandtzaeg, The mucosal immune system and its integration with the mammary glands, The Journal of Pediatrics 156 (2010) S8–S15.

[53] A.J. Hodgkinson, J. Cakebread, M. Callaghan, P. Harris, R. Brunt, R.C. Anderson, et al., Comparative innate immune interactions of human and bovine secretory IgA with pathogenic and non-pathogenic bacteria, Developmental & Comparative Immunology 68 (2017) 21–25.

[54] C. Flore, S. Chinenye, L. Elfstrand, C. Hagman, I. Ihse, ColoPlus, a new product based on bovine colostrum, alleviates HIV-associated diarrhoea, Scandinavian Journal of Gastroenterology 41 (2006) 682–686.

[55] F.O. Kaducu, S.A. Okia, G. Upenytho, L. Elfrstrand, C.H. Florén, Effect of bovine colostrum-based food supplement in the treatment of HIV-associated diarrhea in Northern Uganda: a randomized controlled trial, Indian Journal of Gastroenterology 30 (2011) 270–276.

[56] D.M. Asmuth, Z. Ma, A. Albanese, N.G. Sandler, S. Devaraj, T.H. Knight, et al., Oral serum-derived bovine immunoglobulin improves duodenal immune reconstitution and absorption function in patients with HIV enteropathy, AIDS 27 (2013) 2207–2217.

[57] R.J.J. Neerven, The effects of milk and colostrum on allergy and infection: mechanisms and implications, Animal Frontiers 4 (2014) 16–23.

[58] P.O. Odong, P.J. Angwech, J. Obol, J. Kuule, C. Florén, Management of HIV in children using a bovine colostrum-based food product: an observational field study, World Journal of AIDS 5 (2015) 100–104.

[59] M.L. Godhia, N. Patel, Colostrum - its composition, benefits as a nutraceutical: a review, Current Research in Nutrition Food Science 1 (2013) 37–47.

[60] H.B. MacDonald, Conjugated linoleic acid and disease prevention: a review of current knowledge, Journal of the American College of Nutrition 19 (2000) 111S–118S.

[61] Z. Hofman, R. Smeets, G. Verlaan, P.A. Verstappen, The effect of bovine colostrum supplementation on exercise performance in elite field hockey players, International Journal of Sport Nutrition and Exercise Metabolism 12 (2002) 461–469.

[62] J. Antonio, M.S. Sanders, D. Gammeren, The effects of bovine colostrum supplementation on body composition and exercise performance in active men and women, Nutrition 17 (2001) 243–247.

[63] T. Marchbank, G. Davison, J.R. Oakes, M.A. Ghatei, M. Patterson, M.P. Moyer, et al., The nutriceutical bovine colostrum truncates the increase in gut permeability caused by heavy exercise in athletes, American Journal of Physiology Gastrointestinal and Liver Physiology 300 (2011) 477–484.

[64] A. Mero, H. Miikkulainen, J. Riski, R. Pakkanen, J. Aalto, T. Takala, Effects of bovine colostrum supplementation on serum IGF-I, IgG, hormone, and saliva IgA during training, Journal of Applied Physiology 83 (1997) 1144–1151.

[65] R.R.B. Russell, The application of molecular genetics to the microbiology of dental caries, Caries Research 28 (1994) 69–82.

[66] T. Koga, T. Oho, Y. Shimazaki, Y. Nakano, Immunization against dental caries, Vaccine 20 (2002) 2027–2044.

[67] Y. Shimazaki, M. Mitoma, T. Oho, Y. Nakano, Y. Yamashita, K. Okano, et al., Passive immunization with milk produced from an immunized cow prevents oral recolonization by *Streptococcus mutans*, Clinical and Diagnostic Laboratory Immunology 8 (2001) 1136–1139.

[68] H.A.O. Yu, Y. Nakano, Y. Yamashita, T. Oho, Effects of antibodies against cell surface protein antigen PAc-glucosyltransferase fusion proteins on glucan synthesis and cell adhesion of *Streptococcus mutans*, Infection and Immunity 65 (1997) 2292–2298.

[69] V.J.S. Abreu, A.L. Cardoso, H.F.J. Pena, S.M. Gennari, I. Sinhorini, S.B. Damy, Evaluation of the efficacy of hyperimmune bovine colostrum on *Cryptosporidium parvum* experimental infection of rodents, Brazilian Journal of Veterinary Research and Animal Science 40 (2003) 191–198.

[70] T.M.P. Cruz, D.B. Moretti, W.M. Nordi, J.E.P. Cyrino, R. Machado-Neto, Intestinal epithelium of juvenile dourado Salminus brasiliensis (Cuvier, 1816) fed diet with lyophilized bovine colostrum, Aquaculture Research 47 (2014) 561–569.

[71] A. Huguet, J. Dividich, I. Huërou-Luron, Improvement of growth performance and sanitary status of weaned piglets fed a bovine colostrum-supplemented diet, Journal of Animal Science 90 (2011) 1513–1520.

[72] A.L. Lima, P. Pauletti, I. Susin, R. Machado-neto, Fluctuation of serum variables in goats and comparative study of antibody absorption in new-born kids using cattle and goat colostrum, Brazilian Journal of Zootechnology 38 (2009) 2211–2217.

[73] L.J. Palczynski, E.C.L. Bleach, M.L. Brennan, P.A. Robinson, Giving calves "the best start": perceptions of colostrum management on dairy farms in England, Animal Welfare 29 (2020) 45–58.

[74] N.N. Pandey, A.A. Dar, D.B. Mondal, L. Nagaraja, Bovine colostrum: a veterinary nutraceutical, Journal of Veterinary Medicine and Animal Health 3 (2011) 31–35.

[75] M.J.R. Sturaro, E.M. Pituco, C.M.M. Bittar, L.G.R. Stuaro, R.G. Saraiva, Transfer colostrum of passive immunity by antibody in Murrah calf buffalo, Buffalo Bulletin 39 (2020) 27–33.

[76] L. Kindlein, D.B. Moretti, P. Pauletti, A.R. Bagaldo, A.P.O. Rodrigues, R. Machado-Neto, Bovine colostrum enriched with lyophilized bovine colostrum stimulates intestinal epithelium renewal of Holstein calves in the first days of life, Journal of Animal Physiology and Animal Nutrition 102 (2018) 514–524.

[77] L.J. Palczynski, E.C.L. Bleach, M.L. Brennan, P.A. Robinson, Appropriate dairy calf feeding from birth to weaning: "it's an investment for the future", Animals 10 (2020) 116.

[78] E. Salazar-Acosta, J.A. Elizondo-Salazar, Heat treatment of colostrum increases immunoglobulin absorption in Holstein heifer calves, Agronomía Mesoamericana 30 (2019) 229–238.

[79] M. Rafiei, T. Ghoorchi, A. Toghdory, M. Moazeni, M. Khalili, Effect of feeding heat-treated and unheated colostrum on immunoglobulin G absorption, health and performance of neonatal Holstein dairy calves, Acta Scientiarum. Animal Sciences 41 (2019) e45533.

[80] C.R. Baumrucker, A.M. Burkett, A.L. Magliaro-Macrina, C.D. Dechow, Colostrogenesis: mass transfer of immunoglobulin G1 into colostrum, Journal of Dairy Science 93 (2010) 3031–3038.

[81] P.D. Cockcroft, C. Jenvey, M.P. Reichel, Role for colostrum and whey in testing for bovine TB and Johne's disease? Veterinary Record 175 (2014) 597.

[82] M. Borissenko, Colostrum and the performance horse, Institute of Colostrum Research (2002) 1–13.

[83] J. Hernandez, D.L. Hawkins, Training failure among yearling horses, American Journal of Veterinary Research 62 (2001) 1418–1422.

[84] C. Fenger, T. Tobin, P. Casey, J. Langemeier, D. Haines, Bovine colostrum supplementation does not influence serum insulin-like growth factor-1 in horses in race training, Journal of Equine Veterinary Science 34 (2014) 1025–1027.

[85] L. Kindlein, P. Pauletti, Effects of enriched colostrum supply in intestinal mucosa morphology of newborn calves, Acta Scientiae Veterinariae 36 (2008) 31–34.

[86] E. Peterhans, T. Greenland, J. Badiola, G. Harkiss, G. Bertoni, B. Amorena, et al., Routes of transmission and consequences of small ruminant lentiviruses (SRLVs) infection and eradication schemes, Veterinay Research 35 (2004) 257–274.

[87] D.B. Moretti, W.M. Nordi, A.L. Lima, P. Pauletti, I. Susin, R. Machado-Neto, Goat kids' intestinal absorptive mucosa in period of passive immunity acquisition, Livestock Science 144 (2012) 1–10.

[88] D.B. Moretti, W.M. Nordi, A.L. Lima, P. Pauletti, I. Susin, R. Machado-Neto, Lyophilized bovine colostrum as a source of immunoglobulins and insulin-like growth factor for newborn goat kids, Livestock Science 145 (2012) 223–229.

[89] W.M. Nordi, D.B. Moretti, A.L. Lima, P. Pauletti, I. Susin, R. Machado-Neto, Intestinal histology of newborn goat kids fed lyophilized bovine colostrum, Czech Journal of Animal Science 58 (2013) 232–241.

[90] R. Machado-Neto, M.C.F. Pontin, W.M. Nordi, A.L. Lima, D.B. Moretti, Goblet cell mucin distribution in the small intestine of newborn goat kids fed lyophilized bovine colostrum, Livestock Science 157 (2013) 125–131.

[91] A. Brunse, P. Worsøe, S.E. Pors, K. Skovgaard, P.T. Sangild, Oral supplementation with bovine colostrum prevents septic shock and brain barrier disruption during bloodstream infection in preterm newborn pigs, Shock 51 (2019) 337–347.

[92] S.O. Rasmussen, L. Martin, M.V. Østergaard, S. Rudloff, Y. Li, M. Roggenbuck, et al., Bovine colostrum improves neonatal growth, digestive function, and gut immunity relative to donor human milk and infant formula in preterm pigs, American Journal of Physiology - Gastrointestinal and Liver Physiology 311 (2016) G480–G491.

[93] J. Sun, Y. Li, X. Pan, D.N. Nguyen, A. Brunse, A.M. Bojesen, et al., Human milk fortification with bovine colostrum is superior to formula-based fortifiers to prevent gut dysfunction, necrotizing enterocolitis, and systemic infection in preterm pigs, Journal of Parenteral and Enteral Nutrition 43 (2019) 252–262.

[94] J. Sun, Y. Li, D.N. Nguyen, M.S. Mortensen, C.H.P. van den Akker, T. Skeath, et al., Nutrient fortification of human donor milk affects intestinal function and protein metabolism in preterm pigs, The Journal of Nutrition 148 (2018) 336–347.

[95] M.S. Cilieborg, M. Boye, T. Thymann, B.B. Jensen, P.T. Sangild, Diet-dependent effects of minimal enteral nutrition on intestinal function and necrotizing enterocolitis in preterm pigs, Journal of Parenteral and Enteral Nutrition 35 (2011) 32–42.

[96] P.W. Lin, B.J. Stoll, Necrotising enterocolitis, Lancet 368 (2006) 1271–1283.

[97] Y. Li, X. Pan, D.N. Nguyen, S. Ren, A. Moodley, Bovine colostrum before or after formula feeding improves systemic immune protection and gut function in newborn preterm pigs, Frontiers in Immunology 10 (2020) 3062.

[98] M. Alinaghi, P.P. Jiang, A. Brunse, P.T. Sangild, H.C. Bertram, Rapid cerebral metabolic shift during neonatal sepsis is attenuated by enteral colostrum supplementation in preterm pigs, Metabolites 9 (2019) 13.

[99] B. Balachandran, S. Dutta, R. Singh, R. Prasad, P. Kumar, Bovine colostrum in prevention of necrotizing enterocolitis and sepsis in very low birth weight neonates: a randomized, double-blind, placebo-controlled pilot trial, Journal of Tropical Pediatrics 63 (2017) 10–17.

[100] Y. Li, M.V. Østergaard, P. Jiang, D.E.W. Chatterton, T. Thymann, A.S. Kvistgaard, et al., Whey protein processing influences formula-induced gut maturation in preterm pigs, The Journal of Nutrition 143 (2013) 1934–1942.

[101] A.C.F. Støy, P.T. Sangild, Ā.K. Skovgaard, T. Thymann, Spray dried, pasteurised bovine colostrum protects against gut dysfunction and inflammation in preterm pigs, Journal of Pediatric Gastroenterology and Nutrition 63 (2016) 280–287.

[102] W.M. Nordi, D.B. Moretti, T.M.P. da Cruz, J.E.P. Cyrino, R. Machado-Neto, Cellular activity and development of enteric, hepatic and muscle tissues of juvenile Pacu Piaractus mesopotamicus (Holmberg 1887) with lyophilized bovine colostrum, Aquaculture Research 48 (2015) 1099–1109.

[103] D.B. Moretti, W.M. Nordi, T.M.P. Cruz, R. Machado-Neto, Catalase, superoxide dismutase, glutathione peroxidase and oxygen radical absorbance capacity in the gut of juvenile Pacu *Piaractus mesopotamicus*

and dourado *Salminus brasiliensis* fed bovine first milk secretion, Latin American Journal of Aquatic Research 45 (2017) 717–723.

[104] A.P.O. Rodrigues, P. Pauletti, L. Kindlein, E.F. Delgado, J.E.P. Cyrino, R. Machado-Neto, Intestinal histomorphology in Pseudoplatystoma fasciatum fed bovine colostrum as source of protein and bioactive peptides, Scientia Agricola 67 (2010) 524–530.

[105] P. Pauletti, L. Kindlein, A.R. Bagaldo, A.P.O. Rodrigues, E.F. Delgado, J.E.P. Cyrino, et al., Neto, Growth performance and muscle protein, RNA and DNA contents in juveniles Pseudoplatystoma fasciatum (Teleostei: Pimelodidae) fed lyiphilized bovine colostrum, Journal of Animal Science 85 (2007) 89.

[106] D.B. Moretti, W.M. Nordi, T.M.P. Cruz, J.E.P. Cyrino, R. Machado-Neto, Histochemical distribution of intestinal enzymes of juvenile Pacu (*Piaractus mesopotamicus*) fed lyophilized bovine colostrum, Fish Physiology and Biochemistry 40 (2014) 1487–1493.

[107] D.B. Moretti, W.M. Nordi, A.L. Lima, P. Pauletti, R. Machado-Neto, Enteric, hepatic and muscle tissue development of goat kids fed with lyophilized bovine colostrum, Journal of Animal Physiology and Animal Nutrition 98 (2013) 201–208.

[108] A.P.O. Rodrigues, P. Pauletti, L. Kindlein, J.E.P. Cyrino, E.F. Delgado, R. MacHado-Neto, Intestinal morphology and histology of the striped catfish Pseudoplatystoma fasciatum (Linnaeus, 1766) fed dry diets, Aquaculture Nutrition 15 (2009) 559–563.

[109] R. Machado-Neto, D.B. Moretti, W.M. Nordi, T.M.P. da Cruz, J.E.P. Cyrino, Growth performance of juvenile Pacu (*Piaractus mesopotamicus*) and dourado (*Salminus brasiliensis*) fed with lyophilized bovine colostrum, Aquaculture Research 47 (2015) 3551–3557.

[110] W.M. Nordi, D.B. Moretti, T.M.P. da Cruz, J.E.P. Cyrino, R. Machado-Neto, Enteric, hepatic and muscle tissue development of juvenile dourado *Salminus brasiliensis* fed with lyophilized bovine colostrum, Fisheries Science 82 (2016) 321–326.

[111] B. Falahatkar, K. Eslamloo, S. Yokoyama, Suppression of stress responses in Siberian Sturgeon, Acipenser Baeri, juveniles by the dietary administration of bovine lactoferrin, Journal of World Aquaculture Society 45 (2014) 699–708.

[112] A.M.M. El-Ashram, M.E. El-Boshy, Assessment of dietary bovine lactoferrin in enhancement of immune function and disease resistance in Nile Tilapia (*Oreochromis Niloticus*), The Eighth International Symposium on Tilapia in Aquaculture 2 (2008) 1097–1107.

[113] D.B. Moretti, W.M. Nordi, R. Machado-Neto, Redox balance and tissue development of juvenile piaractus mesopotamicus subjected to high stocking density and fed dry diets containing nutraceutical food, Latin American Journal Aquatic Research 47 (2019) 423–432.

18

Colostrum new insights: products and processes

Alessandra Cristine Novak Sydney[1], Isadora Kaniak Ikeda[1], Maria Carolina de Oliveira Ribeiro[1], Eduardo Bittencourt Sydney[1], Dão Pedro de Carvalho Neto[2], Susan Grace Karp[2], Cristine Rodrigues[2], Carlos Ricardo Soccol[2]

[1]FEDERAL UNIVERSITY OF TECHNOLOGY – PARANA (UTFPR) - CAMPUS PONTA GROSSA, PONTA GROSSA, PARANÁ, BRAZIL [2]FEDERAL UNIVERSITY OF PARANÁ—(UFPR), CURITIBA, PARANÁ, BRAZIL

18.1 Introduction

Colostrum is a mammary secretion produced immediately after parturition (24–48 hours) composed of nutritional components (e.g., carbohydrates, proteins, fat, vitamins, and minerals), immune modulators (e.g., β-lactoglobulin, secretory IgA, lactalbumin, α2-macroglobulin, and cytokines), and growth factors [1–5]. In the firsts hours, it serves for the newborn as an energy source for heat generation, passive immunity (by means of Ig acquiring), and also as a laxative, allowing the meconium exit, being also important for the development of the digestive system [6].

Although all mammals are able to produce it, researches demonstrated that bovine colostrum (BC) is up to 100-fold more potent than human's due to high concentrations of immunoglobulin G (IgG), and lactoferrin, which are responsible for immunological modulation and absorption of iron in the intestine, respectively [5,7]. In addition, the secretion of antigen-specific antibodies in BC was proven a reliable heterologous source of passive immunity against polio in humans during the 20th century [8–10]. The high nutraceutical potential of BC and the fact that it is considered safe with transient, minimal side effects—nausea, vomiting, and flatulence—stimulated its therapeutic applications. Several clinical investigations demonstrated that BC shows positive results in the treatment of allergies [11], respiratory and gastrointestinal infections [12,13], and as a nutritional supplement to increase immunity on athletes [14,15]. These early evidences stimulated the creation of a new market niche of colostrum-based products.

The first applications occurred in the 1990s and were addressed for the fortification of infant formulas with immunoglobulins (Ig) from bovine milk, mainly for neonates which cannot be breast feeding. Colostrum could be responsible for protection against *Escherichia coli*, *Cryptosporidium* spp. and also rotavirus, preventing diarrhea, leading the application of bovine Ig in colostrum as immunological supplements [16]. In the 1990s, researchers indicated the effect of heating on the quality of colostrum [16]. After severe thermal treatment, most of the colostrum and milk's antibodies were denaturized, generating new challenges for the processing of colostrum to make it a safer product without losing its qualities. Until now, hundreds of articles have been published about processing colostrum and how its quality can be maintained or even improved. They have discussed the techniques and equipment, physico-chemical, microbiological, and bioactive properties after processing.

At birth, mammalians have an immature immune system, and until their own immune system develops and becomes effectively protective, the colostrum, rich in Ig, acts as the first barrier by means of passive transfer of immunity. Colostrum can be used to feed newborns who cannot be breastfed, as a supplement when maternal colostrum has low quality or is produced at insufficient volumes, to prevent infections [17]. Calves fed with colostrum instead of milk replacer, for example, showed best performances in growth and health [17,18].

Generally, colostrum replacers are processed to have a low level of bacteria reducing the exposure to pathogens and preventing the failure of Ig passive transfer. It can be further enriched with IgG from whey, milk, or colostrum, from blood or egg, to improve the Ig absorption by newborn [17]. Despite the BC being the most studied, colostrum from other sources such as sows [19], ewe [6], and goat [20] is being studied and considered potential materials for the development of new products. This chapter covers the most recent studies and developments on Colostrum technology, including a review of the patented technologies of the last 3 years and available commercial products based on containing colostrum.

18.2 New insights in colostrum production process and quality control

To obtain good quality colostrum many factors can be modulated. In the last years, studies indicated that the main parameter is to raise the flock appropriately, which also positively impacts the produced volume. Colostrum as a raw material must be processed in order to maintain its properties. Advances in analytical methods are crucial for the development of rapid and precise colostrum quality testing, which may help to take crucial decisions quickly.

18.2.1 Impact of flock managing in colostrum quality

The colostrum's quality control is similar to quality control protocols in the food industry, for example, and can be managed with the Best Practice of Production (BPP). BPP is indicated for the microbiological control of colostrum in three critical points: suckling colostrum from mother, feeding colostrum using bottles and when the colostrum is milked from large herds [21].

Feeding procedures and management have also an important role in colostrum quality. During flock management, they could be exposed to many toxins. Deoxyvalenol (DON) is one of the most common mycotoxins; it is produced mainly by species of fungus *Fusarium* sp., which contaminates wheat, maize, and barley and is especially toxic for pigs. DON and its metabolites can pass by sows to piglets through the intrauterine path and was proven to pass via colostrum, even in trace amounts, in cows, sheeps, and pigs. It causes vomiting and diarrhea, impacting weight gain and general health. Suitable stock and management of food are necessary to avoid these problems [22].

Studies indicated that flock management is also crucial for successful colostrum production. Procedures before and after calving are also relevant to obtain high-quantity and quality colostrum. A study performed in Germany using parturient cows evaluated the effect of oxytocin application after calving and the presence of calves during colostrum milking in the colostrum quality. Researchers observed that the number of parities, calf birth weight, and calving, among others, were related to the quantity of colostrum produced (Table 18−1). It is possible to conclude that first or third (or more) parity and nocturne calving impacted positively in colostrum quantity (4.75 and 4.93 kg, respectively) [23].

Other factors also have influenced the BC quality and quantity (Table 18−2). Among which superior colostrum amount and the bigger time interval between calving and milking

Table 18.1 Impact of parity number and day's period of calving in quantity of colostrum produced.

Parity number	Quantity of colostrum (kg)
1	4.75 ± 0.34
2	3.74 ± 0.37
3 +	4.75 ± 0.38
Day's period of calving	**Quantity of colostrum (kg)**
Night (22:00−6:00 h)	4.93 ± 0.37
Morning (6:00−14:00 h)	4.17 ± 0.38
Afternoon (14:00−22:00 h)	4.14 ± 0.34

Based on: F. Sutter, S. Borchardt, G.M. Schuenemann, E. Rauch, M. Erhard, W. Heuwieser, Evaluation of 2 different treatment procedures after calving to improve harvesting of high-quantity and high-quality colostrum, Journal of Dairy Science 102 (10) (2019) 9370−9381 [23].

Table 18−2 Effect of treatment (oxytocin, presence/absence of calves), parity number and day's period of calving in colostrum quality, in terms of IgG.

Treatment	IgG concentration (mg of IgG/mL)
Application of Oxytocin	57
Presence of calves	56
Absence of calves (control)	50.7
Parity number	**IgG concentration (mg of IgG/mL)**
1	48.5 ± 2.86
2	50.7 ± 2.89
3 +	64.6 ± 2.59
Day's period of calving	**IgG concentration (mg of IgG/mL)**
Night (22:00−6:00 h)	60.4 ± 2.92
Morning (6:00−14:00 h)	51.9 ± 2.98
Afternoon (14:00−22:00 h)	51.3 ± 2.71
Sunday (quiet days)	61.4 ± 3.70

Based on: F. Sutter, S. Borchardt, G.M. Schuenemann, E. Rauch, M. Erhard, W. Heuwieser, Evaluation of 2 different treatment procedures after calving to improve harvesting of high-quantity and high-quality colostrum, Journal of Dairy Science 102 (10) (2019) 9370−9381 [23].

affected negatively in the colostrum quality. Nocturne calving also results in an elevated quality in terms of IgG besides better production in quantity [23].

It's already known that gestation length, dam age, and dam breed can also affect colostrum production. Studies showed that calving during calm days (Sundays) resulted in the best colostrum quality, reinforcing the importance of flock management. Nowadays animal welfare is a constant concerning for the producer providing a quiet place for calving can be both an animal-friendly procedure with profitable returns.

The fluctuations on the metabolic behavior of pregnant females are caused by different gestation phases and reflect the embryo requirements. A study performed with sows in the last gestation phase proved that females not only require constant feeding but also more protein. Feeding gestating sows with different energy levels with fixed or increasing proportions of protein were evaluated in relation to its effect on piglet birth weight, weight gain and blood analysis, and colostrum quality. The feeding with increased protein content resulted in increased fat levels in the prepartum colostrum. This could be used as a strategy to obtain colostrum with best levels of fat, amplifying the range of colostrum use [19]. Another study showed the use of living yeasts as a safe nonanimal protein source to increase the protein content in the feed of late gestation and lactation sows [24].

These studies sustain that many of the flock characteristics can affect the colostrum quality and quantity, independent of the species studied. This information can support medium- and long-term decisions and may support the development of new strategies or methodologies, aiming to produce the better possible colostrum.

18.2.2 Effect in colostrum quality by processing

The colostrum therapeutic properties are best preserved in the dried form. Longer storage time and high temperature (above 25°C) influenced negatively on the content of Ig, thus lower storage temperatures are recommended even in the production of colostrum powder by spray drying [1].

Despite colostrum has been safely used for many years, it is necessary to update the basic toxicity information of natural and processed products. Using bacterial reverse mutation, in vitro chromosomal aberration test, and in vivo mammalian micronucleus test, Thiel et al. identified no adverse effects in organs or mortality of BC after 90 days of exposure, the NOAEL (no-observed-adverse-effect-level) was determined as 4200 mg/kg bw/day [25]. Recent studies involving colostrum processing include enzymatic extraction, protein precipitation, processes to separate protein fractions, and lactose are described in Section 18.3.

Colostrum is commonly stored in frozen form. Before freezing, it undergoes a thermal treatment, such as pasteurization. A study evaluated, pasteurization at 60°C/60 minutes affected negatively in the increase in blood IgG concentration, but it was not clear if the problem was IgG availability or absorption [26]. An Egyptian study indicates that a colostrum treatment by fast pasteurization (80°C/15 seconds) did not completely denatured IgG and IgM. They demonstrated that ultrafiltration could be used in the production of an IgG-rich retentate, which can be used to produce, for example, a rich-IgG cheese [27].

18.2.3 Quality parameters

Serious problems can occur to newborns when colostrum is not available in sufficient quantities or has low quality, including death. Lamb mortality reaches 15%–20% of total lambs birth [6], for example. It is crucial to quickly determine the colostrum quality to take the right decision of replacing or supplementing breastfeeding if necessary, just in time that it is needed, preserving the newborn's life and improving its health.

Refractometry can be a useful tool for determining serum total protein and brix in animal blood. A serum total protein content of 4.2 g/dL is related to the presence of 10 mg/mL of IgG [17], a value that can be used, carefully, for decision-making. The IgG level needed in newborns to avoid infections is not known, but often concentrations below 50 g/L are considered inappropriate and indicate the need of high-quality colostrum ingestion for supplementation [6]. This is a promising strategy that demonstrates the rapid and in loco quality tools can actively help in flock managing. Colostrum could contain toxic compounds that should be removed or neutralized before consumption, such as the bovine leukemia virus (BLV). Studies indicate that spray-drying is effective in inactivating infectious BLV in colostrum, resulting in a reduced risk of BLV transmission. However, the functional properties and antibody stability were not tested to date [28].

The environmental conditions in which mammalians grow are important to colostrum quality and may be evaluated. An increasing area of research is the detection of toxic molecules in colostrum to determine quality. The Persistent Organic Pollutants (POPs), such as hexachlorocyclohexane (HCH), were found in the colostrum of a human Chinese population living in an easternmost island. In this study, 16S rRNA gene pyrosequencing was used and revealed that the microbiome distribution was altered by the level of HCH in the colostrum. The level of HCH was exceeded in half newborn by colostrum ingestion and was detected that the microbiome diversity had the tendency of decreasing in samples with higher HCH levels. The colostrum is the first source of microorganisms for newborns, and they will colonize the infant gut. So, it is important to guarantee the best environment for mothers, with fewer pollutants possible, as it seems to be relevant to colostrum quality [29].

18.2.4 Other applications

Milk, in general, is already used as a cryoprotectant in microbiology and ultra-congelation, avoiding cells to break up during the freeze-thaw process. Mare colostrum was identified as a good extender for freezing stallion semen, improving semen quality. Besides being a good preservative agent, mare colostrum decreased the percentage of DNA fragmentations during the freezing-thawing procedure [30].

It is known that the health of mammalian infants in the first hours and days of life is decisive to their health and performance. New strategies of colostrum feeding include its supplementation with n-3 fatty acids (FA) and α-tocopherol [31] and guarantee the fast first colostrum feeding [32]. Aiming to use colostrum as an antioxidant feeding, researchers supplemented it with fish and flaxseed oil with 200 mg of tocopherol and administrated to newborn calves after 6 hours of birth. This adds increased plasma concentrations of α-linolenic, eicosapentaenoic, and docosahexaenoic acids. At the same time, a decrease in oxidant status index was observed in the first week of the life of calves tested. By day 14, the oxidant status index returned to the level of control calves, who did not receive the supplementation. Colostrum supplement of n-3 FA and α-tocopherol is safe for newborn calves, efficient in reducing oxidant status (in the first week of life), and may be responsible for improving health and performance in medium and long terms [31].

As a consequence of production management, it is possible that newborn calves suffer a delay in their first colostrum feeding. However, different interval times of first feeding (e.g.,: 45 minutes, 6 hours, and 12 hours after birth) can greatly influence the microbiota mucosa associated with ileum and colon in calves [32]. The predominance of *Enterococcus* and *Streptococcus* species and lower abundance of *Lactobacillus* was observed at an interval time of 12 hours. On the other hand, *Faecalisbacterium* was found in lower abundance in calves fed with colostrum instantly after birth.

18.3 Colostrum patents

Products and processes recently patented related to colostrum can be separated into three main categories: human health, animal health, and cosmetics. For this chapter, a survey was made with inventions published around the world since 2018. From the results, screening was conducted based on the importance of colostrum in the invention. The selected patents are here approached, with a brief description of them in their respective category.

18.3.1 Human health

BC has been used as an antibacterial agent until the development of antibiotics. It has been used for human feeding for a long time in India and United States. However, in the 1990s, the specific research increased with the objectives to evaluate and improve the use of BC for human consumption, as well the products development [33]. Some food processing methods comprising BC with aid at its potential bioactive compounds have been proposed. As is described in the Patent CN110235943A [34], which had the purpose of developing a yogurt formulation to help combat aging and to improve immunity. For this, a formulation was elaborated with 70% of BC, 20% of fish collagen peptides, 3% of thickeners, 4% of concentrate juice (noni fruit, cranberry, and blueberry), 1% of tea polyphenols nanoparticles, 0.5% of sweetener, and 0.5% of lactic bacteria inoculum. The obtained results from biological assays indicate antioxidant effects of colostrum as well as of tea polyphenols nanoparticles and of the fruit juice.

Another proposal of the utilization of BC as an ingredient in processed food is disclosed in the Patent AU2018323788A1 [35], in which the object of the invention is a bioactive dairy preparation for the supplementation of IgG, in a liquid or semiliquid state. Its formulation comprises a bioactive powder that should be BC and/or milk powder containing 65%–87% of protein, 20%–27% of IgG, and 1.5%–2% of lecithin, water, citric acid, sweetener, and emulsifier. At the liquid and semiliquid state, the bioactive powder proportion must be 2%–7% and 9%–25%, respectively. The inventors indicate that this invention relates to formulation and processing of ready-to-eat products, as is cheese, ice cream, desserts, yogurts, ready-to-drink products such as carbonated drinks, fermented milk, coffee preparations, and hot chocolate.

In the Western World, the use of colostrum in medicinal form dates from the 18th century, with its use aimed at improvements in the immune system [33]. The high concentration of bioactive compounds in colostrum improves the immune system, as well as actuating in the body development by the presence of growth factors. As a result, its use as a food supplement in recent decades is increasing [36]. Its role as an ingredient, often as majority, in supplement food formulations is very widespread. Border 18−1 presents a summary of the recently published innovations related to the use of colostrum in supplements for health and well-being.

Although it has beneficial properties to human health, the therapeutic use of BC in the past has been limited due to technical factors such as the oxidation sensitivity of colostrum lipid components and the need for cooling during storage [57]. In order to maintain its chemical, nutritional, and microbial stability, some processes have been proposed, like the one presented in Patent CN109349350A [58]. It relates operations of centrifugation for fat separation, followed by heat treatment, concentration, lyophilization, and pulverization. The process disclosed results in a colostrum powder with the maintenance of the nutritional value and the Ig.

In formulations, colostrum can be applied fully or as its derivatives, like protein fractions, Ig, lactoferrin, peptides, and oligosaccharides. In this context, Patent CN110973345A [59] proposes a methodology comprising techniques of chemical and enzymatic precipitations, followed by separation and purification processes. Therefore, protein fractions, Ig, and lactoferrin are obtained. Patent CN107760738A [60] relates a methodology of enzymatic extraction with β-galactosidase in combination with filtration operations in order to obtain oligosaccharides up to 50% of purity from BC (or from whey).

Colostrum may also be subjected to several operations of separation and enzymatic treatments with the objective of modifying its composition partially and increasing its potential as a functional ingredient. The invention CN109287747A [61] proposes a method with various separation process for obtaining a product free of casein and lactose, with less than 10% of humidity, and approximate concentrations of 40% of proteins, 10% of Ig, 5% of lactoferrin, 2% of fat, 23% of vitamin C, 5% of taurine, 5.5% of vitamin E, 1.8% of calcium, 8% of zinc, and 1.6% of iron.

The high concentration of bioactive components in BC such as lactoferrin, lysozyme, leukocytes, Ig, and cytokines guarantee immune protection until the newborn's body is able to develop its own responses [62]. These characteristics have attracted the interest of the pharmaceutical industry to BC components and their applications in human health [33]. Several examples of colostrum application in products directed for the prevention and treatment of diseases can be found. Some innovations have been disclosed proposing solutions for bone treatment. Patent CN108576816A [63] inventors consider bone density as a good indicator of the osteoporosis grade of the patient. For this reason, it proposes a composition to enhance the bone density, taking into account the synergy between its components, as described in Table 18−3.

The ingredient called colostrum basic protein can effectively increase 1,25-dihydroxyvitamin D concentration, a vitamin that regulates active intestinal absorption of calcium and reduces calcium extraction through the urinary tract. Thus it enhances its mineral utilization. According to the inventors, this formulation inhibits calcium loss, promotes bone reabsorption, and presents an effect in increasing bone density. The invention CN110338409A [64] highlights that women are the most affected by bone density loss due to pregnancy, breast feeding, menopause, and inadequate diet,

Border 18–1 Border summary of innovations comprising colostrum use in supplements for health and well-being.

Patent/inventors	Publication year	Worldwide applications	Object of the invention
CN108720000A [37]	2019	CN	Supplement formulation with colostrum mineral salts and calcium, and its obtaining process. It is indicated for the restitution of calcium.
CZ33239U1 [38]	2019	CZ	A nutritional supplement containing granulated colostrum, especially goat and cow obtained in the first 6 h after delivery, with supportive properties for the immune system.
CN109645133A [39]	2019	CN	A kind of strengthen immunity milk powder and preparation method thereof, including colostrum in the formulation in a proportion of 6%–12%.
CN109717249A [40]	2019	CN	Development of a supplement in the tablet form, in which bovine colostrum is the majority ingredient with a proportion of 32%–55%, and its preparation method. It is indicated for immunity strengthening.
CN108124963A [41]	2018	CN	Development of a powder nutritional supplement with numerous functional ingredients, including bovine colostrum as majority in a proportion of 20%–30%, and its preparation method. It is indicated for the enhancement of memory, immunity, and gastrointestinal tract of children.
CN110037257A [42]	2019	CN	Development of a powder supplement with numerous functional ingredients, including bovine colostrum powder, and its preparation method. Indicated mainly for immunity regulation and enhancement of physical strength.
CN107890104A [43]	2018	CN	Development of a formulation indicated for improvement of growth and memory, and its preparation method. It is proposed in the capsule form, being walnut oil and colostrum (11%–20%) the main ingredients.
CN109198601A [44]	2019	CN	Development of a food supplement, and its preparation method. Indicated for immunity improvement, tumor prevention, and to assist in the treatment of cancer. Formulated with various compounds of fungi and protein source, and colostrum is a minority ingredient.
CN109430413A [45]	2019	CN	Development of a food supplements formulation with bovine colostrum in a ratio of 1%–10%, and its preparation method. Indicated for immunity increase.
EP3177636A1 [46]	2019	DK, PT, WO, EP	Process for the preparation of highly pure mixtures of proteins factors from bovine colostrum. It can be used in pharmaceutical and supplement compositions with therapeutic activities, promoting tissue healing and regeneration.
CN109674064A [47]	2019	CN	Development of a formulation rich in vitamins, functional ingredients, herbaceous extract and bovine colostrum (1%–10%), and its preparation method. Indicated for the strengthening of the immune system and the treatment of respiratory diseases.
CN108142925A [48]	2018	CN	Liquid formulation with high protein content with vegetable protein and bovine colostrum protein (15%–20%) within animal proteins, and its method of preparation. Indicated to help in recovery of patients.
CN107980906A [49]	2018	CN	Processing method of a formulation rich in vitamins, minerals, and vegetable and animal proteins, including 60%–70% of bovine colostrum. The proteins are treated with proteases to provide polypeptides easier absorbed by the organism. It helps in recovery of patients.
CN110839698A [50]	2020	CN	A composition formulated mainly with bovine colostrum, sialic acid, and bifidobacteria. It promotes wound healing and it prevents or treats inflammatory infections.

(Continued)

Border 18–1 (Continued)

Patent/inventors	Publication year	Worldwide applications	Object of the invention
CN109730156A [51]	2019	CN	Development of a formulation enriched with bovine colostrum powder and other ingredients to reinforce the immunity of babies and children and to treat gastrointestinal disorders.
CN109645473A [52]	2019	CN	Food supplement formulated mainly with immune enhancers components of bovine colostrum and probiotic bacteria. Indicated to enhance immunity and the gastrointestinal tract of children.
EP3479699A1 [53]	2019	WO	Supplement to improve commensal colonization in the children's intestine, based on the enrichment of IgG β-lactoglobulin and α-lactalbumin fractions from bovine colostrum.
CN109770345A [54]	2019	CN	Formulation indicated for patients with reduced capacity of nutrients absorption. It is highly enriched and supplemented with bovine colostrum powder with 20%–30% of IgG.
CN109907264A [55]	2019	CN	Processing of a formulation to enhance antiviral capacity and improve flu symptoms. It is prepared with plants and roots extracts, and bovine colostrum in immunoglobulin form.
CN109700999A [56]	2019	CN	Processing of a paste formulation. It is prepared with natural ingredients, including bovine colostrum protein. It has the purpose of increasing white cells in the blood and increase the immunity.

Table 18–3 Proposed formulation to enhance bone density.

Ingredients	Range (weight/weight)
Milk mineral salt	11.0–50.0
Oligosaccharide	6.0–16.0
Vitamin D	0.00005–0.0005
Collagen fish	10.0–30.0
Colostrum basic protein	0.6–1,9
Vitamin K2	0.0003–0.02
Vitamin C	0.5–3.0
Vitamin E	0.03–0.3
Magnesium	0.8–2.1
Zinc	0.015–0.1
Manganese	0.02–0.1
Eucommiaulmoides extract	0.5–5.0
Rhizomadrynaria extract	0.8–6.0

Based on: X. Liu, Y. Zhang, Composition capable of increasing bone density, CN108576816A, 2018 [63].

leading to higher possibilities of evaluating osteoporosis and suffering from a fracture. That said, a formulation is proposed for promoting skeletal system health and preventing bone density loss in women, according to Table 18–4. The inventors relate the formulation used in powder or granulated form, associated with food consumption. In this composition, BC is present as peptides, being the components that influence directly the bone density. Other ingredients also play an important role in formulation, like oligosaccharide and casein phosphopeptide, as they promote the solubility and absorption of minerals by the intestine. Different variations of the disclosed formula were tested in a group of 30 women being 30–55 years old, at pregnancy, lactation, and postmenopause periods. After 6 months of consumption, an increase in bone density was observed.

In the case of Patent CN110292177A [65], the invention overcomes the bone development, which problematic in childhood and adolescence, as calcium nutrition at these phases is crucial for bone mass in the human body. During childhood, the absorption rate of calcium achieves 75% due to physical development. And in adolescence, 50% of adult mass bone is obtained. Thus, the insufficient intake of calcium at these periods can lead to growth retardation, osteomalacia, and bone deformation. Therefore, the invention provides a composition for promoting bone growth and development of children, aiming at overcoming the deficiency of karyo systemic health caused by insufficient calcium intake in children in the rapid growth and development stage. Six formulations were developed, which are made of the following raw materials, as described in Table 18–5.

The composition of the present invention can be used for preparing a health food for promoting the health of the skeletal system, that has a wide range of applications, and has good economic and social effects. An assay comprising 40 children with low bone density was

Table 18–4 Proposed formulation for preventing bone density loss.

Ingredients	Range (weigth/weigth)
Skim milk powder	200.0
D-glucosaminehydrochloride	150.0
Oligosaccharide	250.0
Orange fruit powder	50.0
Collagen	150.0
Casein phosphopeptide	250.0
Colostrum basic peptide	24.0
Vitamin D3	1.0

Based on: J. Li, Y. Lu, X. Wu, Composition for preventing and treating bone loss of women and application thereof, CN110338409A, 2019 [64].

Table 18–5 Proposed formulation to promote bone growth and child development.

Ingredients	Range (weigth/weigth)
Whole milk powder	50.0
Milk mineral salt	10.0
D-glucosaminehydrochloride	0–5.0
Oligofructose	10.0
Orange fruit powder	5.0
Casein phosphopeptide	10.0
Colostrum basic peptide	10.0

Based on: J. Li, Y. Lu, X. Wu, Composition for promoting children bone growth and development and preparing method and application thereof, CN110292177A, 2019 [65].

Table 18–6 Proposed formulation to repair osteoarthritis injuries.

Ingredients	Range (%)
Colostrum basic protein	31.0
Freshwater fish collagen peptide powder	29.0
Sea cucumber lyophilized powder	21.0
Yak bone powder	9.3
Maltodextrin	5.0
Vitamin C	3.3
Sorbitol	1.2
Magnesium stearate	0.2

Based on: J. Chen, H. Chunyue, S. Zhiyong, W. Jiguang, J. Zhang, Protein compound preparation for improving bone and joint, CN110354256A, 2019 [66].

conducted. They consumed the proposed formulations for 3 months. It was observed that the arising symptoms from calcium deficiency significantly improved. Another formulation described in the field of bone density is described in Patent CN110354256A [66]. It relates a preparation intended to repair injuries caused by osteoarthritis, such as loss of elasticity in cartilage and decreased joint function. Also, it supplements calcium, promoting bone density increase. The main components of the described formulation are colostrum basic protein, sea cucumber lyophilized powder and yak bone powder, as it is shown in Table 18–6.

The international document WO2019162895A1 [67] discloses a composition for topical use to treat vaginal dryness, vaginitis, and vaginosis. It comprises a combination of an oil mix (sunflower, argan, olive, jojoba, macadamia, and linseed) and a mixture of biological factors (cytokines, growth factors, chemotactic factors, stimulating factors for stem cells, antibacterial/antiviral factors, and Ig) that is provided from BC.

In order to help in the curative or preventive treatment of pathologies of the urinary system pathologies, mainly cystitis and interstitial cystitis, the invention WO2018029558A1 [68] presents a liquid composition for topical use by intravesical administration. It is composed preferably of 1.8%–2% of purified BC with at least 75% of its composition consisting of IgG. Also, it has an addition of at least one other active ingredient from among chondroitin sulfate, or the salts thereof, such as the sodium salt, salicin, and mixtures thereof. Other applications of BC in products and methods for the prevention and treatment of disorders can be observed in Border 18–2.

Innovators methods and products comprising colostrum in its full or derivative form have been proposed as are presented by American Patents United States 9943482B2 [73] and United States 10166259B1 [74]. They aid to offer techniques for improving the absorption of drugs administered orally in chemotherapy therapy. For example, the second invention proposes a methodology of isolation and purification of exosomes from bovine and caprine colostrum powder, in which several centrifugation steps are applied. Yet, it is disclosed a technique of using these exosomes for carrying small molecules, such as curcumin, withaferin A, paclitaxel, and anthocyanidins.

BC has also been used in the development of products for the treatment, prevention, or remission of cancer and infectious disease as an immunotherapy technique, as is disclosed

Border 18–2 Various applications of bovine colostrum in the prevention and treatment of pathologies.

Patent	Publication	Worldwide applications	Object of the invention
United States10464998B2 [69]	2019	WO, United States, EP	This invention relates to methods and compositions for treating fibrosis, by administering compositions comprising anti-LPS immunoglobulin enriched colostrum preparations. In particular, the invention relates to methods and compositions for the treatment of liver fibrosis and/or lung fibrosis. Prophylactic or therapeutic compositions and diagnostic methods are also disclosed and claimed.
EP3127543B1 [70]	2018	EP	The object of the invention is therefore ophthalmic formulations comprising one or more oligosaccharides present in human colostrum or milk, especially 2′-fucosyl-lactose as specified in the claims. The formulations according to the invention can also optionally include hyaluronic acid and salts thereof.
WO2020056521A1 [71]	2020	WO	The invention is based on the development of colostrum-derived stem cell populations capable of providing an antimicrobial effect and therefore useful as supplements or compositions for enhancing mammalian health, and for treating and preventing microbial contamination or infection.
EP3461840A1 [72]	2019	BR; EP; United States; AU; CA; *MX*; DK; WO; *SG*; JP; ES; KR; IL;	The present invention provides a composition comprising an anti-LPS enriched immunoglobulin preparation for use in treatment and/or prophylaxis of a pathologic disorder such as chronic liver disease and cirrhosis. The anti-LPS enriched immunoglobulin preparation may be derived from colostrum or from avian eggs.

in Patent AU2020201700A1 [75]. It relates a methodology that presents colostrum as an activating agent of macrophages when treated enzymatically with β-galactosidase or β-galactosidase and sialidase. This product can be vehiculated as a pharmaceutical composition, quasi-drug, or as a supplement in food and drinks.

18.3.2 Animal health

Animal nutrition is a relevant field, especially due to the importance of the livestock industry, but also in domestic environments with companion animals. Although newborn mammal animals uptake colostrum as a natural mechanism for acquiring passive immunization, sometimes this transfer fails or is even a source of infections when contaminated. Additionally, farm animals are exposed to pathogens that can cause serious problems. These factors can lead to morbidity, mortality, and lower growth rates [76,77]. In this context, efforts have been made to develop supplements or pharmaceuticals to enhance animal health and to overcome pathologies, including colostrum in compositions.

Firstly, we have the inventions that disclose products made from colostrum. The American Patent United States 2019313667A1 [78] presents nutritional compositions comprising fermented colostrum components as well as methods for their manufacture and administration. The main point of this invention is the superior nutritional value of fractions derived from the fermented colostrum compared to the unfermented one. The colostrum used is from a mammalian source and goes through a fermentation process under defined conditions with one or more bacterial strains called Przewalski culture. This culture is referred to as bacteria isolated from Przewalski horse and comprises a mixture of *Ruminococcus flavefaciens, Butyrivibrio fibrisolvens, Fibrobacter succinogenes*, and *Ruminococcus albus*.

The claimed compositions comprise fermented colostrum fractions composed of novel bioactive peptides with therapeutic activities. In summary, the colostrum is fermented followed by an enzymatic process, in which hydrolysis occurs with a protease sequence. Then it is heated and passed through ultrafiltration. The resulting filtrate is concentrated with the desired peptides, so they can be separated from other components. Two important obtained fractions are the proline-rich polypeptides (PRP) and the Bioactive Component 1, which contains at least forty distinctive bioactive polypeptides. In addition to the cited fractions, the compositions may include other nutritious ingredients like vitamins and minerals. These compositions intend to enhance growth and strengthen the immune system in animals, especially immature ones. They promote cell maturation, help nutrient uptake, and aid gut maturation. Thus it stimulates the health and growth of animals at any age. It is possible to use it as a feed complement or as a substitute for mother's milk and for a neonatal diet.

Additionally, another component from the disclosed fermented colostrum presented an unexpected activity: antianxiety effect. Therefore, an anxiolytic product is also claimed in this invention. This product's active ingredient is a decapeptide present in the Bioactive Component 1, which is purified by chromatography. A significant calming effect is observed throughout its administration, being used for ameliorating stress and anxiety in animals in certain high-stress situations. A different approach is used in the Patent JP6384928B2 [79] in which the component derived from colostrum is a protein with its amino acid sequence modified. It is based on the fact that even though colostrum presents many benefits to the animal organism, its protein content is unstable, difficult to collect and difficult to store. Hence, the object of the invention is a modification that provides the protein with higher stability outside the body and its use prevents or resists foreign pathogens. In the modified colostrum protein, the isoleucine at position 33, glutamic acid at position 101, and arginine at position 175 is, respectively, replaced by alanine, cysteine, and cysteine (the whole amino acid sequence can be accessed in the Patent archive). These modifications provide a more stable tertiary structure compared to that of the wild colostrum protein.

Specifically, the modified colostrum protein of the present invention can be obtained by purifying the protein that binds to the peptidoglycan on the bacterial cell wall and modifying its amino acid sequence. It is named Pathological Recognition Protein due to its characteristics. Once the amino acid sequence of Pathological Recognition Protein is obtained, its gene can be cloned and constructed in a yeast expression vector, which can be further introduced into yeast strain. Thus the modified colostrum protein can be produced under ideal fermentation conditions. The oral dosage form can be manufactured such as solid (tablets, granules, powder, or capsules), semisolid, or liquid form. The animal feed should contain 0.01%–0.02% (w/w) of the protein and its administration induces animal immunity by increasing the production of immunoglobulin A (IgA) in the organism. The antibody IgA is capable of binding to multiple epitopes of pathogens, so they cannot bind to mucosal cells and infect the animal. More specifically, it is claimed the use of the feed to treat or prevent diseases that induce mucosal immunity, including porcine reproductive and respiratory syndrome, foot-and-mouth disease, porcine epidemic diarrhea, avian influenza, and human influenza. The pharmaceutical composition, which comprises a pharmaceutical carrier, a vaccine adjuvant, and the modified protein, also intends to treat and prevent the animals from the same diseases described above.

The invention EP2249661B [80] proposes compositions and methods for enhancing animal health relating to a germ-free colostrum and supplements enriched with synthetically multimerized Ig. The production method of this kind of Ig is also disclosed, in which colostrum can be a raw material among others. These compositions are mainly for use of farm animals and promote resistance to pathogenic agents and passive immunization through oral administration. It can replace colostrum, or it can be used as a supplement.

Colostrum is also part of compositions destined for animals. A recent invention assigned by Nestlé S.A. (United States 10010566B2 [81]) provides compositions and methods that promote amelioration of age-related maladies. The compositions comprise a combination of at least one or more of each of the following components: unsaturated fatty acids (UFA), nitric-oxide releasing compounds, antiglycation agents, and colostrum. And the methods present the therapeutically effective amount of the compositions that should be administered to an animal. Therefore, this invention has the objective of enhancing and maintaining the immune system, cognitive function, muscle strength, balance, and reducing and mitigating oxidative stress in animals. In general, the compositions and methods described promote health and wellness, consequently, extending the healthy life years of the subject, especially companion animals.

Although not many details are given specifically about colostrum in terms of its function in the compositions, from what has already been described in this chapter, it is possible to believe it has a great role in the immune system. Thus its uptake together with the other ingredients causes the amelioration of the quality of life. The colostrum used in the invention should be suitable for administration to an animal, being obtained from any suitable source, either synthetic or natural, but preferably being BC. Its administration to an animal through the compositions disclosed can range from nearly 0.001 to about 10 g/kg/day.

The compositions can be administered to the animal in any form like being a part of food composition, as a dietary supplement, pharmaceutical or nutraceutical composition, or as a medicament. Also, many routes can be used, including oral, intranasal, intravenous, intramuscular, intragastric, transpyloric, subcutaneous, and the like.

18.3.3 Cosmetics

The colostrum composition and bioactivity make it very suitable for applications in cosmetology. Asian countries are known for using natural ingredients in the creation of innovative cosmetics.

Thus they appear as the main source of the patents comprising colostrum and derivatives having the objective of using it in the skincare industry as well as replacing synthetic active ingredients. Farm Skin is a South Korean cosmetic brand specialized in manufacturing colostrum cosmetics and it owes three recently published patents involving colostrum fermentation. The first one, International Patent WO2018174669A1 [82], discloses a colostrum fermented product and a cosmetic composition containing it. The fermentation is led by one or more lactic acid bacteria, and through this process, the fermented colostrum shows increased bioactivity. The claims design the fermented product as having elevated antiinflammatory and skin irritation mitigation effects, as well as whitening activity, skin moisturizing, and wrinkle improvement, being helpful for the treatment of psoriasis as well. According to the inventors, the fermented colostrum product exhibits very useful properties different from any components derived from colostrum before and with more efficient effects than fermented milk. Thus it is very profitable to the cosmetic industry.

The process claimed starts with a pasteurized colostrum, which can be from any animal that is then inoculated with the lactic acid bacteria. The chosen bacteria will affect the product effects, so they must be applied according to the desired purpose. The use of a specific bacteria or a combination of them can enhance a colostrum effect acting in synergy with components already present in the raw material, like lactoferrin and TGF-ß. The fermentation occurs at 30°C–50°C for 30–120 minutes. Then the broth is centrifuged followed by an important step of the invention that is the casein removal, a protein that may present negative effects. This is made through an agglomeration process using a coagulant enzyme, like Rennet. Consequently, the functional component is enhanced. After, centrifugation and filtration are applied, resulting in the lactic acid fermented colostrum extract, which passes through filtration (twice or more times) to sterilize the product.

The invention claims the use of this final product to compose cosmetics compositions for the following purposes: antiinflammatory activity, skin irritation relief, skin whitening, moisturizing effect, and wrinkle improvement. Moreover, food and pharmaceutical compositions are claimed, given the same skin effects, with the addition of psoriasis treatment. The Patent KR20190060556A [83] provides an additional property of the fermented colostrum of the prior described invention, using it as an active ingredient in an antiacne cosmetic since it presents good antibacterial activity against *Propionibacterium acnes*. In both inventions, it is stated that the fermentation of colostrum provides a safe product, capable of replacing synthetic compounds usually present in cosmetic formulations intended to achieve the respective effects mentioned above.

Also, Farm Skin published another fermentation method (KR20190082159A [84]) of colostrum, occurring under aerobic conditions. In this manner, a process with high quality and efficiency is provided through simpler requirements, in which oxygen removal is not necessary. Thus it can be a more economical process with easier conditions. The production process disclosed is similar to the anaerobic fermented product one. The difference presented in the claims is the removal of fat colostrum, the time at which the casein is removed (it can be either before or after fermentation), and the presence of glycomacropeptide in the product.

The product obtained in this invention exhibits even better effects on the skin compared to the anaerobically fermented one. Thus it was claimed its use in cosmetics or pharmaceutical formulations as an active ingredient to wrinkle improving, whitening, moisturizing, antiatopic dermatitis, antiacne, antibacterial, antiinflammatory, antioxidant, and skin irritation mitigation purposes.

The object of the Korean invention KR101974405B1 [85] is also about processed colostrum, but in this case, it relates to a cosmetic composition comprising colostrum whey protein hydrolyzed as an active ingredient. The purpose is to create a cosmetic for skin regeneration, which can be achieved since the hydrolysates from colostrum whey protein promote dermal fibroblasts proliferation, keratinocyte differentiation, and fibronectin secretion. These properties showed to be superior in the claimed product compared to colostrum whey protein and milk whey protein hydrolyzed. It is obtained through alkaline protease treatment having a hydrolysis rate of 40%–50%, at pH 8–10, at 50°C–60°C, followed by enzyme inactivation at 80°C–100°C. Thus, cosmetic for skin regeneration is claimed, being one or more of the following formulations: lotions, softeners, toner, astringents, creams, foundations, essences, packs, soaps, cleansing foams, and body cleansers.

The Chinese patent CN107822985A [86] makes use of the grease part of colostrum in the development of a lotion. The formulation claimed comprises 5%–15% (w/) of sheep colostrum fat in addition to the raw materials (w/w): 5%–15% of cetyl alcohol, 1%–5% of carbomer, 1%–10% of sodium alginate, 3%–9% of glycerin, 2%–3% of Tween, 1%–5% of ethanol solution, 11%–19% of magnolia flavor, 5%–7% of green tea extract, 3%–5% of butanediol, 3%–5% of water-soluble jojoba oil, 1%–5% of sodium lactate, 5%–10% of methylparaben, 1%–1.15% of propylene glycol, 4%–5% of emulsifier, 2%–4% of viscosity modifier. This formula is reasonable, easy to produce and intends to make the skin delicate and smooth. It is based on the sheep colostrum fat capability of inducing the skin fibroblasts proliferation, accelerating the epidermis regeneration.

The effects related to the colostrum are likewise useful for a children's soap formulation. A Chinese Patent (CN108085186A [87]) includes the bovine-derived colostrum in soap in order to create a product safe for a child's skin, characterized by natural and mild substances. The use of this raw material aggregates immune regulation and inhibition of various pathogens to the product, while not being irritating to the young skin as commonly used bactericides. The invention claims the soap formulation comprising: 75%–80% of soap granules, 0.05%–0.1% of 2,6-di-tert-butyl-p-cresol, 0.1%–0.5% of EDTA disodium, 0.5%–2% of horse oil, 1%–3% of BC, 1%–3% of glycerin, 0.3%–1% of wormwood oil 0.3, 3%–6% of wormwood extract, and 10%–12% of deionized water. Due to the presence of colostrum in combination with wormwood oil and extract, the soap exhibits immunomodulating, antibacterial, antiviral, dehumidifying, antipruritic, and antiallergic effects.

Properties of colostrum combined with other active ingredients are an object of other inventions as well. The inventors of the Patent KR102076178B1 [88] report the synergistic effect of a mixed composition of colostrum and orthenine in improving wrinkles while showing almost no skin irritation. Thus the object of the invention relates to a cosmetic composition comprising these safe ingredients, which has collagen synthesis enhancing effect and skin barrier improvement, ameliorating skin wrinkles and elasticity. Both colostrum and ornithine must be contained in the cosmetic composition for wrinkles improvement in a ratio of 0.001%–10% (w/w). The formulations claimed various forms for different uses, including simple forms as solutions and suspensions, passing by emulsions, pastes, gels, creams, lotions, and reaching complex formulation as powders, soaps, surfactant-containing cleansings, oils, and formulations used in make-up, as powder foundation, emulsion foundation, wax foundation, and spray.

The use of colostrum along with collagen is also an object of a Korean invention (KR20180114493A [89]). Yet, the main focus is related to the presentation form of the final product. This cosmetic is produced through freeze-drying technology, which means it has no moisture and has no need for preservatives. Due to the presence of sodium alginate in its composition, the product acquires a film form. As a result, it can easily be applied directly to the targeted skin area and also can be easily dissolved. Colostrum enhances the moisturizing power of the cosmetic and enriches its composition, giving better and more persistent effects than collagen alone. The invention claims a composition consisting of an aqueous phase comprising (w/w) 0.5%–10% of natural collagen, 0.1%–5% of sodium alginate, 0.01%–1% of propylene glycol alginate, and 2%–5% of niacinamide; and a solubilizing part containing 1%–10% of colostrum 0.1%–1% of surfactant and purified water until 100% is reached.

The use of encapsulation technology is applied in the KR102023724B1 [90] invention. The patent relates to a single hydrogel capsule containing a protein mixture derived from colostrum, and a fat-soluble component. Consequently, when applied to the skin, it delivers water-soluble protein and a fat-soluble component at the same time, also increasing the moisturizing efficiency and the absorption by the skin. Besides its great effects, the hydrogel capsule enables a stable cosmetic formulation, being in a water-in-oil-in-water form with uniform physical properties. So, the structure claimed in the invention is formed by three layers: an aqueous core comprising the colostrum proteins and a water-soluble polymer, a fat-soluble intermediate layer involving the aqueous core, and a polymer hydrogel capsule located on the intermediate layer. This composition leads to protein protection from denaturation and oxidation and it blocks off the odor derivative from the colostrum. Then, it can be added to cosmetic formulations maintaining its stability and acting for skin moisturizing and improvement.

Different from the inventions described above, the Patent WO2019031976A2 [91] has an European origin and it discloses a product not for skincare, but for nail regeneration. The invention relates a multiingredient preparation, in which colostrum fat is the active ingredient and the method for making it. This product may be used for the treatment and regeneration of finger and toenails stricken by onycholysis or affected by skin conditions such as psoriasis. The invention overcomes the necessity of therapy with long-term success and that carries the active ingredient through the nail to the nail bed and nail root.

Natural proteins present in lyophilized colostrum fat such as Ig, lactoferrin, lysozyme, or lactoperoxidase are able to penetrate nails and efficiently treat nail inflammatory disorders, and especially to successfully treat onycholysis. The commercial lyophilized colostrum (called cream colostrum) contains several highly concentrated compounds that are not found in any other natural product. Due to these components and properties, it becomes an advantageous ingredient for the purpose of the invention.

The multiingredient preparation described comprises 2%–8% (w/w) of the cream colostrum containing 30%–50% of BC fat, emulsifiers, emollient, antioxidants, conditioners, thickeners, perfume, and deionized water. Its formula is based on liquid crystal emulsions, which promote the penetration of the active ingredients. Also, the presence of other valuable oils augments the product regenerating properties.

18.3.4 Colostrum patents overview

From the inventions covered, it is possible to realize the main uses of colostrum and countries with more research being made with it. In Fig. 18–1, a significant presence of China is seen as a source

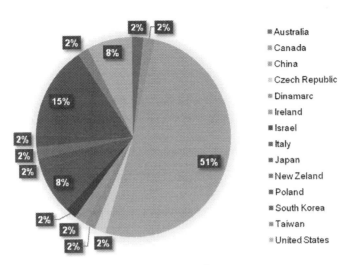

FIGURE 18–1 Source countries of inventions approached in this chapter.

of the Patents described, with a huge concentration of inventions in the human health field. Several inventions disclose colostrum or its components as a good ingredient for supplementing the human body, assisting patients, treating diseases, or just maintaining the subject's health. This could be linked to the Asian tendency of exploring the potential of natural ingredients and also, especially in China, to the application of Traditional Chinese Medicine (TCM). Hence, some of the inventions mention they are applied in the TCM field, proposing compositions made of colostrum in combination with a variety of plants and extracts.

It is also notable South Korea is the main source of innovations comprising the application of colostrum in cosmetics. As it was already mentioned, Asian countries put efforts into developing healthy products with natural additives, and this trend is reflected in the modern cosmetic industry, being Korea a leader in creating innovative cosmeceutical products (a combination of pharmaceuticals and cosmetics) [92]. So, colostrum seems to be a useful resource for these kinds of creations. As a result, human health and cosmetics appear as major categories (Fig. 18–2), since they are dominated by these Asian countries. Other countries demonstrate a more subtle action with this product, with Italy and the United States making a certain contribution.

18.4 Commercial bovine colostrum products

The main source of colostrum to develop products is the BC, because the production yield is related to the animal weight; then, the great amount of colostrum is obtained with the bigger animal (cows, instead of other animals as ewes or pigs), being a competitive factor. Biotest Pharm GmbH (Germany) was the first company to elaborate and test a colostrum-based product, the lactimmunoglobulin. This product was a bulk of Ig isolated from BC that was able to alleviate diarrhea in immunosuppressed patients [93,94]. Later, the Viable Bioproducts Ltd. (Finland) commercialized the Bioenervi, a sterile-filtered colostrum formula, which was able to increase the

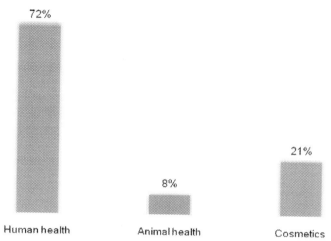

FIGURE 18–2 Distribution of the inventions comprising colostrum approached in this chapter based on the application categories.

serum concentration of insulin-like growth factor I (IGF-1) in male athletes [95]. The Sweden company ColoPlus AB enriched organic support (rice flakes) with sterilized and lyophilized BC. This product was further investigated as a functional food to ameliorate HIV-associated diarrhea in Port Harcourt, Nigeria [96]. Despite the promising results, the lack of approval from international vigilance bodies [e.g., Food and Drug Administration (FDA) and the European Food Safety Authority (EFSA)] has resulted in the low distribution of these products.

It was only in December 2019 that the BC produced by the biotechnology company PanTheryx (Colorado, USA) received the generally recognized as safe classification by the FDA. Although limited to the USA, this certification demonstrates international relevance and increases the expansion of the functional food market. A recent study promoted by Transparency Market Research revealed that the BC market has a projection to increase from $ 2.6 bi in 2019 to $ 4.3 bi in 2027. To date, 13 companies distribute BC-based products directed to human nutritional supplementation (Table 18–7). According to the manufacturers, the colostrum is collected from buffalos and cows without treatment with exogenous antibiotics and hormones. In addition, the processing of the BC is performed under low temperatures and low pressure in order to avoid the loss of Ig and vitamin contents [3].

PanTheryx is nowadays the main supplier of BC worldwide and is focused on BC-based therapies, commercializing the first-in-class therapeutics produced through patented technologies (https://pantheryx.com/). Other companies with prominent market relevance in the segment of colostrum-based products for human nutrition are described as follows. APS BioGroup (Arizona, USA) offers the ColostrumOne Nature's Superfood, bulk colostrum powders for human applications packaged in 20 kg boxes, and colostrum finished products in different forms (capsules, powder-filled containers, individual sachets, immulox spray liquid, specially formulated products). The company uses a batch pasteurization process or a special high-temperature short time pasteurizer with the finishing of the powder carried out by

Table 18–7 Bovine colostrum manufactures and commercially available products.

Company	Country	Product	Composition	Formula
PanTheryx (APS Biogroup)	USA	Colostrum One	Whole, skim or ultra-skim colostrum	Capsules, chewable tablets, and powder
BiostrumNutritech Ltd.	India	—	Whole cow's colostrum	Capsules, chewable tablets, and powder
MIP Colostrum NZ	New Zealand	Maxi Immuno Plus	Whole cow's colostrum	Capsules, chewable tablets, and powder
Sovereign Laboratories	USA	Colostrum-LD	Whole cow's colostrum	Capsules and powder
Jarrow Formulas Inc.	USA	Colostrum Prime Life	Colostrum supplemented with immunoglobulins	Capsules
PuraLife LLC.	USA	PuraLife Colostrum	Whole cow's colostrum	Powder
Agati Healthcare Pvt. Ltd.	India	Colnsta	Whole cow's colostrum	Powder
	India	Immcar-9	Whole cow's colostrum	Powder
		ProElec	Whole buffalo's colostrum	Powder
Fonterra	New Zealand	NZMP	Whole cow's colostrum	Powder
ColoPlus AB	Sweden	ColoPlus	Lyophilized cow's milk added to an organic support (rice flakes)	Granules
Immuno-Dynamics, Inc.	USA	PerCoBa	Whole cow's colostrum	Powder
Colostrum Biotec	Germany	Armacura	Liquid concentrate of cow's colostrum	Liquid
SCCL	Canada	Eterna Gold	Whole cow's colostrum	Powder
Cure Nutraceutical	India	Colostruvita	Spray-dried cow's colostrum	Capsules, chewable tablets, and powder

spray-drying. The agglomeration process is applied in addition to instantizing, to promote homogeneous particle size and better solubility in liquids (https://apsbiogroup.com/colostrum/).

Biostrum Nutritech Pvt. Ltd. (Gujarat, India) commercializes colostrum powders and granules in packages containing 500 mg, 5 kg, or 25 kg (https://www.bcolostrum.com). Other forms of products are also offered by the company such as those of the Immune Strong Products line (pure colostrum powder, especially formulated chocolate and vanilla colostrum powder, colostrum capsules and chewable tablets). MIP Colostrum NZ (Tauranga, New Zealand) offers the MIP Colostrum containing 22%–25% IgG, available in four forms: MIP Colostrum 90 VegeCaps, MIP Colostrum Powder (80 g), MIP Colostrum 100 Chewables, and MIP Day Boost—two-month supply (60 servings), all produced with undiluted BC from New Zealand (https://www.mipcolostrumnz.com/). One of the advantages pointed out by the company is that the colostrum is collected and transported fresh and chilled directly from the farm to the processing facility, which would preserve its active ingredients.

Sovereign Laboratories (Arizona, USA) has launched the Colostrum LD series of dietary supplements, available in the form of powder and capsules (https://www.sovereignlaboratories.com/). It is used as a source of Ig, colostrum polypeptides, growth factors, and lactoferrin. The colostrum is obtained from pasture-raised dairy cows that are healthy and free from bovine spongiform encephalopathy, not supplemented with recombinant bovine somatotropin or treated with antibiotics. The product is tested for the absence of antibiotics, hormones, and herbicides and is flash pasteurized at 161°F for 15 seconds.

The claimed benefits of Colostrum LD include the maintenance of healthy stomach, gastrointestinal tract, and intestinal flora; support to a healthy cognitive function; healthy aging; immune support; and increase in bone and lean muscle mass. The advantages of the product are 1500% increased bioavailability, attributed to the LD Liposomal Delivery system that preserves the components from degradation in the gastrointestinal tract and provides biologically active components at the cellular level; the presence of low molecular weight polypeptides, Ig, and antibodies; and the presence of growth factors such as insulin-like growth factor (IgF-1), growth hormone (GH), and transforming growth factors (TgF).

The American company Jarrow Formulas offers the Colostrum Prime Life in the form of capsules (https://www.jarrow.com/product/207/Colostrum_Prime_Life). The product is obtained from pasture-fed cows raised without the use of recombinant growth hormone, antibiotics, and pesticides. The claimed benefits are the support to the body's defense system, and the promotion of gastrointestinal health. The product contains a minimum of 30% Ig and is spray-dried at low temperature and pressure to avoid denaturation and maintain biological activity. Other formulation ingredients are cellulose, magnesium stearate of vegetable source, silicon dioxide, sunflower lecithin, and medium-chain triglycerides.

Finally, the company PuraLife (Pennsylvania, USA) commercializes the PuraLife Colostrum (https://www.pura-life.com/products/puralifecolostrum), a powder prepared from first-milking colostrum, collected within 16 hours from birth, after the calf is fed for the first time. The colostrum is pasteurized and processed at low temperatures and pressures, by low-heat flash pasteurization and low-heat and low-pressure indirect spray-drying, using a free-flowing powder technology that avoids aggregation and enhances solubility. The final product contains a high level of PRP and is claimed as a booster of strength, stamina, and lean muscle mass production, a promoter of healthy intestinal flora and digestive systems, and an enhancer of cellular growth for healthy skin, bones, muscles, nerves, and cartilage.

18.5 Conclusion and perspectives

Colostrum is a natural product inherent to mammalians' live. Many studies were performed since colostrum awakened the scientific community's attention in the 1990s, elucidating its composition and role in newborn development. Ultimately, the recent colostrum production and processing technologies reside in the improvement of colostrum quantity and quality.

IgG levels play a decisive role in colostrum quality but developing methods to improve its concentration and absorption are great challenges.

The search for a healthier life is growing in the population, and increasing the immune barrier is a crucial point. However, there are still few products that perform this task in a proven way, as is the case of colostrum. The use of products of animal origin requires strict quality control and correct purification. In addition, communicating with the population and demonstrating their ability and safety are crucial factors for the commercial success of a human health product. Considering its health benefits, colostrum-based products are a promising market. Appropriate quality control, especially in terms of IgG content, is crucial for the solidification of a colostrum-based industry. The relationship between patents and scientific studies about colostrum and licensed and commercialized products is still small. This demonstrates the wide gap between what can be exploited commercially and what really is, demonstrating great opportunities for growth in this market that is in expansion.

References

[1] S.G. Borad, A.K. Singh, G.S. Meena, H.V. Raghu, Storage related changes in spray dried colostrum preparations, LWT Food Science and Technology 118 (2020) 108719.

[2] A. Civra, A. Altomare, R. Francese, M. Donalisio, G. Aldini, D. Lembo, Colostrum from cows immunized with a veterinary vaccine against bovine rotavirus displays enhanced in vitro anti-human rotavirus activity, Journal of Dairy Science 102 (6) (2019) 4857–4869.

[3] S. Bagwe, L.J.P. Tharappel, G. Kaur, H.S. Buttar, Bovine colostrum: an emerging nutraceutical, Journal of Complementary and Integrative Medicine 12 (3) (2015) 175–185.

[4] H.S. Buttar, S.M. Bagwe, S.K. Bhullar, G. Kaur, Health benefits of bovine colostrum in children and adults, in: R.R. Watson, R.J. Collier, V.R. Preedy (Eds.), Dairy in Human Health and Disease Across the Lifespan, Academic Press, Boca Raton, FL, 2017, pp. 3–20.

[5] B.A. McGrath, P.F. Fox, P.L.H. McSweeney, A.L. Kelly, Composition and properties of bovine colostrum: a review, Dairy Science and Technology 96 (2) (2016) 133–158.

[6] F.P. Campion, T.F. Crosby, P. Creighton, A.G. Fahey, T.M. Boland, An investigation into the factors associated with ewe colostrum production, Small Ruminant Research 178 (2019) 55–62.

[7] F. Giansanti, G. Panella, L. Leboffe, G. Antonini, Lactoferrin from milk: nutraceutical and pharmacological properties, Pharmaceuticals 9 (4) (2016) 61.

[8] W.L. Hurley, P.K. Theil, Perspectives on immunoglobulins in colostrum and milk, Nutrients 3 (2011) 442–474.

[9] J.M. Rocha, On the historical background of bovine colostrum, EC Nutrition 4 (2016) 980–981.

[10] A.B. Sabin, A.H. Fieldstell, Antipoliomyelitic activity of human and bovine colostrum and milk, Pediatrics 29 (1962) 105–115.

[11] L.H. Ulfman, J.H.W. Leusen, H.F.J. Savelkoul, J.O. Warner, R.J.J. van Neerven, Effects of bovine immunoglobulins on immune function, Allergy, and Infection, Frontiers in Nutrition 5 (2018) 52.

[12] J. Steele, J. Sponseller, D. Schmidt, O. Cohen, S. Tzipori, Hyperimmune bovine colostrum for treatment of GI infections: a review and update on clostridium difficile, Human Vaccines and Immunotherapeutics 9 (7) (2013) 1565–1568.

[13] K. Saad, et al., Effects of bovine colostrum on recurrent respiratory tract infections and diarrhea in children, Medicine (Baltimore) 95 (37) (2016) e4560.

[14] Z. Haimin, Amelioration of decline in immune function in athletes after high-intensity training by bovine colostrum, Current Topics in Nutraceutical Research 17 (2) (2018) 219−222.

[15] A.W. Jones, D.S. March, R. Thatcher, B. Diment, N.P. Walsh, G. Davison, The effects of bovine colostrum supplementation on in vivo immunity following prolonged exercise: a randomised controlled trial, European Journal of Nutrition 58 (2019) 335−344.

[16] E. Dominguez, M.D. Perez, M. Calvo, Effect of heat treatment on the antigen-binding activity of anti-peroxidase immunoglobulins in bovine colostrum, Journal of Dairy Science 80 (12) (1997) 3182−3187.

[17] A.J. Lopez, C.M. Jones, A.J. Geiger, A.J. Heinrichs, Comparison of immunoglobulin G absorption in calves fed maternal colostrum, a commercial whey-based colostrum replacer, or supplemented maternal colostrum, Journal of Dairy Science 103 (5) (2020) 4838−4845.

[18] J.D. Quigley, L. Deikun, T.M. Hill, F.X. Suarez-Mena, T.S. Dennis, W. Hu, Effects of colostrum and milk replacer feeding rates on intake, growth, and digestibility in calves, Journal of Dairy Science 102 (12) (2019) 11016−11025.

[19] U. Krogh, et al., Impact of dietary protein to energy ratio and two different energy levels fed during late gestation on plasma metabolites and colostrum production in sows, Livestock Science 234 (2020) 103999.

[20] Y. Sun, C. Wang, X. Sun, S. Jiang, M. Guo, Characterization of the milk fat globule membrane proteome in colostrum and mature milk of Xinong Saanen goats, Journal of Dairy Science 103 (4) (2020) 3017−3024.

[21] V. Guzman-Carazo, J. Reyes-Vélez, I. Elsohaby, M. Olivera-Angel, Factors associated with microbiological quality of bovine colostrum in Colombian dairy herds, International Dairy Journal 105 (2020) 104670.

[22] K. Stastny, H. Stepanova, K. Hlavova, M. Faldyna, Identification and determination of deoxynivalenol (DON) and deepoxy-deoxynivalenol (DOM-1) in pig colostrum and serum using liquid chromatography in combination with high resolution mass spectrometry (LC-MS/MS (HR)), Journal of Chromatography B 1126−1127 (2019) 121735.

[23] F. Sutter, S. Borchardt, G.M. Schuenemann, E. Rauch, M. Erhard, W. Heuwieser, Evaluation of 2 different treatment procedures after calving to improve harvesting of high-quantity and high-quality colostrum, Journal of Dairy Science 102 (10) (2019) 9370−9381.

[24] X. Peng, et al., Live yeast supplementation during late gestation and lactation affects reproductive performance, colostrum and milk composition, blood biochemical and immunological parameters of sows, Animal Nutrition (2020).

[25] A. Thiel, et al., Toxicological evaluations of colostrum ultrafiltrate, Regulatory Toxicology and Pharmacology 104 (2019) 39−49.

[26] S.L. Gelsinger, A.J. Heinrichs, Comparison of immune responses in calves fed heat-treated or unheated colostrum, Journal of Dairy Science 100 (5) (2017) 4090−4101.

[27] M.M. El-Loly, L.K. Hassan, E.S.A. Farahat, Impact of heat treatments and some technological processing on immunoglobulins of Egyptian buffalo's milk, International Journal of Biological Macromolecules 123 (2019) 939−944.

[28] M. Lomónaco, et al., Efficacy of the spray-drying treatment to inactivate the bovine leukemia virus in bovine colostrum, Journal of Dairy Science (2020).

[29] M. Tang, et al., Hexachlorocyclohexane exposure alters the microbiome of colostrum in Chinese breast-feeding mothers, Environmental Pollution 254 (2019) 112900.

[30] C. Álvarez, V. Luño, N. González, P. Guerra, L. Gil, Effect of mare colostrum in extenders for freezing stallion semen, Journal of Equine Veterinary Science 77 (2019) 23−27.

[31] J. Opgenorth, L.M. Sordillo, M.J. VandeHaar, Colostrum supplementation with n-3 fatty acids and α-tocopherol alters plasma polyunsaturated fatty acid profile and decreases an indicator of oxidative stress in newborn calves, Journal of Dairy Science 103 (4) (2020) 3545−3553.

[32] T. Ma, et al., Altered mucosa-associated microbiota in the ileum and colon of neonatal calves in response to delayed first colostrum feeding, Journal of Dairy Science 102 (8) (2019) 7073−7086.

[33] E.G. dos, S.O. Silva, A.H. do, N. Rangel, L. Mürman, M.F. Bezerra, et al., Bovine colostrum: benefits of its use in human food, Food Science and Technology 39 (Suppl. 2) (2019) 355–362.

[34] Y. Xu, J. Zhang, J. Zhang, Anti-aging yoghurt containing modified starch and preparation method thereof, CN110235943A, 2019.

[35] V. Bhandari, Bioactive dairy products and processes for their manufacture, AU2018323788A1, 2020.

[36] M. Godhia, N. Patel, Colostrum—its composition, benefits as a nutraceutical: a review, Current Research in Nutrition and Food Science Journal 1 (1) (2013) 37–47.

[37] J. Zou, Colostrum mineral calcium supplementing preparation and preparation method thereof, CN108720000A, 2018.

[38] D. Spisar, Nutritional food supplement containing granulated colostrum, CZ33239U1, 2019.

[39] X. Long, X. Zhan, Milk powder for enhancing immunity and preparation method thereof, CN109645133A, 2019.

[40] K. Li, C. Liang, L. Wang, Q. Wang, X. Yang, Bovine coloctrum type solid state shaping preparation capable of enhancing immunity, CN109717249A, 2019.

[41] S. Wu, Bovine colostrum composite nutrition powder and preparation method thereof, CN108124963A, 2018.

[42] M. Chen, Formula for regulating immunity and enhancing physical strength and preparation method thereof, CN110037257A, 2019.

[43] B. Liu, Health product for promoting growth and enhancing memory function and preparation method thereof, CN107890104A, 2018.

[44] D. Kuang, Compound fungus and protein powder nutritious food formula, CN109198601A, 2019.

[45] S. Guo, Y. Tan, S. Wang, Y. Xiao, M. Yang, B. Yi, Composition capable of enhancing immunity, functional food capable of enhancing immunity and preparation method of functional food, CN109430413A, 2019.

[46] S. Biangiolini, J. Chini, G. Cipolletti, L. Vagnoli, Process for the preparation of high purity mixtures of protein factors from bovine colostrum, EP3177636A1, 2019.

[47] J. Feng, H. Li, Z. Qu, C. Xiao, K. Ying, Composite colostrum and preparation method thereof, CN109674064, 2019.

[48] F. Wang, Medical intacted protein-whole nutrition formulated food and preparation method thereof, CN108142925A, 2018.

[49] F. Wang, Polypeptide type total nutrient formula food for medical purpose and preparation method thereof, CN107980906A, 2018.

[50] X. Kong, B. Liu, B. Liu, Composition and food application thereof, CN110839698A, 2020.

[51] W. Li, Z. Li, Antidiarrheal nutrition bag for infants and preparation method thereof, CN109730156A, 2019.

[52] Y. Huang, M. Liu, G. Yin, M. Zhang, Composition for improving gastrointestinal tract and enhancing immunity and application thereof, CN109645473A, 2019.

[53] R. Hickey, M. Marotta, S. Morrin, J. Lane, S. Carrington, A composition and uses thereof, WO2019086613A1, 2019.

[54] L. Gan et al., Total-nutrient formula suitable for enterostomy intestine internal absorption for special medicine, CN109770345A, 2019.

[55] G. Yu, Medicated diet for people with common cold, CN109907264A, 2019.

[56] J. Dai et al., Oral thick paste capable of increasing white blood cell content and preparation method thereof, CN109700999A, 2019.

[57] W.G. Struff, G. Sprotte, Bovine colostrum as a biologic in clinical medicine: a review. Part I: Biotechnological standards, pharmacodynamic and pharmacokinetic characteristics and principles of treatment, International Journal of Clinical Pharmacology and Therapeutics 45 (4) (2007) 193–202.

[58] H. Gao, Bovine colostrum powder and preparing method thereof, CN109349350A, 2019.

[59] M. Diao, T. Zhang, W. Yang, S. Chen, Method for continuously separating and preparing functional lactoprotein in colostrum, CN110973345A, 2020.

[60] W. Chen, Y. Chen, D. Jia, Z. Mu, C. Yang, Method for extracting oligosaccharides, 2018.

[61] W. Yan, Y. Yan, Preparation method of bovine colostrum element, CN109287747A, 2019.

[62] E. Satyaraj, A. Reynolds, R. Pelker, J. Labuda, P. Zhang, P. Sun, Supplementation of diets with bovine colostrum influences immune function in dogs, British Journal of Nutrition 110 (12) (2013) 2216–2221.

[63] X. Liu, Y. Zhang, Composition capable of increasing bone density, CN108576816A, 2018.

[64] J. Li, Y. Lu, X. Wu, Composition for preventing and treating bone loss of women and application thereof, CN110338409A, 2019.

[65] J. Li, Y. Lu, X. Wu, Composition for promoting children bone growth and development and preparing method and application thereof, CN110292177A, 2019.

[66] J. Chen, H. Chunyue, S. Zhiyong, W. Jiguang, J. Zhang, Protein compound preparation for improving bone and joint, CN110354256A, 2019.

[67] G. Zarbo, Composition based on a mixture of biological factors isolated from colostrum and ozonized vegetable oils for use in the treatment of vaginitis and vaginosis, WO2019162895A1, 2019.

[68] A. Ferrari, L. Genova, Composition for the treatment of pathologies of the urinary system, WO2018029558A1, 2018.

[69] Y. Ilan, M. Mizrahi, Anti-LPS enriched immunoglobulin for use in treatment and/or prophylaxis of fibrosis, United States10464998B2, 2019.

[70] C. Bucolo, F. Drago, M. Musumeci, S. Musumeci, Ophthalmic formulations comprising oligosaccharides, EP3127543B1, 2018.

[71] R.E. Burrel, D.H. Gul-Uludag, Colostrum derived stem cells, neural differentiation, compositions and supplements for enhancing mammalian health, WO2020056521A1, 2020.

[72] T. Adar, A. Ben-Ya'acov, Y. Ilan, G. Lalalzar, M. Mizrahi, Anti-lps enriched immunoglobulin preparation for use in treatment and/or prophylaxis of non alcoholic steatohepatitis, EP3461840A1, 2019.

[73] F. Aqil, R. Gupta, J. Jeyabalan, R. Munagala, Milk-derived microvesicle compositions and related methods, United States9943482B2, 2018.

[74] R. Gupta, Isolation of exosomes from colostrum powder and exosomal drug formulations using the same, United States10166259B1, 2019.

[75] H. Hori, T. Inui, K. Kubo, Y. Uto, Bovine colostrum enzyme processed product, production method therefor, composition, and food or beverage, AU2020201700A1, 2020.

[76] A.L. Beam, et al., Prevalence of failure of passive transfer of immunity in newborn heifer calves and associated management practices on United States dairy operations, Journal of Dairy Science 92 (8) (2009) 3973–3980.

[77] E.L. Cuttance, W.A. Mason, R.A. Laven, C.V.C. Phyn, The relationship between failure of passive transfer and mortality, farmer-recorded animal health events and body weights of calves from birth until 12 months of age on pasture-based, seasonal calving dairy farms in New Zealand, The Veterinary Journal 236 (2018) 4–11.

[78] G. Pusillo, Therapeutic and nutritional compositions, United States2019313667A1, 2017.

[79] T.-M. Yu, H.-S. Chang, Modified colostrum protein and method for using the same, JP6384928B2, 2018.

[80] K. Bisgaard-Frantzen, Products and methods to prevent infection, EP2249661B1, 2018.

[81] E. Satyaraj, Compositions and methods useful for ameliorating age related maladies, United States10010566B2, 2018.

[82] Y.D. Ahn, J. Kim, K. Il Kim, T. Kwak, H.R. Seo, S.-A. Seo, Fermented product of colostrum, and cosmetic composition using same, WO2018174669A1, 2018.

[83] M. Jeong, J. Kim, K. Il Kim, T. Kwak, H.R. Seo, S.-A. Seo, A cosmetic composition of fermented colostrum product for anti-acne, KR20190060556A, 2019.

[84] M. Jeong, J. Kim, K. Il Kim, T. Kwak, Aerobic fermented product of colostrum, KR20190082159A, 2019.

[85] C.D. Kim, H. Lee, M.S. Nam, Cosmetic composition comprising hydrolyzate of colostrum whey protein for skin regeneration, KR101974405B1, 2019.

[86] H. Li, A lotion formula, CN107822985A, 2018.

[87] L. Cheng, Soap for children and preparation method thereof, CN108085186A, 2018.

[88] K. Bo-Kyung, P.H. Woo, L. Kun-Kook, K.-S. Lee, Cosmetic composition for improving skin wrinkle containing colostrum and ornithine, KR102076178B1, 2020.

[89] M.J. Jung, Colostrum collagen composite for improving skin using slow speed lyophilization at low temperature and manufacturing method, KR20180114493A, 2018.

[90] L. Ho, Skin-moisturizing hydrogel capsules and manufacturing method thereof, KR102023724B1, 2019.

[91] A. Oleszek, A multi-ingredient preparation for finger and toe nail regeneration and a method of preparation thereof, WO2019031976A2, 2019.

[92] M.L. Juhász, M.K. Levin, E.S. Marmur, The use of natural ingredients in innovative Korean cosmeceuticals, Journal of Cosmetic Dermatology 17 (3) (2018) 305–312.

[93] W. Stephan, H. Dichtelmuller, R. Lissner, Antibodies from colostrum in oral immunotherapy, Journal of Clinical Chemistry and Biochemistry 28 (1) (1990) 19–23.

[94] F.O. Uruakpa, M.A.H. Ismond, E.N.T. Akobundu, Colostrum and its benefits: a review, Nutrition Research 22 (6) (2002) 755–767.

[95] A. Mero, H. Miikkulainen, J. Riski, R. Pakkanen, J. Aalto, T. Takala, Effects of bovine colostrum supplementation on serum IGF-I, IgG, hormone, and saliva IgA during training, Journal of Applied Physiology 83 (4) (1997) 1144–1151.

[96] C.H. Florén, S. Chinenye, L. Elfstrand, C. Hagman, I. Ihse, ColoPlus, a new product based on bovine colostrum, alleviates HIV-associated diarrhoea, Scandinavian Journal of Gastroenterology 41 (6) (2006) 682–686.

Index

Note: Page numbers followed by "*f*" and "*t*" refer to figures and tables, respectively.

A

Achaar, 91
Acidolysis, 215–216
Active microorganisms, whole cells of, 113–116
Aflatoxins, 262, 264, 265*t*
Agricultural products contamination, 261
Agricultural wastes, 301
Agriculture food residue, products from, 304*t*
Agri-food industrial wastes, 301
Agri-food processing residues and valorization, 303–313
    bacterial cellulose (BC), 312–313
    bioactive peptides, 310–311
    bioethanol production, 311–312
    extraction methods, 305–306
    fructooligosaccharides (FOS), 311
    mushroom production, 313
    nutraceuticals and functional food from agri-food waste, 306–308
    organic acids from agro-industrial wastes, 308–309
    polyunsaturated fatty acids (PUFAs), 310
    volatile fatty acids production, 309–310
Agri-food processing residues, 303–305
Agri-food residue to value-added products, 302*f*
Agro-industrial by-products, 116
Agro-industrial waste products, 312–313
Agroindustrial wastes, 173
α-acetolactate decarboxylase, 140*t*
Alpha-amylase, 141
α-galactosidase, 140*t*, 249–250
α-glucan, 338, 340, 343
α-ketoglutaric acid (α-KG), 282
α-linolenic acid (ALA), 15–16, 283–284
Alzheimer disease, 122, 127–128
American company Jarrow Formulas, 417
Amino acids, 58, 240, 331*t*
Ammonium sulfate precipitation, 168
Amylases, 140*t*
Angiotensin-converting enzyme (ACE)-inhibitory BAPs, 54–55, 57
Animal nutrition, 408–409
Anthocyanidins, 2–3, 407
Anthocyanins, 57, 199, 306
Anticoagulants, 157–158
Antimicrobial compounds, production of, 35–36
Antimicrobial peptides, 379–380
Antinutritional compounds, 254
Antinutritional factors (ANFs), 240–241
    classification of, 242*f*
Antinutritional factors, conversion of toxic, 249*f*
Antinutritional factors and adverse effects
    classification of, 241*t*
Antioxidant compounds, production of, 36–37
Antioxidant peptide production, 54
Antioxidants, encapsulation of, 39–40
Apple pomace, 307, 311
AprE34, 167–168
APS BioGroup, 415–416
Arabinofuranosidease, 140*t*
Ascorbic acid, 39–40
Asian fermented shrimp paste, 168–169
Asparaginase, 140*t*
Aspergillus, 149–150
Aspergillus flavus, 262
Aspergillus ibericus, 115
Aspergillus japonicus, 116, 118
Aspergillus niger, 112, 116
Aspergillus parasiticus, 262
ASRA (adaptive substituent reordering algorithm), 148–149
Aureobasidium melanogenum, 119–120
Aureobasidium pullulans, 112, 118–119

B

Bacillus, 79

423

Bacillus amyloliquefaciens, 3–5
Bacillus licheniformis, 3–5
Bacillus natto probiotics, 365–366
Bacillus subtilis, 149–150, 281
Bacterial-amylase, 141
Bacterial cellulose (BC), 312–313
  production of, 313f
Bacterial fermentation, 195, 337
Bacteriocins, 35–36, 53
Baking process, 141
Baking products, 331t
Bamboo Tree, 3–5
Batch and continuous bioreactors, operational stability in, 226–230
Batch operational stability, 227–229, 228t
Berry pomace, 307
Best Practice of Production (BPP), 398
β-D-fructosyltransferase, 111
β-fructofuranosidase, 111, 117, 119–120
β-glucan, 338, 340, 343
Betalains, 307
Bifidobacterium, 79, 250, 330
Bifidobacterium longum, 5
Bifidus factor, 123
Bioactive Component 1, 409
Bioactive compounds, 39
  bioavailability of, 41
  biological activities of, 4f
  microbial-based production of, 3f
Bioactive ingredients, 33–34
Bioactive peptide production, 47–72
  biotechnological progress for industrial production of, 57–62
  in fermented dairy products, 48–52
    health-promoting activities of peptides, 49–52
  in fermented meat products, 52–55
    generation of bioactive peptides, 52–53
    identification and functional activities of bioactive peptides, 53–55
  in fermented vegetables, 55–57
Bioactive peptides (BAPs), 12–14, 47–48, 310–311
  microbial activity, 49
  derived from fermented foods, 48f

  in fermented foods of dairy origin, 50t
  rich functional foods and microbes in production of, 13t
Bioactive phytochemicals, 308
Bioavailability, 39
Biocatalyst immobilization, 217–218
Bio-energy, 311–312
Bioenervi, 414–415
Bioethanol production, 311–312
Biomass, 311–312
BIOPEP database, 60
Bioprocess technologies, 209–238
  lipids, general aspects of, 210–213
  lipids as nutraceuticals, 209–210
  lipids' role in human nutrition, 213–215
    absorption and lipid metabolism, 214–215
    functional/nutritional properties of lipids, 213–214
  structured lipids, definition and types of, 210
  structured lipids, production of, 215–230
    biocatalyst immobilization, 217–218
    case studies on, 219–226
    lipases and phospholipases, 216
    reaction systems, 218–219
Biostrum Nutritech Pvt. Ltd., 416–417
Biotechnological interventions, 158
Biotest Pharm GmbH, 414–415
Blood clotting, mechanism of, 158–159
Bone density, 405t
Bone density loss, 406t
Bone growth and child development, 406t
Boreostereum vibrans, 363
Bovine and human colostrum
  macronutrient composition and bioactive factors in, 379t
Bovine colostrum (BC), 378, 384f
  application for animal therapy, 386–388
  application for human therapy, 383–386
  composition, 378–382
  passive immunity mechanism in human and different animal species, 382–383
Bovine colostrum manufactures
  commercially available products and, 416t
Bovine colostrum products, commercial, 414–417
Bovine lactoferrin, 388

Bovine Tuberculosis (TB), 386
Branched-chain amino acid (BCAA), 58
Bromelain, 140*t*
Bulgaricus, 5
Butter, 142, 330

C

Calcium oxalates, 245
Canavanine, 248
Candida albicans, 6−7
Candida famata, 8
Capric acid, 227−229
Caprylic acid, 227−229
Carbohydrate fermentation, 37, 74
Carbohydrates, 37
Carboxypeptidase, 143
Carotenes, 11−12
Carotenoid pigments, 282
Carotenoids, 10−12, 286−287
Caseins, 48
Casomorphins, 51
Cattle colostrum, 386
Cell conditions, 330
Cerebral ischemia−reperfusion injury, 124
Cheese, 330, 331*t*
Cheonggukjang, 168
Chinese patent CN107822985A, 412
Chinese Patent CN108085186A, 412
Chromatographic techniques, 160−161
Chymosin, 140*t*, 142
Cineromycin B, 359
  antilipid effect of, 359
Citrinins, 262−263
Citrus fruit waste, 307
Citrus peel waste, 307
Cladosporium cladosporioides, 117
Clustered Regularly Interspaced Short
  Palindromic Repeat, 330
Coagulation and fibrinolytic cascades, 159*f*
Coculture of microorganisms, 118−119
Codex Alimentarius Commission (CAC), 268
Collagen, 123−124, 142
  synthesis of, 280
Colonic fermentation, 33
Colostral whey, 386

Colostrum, 377−378, 397, 403−406, 410−411
Colostrum milking, 399
ColostrumOne Nature's Superfood, 415−416
Colostrum patents, 402−414
  animal health, 408−410
  cosmetics, 410−413
  human health, 402−408
Colostrum Prime Life, 417
Colostrum production process and quality
  control, 398−402
  effect in colostrum quality by processing, 400
  impact of flock managing in colostrum quality,
    398−400
  quality parameters, 401
Colostrum quality, 399
Colostrum therapeutic properties, 400
Colostrum's quality control, 398
Commercial probiotics, 93−94
Complex lipids, 212
Computer simulation technology, 149
Condensed tannins, 242−243
Conjugated linoleic acid, 213−214
Conjugated linolenic acid, 385
Corynebacterium glutamicum, 5
CRISPR-associated protein 9, 330
CRISPR-mediated gene editing, 330
Critical humidity point (CHP), 116
Cumbu, 92
Cutting-copying-pasting, 324
Cyanidin-3-O-glucoside, 197*t*
Cyanogens, 239−240, 244−245
Cytokines, 379−381

D

Dahi, 91
Dairy fermented foods, 81−86
  probiotics in, 82*t*
Dairy products, 87, 331*t*
Data sharing, 267−268
Deactivation models, 226
Degree of polymerization (DP), 111
Deoxynivalenol/nivalenol, 265*t*
Deoxyvalenol (DON), 398−399
Dextran, 340−341, 348
Dextransucrase, 341

Dhokla, 91–92
Diabetes mellitus (DM), 90
Diacylglycerols (DAG), 212, 358
2,3-Dibenzylbutane, 192
Dietary fat, 358
Dietary lipids, 362–363
    metabolism of, 357f
Dietary polyphenols, 125, 200
Dietary Supplement Health and Education Act, 1994, 94
1,2-Diglyceride, 142
1,3-Diglyceride, 142
Dihydroxyflavonols, 2–3
1,25-Dihydroxyvitamin D, 403–406
Dimethylallyl pyrophosphate (DMAPP), 286–287
Direct esterification, 215–216
Discovery Studio (DS), 149
Docosahexaenoic acid (DHA), 15–16, 213, 283–284, 310
Doenjang, 167–168
Douche, 164–167
Downstream treatments, 113
Dysbiosis, 92–93, 274–275

E

Eicosapentaenoic acid (EPA), 15–16, 213, 283–284, 310
Ellagitannins, 242–243
Embden–Meyerhof pathway, 74–75
Emergent prebiotics, 125–127, 126t
Encapsulated probiotics, 93
Encapsulation techniques, 38–40, 413
Endo-inulinases, 120
Enterococcus, 77
Environmental and genetic factors, 355–356
Enzymatic synthesis of TAG, 225t
Enzyme-assisted extraction, 306
Enzyme-catalyzed interesterification, 224
Enzyme engineering, 149
Enzyme inhibitors, 252–254, 380–381
    amylase inhibitors, 254
    protease inhibitors, 252
    trypsin inhibitors, 252–254
Enzyme purification methods, 146
Enzymes, role of
    in food industries, 140t
Ergots, 263
Ergot sclerotia, 263
Escherichia coli, 5
Ethnic fermented probiotics products, 90–92
    achaar, 91
    dahi, 91
    dhokla, 91–92
    idlis and dosa, 91
    kaanji, 91
    koozh, 92
    toddy, 92
Ethyl esters, 218
Exopolysaccharides (EPS), 37–38, 331t, 338–339
    as antioxidant, 348
    as antitumor agent, 348–349
    fermented food containing, 345–346
        kefir, 346
        sour dough, 345–346
        yoghurt, 346
    health benefits of, 347–349
    in functional food product development, 345f
    as immunomodulator, 348
    as prebiotic, 348
    technological advantages of, 347
Exopolysaccharides producing bacterial genera, 339–344
    lactic acid bacteria species, 343–344
    Lactobacillus, 339–341
    Leuconostoc, 342
    non-lactic acid bacteria species, 344
    Pediococcus, 342–343
    Weissella, 341–342
Exopolysaccharides production with associated health benefits, 37–38
Extrinsic pathway, 158–159

F

Farm Skin, 411
Fat-soluble vitamins, 7–8, 365
Fatty acid distribution, 221
Favism, 248
Feeding procedures and management, 398–399
Fermentation, 73
Fermentation engineering, 150–151

Fermentative approach, 171—174
  nutrients, 173—174
  optimization of temperature, 173
  pH, 173
  statistical optimization of media, 174
  substrate selection, 173
Fermented colostrum product, 411
Fermented dairy products, 49
Fermented food, for antiviral therapy, 40—41
Fermented food containing exopolysaccharides, 345—346
  kefir, 346
  sour dough, 345—346
  yoghurt, 346
Fermented food products, 9—10, 12, 36—37, 239—240, 323—324, 332—333
Fermented functional foods, 337
Fermented Indian products, 170
Fermented meat products
  generation of bioactive peptides in, 52—53
  identification and functional activities of bioactive peptides in, 53—55
Fermented milk, inhibitory effect of, 366—367
Fermented skim milk, inhibitory effect of, 366—367
Fermented soybean products, 3—5
Fermented soybean seasoning (FSS), peptide enriched, 14
Fermented vegetables, 331t
Fibrin, 159—160
Fibrinogen, 161—162
Fibrinogenesis, mechanism of, 158—159
Fibrinogenolytic activity of FEs, 161—162, 163t
Fibrinolysis, 159—160
Fibrinolytic activity of FEs, 161—162, 163t
Fibrinolytic enzymes (FEs), 157—188
  applications, 174—177
    clinical, 174—177
  biochemical characterization of, 161—163
    amidolytic properties of, 162—163
    physiochemical properties of, 161—162
  classification of, 160
  mechanism of action of, 159—160
  mechanism of blood clotting and fibrinogenesis, 158—159
  production of, 170—174
    biotechnological approaches, 170—171
    fermentative approach, 171—174
    cloning and expression parameters used for, 172t
  purification of, 160—161
  sources of, 163—170, 165t
    Asian fermented shrimp paste, 168—169
    cheonggukjang, 168
    doenjang, 167—168
    Douche, 164—167
    fermented Indian products, 170
    Jeotgal, 164
    kimchi, 169
    kisk, 169—170
    Meju, 170
    natto, 168
    skipjack, 169
    Tempeh, 167
  statistical methods used for optimal production of, 175t
Fibrinolytic mechanism, 158
Field-based residues, 303
First-order deactivation kinetics model, 226
Flavan-3-ols, 2—3
Flavanone-3-hydrolase, 285—286
Flavanones, 2—3, 5, 197t
Flavanonols, 2—3
Flavins, 10—11
Flavones, 2—3
Flavonoids, 2—3, 33, 191, 243
Flavonols, 2—3, 5, 197t
Flavonol synthase, 285—286
Fluorescence-activated cell sorting (FACS), 147—148
Folate, 280—281
Folate metabolism, 277
Fold assignment techniques, 61—62
Food and Agriculture Organization (FAO), 266
Food and Agriculture Organization estimation, 302
Food antioxidants, 36
Food biotechnology, 323—324
Foodborne disease, 35
Food-derived bioactive peptides, 40

Food enzyme fermentation process
  optimization of, 150–151
Food enzymes production
  advanced technologies for, 147–151
    development of safe and efficient expression hosts, 149–150
    high-throughput screening technology, 147–148
    optimization of food enzyme fermentation process, 150–151
    strategies for efficient modification of food enzymes, 148–149
  applications of, 141–144
    dairy industry, 142
    food analysis and testing, 143
    food processing of grains and oils, 141–142
    functional foods, 143–144
    meat products processing, 142–143
  current progress and challenges in, 144–147
  procedures and related technologies, 144f
  progress and prospects of, 139–156
  strategies for, 148f
Food fermentation, 31–32, 189–190
Food-grade microorganisms
Food-grade organic solvents, 219
Food ingredients, in global market
  based on bioactive peptides derived from whey proteins, 59t
Food processing, 251–252
Foods for Specified Health Use (FOSHU), 209–210
Food substrates, 31–32
Food waste, 301–302
Free fatty acids (FFA), 212
Free radicals, 36
French paradox, 191–192
Fructans, 109–112
Fructooligosaccharides (FOS), 15, 110–111, 114t, 121–123, 143–144, 278–279, 283, 311
  biotechnological strategies for production of, 110f
  manufacturers of, 112
  prebiotic effects of, 123
  production of, 112–120

cloning fructosyltransferase gene for, 119–120
coculture of microorganisms, 118–119
fructans occurrence, chemical structure, technological properties, and market, 110–112
mixed enzymes and cell systems, 117–118
whole cells of active microorganisms, 113–116
whole cells of inactive microorganisms, 116–117
Fructosidase, 143–144
Fructosylfuranosidades, 120
Fructosyltransferase, 119–120, 140t
Fruit processing wastes, 302–303
2-Fucosyllactose (2'-FL), 283
Fumonisin B1, 265t
Fumonisins, 264
Functional foods, 2, 143–144, 210
Functional Food Science in Europe (FUFOSE), 209–210
Fungalamylase, 141
Fungal isolates, as pancreatic lipase inhibitors, 362f
Fusarium mycotoxins, 264

G

Galactooligosaccharides (GOS), 15, 37, 109, 123–124, 143–144, 278–279
  prebiotic effects of, 123
Gallotannins, 242–243
Gamma-aminobutyric acid (GABA), 9–10, 279–280
  GABA-enriched yogurt, 9–10
  GABA-rich fermented soy protein (GABA-RFSP) biofilms, 9–10
Gamma-linolenic acid (GLA), 213–214
Gastric lipase, 358
Gastrointestinal absorption, 189–190
Gastrointestinal tract (GI), 355–356
GATA binding protein family (GATA), 359
Gene Gun, 324
Gene–nutrients interactions, 277
Generally recognized as safe (GRAS) species, 74
Genetically modified microorganisms, 323–336

advantage of, 325–326
advantages of genetic modification, 324
characteristics of, 327–330
   cloning of gene and heterologous expression, 328–329
   clustered regularly interspaced short palindromic repeats-Cas9 (CRISPR-Cas9), 330
   development of recombinant enzymes, 330
   engineering regulatory networks, 329
   expression vectors and transformation process, 328
   gene insertion and deletion, 329
   genetic knockout of loci, 330
   redirecting metabolic pathway, 329
   selection of host organism, 328
   stimulation by precursors, 329
and fermented food products, 326
fermented food products and their health benefits, 332–333
and industrial food enzyme production, 326–327
and metabolites production, 327
microorganisms as source of food enzymes, 325
need for genetically modified food products, 324–325
products development using, 330–331
techniques used for genetic modification, 324
Genetically modified microorganisms and different food products
for enhancing nutritional properties, 327f
Genetically modified organism (GMO), 323–324
Genetic manipulation, 8
Genistein, 5
Gentamycin, 356–357
Germination, 253
Global fermented food products, principal groups of, 240t
Glucan, 338
Glucanase, 140t
Glucoamylase, 140t
Glucooligosaccharides (GOS), 283
Glucosamine (GlcN), 281
Glucose isomerase, 118
Glucose oxidase (GOD), 140t, 141
Glucoside hydrolase (GH), 338
Glutaminase, 140t
Glutathione (GSH), 281–282
Glutathione reductase, 10
Glycerol-based lipids, 212
Glycolipids, 212
Grain-based food products, 87
Grape seed extract, 306–307
Grape seed oil, 306–307
Green extraction technique, 305–306
Green technologies, 58
Growth hormone (GH), 385, 417
Gut bacteria, 275
Gut–brain axis, 359–360
Gut microbes, 189–190, 273–300, 359–360
   developments in nutraceuticals production using, 278–279
   glutathione, carotenoids, and other nutraceuticals production by yeasts, 281–282
   market trends in production of nutraceuticals, 287–288
   metabolic engineering of microbes for production of nutraceuticals, 282–287
      carotenoids, 286–287
      mannitol and sorbitol production by lactic acid bacteria, 284–285
      polyamino acids, 286
      polyphenols, 285–286
      prebiotics, 282–284
   metabolites of, 361f
   microbial strategies for production of nutraceuticals, 279
   nutraceuticals production by *Bacillus subtilis*, 281
   nutraceuticals and, 276
   nutrigenomics, 276–278
   production of gamma-aminobutyric acid and hyaluronic acid by lactic acid bacteria, 279–280
      gamma-aminobutyric acid (GABA), 279–280
      hyaluronic acid (HA), 280
   role of, 274–275
   vitamin B12 and folate production by *Propionibacteria*, 280–281

Gut microbes (*Continued*)
    folate, 280−281
    vitamin B12, 280
Gut microbial metabolism, 200
Gut microbiota, 276
    composition of, 356−357
    modification, 122−123

H

Halophiles, 164
Haloterrigena turkmenica, 11−12
Health-promoting bioactive compounds, 31−32, 32f
Health-promoting functional foods, 62
Heat treatment, 251, 388
Hemicellulases, 140t
Heterofermentative metabolism, 75
Heteropolysaccharide (HePS), 338, 343
Hexachlorocyclohexane (HCH), 401
HMFS production
    by batch acidolysis/interesterification, 223t
Homofermentative bacteria, 74−75
Homofermentative metabolism, 74−75
Homologous modeling technology, 149
Homopolysaccharide (HoPS), 338−339, 343
HotSpot Wizard, 148−149
Human angiotensin-converting enzyme 2 (hACE2), 40
Human gut microbiome, impact of probiotics on, 87−90
    probiotics in prevention and treatment of clinical diseases, 88−90
        diabetes mellitus (DM), 90
        obesity, 89−90
        probiotics and COVID-19, 89
Human Microbiome Project, 274−275
Human milk fat (HMF), 220−221
    substitutes, 220−223
Hyaluronic acid (HA), 280
Hybrid methods, 61f
Hydrogen peroxide, 141
Hydrolysable tannins, 242−243
Hydroxybenzoic acid, 5, 191
Hydroxycinnamic acids, 191

I

Idlis and dosa, 91
Immobilized enzyme technology, 146−147
Immobilized lipases, 217
Immune system stimulation, 311
Immunoglobulin concentration, 380
Immunoglobulin G (IgG), 397
Indigo, 10−11
Industrial growth, in agri-food sector, 303
Infection resistance, 311
Ingested bacteria, 88
Inhibitors, 161
Innovators methods and products, 407
Inorganic selenium, 6−7
In silico methods, 60
In silico tools, 148−149
Insulin-like growth factors, 385
Interesterification, 215−216, 223−224
Interesterified fat blends, 223
Interesterified trans-free fat blends and triacylglycerols, 223−226
International Scientific Association for Probiotics and Prebiotics (ISAPP), 109
Intrinsic pathway, 158−159
Inulin, 111, 278−279, 283
Inulooligosaccharides, 120
Invertase, 140t
Isoflavones, 2−3
Isomaltooligosaccharides (IMO), 125−127, 143−144
Isopentenyl pyrophosphate (IPP), 286−287
    derived fat-soluble pigments, 11−12
Itaconic acid, 309

J

Japanese knotweed, 191−192
Jeotgal, 164

K

Kaanji, 91
Kaempferol, 197t
Katsuwokinase (KK), 169
Kefir, 346
Kefiran, 340−341, 346
Kezhvaragu, 92

Kimchi, 169
Kisk, 169—170
Kombucha tea, 36—37
Koozh, 92
Krüppel-like factor (KLF) family, 359

L

Laccase, 140*t*, 141
Lactase, 140*t*
Lactate dehydrogenase (LDH), 284
Lactic acid bacteria (LAB), 6—7, 9—10, 49, 74—79, 149—150, 250, 325, 330, 337, 343—344
    classification of, 75—79
        Bacillus, 79
        Enterococcus and *Streptococcus*, 77
        Lactobacillus, 76
        Lactococci, 78—79
        Leuconostoc, 78
        Pediococcus, 76—77
    metabolic pathway, 75*f*
    metabolism, 74—75
        heterofermentative, 75
        homofermentative, 74—75
    proteolytic system, 49
Lactic acid fermentation, 251
Lactic acid fermented foods, 35—36
Lactobacilli, 76
Lactobacillus, 76, 330, 339—341
Lactobacillus acidophilus, 5, 51, 251—252, 366—367
Lactobacillus brevis, 366—367
Lactobacillus delbrueckii, 366—367
Lactobacillus delbrueckii ssp, 5
Lactobacillus helveticus, 51, 366—367
Lactobacillus johnsonii, 6—7
Lactobacillus paracasei, 5
Lactobacillus plantarum, 3—7, 54, 251—252, 275, 332—333, 366—367
Lactobacillus probiotic species, 77*t*
Lactobacillus sakei, 54—55
Lactobacillus spp., 74
Lactococci, 78—79
Lactococcus, 278—279, 330
Lactococcus lactis, 5, 40, 366—367
Lactococcus spp., 343—344

Lactoferrin, 379—380, 385, 397
Lactoperoxidase, 385
Lactose, 142
Lactulose, 37, 109, 124—125
Legumes, 3—5, 252
Leuconostoc, 78, 330, 342
Leuconostoc mesenteroides, 342
Lignans, 2—3, 192
Lignin-rich waste materials, 305
Lingual lipase, 358
Linoleic acid, 15—16
Lipase-catalyzed process, 226
Lipase inhibitor, 364
Lipases, 216, 358
Lipids
    general aspects of, 210—213
    metabolism, 357—358
    as nutraceuticals, 209—210
    role in human nutrition, 213—215
        absorption and lipid metabolism, 214—215
        functional/nutritional properties of lipids, 213—214
    structured lipids, definition and types of, 210
    structured lipids, production of, 215—230
        biocatalyst immobilization, 217—218
        case studies on, 219—226
        lipases and phospholipases, 216
        reaction systems, 218—219
Lipopolysaccharides (LPS), 359
Lipoxygenase, 140*t*, 141
Lipstatin, 362—364
Living cells, encapsulation of, 39
Long-chain fatty acids (LCFAs), 212—215, 219—220
Low-calorie sugar, 285
Low-calorie triacylglycerols (TAG), 219—220
Lumbrokinase, 160
Lycopene, 282, 286—287
Lyophilized bovine colostrum (LBC), 386—387
Lysozyme, 140*t*

M

Macroantioxidants, 200
Malt-amylase, 141
Mammals, 382

Mannitol, 284–285
Mannitol-phosphate dehydrogenase (MPDH), 284
Meat processing industrial waste, 305
Meat products, 331t
Medium-chain fatty acids (MCFAs), 212–213, 215, 219–220
Meju, 170
Melanin, 10–11
Membrane techniques, 113
Menaquinone (MK), 8–9
Metabolic engineering, 190, 196–198, 288
Metabolomics, 3–5
Metalloprotease, 160
Methyl esters, 218
Microbial-based polyphenols, 5
Microbial carotenoids, 11–12
Microbial enzymes, 170
Microbial expolysaccharides, 280
   classification of, 339f
Microbial fermentation, 195–196, 326–327
   bottleneck of, 60
Microbial fermentation, for reduction of antinutritional factors, 239–260
   neutralizing antinutritional factors, in fermented foods, 248–254
     cyanogenic glycosides, 251
     enzyme inhibitors, 252–254
     glucosinolates, 254
     lectins, 252
     oxalates, 251–252
     phytic acid, 250
     saponins, 251
     tannins, 251
   nutritive and antinutritive properties in food, 240–248
     alkaloids, 243
     cyanogenic glycosides/glucosides, 244–245
     enzyme inhibitors, 246–248
     glucosinolates, 248
     lectins, 245–246
     oxalates, 245
     phytates, 241–242
     polyphenols, 242–243
     saponins, 243–244
     toxic amino acids, 248
Microbial hydrolysis, 196
Microbial metabolites beneficial in regulation of obesity, 355–376
   antilipid effect of Cineromycin B, 359
   Bacillus natto probiotics, 365–366
   composition of gut microbiota, 356–357
   gut–brain axis, 359–360
   inhibitory effect of fermented skim milk, 366–367
   new players for obesity treatment, 362–364
   obesity and lipid metabolism, 357–358
   orlistat, 364–365
     mechanism of action, 364–365
     side effects of, 365
   pancreatic lipase, 358
   short-chain fatty acids (SCFAs), 360–362
Microbial production of nutraceuticals and functional foods, 2–16
   bioactive peptides, 12–14
   gamma-aminobutyric acid (GABA), 9–10
   oligosaccharides, 14–15
   pigments, 10–12
   polyphenols, 2–5
   polyunsaturated fatty acids (PUFAs), 15–16
   selenium, 6–7
   vitamins, 7–9
     riboflavin (B_2), 8
     vitamin K, 8–9
Microbial transformation, 31–46
   bioactive compounds as functional food ingredients, 33–34
   delivery of bioactive compounds targeting enhanced food functionality, 38–40
     antioxidants, encapsulation of, 39–40
     living cells, encapsulation of, 39
   fermented food, for antiviral therapy, 40–41
   health-promoting bioactive compounds synthesized via, 34–38
     antimicrobial compounds, production of, 35–36
     antioxidant compounds, production of, 36–37
     exopolysaccharides, production of, 37–38
     oligosaccharides, production of, 37

microbes and food production over centuries, 31–33
Microbiome, 123–124
Microencapsulation, 93
Microwave-assisted product extraction, 306
Milk, 330, 401
    composition, 48
Mineral carriers, 116
Mitogen-activated protein kinase (MAPK) pathway, 123–124
MLM-type structured lipid production
    by batch acidolysis or interesterification, 221t
Modern molecular-based techniques, 95
Modern molecular techniques, 170
Molecular weight, 161
Monascins, 10–11
Monascus pigments (MPs), 11
Monoacylglycerols (MAG), 212
Mushroom cultivation, 313
Mushroom production, 313
Mycotoxicoses, 261–262
Mycotoxin contamination, 267
Mycotoxins, in foods, 261–272
    economic implications of, 266
    food contamination, 268–269
    impact on health, 264–266
    mitigation and control of, 266–268
        adequate monitoring of mycotoxin management, 267
        development and application of innovative and resourceful technologies, 267
        development and formulation of good policies and regulations, 268
        promoting and protecting human lives, 267
        transparent and responsible data sharing, 267–268
    molecular structure of, 263f
    legislations on, 268
    staple grains and seeds, 265t

N

N-acetylglucosamine (GlcNAc), 281
N-acetylneuramic acid, 283
Naringenin, 5, 197t, 285–286
Natto, 14, 168

Nattokinase (NK), 160, 176
Natural BAP production, 58–60
Natural coloring agents, 307
Natural proteins, 413
Necrotizing enterocolitis (NEC), 387
Neutralizing antinutritional factors, in fermented foods, 248–254
    cyanogenic glycosides, 251
    enzyme inhibitors, 252–254
    glucosinolates, 254
        amylase inhibitors, 254
        protease inhibitors, 252
        trypsin inhibitors, 252–254
    lectins, 252
    oxalates, 251–252
    phytic acid, 250
    saponins, 251
    tannins, 251
Next-Generation Probiotics (NGP), 94
NF-E2-related factor 2 (Nrf2), 124
NHEJ DNA repair system, 330
Nisin, 40
Noncarbohydrate plant bioactive compounds, 125
Nondairy fermented foods
    probiotics in, 83t
Nonextractable bound phenolic compounds, 200
Nonextractable polyphenols, 200
Nonfermented skimmed milk (NFSM), 367
Nonflavonoids, 2–3
Non-lactic acid bacteria species, 344
Nonpolar lipids, 211–212
Nonregioselective lipase, 224f
Noreugenin, 5
Nucleosides, 381–382
Nucleotide concentration, 381–382
Nutraceutical production, using gut microbes developments in, 278–279
Nutraceuticals, 1, 273–274
Nutraceuticals production, from engineered strains, 288t
Nutrient-rich legumes, 254
Nutrients, 173–174
Nutrigenomics, 276–278
Nutritional value, 240

Nutritive and antinutritive properties in food, 240–248
  alkaloids, 243
  cyanogenic glycosides/glucosides, 244–245
  enzyme inhibitors, 246–248
    amylase inhibitors, 247–248
    lipase inhibitors, 247
    protease inhibitors, 246
    trypsin inhibitor, 246–247
  glucosinolates, 248
  lectins, 245–246
  oxalates, 245
  phytates, 241–242
  polyphenols, 242–243
    flavonoids, 243
    tannins, 242–243
  saponins, 243–244
  toxic amino acids, 248

O

Obesity, 89–90, 355, 357–358, 362–363
Obesity treatment, new players for, 362–364
Ochratoxin A, 262, 265t
Ochratoxin B, 262
Ochratoxin C, 262
Oenococcus oeni, 344
Oleo-chemical industry, 310
Oleogels, 210
Oligosaccharides, 14–15, 37
  production with prebiotic potential, 37
Omega-3 fatty acids, 15–16, 283–284
Omega-3 linolenic acid, 213
Omega-3 PUFA, 224
Omega-6 fatty acids, 15–16, 283–284
Omega-6 linoleic acid, 213
Operational stability, in batch and continuous bioreactors, 226–230
Optimization of temperature, 173
Optimum pH, 161
Organic acids, 331t
Organic waste, strategic valorization of, 302–303
Orientase (OR), 311
Orlistat, 364–365
  mechanism of action, 364–365
  side effects of, 365

Osteoarthritis injury
  proposed formulation to repair, 407t
Overexploitation, 311–312
Oxidative stress, 348

P

Pancreatic lipase, 358
PanTheryx, 415–416
Papain, 140t
Parmigiano-Reggiano (PR) cheese, 12–14
Patent KR102076178B1, 412
Patent WO2019031976A2, 413
Pathological Recognition Protein (PRP), 409–410
Patulin, 263
P-coumaric acid, 197t, 199, 285–286
Pectin, 307
Pectinase, 140t
Pediococcus, 330, 342–343
Pediococcus acidilactici, 51
Pediococcus pentosaceus, 54–55, 343, 366–367
Pediococcus strains, 76–77
Penicillium camemberti, 262–263
Penicillium strains, 115
Peptidase, 140t
Peptide production, development of, 60
Peptidomics profiling, 12–14
Persistent Organic Pollutants (POPs), 401
Petrovac Sausage, 54
pH, 173
Phenazines, 10–11
Phenolic acids, 2–3, 191, 307
Phenolic compounds, 305
Phenylalanine ammonia lyase (PAL), 285–286
Phenyl methyl sulfonyl fluoride (PMSF), 160
Phenylpropanoid, 198
Phosphoglycerides, 212
Phosphoketolase pathway, 75
Phospholipases, 140t, 216
Phospholipases A1 and A2, 216
Phospholipids, 212
Phospholipid values, 381
Phylloquinone (PK), 8–9
Physically structured oils, 210
Phytates, 241–242
Phytoalexin, 191–192

Phytoanticipins, 244–245
Phytochemicals, 278
Pickled vegetables, 330
Pigments, 10–12
Pinosylvin, 197t
Plant-derived pathways, 197t
Plant oils, 142
Plant polyphenols, 198, 276
  effect of microbial fermentation for release of, 194t
Plant secondary metabolites, 240–241
Plasmin, 159–160
Plasminogen activation activity of FEs, 161–162, 163t
Plasminogen activators (PAs), 160
Pleurotus ostreatus, 313
Pleurotus pulmonarius, 313
Pleurotus sajor-caju, 313
Plum pomace, 307
Plum seed oil, 307
Polar lipids, 211–212
Polyamino acids, 278, 286
Polygonum cuspidatum, 191–192
Polyphenolic compounds, 305–306
Polyphenols, 2–5, 33, 242–243, 275, 285–286
  bioavailability of, 189–190
  effect of gut microbes on, 200
  effect of microbial fermentation on, 193f
  flavonoids, 243
  microbial production and enhancement of, 192–199
    addition of polyphenols in fermented food products, 199
    microbial fermentation for recovery of polyphenols from agri-food industry, 195–196
    polyphenol production by metabolic engineering, 196–199
    through food fermentation, 193–195
  microbial production and transformation of, 189–208
  tannins, 242–243
  types and health benefits of, 190–192
    flavonoids, 191
    lignans, 192
    phenolic acids, 191
    stilbenes, 191–192
Polysaccharides, 278
Polyunsaturated fatty acids (PUFAs), 15–16, 310
Prebiotics, 37, 282–284, 356–357
  in control of human diseases, 120–127
  emergent, 125–127
  fructooligosaccharides as, 121–123
  galactooligosaccharides as, 123–124
  global market for, 112
  lactulose as, 124–125
Proanthocyanidins, 242–243
Proanthocyanins, 199
Probioactive, 86
Probiotic bacteria, 74–80
  Bifidobacterium, 79
  lactic acid bacteria (LAB), 74–79
  yeast, 79–80
Probiotic Bifidobacteria, 6–7
Probiotics, 39, 337, 356–357, 365
  defined, 73–74
  in fermented foods, 80–86
    dairy fermented foods, 81
    history of fermented foods, 80–81
    nondairy fermented foods, 81–86
Probiotics and components of fermented foods, interaction between, 86–87
Probiotics and COVID-19, 89
Probiotics bacteria, 34
Probiotics industry and future prospects
  challenges in current status of, 92–95
    safety and associated risks, 93–94
    unregulated market, 94–95
    viability and stability of probiotics, 93
Probiotic strains, 12–14
Process state variables, 151
Prodigiosin, 10–11
Production of peptides, via fermentation, 56f
Proline-rich polypeptides (PRP), 409
Propionibacteria, 280
ProSAR, 148–149
Proteases, 140t, 141–142
Protease XXIII (PR) enzymes, 311
Protein, proteolysis of, 12–14

Protein Digestibility Corrected Amino Acid Score, 58
Protein expression analysis, 147–148
Pullulan, 344
Punjabi drink, 91
PuraLife, 417
PuraLife Colostrum, 417
Purification techniques, 168, 170

Q
Quantitative structure–activity relationship, 48
Quantitative structure–activity relationship method, 148–149
Quercetin, 197t
Quinones, 10–11

R
Refractometry, 401
Resveratrol (3,5,4-trihydroxystilbene), 191–192, 197t, 198–199, 277, 285–286
Rhizopus oryzae lipase, 229–230
Rhodotorula mucilaginosa, 11–12
Riboflavin (B_2), 8

S
Saccharomyces, 79–80
Saccharomyces cerevisiae, 118–119, 145–146
Sadana deactivation kinetics, 227
SALATRIM, 220
SCHEMA, 148–149
Secondary metabolites, 361f
Se-enriched *Candida utilis*, 6
Selenite, 6
Selenium (Se), 6–7
Selenium-enriched functional foods, 7
Selenocysteine (SeCys), 6–7
Selenomethionine (SeMet), 6–7
Se-nanoparticles (SeNPs), 6–7
SeNPs-enriched probiotics, 6–7
Series-type deactivation kinetics model, 227
Serine metalloprotease, 160
Serine protease, 160
Serrapeptase, 176
Shaking condition and other factors, 174

Short-chain fatty acids (SCFAs), 121, 212–213, 276, 360–362
Simple lipids, 212
Site-directed mutagenesis, 170–171
Skipjack, 169
Sn-1,3 regioselective lipase, 216, 227f
Soaking, 253
Solid-state fermentation (SSF), 14, 112
Solvent-free systems, 219
Sorbitol, 284–285
Sour dough, 345–346
Sovereign Laboratories, 417
Soybean, 14
Soy genistein, 277
Soymilk, 8
Sphingomonas paucimobilis, 344
Spray-drying techniques, 39
Srebp-1c, 366
Stachyose, 127
Staphylococcus carnosus, 54–55
Staphylococcus xylosus, 54–55
Staphylokinase, 176
Statistical optimization of media, 174
Stilbenes, 191–192
Stilbene synthase (STS), 285–286
Stilbenoids, 2–3
Streptococcus, 77, 278–279
Streptococcus thermophilus, 5
Streptokinase, 177
Streptomyces aburaviensis, 363
Streptomyces albolongus, 363
Streptomyces cerevisiae, 5
Streptomyces lavendulae, 363
Streptomyces toxytricini, 362–363
Streptomyces venezuelae, 5
Structured lipids (SL), 210
  main characteristics and functional properties of, 211t
Structured phospholipids (SPL), 218
Subcritical water extraction, 305–306
Submerged fermentation (SmF), 112
Substrate selection, 173
Subtilisins, 170–171
Succinic acid, 308
Succinoglucan, 344

Superbugs, 88–89
Supercritical fluid extraction, 305–306
Superoxide dismutase, 10
Surfactants and antimicrobial properties, 177
Sustainable fuels, 311–312
SWISS-MODEL, 148–149
Synbiotics, 92–93, 356–357
Synthetic materials, 116

T
Tagatose, 284–285
Tagatose-6-phosphate pathway, 284–285
Tannase, 140t
Tannins, 242–243
Tempeh, 167
Temperature, 161
Terpenoids, 33, 278–279
Tetrahydrolipstatin (THL), 362–363
Tetraterpene carotenoids, 278–279
Thrombolysis, 159–160
Thrombolytic therapy, 158
Thrombus, 174
Time-delay inactivation model, 226
Toddy, 92
TPA (tissue plasminogen activator), 159–160, 159f
TPase, 167
Traditional Chinese Medicine (TCM), 413–414
Traditional fermentation, 171
Traditional fermented food products, 347
Traditional immobilization methods, 146–147
Transesterification, 215–216
Transforming growth factors (TgF), 417
Transglucosidase, 140t
Transglutaminase (TG), 140t, 141–143
Trehalose, 285
Triterpene saponins, 243–244
Tryptophan, 58
Tuna-based industry, 311
Tyrosine ammonia-lyase (TAL), 285–286

U
Ultrasound-assisted extraction, 306
UPA (urokinase plasminogen activator), 159–160, 159f
Urease, 140t

V
Valorization of wastes, 303–305
Vancomycin, 356–357
Vegetable fermentation, 57
Vegetable products, fermentation of, 56
Veratrum grandiflorum, 191–192
Vitamins, 7–9, 331t
    riboflavin (B_2), 8
    vitamin B12, 280
    vitamin K, 8–9
Volatile fatty acids (VFAs), 309
    production, 309–310

W
Water adsorption index (WAI), 116
Water-soluble vitamins, 7–8
Weissella, 78, 341–342
Weissella cibaria, 341
Weissella confuse, 341
Whey proteins, 48

X
Xanthan gum, 344
Xanthomonas campestris, 344
Xanthophylls, 11–12
Xylanases, 140t, 141
Xylooligosaccharides (XOS), 15, 143–144

Y
Yarrowia lipolytica, 120
Yeast cells, 6
Yoghurt, 330, 331t, 346

Z
Zearalenone, 262, 264, 265t

Printed in the United States
by Baker & Taylor Publisher Services